Vibrational Spectroscopy with Neutrons

With Applications in Chemistry, Biology, Materials Science and Catalysis

Series on Neutron Techniques and Applications – Vol. 3

Vibrational Spectroscopy with Neutrons

With Applications in Chemistry, Biology, Materials Science and Catalysis

PCH Mitchell
SF Parker
AJ Ramirez-Cuesta
J Tomkinson

World Scientific

NEW JERSEY · LONDON · SINGAPORE · BEIJING · SHANGHAI · HONG KONG · TAIPEI · CHENNAI

Published by

World Scientific Publishing Co. Pte. Ltd.

5 Toh Tuck Link, Singapore 596224

USA office: 27 Warren Street, Suite 401-402, Hackensack, NJ 07601

UK office: 57 Shelton Street, Covent Garden, London WC2H 9HE

British Library Cataloguing-in-Publication Data
A catalogue record for this book is available from the British Library.

Series on Neutron Techniques and Applications — Vol. 3
VIBRATIONAL SPECTROSCOPY WITH NEUTRONS
With Applications in Chemistry, Biology, Materials Science and Catalysis

ISBN 981-256-013-0

Printed in Singapore by World Scientific Printers (S) Pte Ltd

Foreword

This volume in the *Series on Neutron Techniques and Applications* is the first account in book form since the late 1960s of the principles and applications of inelastic neutron scattering (INS) as a vibrational spectroscopic technique. Our aim has been to provide a hands-on account of inelastic neutron scattering concentrating on how neutron scattering can be used to obtain vibrational spectroscopic and bonding information on a range of materials.

The practical aspects of neutron scattering are described in detail—neutron sources, modern instrumentation, sample handling, analysing INS spectra, extracting information from spectra and common pitfalls.

The theory is carefully presented with emphasis on how theory and experiment connect in the interpretation of INS data and in the computational modelling of INS spectra. A particular emphasis is on solid state vibrational spectroscopy and the exploitation of *ab initio* computational methods and modern computational techniques generally.

The application of INS spectroscopy is described in studies of dihydrogen and its compounds, hydrides, hydrogen-bonded systems including water and ice, carbon, polymers and biological systems, in surface chemistry and catalysis and to organic and organometallic compounds. Many spectra are presented and a comprehensive listing of compounds and materials studied is given. Our book should be the first port of call for those seeking not only an understanding of the INS technique but information on the spectra of particular substances.

Our book will meet a longstanding need of users and potential users of neutron scattering spectroscopy—academics, staff of neutron scattering institutes world wide, research workers and graduate students—for an accessible and comprehensive one volume primary text and reference source. Moreover, publication is especially timely in view of the increasing use of INS in chemistry, materials science and biology, the introduction of a second target station at ISIS and the coming on stream of new pulsed neutron sources in Japan (J–PARC), the United States (SNS), and the proposed new European spallation source (ESS).

The benefits of using neutrons for vibrational spectroscopy—strong scattering by hydrogen, high penetration, straightforward and accurate modelling of the spectra including intensities—and comparisons with infrared and Raman spectroscopy are fully explored.

To summarise: a neutron scattering spectrum contains a wealth of information on the forces acting internally between atoms within molecules and, externally, between atoms and molecules in extended lattices; the spectra, energies and intensities, can be accurately modelled with modern *ab initio* computational methods. Inelastic scattering is a well-established technique which is being applied to an increasingly wide range of materials and problems. We have covered the reasons for choosing INS and how to apply it, and provided a detailed review of its applications and we commend our book to all present and future workers in the field of *Vibrational Spectroscopy with Neutrons*.

Philip C.H. Mitchell
School of Chemistry
University of Reading
Reading
RG6 6AD UK

Stewart F. Parker
ISIS Facility
CCLRC Rutherford Appleton
Laboratory
Chilton
OX11 0QX UK

Anibal J. Ramirez-Cuesta
ISIS Facility
CCLRC Rutherford Appleton
Laboratory
Chilton
OX11 0QX UK

John Tomkinson
ISIS Facility
CCLRC Rutherford Appleton
Laboratory
Chilton
OX11 0QX UK

October 2004

Contents

Acknowledgements

The authors wish to acknowledge the help and support of friends and colleagues at the ISIS Facility, Rutherford Appleton Laboratory, in the University of Reading, and in the neutron scattering community. We have benefited from discussions of particular topics with Gustavo Barrera, Daniele Colognesi, François Fillaux and Peter Trommsdorff. We are grateful to Luc Daeman, Jurgen Eckert, Bruce Hudson, Mark Johnson, Helmut Schober and Bill Stirling for providing data and spectra and to publishers and authors who have allowed us to reproduce published spectra. We have redrawn diagrams and spectra of which we were unable to obtain adequate copies. Adrian Champion helped with some of the figures and the literature survey for Appendix 4. Marina Taylor read and made helpful comments on draft chapters and checked the references. We thank the ISIS Facility, Rutherford Appleton Laboratory for providing time, library and computing facilities and likewise the School of Chemistry, University of Reading. The Leverhulme Trust is acknowledged for the award of a Leverhulme Emeritus Fellowship (to PCHM) during the tenure of which most of the work on the book was carried out. Finally, any errors and omissions are, as is customary, to be blamed on the authors—who will be grateful to those readers who bring them to their attention. We can be contacted at: insbook@rl.ac.uk.

Abbreviations, Acronyms and Computer Programs

We have restricted our use of abbreviations and acronyms to specific sections in the text; they are defined early within the section where they appear. The names of INS spectrometers are found in Table 3.2.

Below we provide the full form (and where appropriate a very brief description) of those abbreviations, acronyms and named computer programs that are found more generally throughout the book. References are in the text.

Item	Meaning
ABINIT	Plane wave DFT program, appropriate mostly for extended solids.
ACLIMAX	Program that calculates the INS spectrum (on a low-bandpass spectrometer) from the output of an *ab initio* calculation.
CASTEP	Plane wave DFT program, appropriate mostly for extended solids.
CHARMM	Chemistry at HARvard Molecular Mechanics program, appropriate for macromolecular systems.
CPMD	Car-Parrinello Molecular Dynamics, molecular dynamics plane wave DFT program, used mostly for extended solids.
DFT	Density Functional Theory.
DMOL3	DFT program, using localised numerical basis sets.
FRM-II	Forschungsneutronenquelle, Garching, Germany.
GAUSSIAN	Quantum mechanics computational program including DFT..
HREELS	High Resolution Electron Energy Loss Spectroscopy.
IBR-2	Frank Laboratory of Neutron Physics, Dubna, Russia.
ILL	Institut Laue Langevin, France.
INS	Inelastic Neutron Scattering.
IPNS	Intense Pulsed Neutron Source, USA.
ISIS	The ISIS Facility, Rutherford Appleton Laboratory, UK.
J-PARC	Japan Proton Accelerator Research Complex, Tokai, Japan.
KENS	Neutron Science Laboratory, KEK, Tsukuba, Japan.
LA	Longitudinally polarised acoustic phonon branch.
LAM	Longitudinal acoustic type (or accordion like) mode in alkanes.
LANSCE	Los Alamos Neutron Science Centre, Los Alamos, USA.
LLB	Laboratoire Leon Brillouin, Saclay, France.
LO	Longitudinally polarised optical phonon branch.
LO-TO	Longitudinal and transverse polarisations taken together.
NIST	National Institute for Standards and Technology, Gaithersburg, USA.

Item	Meaning
SHELL	Molecular Mechanics program, appropriate for extended solids.
SINQ	Paul Scherrer Institut, Villigen, Switzerland.
SNS	Spallation Neutron Source, Oak Ridge, USA.
TA	Transverse polarised acoustic phonon branch.
TAM	Transverse acoustic type mode in alkanes.
TO	Transverse polarised optical phonon branch.

Glossary of Symbols

Symbol	Equation (a)	Meaning	Units (b)
a_1, a_2, a_3	4.41	lattice translation vectors	Å
a and b	6.20	rotational field strength parameters	B_{rot}
a, b, c	§10.1.1	crystallographic axis labels	
$\hat{a}^+, \hat{a}^-$	A2.60	creation, annihilation operators	
$A_{disorder}$	5.20	apparent mean square displacement tensor due to static disorder	$Å^2$
$A_{int, ext}$	5.19	mean square displacement tensor, summed over internal modes, external modes	$Å^2$
A_l	§2.4	mean square displacement tensor of atom l, summed over all, usually, internal modes	$Å^2$
A_t	5.18	mean square displacement tensor summed over all modes, internal and external	$Å^2$
A	A3.10	moderator constant	$\hbar^2$ (e)
b	2.3	scattering length	fermi (c)
b_{coh}	2.2	coherent scattering length	fermi (c)
b_{inc}	2.3	incoherent scattering length	fermi (c)
b_+	2.6	scattering length, neutron with proton, unpaired spins	fermi (c)
b_-	2.6	scattering length, neutron with proton, paired spins	fermi (c)
$B, {}^{\nu}B_l$	2.37	mean square displacement tensor, atom l in specific vibrational mode ν.	$Å^2$
B_{rot}, B_{rot}^{HH}	6.9	rotational constant of a molecule, hydrogen	cm^{-1} (d)
$(B_{rot}^{HH})_e$	§6	spectroscopic rotational constant	cm^{-1} (d)
B	§5.2.1.2	isotropic crystallographic thermal paramete	$Å^2$
c	2.15	speed of light, 2.998×10^8	m s^{-1}
C	2.1	concentration	mol m^{-3}
d	2.21	distance,	m (e)
$d_i, d_s, d_f, d_m,$ d_F, d_t	§3.4	neutron flight path length, initial, in sample, final, moderator to Fermi chopper, Fermi chopper to sample, total	

Symbol	Equation (a)	Meaning	Units (b)
$d^2\sigma/dE_f\,d\Omega$	2.28	double differential scattering cross section	barn $(cm^{-1})^{-1}$ sr^{-1}
dA	§2	small element of area	m^2
$d\Omega$	2.27	small element of solid angle	sr
$e, e_x, e_y, e_z,$ e_Q	2.37	unit vector, the Cartesian unit vectors, that along Q	varies
E		energy	J
E_{elec}	4.2	typical energy of electronic transitions in a molecule	cm^{-1} (d)
$E_i, E_f,$	2.15	neutron energy, initial, final	cm^{-1} (d) (e)
E_k	2.78	kinetic energy	cm^{-1} (d) (e)
E_k^{mol}	4.4	kinetic energy of the molecule	varies
E_r	2.77	recoil energy	cm^{-1} (d)
E_{rot}	4.3	typical energy of rotational transitions in a molecule	cm^{-1} (d)
E_t	3.14	transferred neutron energy	cm^{-1} (d) (e)
E_{vib}	4.2	typical energy of vibrational transitions in a molecule	cm^{-1} (d)
f, f_{ads}	4.10	force constant, for adsorbed molecules	mdyn $Å^{-1}$
F_l	4.61	force acting on atom l from other atoms in the assembly	J
F		Helmholtz free energy	J
$g(\omega)$	2.55	density of vibrational states	$(cm^{-1})^{-1}$
h	2.16	Planck constant, 6.626×10^{-34}	J s
$\hbar$	2.21	$h/2\pi$	J s
$\hat{H}$	§6.21	Hamiltonian operator	J
i		$\sqrt{-1}$	
I_H, I_n		spin quantum number, proton, neutron	
I_i, I_j, I_k		moments of inertia, about the principal axes, $i, j, k,$	amu $Å^2$ (f)
$I_n\{x\}$		Bessel function of the first kind	
j	2.55	external mode index	
J	6.7	principal rotational quantum number	
J_i, J_f	2.1	neutron flux, initial, final	varies
$\hat{J}, \hat{J}_z$	6.6	rotational momentum operators	

Symbol	Equation (a)	Meaning	Units (b)
k	2.16	wavevector,	Å^{-1}
k_i, k_f		neutron momentum, initial, final	
k		$\lvert k \rvert$, magnitude of k	Å^{-1}
k_ν	10.2	molecular wavevector of the mode ν	rad (g)
k_B	2.21	Boltzmann constant, 1.381×10^{-23}	J K^{-1}
$k(\omega_\nu)_{max}$	4.45	molecular k value at which $(\omega_\nu)_{max}$ occurs	rad (g)
L	4.13	Lagrangian of the system	J
$^\nu L_l^{mol}(k)$	4.20	normalised molecular vibrational amplitude for atom l, mode ν, at k.	Å^2
$^\nu L_{l,}^{crys}$	4.56	wavevector independent normalised crystalline vibrational amplitude for atom l, mode ν	Å^2
m	§3.2.2	critical angle of reflection, referred to that of nickel	
m_n, m_l	2.9	rest mass, neutron, general atom l	amu (f)
$(m_H)_{eff}$, $(m_l)_{eff}$,		effective mass, hydrogen, general atom l	amu (f)
$M_{S\text{-}T}$	2.73	mass tensor of Sachs and Teller,	amu (f)
M_{lib}, M_{trans}		librational, translational components of $M_{S\text{-}T}$	
M	3.25	molar mass	kg mol^{-1}
M	6.7	magnetic rotational quantum number	
n		quantum number of final molecular vibrational state, order of a vibrational transition, the number of formulae units in a chain, generally any integer	
N_A		Avogadro number, 6.022×10^{23}	mol^{-1}
N_{atom}, N_{mol}		number, atoms, molecules	
p_{atom}	2.75	atomic momentum	Å^{-1}
p		occasionally represents an external vibration (phonon) ω_j	
P_i, P_n	A2.44	initial, final, vibrational state probability	
P_s	2.5	total spin	
Q	4.21	normal mode coordinate	
Q	§2.3	neutron momentum transfer vector	Å^{-1}
Q	2.23	$\lvert Q \rvert$, magnitude of Q	Å^{-1}
$Q_{01,\ 02}$	5.10	value of Q for a fundamental 01, overtone 02	Å^{-1}
q	4.5	mass weighted atomic displacement	amu^{½} Å (f)

Symbol	Equation (a)	Meaning	Units (b)
r_l, r_x, r_y, r_z	2.28	position of atom l, its Cartesian components	Å
$r_l(t)$	A2.41	time dependent position of atom l	Å
$S(Q,\omega)_l$	2.30	scattering law (or function) of atom l	$(cm^{-1})^{-1}$
$S''(Q,\omega)$	2.31	scaled scattering law, weighted by scattering cross sections	barn $(cm^{-1})^{-1}$
$S''(Q,\omega_v)^n$	2.32	scaled scattering law of vth mode to order n	barn $(cm^{-1})^{-1}$
$S(Q,\omega)_{int}, S(Q,\omega)_{ext}$	A2.53	scattering law for the internal (int) and external (ext) modes	barn $(cm^{-1})^{-1}$
S_O	2.64	that part of the scattering law (intensity) remaining at the band origin	barn $(cm^{-1})^{-1}$
S_W	2.65	that part of the scattering law (intensity) in the phonon wing	barn $(cm^{-1})^{-1}$
$S_{1,2}$	5.6	that part of the scattering law (intensity) in the 1st (2nd) phonon wing	barn $(cm^{-1})^{-1}$
$S_{1\leftarrow0, 2\leftarrow0}$	5.10	scattering law (intensity) in the fundamental ($1\leftarrow0$), overtone ($2\leftarrow0$)	barn $(cm^{-1})^{-1}$
t	3.12	time	μs
t_{chop}	3.23	neutron flight time from moderator to chopper	varies (e)
t_i	3.13	neutron flight time to cover the incident fight path	varies (e)
t_F	3.13	neutron flight time to the Fermi chopper	varies (e)
t_f	3.13	neutron flight time to cover final fight path	varies (e)
t_t	3.13	total time, neutron flight time to cover the total flight path	varies (e)
T	2.21	temperature	K
T_{eff}	3.3	effective temperature	K
$u(t)_{int}, u(t)_{ext}$	2.29	time dependent atomic displacement vector, internal mode, external mode	Å
$^v u_l$	2.32	displacement vector, atom l in mode v	Å
u		$\mid u \mid$ magnitude of u	Å
$(^j u_l)_k$	4.44	atomic displacement vector, atom l in external mode j, for the wavevector k	Å
v, v_i, v_f	2.15	velocity, initial, final	$m\ s^{-1}$
v		$\mid v \mid$ magnitude of v	$m\ s^{-1}$
v_n, v_N	A2.17	volume element of neutron, system of N atoms	m^3
V	4.23	potential energy	J

Symbol	Equation (a)	Meaning	Units (b)
$V(\theta, \phi)$	6.20	potential energy field in polar coordinates	J
V^{mol}, V_0^{mol}	4.7	potential energy of the molecule, its minimum value	J
V_F	A2.27	Fermi scattering potential, Fermi pseudo potential	J m^3
$2W$	A2.83	argument in the Debye-Waller factor	
$W_{i \to f}$	A2.16	transition rate, initial to final state	
x, y, z		Cartesian coordinates, mole fractions in chemical formulae	varies
y	2.31	scaling factor	varies
$Y_{JM}(\theta, \phi)$	6.12	spherical harmonics	
Z	§10.1.2.2	number of molecules in the crystallographic cell	
α_{tran}	5.6	translational part of $(\alpha^v_l)_{ext}$	Å^2
α_T	5.6	mean square displacement at temperature T	Å^2
α_0	5.6	minimum mean square displacement, zero point motion	Å^2
α^v_l	2.41	powder averaged total mean square displacement, atom l in the one quantum excitation of mode v	Å^2
$(\alpha^v_l)_{ext}$	2.61	powder averaged total mean square displacement, atom l, due to the external modes but determined in the direction of the internal mode v	Å^2
β	A2.11	amplitude of the wavefunction	
β^v	A2.96	two quantum version of α^v_l, usually taken as isotropic.	Å^2
β	A2.66	$\hbar \omega_v / 2 k_B T$	
γ_c	3.4	critical angle of neutron reflection	rad
Γ	§4.3.3	gamma point, centre of the Brillouin zone	
Γ	2.78	standard deviation of a Gaussian function	varies
$\delta(\omega)$	A2.24	delta function, equals zero unless argument is zero	
Δ	2.78	full width at half height of a Gaussian function	varies

Symbol	Equation (a)	Meaning	Units (b)
$\Delta d_i, \Delta d_f$	3.14	uncertainty in distances, initial, final	varies (e)
ΔE_{chop}	3.22	uncertainty in energy arising from the chopper	varies (e)
ΔE_f	3.14	uncertainty in final neutron energy	varies (e)
$\Delta E/E_i$	3.22	relative incident energy uncertainty (instrumental resolution of direct geometry spectrometers)	
ΔE_m	3.23	uncertainty in energy arising from the moderator	varies (e)
$\Delta E/E_t$	3.14	relative energy transfer uncertainty (instrumental resolution of indirect geometry spectrometers)	
Δt	§4.4.1	time step	ms
Δt_t	3.14	uncertainty in total time	varies (e)
Δt_{ch}	A3.10	uncertainty in time arising from time channel widths	varies (e)
Δt_{chop}	3.23	uncertainty in time arising from the chopper	varies (e)
Δt_m	3.23	uncertainty in time arising from the moderator	varies (e)
$\Delta \omega_{integral}$	§5.4.3	energy band within which to integrate observed intensity across Q	cm^{-1} (d)
ε	4.16	phase of the normal mode	rad
θ, θ_B	2.24, 2.25	scattering angle, polar angle (colatitude), Bragg angle	varies
κ	4.1	Hamiltonian expansion parameter	
λ	2.16	wavelength	Å
$\tilde{\lambda}$	4.17	eigenvalue solutions to the secular determinant	hartree $(bohr^2 \; amu)^{-1}$ (f)
$\mu, {}^{\nu}\mu_l$	2.9	reduced mass, of atom l in mode ν	amu (f)
ν	2.32	internal mode index, expanded to include the external modes in crystal calculations	
$\tilde{\nu}$		wavenumber	cm^{-1} (d)
$\varXi_n$	3.3	integrated neutron intensity across the Maxwellian	
ρ_N	3.4	atom-number density	m^{-3}

Symbol	Equation (a)	Meaning	Units (b)
ρ_f	A2.16	density of final states	
σ_{bound}	2.10	scattering cross section, atom bound	barn
σ_{coh}	2.4	coherent scattering cross section	barn
σ_{free}	2.10	scattering cross section, atom free	barn
σ_{inc}	2.4	incoherent scattering cross section	barn
σ_{total}, σ_l	2.12	total incoherent scattering cross section, of atom l	barn
ϕ	2.33	angle between two scattering vectors, azimuthal polar angle	varies
φ	§2.6.1.1	phase angle in the dispersion relation, (E, φ)	rad
ψ_{rot}, ψ_{vib}	4.24	rotational, vibrational molecular wavefunctions	
$\psi_{n, i}$, $\psi_{N, f}$ $\psi^*_{n, i}$, $\psi^*_{N, f}$	A2.2	wavefunctions, neutron -initial, scattering system -final, & their complex conjugates	
Ω	2.27	solid angle	sr
ω, ω_ν	2.15	circular frequency, eigenvalue of the νth mode	cm^{-1} (d)
$\omega_\nu(k)$	4.45	eigenvalue, of the νth mode, at the wavevector value k	cm^{-1} (d)
$(\omega_\nu)_{min}$, $(\omega_\nu)_{max}$ $(\omega_j)_{min}$, $(\omega_j)_{max}$	2.55	minimum, maximum eigenvalue of the mode ν or j. This covers the frequency range over which the mode may be dispersed and refers to the one phonon states for j, and the band origin region for ν	cm^{-1} (d)
$(\omega_E)_j$	2.56	a single, characteristic frequency representing all vibrations in the jth type of external mode. External modes are occasionally represented by p to simplify the symbolism and avoid confusion	cm^{-1} (d)
$(\omega_\nu)_{ideal}$	§10.1.2	energy of mode ν for an infinite, ideal chain	cm^{-1} (d)
ϖ_E	2.57	single weighted mean frequency (the Einstein frequency) representative of all the external vibrations	cm^{-1} (d)
$\mathfrak{I}(Q, t)$	A2.50	intermediate scattering function	
$\wp(\omega)$	4.63	power spectrum of the velocity autocorrelation function	

(a) The number of the equation, or section, where the symbol is first used.

(b) We have generally followed IUPAC recommendations (§1.6) for quantities

(c) 1 fermi = 10^{-15} m

(d) To be consistent with spectroscopic usage we have adopted the wavenumber as the energy unit; 1 kJ mol^{-1} = 83.59 cm^{-1} In INS literature the energy unit is meV; 1 meV = 8.066 cm^{-1}.

(e) The units of the experimental uncertainties that appear in detailed calculations of instrumental resolution are usually specific to a given neutron source and spectrometer. Those of a typical low-bandpass spectrometer, TOSCA, are given in Table A3.1.

(f) Unified atomic mass unit, one-twelfth mass of ^{12}C atom: m_u = 1 u = 1.661 × 10^{-27} kg. Also called the dalton (Da). We have used amu rather than u to avoid confusion with other uses of u in neutron scattering. Note m_H = 1.0073 amu; m_n = 1.0087 amu. (See §1.6 and references.)

(g) Molecular wavevectors are given in radians and neutron wavevectors are given in Å^{-1}.

1

Introduction

Vibrational spectroscopy with neutrons is a spectroscopic technique in which the neutron is used to probe the dynamics of atoms and molecules in solids. In this introductory chapter we provide a descriptive account of the discovery and properties of the neutron, the development of neutron scattering, how inelastic neutron scattering spectroscopy compares with infrared and Raman spectroscopy and the benefits of using the neutron as a spectroscopic probe.

1.1 Historical development of neutron scattering and key concepts

The neutron was discovered by J. Chadwick (1932) who showed that when beryllium was bombarded with α-particles (helium nuclei) neutral particles were emitted having a mass close to the proton mass [1,2]. The particles were called neutrons, designated $^1_0 n$.

$$^9_4 Be + ^4_2 He \rightarrow ^{12}_6 C + ^1_0 n$$

The neutron is an elementary particle with zero charge, rest mass close to that of the proton, and a magnetic moment (spin ½). Properties of the neutron are listed in Table 1.1. The neutron displays wave-particle duality; whether we treat the neutron as a particle or a wave depends on the phenomenon observed. In incoherent inelastic neutron scattering it is regarded as a particle, although the scattered neutron is treated theoretically as a spherical wave. The wave properties of the neutron are revealed by observing how the phenomena of interference, diffraction, determine the way in which a beam of neutrons propagates and spreads. The de Broglie wavelength of a thermalised neutron (*ca* 1–5 Å) is

1

comparable to interatomic and intermolecular distances, e.g. C–H, 1.09 Å, and its energy (30–700 cm^{-1}) is comparable to molecular vibrational energies. This is why neutron scattering experiments can yield, simultaneously, both structural and dynamic information on the scattering system.

Table 1.1 Properties of the neutron [3].

Property	Value
Rest mass	
m_n /kg	1.674 928 6(10) x 10^{-27}
m_n /u (a)	1.008 664 904(14)
$m_n c^2$ /MeV (b)	939.565 63(28)
Spin, I	½
Charge number, z	0
Mean life /s	889.1(21)

(a) On the unified atomic mass scale, relative to the mass of carbon-12 defined as 12. (b) Energy equivalent.

Early studies (1936–1950) of neutron scattering used radium-beryllium neutron sources but their low neutron flux prevented exploitation of neutron scattering as a spectroscopic technique [4]. Today neutrons are either extracted from a nuclear reactor or generated at a pulsed, accelerator-based spallation source. The exploitation of neutrons from nuclear reactors in structural studies and spectroscopy dates from the 1950s and from pulsed sources from the 1970s. A useful summary of the development of neutron sources is given in [5].

Neutrons are scattered by nuclei whereas photons (X-rays) and electrons are scattered by electrons. There is a certain probability that a neutron passing through a substance will be scattered with no loss of energy; the scattering is *elastic*. The scattered neutron waves may, or may not, undergo interference. *Coherent scattering* arises when the scattered waves from different nuclei of the same type interfere. Coherent elastic scattering is measured in diffraction experiments and tells us the relative positions of atoms, structure. *Incoherent scattering* arises when the natural isotopic and spin mixture of the sample destroys local order and reduces interference between the scattered waves, sometimes completely. (Neutrons are sensitive to the isotopic make-up of samples, which is exclusively a nuclear property, because they interact with the nucleus.) When the neutron exchanges energy with the sample the scattering is *inelastic*. This is the type of scattering that provides the spectroscopic information that is the subject matter of this book. We have limited our coverage to *inelastic neutron scattering from incoherent systems*, principally hydrogenous molecules, solids and surfaces and use the abbreviation INS for this type of scattering.

The first experiments on neutron diffraction were carried out in 1936. The use of neutron diffraction as a structural technique, in particular for the location of hydrogen atoms, developed from studies of potassium dihydrogenphosphate, KH_2PO_4, single crystals. The early work on neutron diffraction is described in Bacon's classic text (1962) [6].

Energy transfer resulting from inelastic scattering of neutrons by polycrystalline materials was demonstrated by P.A. Egelstaff (1951) [7] and by B.N. Brockhouse and D.G. Hurst (1952) [8]. It was early realised that the study of the vibrations of hydrogen atoms in compounds was particularly well suited to neutron scattering techniques: zirconium hydride was the first hydrogenous substance to be studied [9]. The theory of inelastic neutron scattering and early studies of inorganic and organic hydrogenous compounds are summarised in [10]. For molecules see also [11,12]. Recently the Institut Laue-Langevin (the ILL) has published a pocket-size *Neutron Data Booklet* covering many aspects of neutron production, detection, and scattering [13].

Finally, in this historical introduction, we should recall that neutrons have been the subject of two Nobel Prizes: to J. Chadwick for the discovery of the neutron (1935) and to B.N. Brockhouse and C.G. Shull (1994) for their pioneering contributions to the development of neutron scattering techniques (inelastic scattering and diffraction respectively).

1.2 Inelastic neutron scattering (INS)—a spectroscopic technique

The most common methods for studying molecular vibrations are the well established optical techniques of infrared and Raman spectroscopy. It is through a direct comparison with these techniques that the advantages of INS can be most readily grasped.

INS spectra are readily and accurately modelled.
Measured INS intensities are straightforwardly related to the atomic displacements of the scattering atom, which can often be obtained from simple classical dynamics. Any complications arising from the electro-optic parameters are avoided. Indeed the band positions and intensities of most molecular systems can be accurately calculated using modern *ab initio* computational methods. This is especially valuable since these methods are a well established part of the modern chemist's approach to understanding molecular structure and dynamics. The manipulation of an INS spectrum, e.g. subtracting a background, is straightforward.

INS spectra are sensitive to hydrogen atom vibrations.
Optical (i.e. infrared and Raman) techniques are, generally, most sensitive to vibrations involving the heavier atoms, because of the number of their electrons. The neutron incoherent scattering cross section of hydrogen is uniquely high, and makes it about ten times more visible than any other atom.

INS spectra are not subject to the rules of optical selection.
All vibrations are active in INS and, in principle, measurable. This stems from the mass of the neutron (*ca* 1 unified atomic mass unit). When scattered the neutron transfers momentum to the atom and INS measurements are not limited to observation at the Brillouin zone centre, as are photon techniques. The measured INS intensities are, *inter alia*, proportional to the concentration of the elements in the sample.

Neutrons are penetrating—photons are not.
Neutrons penetrate deeply, of the order of millimetres, into typical samples and pass readily through the walls of containment vessels, generally aluminium or steel. INS results are thus naturally weighted to the measurement of bulk properties.

Wide spectral range
INS spectrometers cover the whole molecular vibrational range of interest (16–4000 cm^{-1}, see Fig. 1.1). The lower energy range (below 400 cm^{-1}) is readily accessible, a region that is more difficult experimentally for infrared and Raman spectroscopies. With modern instrumentation, the quality of INS spectra approaches that of infrared and Raman spectra obtained from the same system under the same conditions.

1.3 INS spectra

In an INS experiment we observe how the strength of neutron scattering varies with the energy transfer and momentum transfer. The spectrum is typically in neutron energy loss, where energy is transferred from the incident neutrons to the scattering atoms.

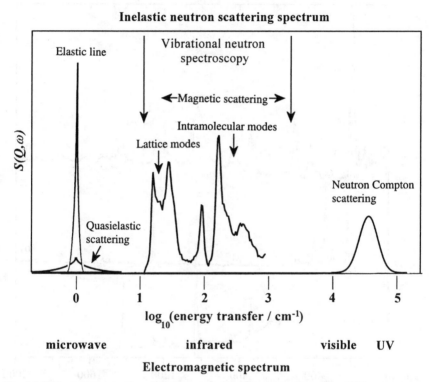

Fig. 1.1 INS spectroscopy in relation to optical (photon) spectroscopies. INS spectra are within the wavenumber range 16–4000 cm^{-1}, the same as the mid-infrared range. The vertical axis is the neutron scattering intensity expressed as the scattering law (or function), see text.

The atoms are embedded in the molecule and can only gain energy in the vibrational quanta characterised by the molecular structure. As an example we show in Fig. 1.2 the INS spectrum of a solid molecular system, Zeise's salt, $K[Pt(C_2H_4)Cl_3]$, and make comparison with its infrared and Raman spectrum.. Several points are apparent: the most basic is that when the same mode is observed by different types of spectroscopy it occurs at the same energy. This must be the case since the vibrational energies are determined by the molecule not the technique. It is also apparent that to obtain a complete description of the spectrum all three forms of vibrational spectroscopy are needed.

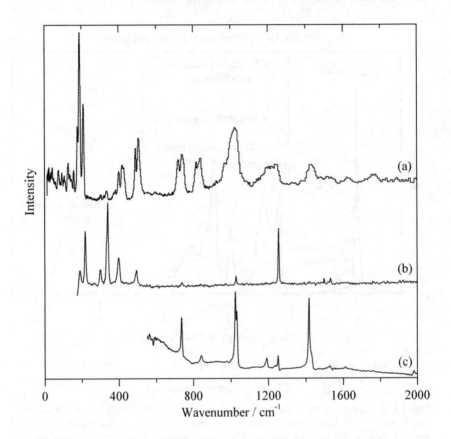

Fig. 1.2 (a) INS (ISIS, TOSCA spectrometer), (b) Raman,(c) infrared spectra of Zeise's salt, $K[Pt(C_2H_4)Cl_3]$. Relative intensities are not to scale. See also Fig. 7.18.

In the INS spectrum of Zeise's salt vibrations involving hydrogen atom displacements have high intensity, for example the torsion of the ethene ligand at 190 cm^{-1} is particularly strong, although all the modes above 100 cm^{-1} involve ethene. The scattering from other atoms, chlorine in this case, appears weakly. This is exemplified by the Pt–Cl stretch at 336 cm^{-1} that gives a very intense Raman band but only a weak INS feature.

Some peaks in the INS spectrum are not seen in the infrared or Raman spectra: for example the factor group splitting of all the modes in the INS is readily apparent because of the absence of selection rules. In the infrared and Raman spectra, some of the factor group components are either forbidden or have zero intensity.

Access to the low energy region (<300 cm^{-1}) is the usual situation in INS spectroscopy. For infrared and, to a lesser extent, Raman spectroscopy, this is not the case and for instrumental reasons, 200–400 cm^{-1} is the usual cut-off. Even when the region can be observed, it is often found that at least some of the librational and translational modes are very weak in an infrared or Raman spectrum.

1.4 Information content of an INS spectrum

To make the most of the INS technique and to avoid pitfalls some knowledge of the physics of neutron scattering is desirable. The theory of neutron scattering can be presented at various levels of sophistication. In the classic texts [14, 15] the formalism and the accompanying symbolism can look formidable indeed. In this book, while the presentation of the necessary theory is accurate and rigorous, our concern is the information content of the equations rather than their derivation. Our aim is to acquaint the reader with the basic concepts of neutron scattering, to make the connection between theory and experiment and to show how useful information can be extracted from the experimental data.

The INS technique can be simply summarised. The observed positions of the transitions (the eigenvalues) are a function of the molecule's structure and the intramolecular forces, as in optical spectroscopy, and correspond to the energies lost by the neutron. The strength of the observed transition is a function of the atomic displacement occurring during that vibration (the eigenvector) and the momentum lost by the neutron. The atomic displacements are again determined by the molecule's structure and the intramolecular forces but the momentum transferred is determined by the neutron spectrometer. INS spectroscopy gives direct access to both the vibrational eigenvalues and eigenvectors and no other technique gives it so straightforwardly.

1.5 When to use neutrons

Neutron scattering as a spectroscopic technique is demanding of time, effort and commitment. To use the technique the experimentalist has to travel to a neutron facility with their sample and, sometimes, with their equipment. So, knowing when to use neutrons, is important.

We might first remind ourselves of when the use of a spectroscopic technique is indicated. Most molecular properties are rationalised through an appeal to an equilibrium molecular structure. Detailed understanding of a molecular structure is the objective of much of the published chemical literature and is best achieved by the use of diffraction techniques. These offer a direct approach to molecular structure. However, many systems of interest and importance to the chemical and materials science communities do not appear as single crystals. Wherever long range order is absent, or seriously confused, diffraction techniques falter and recourse must be had to alternative approaches. All spectroscopies offer an indirect method of obtaining structural information. In molecular vibrational spectroscopy the vibrational resonances that are characteristic of the molecular structure are studied. The spectra are measured and compared with the calculated responses of putative molecular structures. Neutron vibrational spectroscopy is simply another variant on this theme but one that is uniquely advantaged by our ability to calculate the spectral band positions and intensities simply and with confidence.

No one could hope to prescribe which systems should, or should not, be studied by INS; therefore, in this book we offer a review of all the fields of research that have profitably exploited the INS technique. We have been particularly careful to make the results accessible to the broader chemical and materials science community by simplifying, where desirable, the theoretical approaches, in using generally available *ab initio* methods, accepting the conventions of optical spectroscopy and retaining the wavenumber as our energy unit.

The reader is invited to study the contents pages to appreciate the broad range of subjects that have been successfully tackled by INS. Then having read through those sections of relevance to their own speciality decide for themselves when, and how, to apply the INS technique to best effect.

1.6 A note on units, symbols and chemical names

Throughout this book, we have, as far as possible, used SI units or derived units, and the symbols for physical quantities and the printing conventions recommended by the International Union of Pure and Applied Chemistry (IUPAC) [3]. The single significant deviation from this approach is the use of the (Å) as the unit of length.

The reciprocal Ångstrom unit is also conventionally used for momentum (Å^{-1}) throughout the neutron scattering literature. There was neither a common optical spectroscopic alternative (as there was to justify the adoption of the wavenumber, cm^{-1}, for the unit of energy) nor chemical reason (as there was to justify the adoption of the atomic mass unit, u or amu, for the unit of mass). For quantities not defined by IUPAC we have mostly used symbols consistent with the neutron scattering literature [14,15] but have, on rare occasions, been forced to invent our own symbol for the sake of clarity. We provide a table of symbols and units (p. xix).

In handling physical quantities and their units we have used the method of quantity calculus [3] (also called dimensional analysis). The value of a physical quantity is expressed as the product of a numerical value and a unit

physical quantity = numerical value × unit (1.4)

In tabulating numerical values of physical quantities and labelling the axes of graphs, we use the quotient of a physical quantity and a unit in such a form that the values are pure numbers, for example

$\omega\ /\ \text{cm}^{-1} = 500$ (1.5)

Finally, the reader should be beware that the neutron scattering literature, certainly the older literature, often fails to use systematic chemical names. While not wishing to be accused of pedantry, we have chosen to use the systematic name where to do so would be unlikely to cause a problem of understanding (e.g. ethene for ethylene, ethyne for acetylene), and, otherwise, the common name with, at first use, the correct name in parenthesis (e.g. the $[HF_2]^-$ or $[HFH]^-$ ion, commonly bifluoride in the neutron scattering literature but, correctly, hydrogendifluoride or even difluorohydrogenate(1−) [16]. We write the systematic names and formulae of coordination compounds and organometallics following the latest (provisional) IUPAC convention: listing the ligands in alphabetical order 17.

1.6.1 Spectrometers—accuracy and precision of reported results

Vibrational spectroscopy with neutrons has been dominated by one particular type of instrument, which combined a broad energy range and good spectral resolution. The principal technique used to obtain these advantages was to restrict the final neutron energy to low values (< 40 cm^{-1}). This opened the spectral range to the full energy range available from the neutron source. Historically, the initial emphasis was to increase the spectral range available (reactor with thermal source → reactor with hot source → spallation source). More recently the emphasis has shifted to spectrometers with improved resolution, whilst retaining the broad range (beryllium filter → beryllium-graphite filter → crystal analyser). Throughout the book we use the term *low-bandpass spectrometer* to mean an indirect geometry spectrometer with low final energy, broad spectral range and good spectral resolution. Such spectrometers are described in §3.4.2. Direct geometry chopper instruments have been much less used in this field and this is reflected in the number of results obtained by this technique.

We report the published values of the spectral transitions uncritically as given by the original authors, except for any necessary change to wavenumber units. Generally no more than three significant figures is justified but we confess to not having been consistent in rounding published values.The accuracy of these values will have suffered more or less from several sources of experimental error but especially instrumental calibration errors. Their precision is strongly dependent on the instrumental resolution and appropriateness of subsequent fitting procedures, if any, that were used. Generally the spectral accuracy will not be better than the instrumental resolution (§3.4.2.3.1, §3.4.3.1).

In the figure captions we follow the convention of attributing the spectra to an identified instrument at its Institution. If no such attribution is given, the reader should assume that the spectrum was recorded using TOSCA (or its predecessor TFXA) at ISIS.

1.7 References

1   J. Chadwick (1932). Nature, 120, 312. Possible existence of a neutron.
2   G.E. Bacon (1969). *Neutron Physics*, Wykeham Publications, London and Winchester, Chapter 2. Discovery of the neutron.
3   I. Mills, T. Cvitaš, K. Homann, N. Kallay & K. Kuchitsu (1993). *Quantities, Units and Symbols in Physical Chemistry*, 2nd Edition, International Union of Pure and Applied Chemistry (IUPAC), Blackwell Scientific Publications, Oxford.
4   P.A. Egelstaff (Ed.) (1965). *Thermal Neutron Scattering*, Academic Press, London and New York.
5   C.C. Wilson (2000). *Single Crystal Neutron Diffraction from Molecular Materials*, World Scientific, London. Chapter 2. Neutron Scattering
6   G.E. Bacon (1962). *Neutron Diffraction*. 2nd Edition. Clarendon Press, Oxford.
7   P.A. Egelstaff (1951). Nature, 168, 290. Inelastic scattering of cold neutrons.
8   B.N. Brockhouse & D.G. Hurst (1952). Phys. Rev., 88, 542–547. Energy distribution of slow neutrons scattered from solids.
9   G. Dolling & A.D.B. Woods (1965). In *Thermal Neutron Scattering*, (Ed.) P.A. Egelstaff, Chapter 5, Academic Press, London and New York. Thermal vibrations of crystal lattices.
10   H. Boutin & S. Yip (1968). *Molecular Spectroscopy with Neutrons*, The M.I.T. Press, Cambridge, Massachusetts.
11   J.A. Janik & A. Kowalska (1965). In *Thermal Neutron Scattering*, (Ed.) P.A. Egelstaff, Academic Press, London and New York, Chapter 9. The theory of neutron scattering by molecules.
12   J.A. Janik & A. Kowalska (1965). In *Thermal Neutron Scattering*, (Ed.) P.A. Egelstaff, Chapter 10, Academic Press, London and New York. Neutron scattering experiments on molecules.
13   A-J. Dianoux & G. Lander (Ed.) (2001). *Neutron Data Booklet*, Institut Laue-Langevin, Grenoble.

14   S. Lovesey (1984). *Theory of Neutron Scattering from Condensed Matter*, Vol. 1
      and 2, Clarendon Press, Oxford.
15   G.L. Squires (1978). *Introduction to the Theory of Thermal Neutron Scattering*,
      Dover Publications, New York.
16   G.J. Leigh (Ed.) (1990). *Nomenclature of Inorganic Chemistry*. International
      Union of Pure and Applied Chemistry (IUPAC), Blackwell Scientific Publications,
      Oxford.
17   N.G. Connelly (Ed.) (2004). *Nomenclature of Inorganic Chemistry*. International
      Union of Pure and Applied Chemistry (IUPAC), Chemical Nomenclature and
      Structure   Representation   Division.   Provisional   Recommendations.
      http://www.iupac.org/reports/provisional/.

2

The Theory of Inelastic Neutron Scattering Spectroscopy

There have been many excellent publications that have addressed the theoretical basis for our understanding of neutron scattering experiments and it is not our aim to make an original contribution to this well-established and non-controversial field. Unfortunately, whilst some publications are approachable [1,2], most are written for a readership comfortable with mathematical manipulation, and aim to treat all the possible physical cases. Our book, however, is focused on a materials chemistry theme using the technique of molecular vibrational spectroscopy and, moreover, almost exclusively from hydrogenous systems. It is, therefore, appropriate to restrict the scope and style of this theoretical presentation to reflect the theme of the book and the interests of its readership.

In this chapter we report only the essential equations and consign detailed theoretical considerations to Appendix 2. The scattering law equations that are given will be preceded by a brief description of the basic physical assumptions and mathematical methods used in their derivation (§2.5). They should be recognisable in the published literature, since we have remained, mostly, faithful to conventional symbols and units. This chapter is also the appropriate place to introduce and expand on some of the less well known, if straightforward, concepts used in the field of neutron scattering. Under this heading we group: scattering cross sections (§2.1), energy and momentum transfer (§2.3), thermal ellipsoids (§2.4), vibrational dispersion (§2.6.1), the density of vibrational states (§2.6.2), the impact of phonon wings (§2.6.3) and the effects of molecular recoil (§2.6.5).

We begin immediately by describing a simple neutron scattering measurement, shown diagrammatically in Fig. 2.1.

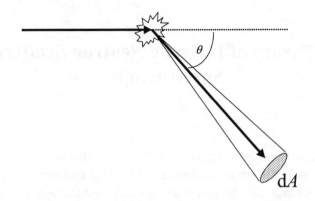

Fig. 2.1 A diagrammatic representation of a simple scattering measurement. The incident neutrons strike a sample and some neutrons are scattered through an angle θ into the detector of area dA a distance d_f from the sample.

A beam of monoenergetic neutrons falls on a sample and is scattered onto a detector of small area, dA, a distance d_f from the sample. This is positioned at the scattering angle, θ, which is defined with respect to the unscattered, straight through, beam. It is found empirically that neutrons are scattered differently by different elements. Some substances scatter neutrons evenly in all directions whilst others scatter most neutrons in specific directions. The latter is of course a consequence of diffraction, more familiar through the scattering of X-rays, and is a coherent event.

The number of neutrons scattered out of the beam is constant for a fixed mass of a given element. It follows a law similar to the Beer-Lambert law for optical absorption and linear approximations to this law can be exploited [3] (for simplicity we shall assume that the element does not absorb neutrons). The initial neutron intensity, or flux on the sample, is J_i and the unscattered flux after the sample is J_f. Then, since the mass of the sample is the product of its thickness, d_s, and concentration, C, we can write:

$$J_f = J_i \exp(-d_s C \sigma) \tag{2.1}$$

The strength of the scattering, σ, has units of cross sectional area (m^2) and is called the cross section of that atom. Neutron scattering is weak (§2.1.4) and so the cross sections are small. Conventionally they are given in units of barn, where 1 barn = 10^{-28} m^2. It is worth noting that the neutron scattering lengths, tabulated in Appendix 1, are exclusively derived from the experimentally measured cross sections. There is no practical theoretical route for the calculation of the scattering lengths from other nuclear properties.

2.1 The atomic cross sections

Atoms both scatter and absorb neutrons; here we shall consider only the scattering cross sections and leave the practical consequences of absorption to be discussed latter (§3.3.2).

2.1.1 The coherent and incoherent scattering strengths

The concepts of coherence and incoherence are related to the way in which the neutron, both as a wave and as a particle, interacts with the scattering sample. Wave-like representations of the neutron view its interaction with solids as occurring simultaneously at several atomic centres; these atoms become the sources of new wavefronts. Since the scattering occurs simultaneously from all of these atoms the new wavefronts will spread out spherically from each new source and remain in phase. Provided the lattice is ordered, the coherence of the incident wave has been conserved. Constructive interference between the new wavefronts leads to the generation of distinctive diffraction patterns with well-defined beams, or reflections, appearing only in certain directions in space and no intensity in other directions.

Diffraction patterns are a defining characteristic of coherent scattering. The number of reflections is limited and each is therefore intense. Since the incident neutron wave interacted with many atoms the momentum transferred (§2.3) to the sample was not absorbed by any single atom but, in the limit, by the whole crystal with its network of bonds.

On the other hand, the particle representation of the neutron has it interacting with a single scattering centre in the sample. The scattered neutron is represented as a spherical wave emanating from this unique source and there is no possibility of interference between wavefronts from many sources: it is incoherent. This type of scattering is characterised by an even distribution of intensity appearing in all directions. The neutrons are scattered in all directions and the strength of scattering in any one direction is weak. Since the neutron is scattered at a single centre, only the local network of bonds is involved and at some value of momentum transfer this local network must fail.

The degree to which a sample can be regarded as a coherent or incoherent scatterer depends on the sample's composition. Take the example of an idealised diatomic solid AE with the rock-salt structure with all the A atoms on one sublattice and the E atoms on the other. All the scattering will be coherent since the structure shows true translational symmetry and long range order. However, if the AE sample has its atoms arranged randomly amongst the sublattices then translational symmetry is maintained only in an average sense. Long-range order persists but only for the pseudo-monatomic lattice, where the pseudo-atom is the average atom, Æ. The sharp and intense diffraction (coherent) features are still present, because the long range order of Æ atoms is well preserved, but they are now somewhat weaker. In local areas of the crystal the short-range order varies; these fluctuations reduce the coherent scattering and produce a level (incoherent) background. All solids that are composed of a single nuclear type will give a purely coherent signal. Note, however, that elemental solids do not necessarily give a purely coherent signal because of the possibility of different isotopes. As an example we discuss neutron scattering by chlorine.

To an incident beam of neutrons the atoms of chorine are a random mixture of ^{35}Cl and ^{37}Cl. We calculate the proportions of neutrons scattered coherently and incoherently from chlorine (75% ^{35}Cl, 25% ^{37}Cl) with coherent scattering lengths, b_{coh} (m) from Appendix 1. The average scattering length, which provides the coherent signal, is

$$b_{coh}(\langle Cl \rangle) = \left[0.75\, b_{coh}(^{35}Cl) + 0.25\, b_{coh}(^{37}Cl) \right] \tag{2.2}$$

$$= [0.75 \times 11.8 \text{ m} + 0.25 \times 2.60 \text{ m})] \times 10^{-15} = 9.50 \times 10^{-15} \text{ m} = 9.50 \text{ fermi}.$$

The incoherent scattering length, the fluctuation from the average, is given by the difference between the average of the squared lengths and the square of the average length.

$$\langle b_{inc} \rangle^2 = \langle b^2 \rangle - \langle b \rangle^2$$

$$\left(b_{inc}(\langle Cl \rangle) \right)^2 = \left(0.75 \times 11.8^2 + 0.25 \times 2.60^2 - 9.50^2 \right) \times 10^{-30}\, m^2 \quad (2.3)$$

$$b_{inc}(\langle Cl \rangle) = \sqrt{15.9}\ \text{fermi} = 3.98\ \text{fermi}$$

2.1.2 Spin incoherence

Neutrons are sensitive not only to isotopic variation but to the breakdown of all spatial correlation. This includes the correlation of nuclear spins, through the neutron's own spin $I_n = 1/2$. Without this extra incoherence, the scattering cross sections of chlorine would be

$$\sigma_{coh} = 4\pi \langle b_{coh} \rangle^2; \quad \sigma_{inc} = 4\pi \langle b_{inc} \rangle^2$$

$$\sigma_{coh}(Cl) = 11.5\ \text{barn} \qquad \sigma_{inc}(Cl) = 1.80\ \text{barn} \quad (2.4)$$

However, because of the nuclear spin incoherence of ^{35}Cl, the total incoherent cross section of chlorine is σ_{inc} (Cl) = 5.30 barn. Spin incoherence is particularly important for an understanding of the neutron scattering cross section of hydrogen nuclei, $I_H = 1/2$.

Except for a few small molecules, like methane and dihydrogen (§6.1), the spin ground state of a hydrogenous molecule is energetically very close to the next state. Samples must be cooled well below 1K before a molecule can stay in its spin ground state. Under most practical experimental conditions the proton spins of a molecule are uncorrelated. The interaction of neutrons with unbound protons yields one of two possible total spin states, $I_\pm$, governed by the two scattering lengths. Each spin state has its own value of b: for neutron and hydrogen spins parallel (triplet) $b_+ = 0.538 \times 10^{-15}$ m; for the spins antiparallel, (singlet) $b_- = -2.37 \times 10^{-15}$ m.

$$I_{total} = I_H \pm I_n \qquad I_- = 0 \quad \text{or} \quad I_+ = 1 \qquad \text{and the total spin, } P, \text{ is}$$

$$P_\pm = 2I_\pm + 1 \quad \text{so} \quad P_- = 1 \quad \text{and} \quad P_+ = 3 \quad (2.5)$$

$$P_s = P_- + P_+ = 4$$

The coherent and incoherent scattering lengths are:

$$b_{coh}(\langle H \rangle) = \frac{P_+}{P_s}b_+ + \frac{P_-}{P_s}b_-$$

$$= [0.75 \times 0.538 + 0.25 \times (-2.37)] \times 10^{-15}\,m = -0.189 \times 10^{-15}\,m \quad (2.6)$$

$$(b_{inc}(\langle H \rangle))^2 = \frac{P_+}{P_s}|b_+|^2 + \frac{P_-}{P_s}|b_-|^2 - (b_{coh}(\langle H \rangle))^2 \quad (2.7)$$

$$= \{[0.75 \times 0.538^2 + 0.25 \times (-2.37)^2] - (-0.189)^2\} \times 10^{-30}\,m^2$$

and the free atom cross sections are,

$$\sigma_{coh} = 0.439\,barn; \sigma_{inc} = 20.1\,barn \quad (2.8)$$

It is more convenient theoretically to work with the bound scattering lengths, which represent the scattering length the atom would have if its mass were infinite. This has the advantage of placing the scattering atom at the origin of coordinates and allows the equations describing the scattering to be expressed without reduced masses. The incident neutron, of rest mass m_n, has a reduced mass, μ_n, which depends on the scattering atom mass, m. (Strictly, $m_n = 1.675 \times 10^{-27}$ kg and $m_H = 1.672 \times 10^{-27}$ kg, but given our required accuracy we can take $m_n = m_H = 1$ amu $= 1.661 \times 10^{-27}$ kg):

$$\frac{1}{\mu_n} = \frac{1}{m} + \frac{1}{m_n} \quad (2.9)$$

The reduced mass enters quadratically into expressions for the atomic cross sections.

$$\frac{\sigma_{bound}}{m_n^2} = \frac{\sigma_{free}}{\mu_n^2} \quad (2.10)$$

The bound cross sections for hydrogen, $m_H = 1$, are (see Appendix 1)

$$\sigma_{coh}(H) = 1.76\,barn \quad \sigma_{inc}(H) = 80.3\,barn \quad \sigma_{total}(H) = 82.0\,barn \quad (2.11)$$

It is easy to see that neutron diffraction experiments on hydrogenous materials are difficult because the coherent scattering is weak and the incoherent scattering is strong. Fortunately the heavier isotope, deuterium (^{2}H), has a much more favourable ratio of bound scattering cross sections, see Appendix 1.

$$\sigma_{coh}(^2H) = 5.59 \, barn \quad \sigma_{inc}(^2H) = 2.05 \, barn \quad \sigma_{total}(^2H) = 7.64 \, barn \quad (2.12)$$

Neutron cross sections themselves follow no trend. Moreover, as was shown above, quite different values are found for different isotopes of the same element. This is extensively exploited in diffraction studies of liquid solutions [4]. Occasionally, when isotopes of the same element have scattering lengths of opposite sign, the average coherent cross section can be reduced to zero by making a sample with the correct proportions of each isotope. Such 'null-scattering' samples produce only incoherent signals.

2.1.3 The incoherent approximation

As the magnitude of the momentum transfer vector (§2.3) increases, the strength of elastic scattering (no energy exchange) falls away and most scattering occurs with an exchange of energy. Eventually, at the highest momentum transfers, all scattering will involve exchanging the recoil energy of the scattering atom (§2.6.5). The neutrons involved in recoil scattering have wavelengths (§2.3) less than the interatomic spacings in the crystal. The dramatic angular variations in detected intensity associated with crystalline diffraction are thus no longer visible and the response is smooth and incoherent-like. The dynamic response can then be calculated simply by treating this scattering as if it were incoherent and using the total scattering cross section of the atom: this is the incoherent approximation.

The approximation works very well within the limits originally set for its use [1]. However, in the molecular spectroscopy of hydrogenous molecules a further version of the approximation is made. Here the cross section used in the calculation of the observed intensities is the total bound scattering cross section of each atom and the dynamics are always treated as incoherent, irrespective of the momentum transferred. This is discussed further with specific examples in §11.1.

2.1.4 Comparison with photon scattering cross sections

Whilst photons have wavelengths similar to interatomic distances for crystallographic work, or energies similar to molecular vibrational energies for spectroscopic work, a neutron has both, i.e. neutrons are suitable for simultaneous elastic and inelastic measurements.

Neutrons interact with materials through their nuclear properties but photons interact with materials through their electrical properties and much more strongly than can neutrons. Therefore, the results from photon experiments are heavily weighted to observing the surface of samples. The use of thin samples in infrared absorption spectroscopy is simply an acknowledgement of this. The results from neutron experiments are weighted to the bulk provided that the sample is not specifically prepared with a high surface area. Damage to the sample from the experimental probe is also much more likely for photons than for neutrons.

Photons are scattered by the electrons around an atom; heavier atoms, with more electrons, scatter much more strongly than light atoms. To emphasise this trend some X-ray scattering cross sections (based on the Thomson free electron cross section, $\sigma_T = 0.7$ barn) are given below. These can be compared with the neutron cross sections given in Appendix 1.

$$\sigma_{X\text{-ray}}(H) \approx 1\,\text{barn} \quad \sigma_{X\text{-ray}}(Cl) \approx 11\,\text{barn} \quad \sigma_{X\text{-ray}}(Pb) \approx 5000\,\text{barn} \qquad (2.13)$$

Photons exchange energy with materials in two principal ways. First, low energy photons, in the infrared region of the spectrum, are absorbed or emitted. The process is mediated by the dipole moment change induced in a molecule by a given vibration. Second, high energy photons, in the visible range of the spectrum, are Raman (inelastically) scattered and this process is mediated by the polarisability change induced by a vibration. If the molecular vibration fails to change the dipole moment, or polarisability, photons cannot exchange energy and the mode is inactive. Different selection rules allow some modes which are infrared inactive to be Raman active and *vice versa*; however, vibrations which are silent in both spectroscopies are common and the higher the molecular symmetry the more modes remain silent. Optical spectroscopy is well adapted for the study of most molecular systems although its inconveniences are sometimes forgotten. A particularly striking example of this is seen in the optical spectroscopy of C_{60}, where 70% of the vibrational modes are optically inactive (§11.2.3.1). Approximate cross sections for the optical processes are, in barn per molecule.

$$\sigma_{Raman} \approx 10^{-4}; \qquad \sigma_{infra\ red} \approx 10^6 \tag{2.14}$$

From one vibration to another there can be considerable variation and, generally, the displacement of electron rich atoms will influence the spectral intensities more than electron poor atoms. Thus, since polarisability increases with atomic size, Raman spectra involving the heavier atoms are easier to observe, as demonstrated by the weakness of hydrogen's vibrational features in metal hydridocarbonyls (§11.2.6). It also follows that the optically strong features of a spectrum may mask weaker vibrations of interest, a common problem in intercalate and catalytic systems.

2.2 Some practical consequences

Molecular vibrational spectroscopy with neutrons relies on the large total cross section of hydrogen (82.0 barn) compared with the cross sections of the other elements commonly found in molecules: carbon, 5.55 barn; nitrogen, 11.5 barn; oxygen, 4.23 barn; sulfur, 1.03 barn. Therefore, in typical molecules the dynamics of a single hydrogen atom will clearly show more strongly than the vibrations of all of the heavy atoms. Alternatively, the dynamics of individual hydrogenous molecules can be observed in the presence of a non-hydrogenous matrix.

By the same token, some of the best known group frequency vibrations of molecular spectroscopy, like the strong carbonyl stretch at about 1700 cm^{-1} in the infrared, are almost invisible in neutron spectroscopy. However, many of the techniques of optical spectroscopy retain much of their significance and indeed the technique of isotopic substitution can be dramatically exploited in neutron spectroscopy.

2.2.1 Effects of deuteration

In optical spectroscopy advantage is taken of the change in atomic mass when deuterium is substituted for hydrogen. Those transitions that involve significant hydrogen motion will change their energies upon deuteration because the mass of the oscillator has changed. This is treated further in Chapter 5. As the mass increases in going from H to D, for a fixed vibrational force constant, the transition energy falls. The greatest decrease occurs for modes where only the hydrogen atom is moving during the vibration. Then the effective mass is doubled and the energy falls by $\sqrt{2}$.

In the optical techniques the spectra of the deuterated and hydrogenated compounds are compared and the bands that have shifted on deuteration are identified. This exposes those modes involving hydrogen atom displacements but it is not necessarily a simple exercise, especially if the spectrum is rich. In neutron spectroscopy the same effects occur and the band shifts are the same. What has changed, however, is that in neutron spectroscopy the cross sections of the hydrogenated and deuterated molecules are quite different. Those modes involving the motion of hydrogen atoms that have been replaced by deuterium have a scattering cross section reduced by an order of magnitude (82.0 barn changed to 7.64 barn) and appear only weakly in the spectrum. Often the bands seem to disappear altogether from the spectrum [5]. This has two important advantages over the optical technique: most obviously, the assignment of the modes that involve the remaining hydrogen atoms is trivial, and the the spectrum is less congested.

It is the latter advantage that is valuable since it allows uniquely important, individual hydrogen atoms to be studied free from the interference of other hydrogen atom vibrations. This is why neutron spectroscopy is outstandingly suited to the study of hydrogen bonds.

It is useful to summarise here some of the less well-publicised aspects of the isotopic substitution technique. Most obviously, the vibrations involving the substituted atoms will still change their energies and so change the neutron spectrum, although this can only be seen under suitable experimental resolution [5]. Less obviously is that the vibrations of modes involving the deuterium atoms may still involve a significant cross section compared with other atoms and their intensity does not totally disappear. This is especially troublesome if the weak features of a complex neutron spectrum must be analysed.

2.3 Energy and momentum transfer

Here we shall define the energy, E, and momentum, k, of a neutron and address their manipulation. We recall the expression for the de Broglie energy, E, and wavelength, λ, of any quantum particle

$$E = \hbar \omega = \tilde{\nu} h c = \frac{m_n |v|^2}{2} \tag{2.15}$$

where $\hbar$ is Planck constant/2π; ω, angular frequency; $\tilde{\nu}$, wavenumber; c, speed of light; m_n, neutron mass, v, neutron velocity.

Also

$$\lambda = \frac{h}{k} = \frac{h}{m_n v} \quad \text{or} \quad \frac{2\pi}{\lambda} = \frac{m_n v}{\hbar} \quad \text{and} \quad \frac{2\pi}{\lambda} = \frac{m_n v}{\hbar} = k \tag{2.16}$$

Although the energy is strictly $\tilde{\nu} h c$, to be consistent with the neutron scattering literature we use ω as the symbol for energy with the spectroscopic wavenumber (cm^{-1}) unit (see box).

A note on units of energy and momentum

Energy. Although the SI unit of energy is the joule, in neutron scattering it is conventional to express neutron energy in meV or THz units. In molecular spectroscopy the wavenumber, cm^{-1}, is the common energy unit and this is the unit we use. We have: 1 meV = 0.2418 THz = 1.602×10^{-22} J = 8.066 cm^{-1}

Momentum. Although the SI unit of length is the metre and the units of momentum are kg m s^{-1}, it is conventional in neutron scattering to express length in Ångstrom units: 1Å = 10^{-10} m. Consequently the conventional unit for the neutron wavelength is Å and for the neutron wavevector Å^{-1}.

The energy and momentum of the neutron are related through the neutron velocity. We evaluate the energy and momentum of a neutron, of mass 1.67×10^{-27} kg, travelling with a speed, $v = 1000$ m s^{-1}.

$$E = \frac{m_n v^2}{2} = 0.5 \times 1.67 \times 10^{-27} \text{ kg} \times 10^6 \text{ m}^2\text{s}^{-2}$$

$$= 0.838 \times 10^{-21} \text{ J or } 5.23 \text{ meV or } 42.2 \text{ cm}^{-1} \tag{2.17}$$

$$|k| = m_n v = \left(1.67 \times 10^{-27} \text{ kg}\right) \times \left(1000 \text{ m s}^{-1}\right) = 1.67 \times 10^{-24} \text{ kg m s}^{-1} \tag{2.18}$$

$$\lambda = \frac{h}{m_n v} = \frac{\left(6.63 \times 10^{-34} \text{ J s}\right)}{\left(1.67 \times 10^{-27} \text{ kg} \times 1000 \text{ m s}^{-1}\right)} \tag{2.19}$$

$$= 3.96 \times 10^{-10} \text{ m} = 3.96 \text{ Å}$$

$$k = \frac{2\pi}{\lambda} = \frac{6.28}{3.96 \text{ Å}} = 1.59 \text{ Å}^{-1} \tag{2.20}$$

In terms of the conventional units used in this book for energy, E (cm^{-1}); wavelength, λ (Å); momentum, k (Å^{-1}); time, t (μs); temperature, T (K) and distance, d (m) the conversions are:

$$E = \frac{\hbar^2 |k|^2}{2 m_n} = \left(16.7 \frac{\text{cm}^{-1}}{\text{Å}^{-2}}\right) |k|^2$$

$$= \frac{h^2}{2 m_n}\left(\frac{1}{\lambda^2}\right) = \left(660 \frac{\text{cm}^{-1}}{\text{Å}^{-2}}\right)\frac{1}{\lambda^2}$$

$$= \frac{m_n}{2}\left(\frac{d}{t}\right)^2 = \left(42.2 \times 10^6 \frac{\text{cm}^{-1} \mu\text{s}^2}{\text{m}^2}\right)\left(\frac{d}{t}\right)^2 \tag{2.21}$$

$$= k_B T = \left(0.695 \frac{\text{cm}^{-1}}{\text{K}}\right) T$$

Readers who are familiar with the optical techniques of molecular vibrational spectroscopy will normally ignore momentum. This is because the momentum of a photon is very small; for example, photons of energy 1000 cm^{-1} have a wavelength about 10 microns ($= 10^{-6}$ m or 10^4 Å) and, travelling at the speed of light (3.00×10^8 m s^{-1}), a momentum of 6.28×10^{-4} Å^{-1}. Therefore, those photons used in optical spectroscopy have no significant momentum to exchange with samples and can explore only the zero-momentum-transfer region. This has important consequences, mostly for the study of the external vibrations of molecules. Note, however, that the photons used in X-ray crystallography can, and do, transfer momentum to a substance. However, these photons are very much more energetic than those used in molecular vibrational spectroscopy; if an X-ray had a wavelength *ca* 1.5 Å its energy would be close to atomic ionisation potentials, 10 keV.

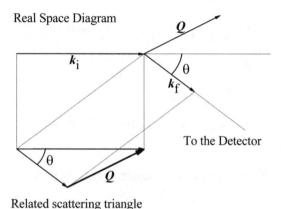

Fig. 2.2 A diagram of the momentum vectors in a scattering event shown in real space, above, and their relationship to the scattering triangle, shown below. The incident, i, and final, f, neutron momenta, k and the transferred momentum Q, are shown in the figure.

In the scattering process the neutron transfers momentum to the sample by virtue of any change in its direction and, or, energy. This is treated classically and is shown diagrammatically in Fig. 2.2. In the resulting vector triangle two sides are found from the incident and final neutron momenta, and the neutron momentum transfer vector, Q, completes the 'scattering triangle'. The value of Q is determined by the application of the well known cosine rule for triangles.

2.3.1 Worked examples—calculating momentum transfer

Here we work through two straightforward examples of the calculation of momentum transfer. The first refers to a typical problem on a direct geometry spectrometer (§3.4.3) and the second outlines considerations on an indirect geometry spectrometer (§3.4.2).

2.3.1.1 Direct geometry

Incident neutrons, of fixed energy 1000 cm^{-1}, are used to excite a vibrational transition at 560 cm^{-1} and the detector is at a scattering angle, θ, of 30° to the incident beam direction. The final energy is:

$$E_f = E_i - \hbar\omega = 1000\ \text{cm}^{-1} - 560\ \text{cm}^{-1} = 440\ \text{cm}^{-1} \tag{2.22}$$

$$k_i = \sqrt{\frac{1000\ \text{cm}^{-1}}{16.7\dfrac{\text{cm}^{-1}}{\text{\AA}^{-2}}}} = 7.74\ \text{\AA}^{-1} \quad k_f = \sqrt{\frac{440\ \text{cm}^{-1}}{16.7\dfrac{\text{cm}^{-1}}{\text{\AA}^{-2}}}} = 5.13\ \text{\AA}^{-1} \tag{2.23}$$

$$Q^2 = k_i^2 + k_f^2 - 2k_i k_f \cos\theta$$
$$= 59.9\ \text{\AA}^{-2} + 26.3\ \text{\AA}^{-2} - 2\times 39.7\ \text{\AA}^{-2} \times 0.866 = 17.4\ \text{\AA}^{-2}$$
$$Q = 4.18\ \text{\AA}^{-1} \tag{2.24}$$

(At $\theta = 50°$ the momentum transfer rises to 5.93 Å^{-1}.)

2.3.1.2 Indirect geometry

The same as in §2.3.1.1 but here it is the wavelength of the scattered neutrons that is fixed, by Bragg reflecting them from a crystal of pyrolitic graphite, PG, into a detector. (The Bragg angle, θ_B, is 45° and the graphite interplanar spacing, d_{PG}, is 3.35 Å.):

$$\lambda_f = 2d_{PG}\sin\theta_B = 2\times 3.35\ \text{\AA} \times 0.707 = 4.74\ \text{\AA} \tag{2.25}$$

$$E_f = \frac{660\ \text{cm}^{-1}/\text{\AA}^{-2}}{\lambda_f^2} = \frac{660\ \text{cm}^{-1}/\text{\AA}^{-2}}{(4.74\ \text{\AA})^2} = 29.4\ \text{cm}^{-1} \text{ so } k_f = 1.32\ \text{\AA}^{-1}$$

$$E_i = E_f + \hbar\omega = 29.4\ \text{cm}^{-1} + 560\ \text{cm}^{-1} = 589\ \text{cm}^{-1} \text{ so } k_i = 5.94\ \text{\AA}^{-1}$$

$$Q^2 = 35.3\ \text{\AA}^{-2} + 1.76\ \text{\AA}^{-2} - 2\times 1.32\ \text{\AA}^{-1} \times 5.94\ \text{\AA}^{-1}\cos(30°)$$
$$= 23.4\ \text{\AA}^{-2} \tag{2.26}$$

$$Q = 4.84\ \text{\AA}^{-1}$$

Here we have used the well-known Bragg's Law [1] to calculate the final wavelength.

2.4 Thermal ellipsoids

Thermal ellipsoids will be used to introduce concepts associated with the representation of atomic displacements in crystals. They are commonly referred to in the crystallographic literature as thermal parameters, or mean square displacements, but modern authors prefer anisotropic displacement parameters. Pictorially they are most familiar through drawings of molecular structures as determined from crystallographic data [6].

The atoms are not represented as points but rather as probability volumes. The volume commonly, but not exclusively, used is generated from three ellipses each centred at the equilibrium position of the atom and each mutually perpendicular. Conventionally there is a 50% chance that the atom will be found within the ellipsoidal volume. One major (and one minor) axis of each ellipse is shared with another. This is the geometric representation of a tensor, a single physical object that has different magnitudes depending on the specific direction in which it is measured. Mathematically these shapes are commonly represented by a 3×3 matrix, see Eq. (2.40). For a harmonic system in a Cartesian frame centred on the atomic position and oriented along the major and minor axes of the ellipsoids the matrix representation can have terms only along its diagonal: these are the principal values. A matrix showing off-diagonal terms is thus referred to a Cartesian frame that is not aligned with the principal axes, or represents an anharmonic system. The sum of the diagonal values of a matrix, its trace, remains constant irrespective of its orientation to the Cartesian frame. Whenever matrices are manipulated together they must be referred to the same axis system. The geometric and matrix representations of a typical anisotropic displacement parameter tensor are shown in Fig. 2.3.

Anisotropic displacement parameters are a measure of the uncertainty of the atomic position and this arises from two sources: the vibrational motion of the atoms (both internal and external vibrations, also called dynamic disorder) and the effects of disorder (both atomic and molecular disorder, also called static disorder). anisotropic displacement parameters, as determined by crystallography, are the sum of all such contributions.

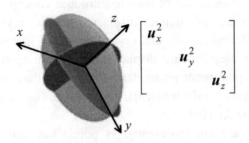

Fig. 2.3 The geometric and matrix representation of the anisotropic displacement parameter tensor of an atom. The u^2 are the mean square displacements of the atom along the Cartesian coordinate axes.

The intensity of a Bragg reflection depends on the size of the anisotropic displacement parameters of those atoms that contribute to the reflection [6] and so individual parameters can be extracted from crystallographic data. As extracted they are referred to the crystallographic axes and are not necessarily simply oriented with respect to significant molecular directions. These parameters are of some interest in the study of molecular vibrations with neutrons because of the dynamical contribution to their value (but not the disorder contribution). The dynamic contribution plays an important role in determining the observed transition intensities, through the Debye-Waller factor (§2.5.1.2).

In crystals with no static disorder the anisotropic displacement parameters measured by crystallographic methods must equal those measured by INS spectroscopy (provided that they were determined under the same conditions, and here temperature is particularly important).

2.5 The theoretical framework of neutron scattering

The theory of thermal neutron scattering has for its objective the interpretation of the experimental observables in terms of the microscopic, quantum-like, properties of the scattering sample. It is as well then to understand exactly what is being measured in neutron spectroscopy. (The theory is fully developed in Appendix 2.)

In its simplest form a neutron spectrometer consists of a single detector (of modest area, A) held at some distance, d_f, with respect to the sample and at some angle, θ, with respect to the direction of the incident neutron beam see Fig. 2.1. These quantities are its polar coordinates (d_f, θ). The solid angle, $d\Omega$, (subtended by an angular element, $d\theta$) is $2\pi \sin\theta\, d\theta$. Depending on the detailed design of the spectrometer the energy spectrum of the sample is scanned in some way and, as a function of this scan, the number of neutrons measured in the detector is recorded, per second (the final, or detected, neutron flux, J_f). This must be normalised to the number of incident neutrons reaching the sample, per second (the incident flux, J_i). The strength of the sample's response is, therefore, known as a function of energy, E, and, since the polar coordinates of the detector can be changed, also as a function of solid angle.

The observable is the rate of change of the cross section with respect to the final energy, E_f, and solid angle, $d\Omega$. This is the double differential scattering cross section, $(d^2\sigma/dE_f\, d\Omega)$, with units, barn eV^{-1} steradian^{-1}, where:

$$d\Omega = \frac{A}{d_f^2} \tag{2.27}$$

The changes that occur during the scattering process must now be related to how the initial state of the system (scattering atom plus neutron) is transformed into the final state of the system.

The nature of the neutrons in the incident beam is represented by their initial wavefunction, $\psi_{n,i}$, which at long distances from the scattering nucleus will be a simple plane wave. At short distances from the nucleus the influence of its nuclear potential perturbs the incident wavefunction into the final wavefunction, $\psi_{n,f}$. The form of the final wavefunction is dependent on the energy of the neutron and is determined by solving the

Schrödinger equation. The final wavefunctions can be isotropic in nature, and have spherical (S) wave solutions, or strongly anisotropic, like the P-wave and D-wave solutions (similar to the electron wavefunctions in the Coulombic field of the nucleus). However, the energies of thermal neutrons, about 200–4000 cm^{-1} are too low to require P- or D-wave solutions and only spherical wave solutions need be considered. The sample, of N atoms, is represented by its initial state, i (final state, f) wavefunctions, $\psi_{N,i\ (f)}$. What is needed is an understanding of how the final states, $\langle \psi_{n,f} \psi_{N,f}|$, in changing from the initial states, $\langle \psi_{n,i} \psi_{N,i}|$, relate to $d^2\sigma/dE_f\,d\Omega$.

Fortunately, the solution to this problem was one of the early successes of quantum mechanics and is so central to modern physics that it carries its own name, Fermi's Golden Rule. The transition *rate* between the initial and final states of a system depends on the strength of the coupling between the initial and final states and on the number of ways the transition can happen, the density of the final states [1]. This rule only applies to scattering problems where the incident wavefunction is weakly perturbed by the presence of the sample. If the scattering is not strong we can suppose that the total wavefunction does not differ substantially from the incident wavefunction; this supposition is the Born approximation.

However, we know, from high-energy physics experiments, that the interaction potential between a neutron (of typical energy *ca* 10^{-9} MeV) and the nucleus is *not* weak but involves the strong nuclear force of *ca* 36 MeV. We can, however, avoid the nuclear force problem by searching for a new form for the interaction potential. The required potential must give S-wave solutions to the final neutron wavefunction when the Born approximation is applied.

Only one form for the interaction potential satisfies these conditions, the Fermi pseudo-potential. Its application to neutron scattering can be regarded as justified by its success in interpreting the experimental measurements. Alternatively, we can posit that it is the average strength of the scattering potential that should be considered and, since the strong nuclear force only operates at an extremely short range, *ca* 2 fm, its overall impact at the macroscopic scale of the sample, *ca* 1 mm, is indeed weak. The form of the double differential scattering cross section is then, where the subscript, l, refers to one specific atom of the molecule, Eq. (A2.48):

$$
\left(\frac{d^2\sigma}{dE_f\,d\Omega}\right)_l
$$

$$
= \frac{\sigma_l}{4\pi}\frac{k_f}{k_i}\frac{1}{2\pi\hbar}\sum_l\int_{-\infty}^{\infty}dt\,\langle\exp(-\,i\,Q.r_l(0))\exp(i\,Q.r_l(t))\rangle\exp(-i\omega t)
$$

(2.28)

This form, involving the Fourier transform of time dependent vector operators, is specifically related to the main theme of this book; namely, the vibrational spectroscopy of hydrogenous molecular crystals. Other forms, more appropriate to different disciplines and systems, can be found in the specialist literature (see also Appendix 2).

Lattice modes, or external modes, as well as internal modes produce harmonic displacements of the atoms. The internal modes result in local molecular deformation, whilst external modes are supposed to entrain the atoms when a molecule is rigidly displaced from its equilibrium position in the lattice. The time-dependent atomic position vector r can be expressed in terms of the internal displacement vector u_{int}, taken with respect to the molecular centre of mass, and the external displacement vector, u_{ext} the displacement vectors have units of length, Å, Eq. (A2.52).

$$
\exp(i\,Q.r(t)) = \exp(i\,Q.u(t)_{ext})\exp(i\,Q.u(t)_{int})
$$

(2.29)

This separation of internal and external dynamics will have consequences later.

2.5.1 The scattering law

We rewrite the double differential of Eq. (2.28) in terms of a function, S, which emphasises the dynamics of each individual atom in the sample. This is conventionally called the 'scattering law' and is directly related to the observed intensities when summed over all the atoms in the sample, Eq. (A2.49).

$$
S(Q,\omega)_l = \frac{4\pi}{\sigma_l}\frac{k_i}{k_f}\left(\frac{d^2\sigma}{dE_f\,d\Omega}\right)_l
$$

(2.30)

Most modern neutron spectroscopic data are reduced to the scattering law since its relationship to the common model of molecular vibrations is straightforward. The scattering law is the natural meeting point of experimental data and the dynamical models that have been developed for its understanding. However, it must be admitted that the literature is inconsistent on this point and many different functions can be found. Fortunately many of them are simply scaled versions of the scattering law.

Whilst loosely referring to the scattering law all of the spectroscopic intensities reported in this book are $S^{\bullet}(Q,\omega)$. This function is closely related to the 'amplitude and cross section weighted density of states' often referred to in condensed matter physics studies of molecular systems. For a single molecule containing, N_{atom}, atoms

$$S^{\bullet}(Q,\omega)_{\text{total}} = y \sum_{l=1}^{N_{\text{atom}}} S(Q,\omega)_l \, \sigma_l = 4\pi \frac{k_i}{k_f} \left(\frac{d^2\sigma}{dE_f \, d\Omega} \right)_{\text{observed}} \qquad (2.31)$$

The $S^{\bullet}(Q,\omega)$ are reported in scaled units, which are related to the correct units, barn $(\text{cm}^{-1})^{-1}$, by a linear factor, y. Since the spectra from hydrogenous molecules are dominated by the hydrogen scattering this cross section can be subsumed within y. This practice stems not from a wish to avoid academic rigor but from an acceptance of experimental reality. As with other spectroscopic techniques, neutron spectroscopy is difficult to perform on the basis of absolute measurements and it results in few tangible benefits. Therefore, only the relative strengths of spectral intensities are measured.

Similarly the number of molecules in the neutron beam, N_{mol}, is also subsumed into the scaling factor. What remains has all of the properties of the scattering law since the cross section has effectively disappeared from the formulation. Whenever the expression scattering law is used in the text it will refer to $S^{\bullet}(Q,\omega)$, unless otherwise stated. In Appendix 2 the calculated intensity in a particular mode, labelled ν, involving the atomic displacement $^{\nu}u$ is given (for a *single crystal* sample) by Eq. (A2.80):

$$S^{\bullet}(Q,\omega_v)_l^n = y\,\sigma_l\frac{\left[\left(Q\cdot{}^v u_l\right)^2\right]^n}{n!}\exp\left(-\left(Q\cdot\sum_v {}^v u_l\right)^2\right) \qquad (2.32)$$

Scattering law units. The scattering law, Eq. (2.30), has units of (energy)$^{-1}$, J^{-1}, which in our convention is $(cm^{-1})^{-1}$. For $S^{\bullet}(Q,\omega)$, Eq. (2.31), the units are barn$\times(cm^{-1})^{-1}$, although the scaling factor will effectively hide this. Since neutron momentum transfers are in Å^{-1}, the product $Q.u$ is dimensionless, as it must be if it is to appear as an argument to a function.

This equation, Eq. (2.32), gives the calculated relative intensity of the vth mode determined at a momentum transfer Q and at an energy transfer ω_v (in the low temperature limit)(Fig. 2.4). This is $S^{\bullet}(Q,\omega_v)$ for the atom, labelled l, with a total scattering cross section σ_l. There are two major terms in Eq. (2.32), the pre-exponential factor and an exponential term.

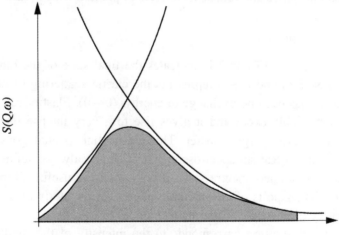

Momentum transfer / Å^{-1}

Fig. 2.4 Diagrammatic representation of the scattering law value, *S*, as it varies with momentum transfer *Q*, for a given vibrational mode, *v*, at an energy transfer of *ω*. At low momentum values the intensity increase is driven by the pre-exponential factor but at high values of momentum the intensity is suppressed by the Debye-Waller factor.

2.5.1.1 The pre-exponential term

The pre-exponential term, Q^2u^2, increases as the momentum transferred, or the scattering atom displacement increases. Since the value of Q is fixed by the spectrometer the determination of $S^*(Q,\omega)$ allows the atomic displacements (the eigenvectors of the vibration) to be extracted.

2.5.1.2 The exponential, or Debye-Waller factor

The exponential term in Eq. (2.32) is called the Debye-Waller factor and it always decreases more rapidly with Q^2u^2 than the pre-exponential factor increases. The argument Q^2u^2 has significant temperature dependence (§A2.4), through the atomic displacement, and samples should always be studied at cryogenic temperatures to increase the observed intensities and reduce the effects of phonon wings (§2.6.3).

2.5.1.3 The orders, n

The value of n (Table 2.1) indicates the final state of the mode that has been excited and $n = 0$ represents the elastic scattering of neutrons where there has been no exchange of energy, (0←0). Elastic scattering is the most probable event and it gives rise to a very intense line in the spectrum at zero energy transfer. This elastic line is analogous to the Rayleigh line in Raman spectroscopy. It is almost always omitted from published vibrational spectra since it contains little useful information. However, its shape is of considerable interest in the study of molecular diffusion by quasielastic scattering techniques [7].

The $n = 1$ solution corresponds to the intensity of the fundamental transition (1←0) and $n = 2$ is the first overtone (2←0) intensity. In optical spectroscopy, the spectra are dominated by the fundamental transitions since overtones, $\Delta v \neq 1$, are formally forbidden and appear only weakly, if at all. However, in neutron spectroscopy overtones are commonly observed and their contribution, at least to $n = 10$, should be calculated. According to Eq. (2.32) the overall calculated spectrum consists of a number of infinitely sharp lines, corresponding to the elastic line, the fundamentals and the overtones of all internal modes, see Table 2.1.

Table 2.1 The first four contributing orders to neutron spectra.

Spectroscopic Description		delta-function representation (a)
Neutron scattering convention	Optical convention	
elastic line $(0 \leftarrow 0)$; n = 0 0 quantum events	Rayleigh line $\Delta v = 0$	$[\, \delta(\omega) +$
fundamentals $(1 \leftarrow 0)$; n = 1 1 quantum events	fundamentals $\Delta v = 1$ 1st harmonic	$+ \delta(\omega\text{-}v_1) + \delta(\omega\text{-}v_2) +.. \delta(\omega\text{-}v_{\text{Natom-6}})$
multi-scattering (b) $(2 \leftarrow 0)$; n = 2 2 quantum events	1st overtone $\Delta v = 2$ 2nd harmonic	$+ \delta(\omega\text{-}2v_1) + \delta(\omega\text{-}2v_2) +.. \delta(\omega\text{-}2v_{\text{Natom-6}})$
multi-scattering (b) $(3 \leftarrow 0)$; n = 3 3 quantum events	2nd overtones $\Delta v = 3$ 3rd harmonic	$+ \delta(\omega\text{-}3v_1) + \delta(\omega\text{-}3v_2) +.. \delta(\omega\text{-}3v_{\text{Natom-6}}) \,]$

(a)see Appendix 2. (b)Compare §3.5.1.

2.5.2 Powder averaging

The reader will appreciate that the scattering law of a particular vibrational transition, given by Eq. (2.32), is written in terms of the scalar (or inner, or dot) product of the neutron momentum transfer vector, $\mathbf{Q}$, and the atomic displacement vector, $\mathbf{u}$. This product is a function of the angle between the vectors and allows some interesting single crystal effects to be examined.

2.5.2.1 Single crystal effects

Occasionally molecular crystals of suitable size are found that have all of their molecules aligned mutually parallel [8]. It is worth investigating how the scattering law of one particular atom, vibrating in the normal mode, v, would change as a function of the relative orientation, ϕ, of the two vectors. Since $\boldsymbol{Q} \cdot \boldsymbol{u} = Q u \cos \phi$ with $Q = |\boldsymbol{Q}|$ and $u = |\boldsymbol{u}|$.

$$S^*(Q,\omega_v) = \left[Q\cos\phi \, {}^v u\right]^2 \exp\left(-\left(Q\cos\phi \sum_v {}^v u\right)^2\right) \tag{2.33}$$

The value of the pre-exponential factor is strongly dependent on the relative orientation of the two vectors. If they are parallel, $\cos\phi = 1$, the scattering strength reaches its maximum value but if they are perpendicular, $\cos\phi = 0$, the intensity in this transition disappears. (Except that, due to the finite thickness of samples, the observed intensity never falls to zero. This is similar to the effects of 'leakage' in single crystal Raman spectroscopy.) In favourable cases the angular dependence of the intensity variation can be used to examine some vibrations in isolation from others, or alternatively to determine the relative orientations of different vibrational displacements.

The angular variation in the exponential of the Debye-Waller factor is not as striking as that for the pre-exponential. This function depends on the sum over all v, and this will always provide some component of atomic displacement parallel to the momentum vector.

Large, aligned single crystal samples are rare and, even if available, a molecular geometry that is too complex will render studies as a function of $\cos\phi$ generally unproductive. Almost all neutron spectroscopy is performed on powders and it is simpler to analyse the spectrum of a powder than that of a partly ordered sample. The experimental results obtained from these powdered samples must be compared with calculations of the scattering law made over all directions of the momentum transfer vector in Eq. (2.32). The powder averaged scattering law is

$$S(Q,\omega)_{\text{powder}} = \frac{1}{4\pi} \int S(\boldsymbol{Q},\omega)\mathrm{d}\boldsymbol{Q} \tag{2.34}$$

2.5.2.2 The powder average of an isotropic system

Here we shall consider the simple case of a single hydrogen atom confined in a box. Such a system is akin to an octahedral interstitial metal hydride (§6.6.1), where the six metal atoms surrounding each hydrogen atom represent the walls of the box.

The great difference in mass between the metal and the hydrogen simplifies the vibrational dynamics. The hydrogen atom has three mutually perpendicular vibrations, labeled $\nu = 1$, 2 and 3. These are of equal energy and displacement, u, and the INS spectrum consists of a single F mode, there are no other hydrogen vibrations. The displacements are conveniently arranged along the Cartesian axes, x, y and z, and the system is isotropic.

Depending on the direction of the momentum transfer vector, Q, it will generate components, q_x along x, q_y along y and q_z along z. The intensity resulting from the $\nu = 1$ hydrogen displacement u is proportional only to $u.q_x$ since there are no components of hydrogen displacement along either y or z. Likewise, for $\nu = 2$ (or 3) only the component q_y (or q_z) is effective. For equal components along each axis, then $q_x = q_y = q_z = Q/\sqrt{3}$, thus $\cos\phi = 1/\sqrt{3}$. Substituting this value into Eq. (2.32) the total signal arising from the three fundamental, $n = 1$, vibrations is

$$S(Q,\omega)_{\text{total}} = \left\{ \left(\frac{Qu}{\sqrt{3}} \right)^2 + \left(\frac{Qu}{\sqrt{3}} \right)^2 + \left(\frac{Qu}{\sqrt{3}} \right)^2 \right\} \exp\left(-\sum_{1,2,3} \left(\frac{Qu}{\sqrt{3}} \right)^2 \right)$$

$$= Q^2 u^2 \exp\left(-Q^2 u^2 \right) \tag{2.35}$$

Because the three vibrations are degenerate, if the q components along one direction are reduced they are compensated by the larger components along other directions. Thus, for example, if $q_y = q_z = 0$ then $q_x = Q$ and substituting this into Eq. (2.32) gives the same result. The direction of the momentum transfer vector is irrelevant and it may be represented by its magnitude alone, Q.

The directionally averaged total intensity is then given by a very straightforward equation.

$$S(Q,\omega)_{\text{total}} = Q^2 u^2 \exp\left(-Q^2 u^2 \right) \tag{2.36}$$

The simplicity of this equation embodies the great strength of INS spectroscopy. The observed intensity, S, is a simple function of the magnitude of the momentum transferred in the scattering event Q (which is determined only by the spectrometer used) and the magnitude of the hydrogen atom displacement u (which is determined only by the forces and the masses in the molecule studied).

This equation, Eq. (2.36), is a popular approximation in condensed matter texts where it is used exclusively to discuss the scattering from incoherent systems. However, in chemistry there are but few examples where the equation should be applied, indeed strictly only when the atomic symmetry is locally O_h. The local symmetry of a scattering atom is usually non-isotropic and is almost never the same as the point group symmetry of the molecule. Moreover, further complexities are introduced by the need to handle the large numbers of different atoms and transitions found in molecular vibrational spectroscopy.

2.5.2.3 The powder average of an anisotropic system

To proceed further we need to develop our representation of the mode specific atomic displacement, presently given as a vector, $^v\!u$. We shall find it most convenient to express this as a tensor in matrix form, $^v\!B$. Where the unit vectors along the Cartesian axes are e_x, e_y, e_z.

$$^v\!u = {}^v\!\left(u_x e_x + u_y e_y + u_z e_z\right)$$

$$^v\!B = {}^v\!u^T\,{}^v\!u = {}^v\!\left(u_x,\ u_y,\ u_z\right)^T\,{}^v\!\left(u_x,\ u_y,\ u_z\right) \tag{2.37}$$

$$= {}^v\!\left(\begin{pmatrix} u_x \\ u_y \\ u_z \end{pmatrix}\begin{pmatrix} u_x & u_y & u_z \end{pmatrix}\right) = {}^v\!\begin{pmatrix} u_x u_x & u_x u_y & u_x u_z \\ u_y u_x & u_y u_y & u_y u_z \\ u_z u_x & u_z u_y & u_z u_z \end{pmatrix}$$

The unit vectors are implicit and only the components, u_x, u_y, u_z, are retained. In this form we may rewrite the displacements of the metal hydride treated above (§2.5.2.2) and Eq. (2.37) becomes:

$$^1\!B_H = (u,0,0)^T\,(u,0,0) = \begin{pmatrix} u^2 & 0 & 0 \\ 0 & 0 & 0 \\ 0 & 0 & 0 \end{pmatrix}$$

$$^2\!B_H = (0,u,0)^T\,(0,u,0) \qquad ^3\!B_H = (0,0,u)^T\,(0,0,u) \tag{2.38}$$

The sum over all the individual mode tensors gives the total atomic displacement tensor, due to the internal modes, A_{int}.

$$\left(A_l\right)_{int} = \sum_v {}^v B_l \tag{2.39}$$

or, specifically for the metal hydride example:

$$\left(A_H\right)_{int} = {}^1 B_H + {}^2 B_H + {}^3 B_H = \begin{pmatrix} u^2 & & \\ & u^2 & \\ & & u^2 \end{pmatrix} \tag{2.40}$$

(Eventually we shall include all the vibrational contributions, both internal and external, and the resulting A_{total} should then equal the anisotropic displacement parameters discussed above (§2.4).)

We are now ready to introduce the 'almost isotropic' approximation to the powder averaged intensity for fundamental transitions; this is derived (along with specific expressions for the combinations and overtones) in Appendix 2, Eq. (A2.93).

$$S(Q, \omega_v)_l = \frac{Q^2 {}^{Tr}\left({}^v B_l\right)}{3} \exp\left(-Q^2 \alpha_l^v\right) \tag{2.41}$$

$$\text{where} \quad \alpha_l^v = \frac{1}{5}\left\{ {}^{Tr} A_l + 2\left(\frac{{}^v B_l : A_l}{{}^{Tr}\left({}^v B_l\right)}\right) \right\}$$

When represented in matrix notation the Trace (^{Tr}) operation, or Spur, of a tensor is the sum of its diagonal components, its scalar value. The contraction operator : implies that the two matrices on either side should be multiplied and the trace of the resulting matrix obtained. This equation, Eq. (2.41), is less troublesome than it first appears, as we shall demonstrate by substituting the results of Eq. (2.38) and (2.40) for the vibration, $v = 1$. First let us evaluate the α_l^v term.

$$\alpha_{\mathrm{H}}^1 = \frac{1}{5}\left\{ {}^{\mathrm{Tr}}\begin{bmatrix} u^2 & & \\ & u^2 & \\ & & u^2 \end{bmatrix}_{\mathrm{H}} + 2\frac{\left({}^1\begin{bmatrix} u^2 & & \\ & 0 & \\ & & 0 \end{bmatrix}_{\mathrm{H}} : \begin{bmatrix} u^2 & & \\ & u^2 & \\ & & u^2 \end{bmatrix}_{\mathrm{H}}\right)}{{}^{\mathrm{Tr}}\left({}^1\begin{bmatrix} u^2 & & \\ & 0 & \\ & & 0 \end{bmatrix}_{\mathrm{H}}\right)} \right\}$$

$$\alpha_{\mathrm{H}}^1 = \frac{1}{5}\left\{\left(3u^2\right) + 2\left(u^2\right)\right\} = u^2 \tag{2.42}$$

The effect of the contraction operator is to select the size of the ellipsoid A, see Fig. 2.3, measured in the direction of the vibration under consideration. This is brought out more fully in the worked example, below. Thus the Debye-Waller factor is determined by an α_l^v value weighted to contributions in the same direction as the vibration of interest. The pre-exponential factor is:

$$\frac{Q^{2\,\mathrm{Tr}}\left({}^1 B_{\mathrm{H}}\right)}{3} = \frac{Q^2 u^2}{3} \tag{2.43}$$

Substituting these values into Eq. (2.41), along with the results for the other two vibrations, $v = 2$ and 3, gives:

$$S(Q, \omega_{1,2,3})_{\mathrm{H}} = Q^2 u^2 \exp\left(-Q^2 u^2\right) \tag{2.44}$$

The results of Eq. (2.36) are recovered.

2.5.2.4 Combinations—the other consequence of working with powders

In powdered samples there are always some crystallites such that components of Q fall parallel to a specific vibrational displacement, so all vibrations are observed. By the same token, all the crystallites in the sample can never be simultaneously aligned and so the intensity observed from a powder is always less than the maximum strength a transition could have in a single crystal study (see §2.5.2.1). This intensity reduction is quite severe and only about one third of the maximum intensity remains, as was shown above. Also, intensities can be found for transitions that involve the simultaneous excitation of different internal modes. All combinations, e.g. $(1 \leftarrow 0)$ $(1' \leftarrow 0)$, between modes involving the displacement of a common scattering atom are possible (provided that momentum and energy are conserved). However, they are not equally probable and so some combinations are stronger than others. The higher quantum events include contributions like:

two quantum
events
$$\left\{ \begin{array}{l} \delta(\omega - 2v_1) + \delta(\omega - 2v_2) + \delta(\omega - 2v_3) + \dots \\ \delta(\omega - v_1 - v_2) + \delta(\omega - v_1 - v_3) + \dots \end{array} \right.$$

and

three quantum
events
$$\left\{ \begin{array}{l} \delta(\omega - 3v_1) + \delta(\omega - 3v_2) + \delta(\omega - 3v_3) + \dots \\ \delta(\omega - 2v_1 - v_2) + \delta(\omega - 2v_1 - v_3) + \dots \\ \delta(\omega - v_1 - v_2 - v_3) + \delta(\omega - v_1 - v_2 - v_4) + \dots \end{array} \right.$$

The process of powder averaging has removed vectors from the problem, as it must, but has reintroduced a degree of mode specificity into the Debye-Waller factor, through α^v_l. The exact value of α^v_l is determined, in part, by the direction of the scattering atom's displacement during the vibration under consideration. This is not an over dominant input and all the modes make some contribution through $^{Tr}A_l$. It is because all vibrations contribute to the argument in the Debye-Waller factor that the simple isotropic approximation to α^v_l is sometimes used in unsophisticated calculations. In the isotropic approximation

$$\alpha^v_l = \frac{^{Tr}A_l}{3} \tag{2.45}$$

We shall now evaluate the intensity of the vibrational bands in a simple system.

2.5.3 Worked example—hydrogendifluoride (bifluoride) ion [HF₂]⁻

The bifluoride ion (hydrogendifluoride, $[HF_2]^-$), offers a simple example that permits the mathematics to be worked through in detail.

The displacement vectors, Å, of the atoms in the internal modes of the ion are given in Table 2.2, from Eqs. (4.38) and (4.40), §4.2.5. Since the *ab initio* calculations, from which these values are drawn, treat the individual components of degenerate modes separately, so shall we.

Table 2.2 The non-zero displacement vectors, Å, of the atoms in the [FHF]⁻ ion.

	Symmetric stretch 575 cm^{-1}	Degenerate bend 1235 cm^{-1}		Antisymmetric stretch 1370 cm^{-1}
	z	x	y	z
F	0.028	-0.003	-0.003	-0.003
H	0.000	0.115	0.115	0.109
F	-0.028	-0.003	-0.003	-0.003

Then, using the hydrogen displacements in the x component of the bending vibration at 1235 cm^{-1}, as an example, the displacement tensors and their trace values, see Eq. (2.38), are calculated.

$$^{1235}B_H = \begin{bmatrix} 0.115 \\ 0.0 \\ 0.0 \end{bmatrix} \begin{bmatrix} 0.115 & 0.0 & 0.0 \end{bmatrix} = {}^{1235}\begin{bmatrix} 0.0132 & & \\ & 0.0 & \\ & & 0.0 \end{bmatrix}_H \quad (2.46)$$

$$^{Tr}\left(^{1235}B_H\right) = 0.0132$$

and using the mode at 575 cm^{-1} to illustrate the fluorine motions

$$^{575}B_F = \begin{bmatrix} 0.0 \\ 0.0 \\ 0.028 \end{bmatrix} \begin{bmatrix} 0.0 & 0.0 & 0.028 \end{bmatrix} = {}^{575}\begin{bmatrix} 0.0 & & \\ & 0.0 & \\ & & 0.0008 \end{bmatrix}_F$$

$$^{Tr}\left(^{575}B_F\right) = 0.0008 \quad (2.47)$$

Readers may wish to convince themselves that they can obtain the remaining vibrational tensors in a similar manner. The trace values of the individual tensors are given in Table 2.3. From the individual tensors the total displacement tensors, A, are generated by addition. The example is given for one of the fluorine atoms.

$$A_F = {}^{575}B_F + \left({}^{1235}B_F\right)_x + \left({}^{1235}B_F\right)_y + {}^{1370}B_F$$

$$= \begin{bmatrix} 0.000009 & & \\ & 0.000009 & \\ & & 0.0008 \end{bmatrix}_F \qquad (2.48)$$

$${}^{Tr}\left(A_F\right) = 0.00082$$

Table 2.3 The trace values, Å^2, of the vibrational displacement tensors for each of the vibrations of the bifluoride ion.

The atom and mode specific tensor	The trace value / Å^2
${}^{Tr}\left({}^{575}B_H\right)$	0.0
${}^{Tr}\left({}^{575}B_F\right)$ (a)	0.0008
${}^{Tr}\left({}^{1275}B_H\right)$	0.0132 (b)
${}^{Tr}\left({}^{1275}B_F\right)$ (a)	0.000009 (b)
${}^{Tr}\left({}^{1370}B_H\right)$	0.0119
${}^{Tr}\left({}^{1370}B_F\right)$ (a)	0.0

(a) for each fluorine. (b) the x and y components are of equal magnitude

The mode-independent total displacement tensor value for the hydrogen atom is:

$$A_H = {}^{575}B_H + \left({}^{1235}B_H\right)_x + \left({}^{1235}B_H\right)_y + {}^{1370}B_H = \begin{bmatrix} 0.0132 & & \\ & 0.0132 & \\ & & 0.0119 \end{bmatrix}_H$$

$${}^{Tr}\left(A_H\right) = 0.0383 \qquad (2.49)$$

For the calculation of α, see Eq. (2.41) and (2.42). We shall use the example of the hydrogen motion in the y component of the bending mode

$$\left({}^{1235}B_H\right)_y : A_H = {}^{Tr}\left(\left({}^{1235}B_H\right)_y \times A_H\right)$$

$$
\overset{Tr}{=} \left(\begin{bmatrix} 0.0132 & & \\ & 0.0132 & \\ & & 0.0119 \end{bmatrix} \begin{bmatrix} 0.0 & & \\ & 0.0132 & \\ & & 0.0 \end{bmatrix} \right) \tag{2.50}
$$

The value of the contracted tensors is 0.000174 Å², and then:

$$
\begin{aligned}
\alpha_H^{1235} &= \frac{1}{5} \left\{ \overset{Tr}{(A_H)} + 2 \frac{A_H : \overset{1235}{B_H}}{\overset{Tr}{(\overset{1235}{B_H})}} \right\} \\
&= \frac{1}{5} \left\{ 0.0383 + 2 \left(\frac{0.000174}{0.0132} \right) \right\} = 0.0129
\end{aligned} \tag{2.51}
$$

The other α values, in Å², are given in Table 2.4.

Table 2.4 The α values, in Å², for the vibrations of the atoms in the [FHF]⁻ ion.

atom	α^{575}	α^{1235}	α^{1370}	α_{iso} (a)
F	0.000484	0.000168	-	0.00027
H	-	0.0129	0.0124	0.0128

(a) In the isotropic approximation, $\alpha = {}^{Tr}A/3$

We shall assume that any value of Q can be achieved and choose $Q = 7$ Å⁻¹. Then, using the symmetric stretch as our example, we calculate the scattering law values (intensities) of the transitions.

$$
S_F = \sigma_F \frac{Q^2 \overset{Tr}{(\overset{575}{B_F})}}{3} \exp(-Q^2 \alpha_F^{575}) =
$$

$$
= 4.0 \frac{49(0.008)}{3} \exp(-49(0.000484)) = 0.0511 \tag{2.52}
$$

$$
S_H = \sigma_H \frac{Q^2 \overset{Tr}{(\overset{575}{B_H})}}{3} \exp(-Q^2 \alpha_H^{575})
$$

$$= 82.0 \frac{49(0.0)}{3} \exp(-49(0.0077)) = 0.0 \qquad (2.53)$$

The total intensity for this mode at 575 cm^{-1} is:

$$\left(S\right)_{\text{total}}^{575} = 2S_{\text{F}} + S_{\text{H}} = 0.102 \qquad (2.54)$$

The remaining results are given in Table 2.5.

Table 2.5 The calculated intensities, S, of the vibrational bands of the [FHF]$^-$ ion.

Atomic contributions	S^{575}	S^{1235}	S^{1370}
F	0.051	0.0012 (b)	0.0
H	0.0	18.76 (b)	8.7
Total	0.102 (a)	18.76	8.7

(a) This includes contributions from both fluorine atoms (b) This includes contributions from both bending components

Here we note that the ratio of the two principal intensities at 1275 and 1370 cm^{-1} is 2.16, which corresponds approximately to the ratio of the degeneracies of their respective modes, 2. This is a consequence of the simple geometry of the bifluoride ion: the proximity of the two vibrational frequencies and the fact that the same Q value was used to calculate both values of S. This is not a common occurrence, if only because molecules usually contain more than a single type of hydrogen.

2.6 Band shaping processes in neutron spectroscopy

There are several band shaping processes that change the appearance of observed neutron spectra from the idealised series of delta sharp transitions to more realistic representations of observations. We reserve instrumental considerations for Chapter 3 and here we focus on dynamical processes occurring in the crystal.

2.6.1 Vibrational dispersion

Vibrational dispersion is, for vibrations of the same type, the variation of the relative phase of atomic displacements with their frequency. Although it is a well known effect in molecular spectroscopy it is not usually presented as such. Rather there is an emphasis on treating each vibration individually and the deep similarities between vibrations of the same type are often obscured. Dispersion in molecules is normally seen in its discrete form and the effect of continuous dispersion is only observed in the spectroscopy of polymers (§10.1.1.1). The connection between the discrete and continuous forms is most easily seen by considering the vibrations of molecules composed of simple chemical motifs repeated several-fold. We have chosen to demonstrate this with the highest frequency vibrations in benzene, the stretches of the six C–H repeat units.

2.6.1.1 Discrete dispersion

The six individual carbon-hydrogen stretching oscillators in a benzene molecule combine to give the six C–H stretching normal modes of the molecule, as follows; A_{1g} (v_1, 3062 cm^{-1}), E_{1u} (v_{12}, 3063 cm^{-1}), E_{2g} (v_{15}, 3047 cm^{-1}) and B_{1u} (v_5, 3068 cm^{-1}). In any one of these modes the C–H oscillators vibrate at the same frequency, undergo the same displacements and have a constant phase angle, φ, with respect to their neighbours. The phase angles determine the relative positions of neighbouring hydrogen atoms on their respective displacement paths. A sinusoidal wave represents these phase differences and the longer its wavelength the less the phase difference between neighbours and the smaller the value of φ, see also (§4.3).

Since benzene is a ring, only whole wavelength solutions are relevant and the simplest vibration to understand is v_5 (all anti-phase). Here, three short wavelengths represent the relative phases and there are six nodes that sit between the hydrogen atoms, as shown in Fig. 2.5.

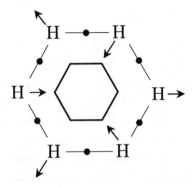

Fig. 2.5 The hydrogen displacement vectors for the ν_5 C–H stretching mode in benzene. All nearest neighbours are in antiphase. The three maximal antinodes correspond to the hydrogen atoms moving out from the molecular centre and the three minimal antinodes to those moving inward. The positions of the six nodes are shown by the • symbol.

The other modes follow accordingly. The mode ν_{15} has two whole wavelengths with four nodes, two aligned with opposite carbons and two aligned with the centres of opposite C–C bonds. The other component of this doubly degenerate mode has the same nodal distribution but with the pattern rotated a one-eighth turn around the C_6 ring. The mode ν_{12} has one whole wavelength with two nodes, aligned with the centres of opposite C–C bonds. The mode ν_1 has all the C–H oscillators in phase, the wavelength is infinite and there are no nodes. The relevant phase angles are: ν_1, $\varphi = 0$; ν_{12}, $\varphi = \pi/3$; ν_{15}, $\varphi = 2\pi/3$ and ν_5, $\varphi = 3\pi/3$ radians.

The number of different phase angles available to any given type of oscillation of a molecule depends on the number of like oscillators involved. The relationship between the energy, E, and the phase angle, φ, of a given type of vibration is its dispersion curve, (E, φ). This is explored further in §4.3.2, where the concept of the vibrational wavevector is introduced and a practical example is presented in §10.1.2. If a larger, flat, conjugated CH system than benzene could be studied the increased number of stretching transitions would again fall, more or less, within the same energy band. The increased number of phase angles would create a denser network of local vibrational states.

2.6.1.2 Continuous dispersion

The phase relationships describing the internal vibrations of a molecule can disappear in larger systems and the normal mode displacements, which in theory are generalised over the whole molecule, become localised on one part or another. The forces carrying the phase information are too weak and it is lost. What can be true in large molecular systems is certainly found in lattices.

Most of the internal vibrations of a molecule in a lattice are rarely in any phase relationship with those of another molecule in the lattice. The energy of these vibrations is the same for all molecules and so independent of any phase. The dispersion curve is flat and the vibrations are localised on individual molecules. This is because the intermolecular forces of the lattice are making insufficient contribution to the potential energy stored during a molecular distortion to carry phase information. As the strength of the intramolecular forces falls the significance of the intermolecular forces rises, phase information is carried and the lowest vibrations of molecules are often dispersed. Under these circumstances the mode is distributed amongst the many molecules involved in the phase relationship and it is, therefore, extended in space and not localised.

This is exactly like the six C–H stretches of benzene but here there are as many oscillators as there are molecules in the crystal, millions. The dispersion relationship now has many points on the (E, φ) plot and it is usually described as a continuous curve. (Strictly it remains discrete and the phase angles are evenly spread along the φ axis.) The intensity, S, arising from such an internal mode no longer appears at a unique energy transfer but is spread over the energy band spanned by the dispersion. The exact band shape produced by this process and its width in energy is varied and depends upon the details of the dispersion. Generally dispersion has an impact on INS spectra of internal vibrations similar to that seen in optical spectroscopy for site-group and factor-group effects. The dispersion of the external modes is covered in greater detail elsewhere (§4.3).

2.6.1.3 Davydov and LO-TO splitting

In Davydov (or factor group) splitting vibrational phase relationships between molecules in the same unit cell give rise to an in-phase and an out-of-phase splitting. Each positive (or negative) component of the split pair falls on its own respective positive (negative) dispersion curve, the two curves are of similar shape but distinct.

The very strong electrostatic forces that are found in some ionic crystals also give rise to band splitting processes. Two wave types are used to describe the phase relationships between molecules in three dimensional lattices, transverse and longitudinal. Transverse waves show the phase relationships between vibrations involving molecular displacements normal to the propagation vector of the wave, whilst longitudinal waves refer to displacements parallel to the propagation vector. In ionic crystals these displacements can have significantly different energies if they change the separation between opposite charges. Vibrations that involve opposite charges are infrared active, hence their description as optic modes. The two types are thus Longitudinal Optic, LO, and Transverse Optic, TO, modes. Since there are always two transverse directions for any longitudinal direction then, at the Brillouin zone centre (§4.3.2), the TO modes should be twice the intensity of the LO modes. However, away from the zone centre the influence of dispersion could change this intensity ratio significantly.

2.6.2 Density of vibrational states

In molecular vibrational spectroscopy the density of vibrational states, $g(\omega)$, is the number of vibrations per wavenumber. It plays an important role in all solid state spectroscopy, especially INS. At the transition itself, the density of states is the degeneracy of the mode, see Table 2.6. There are only $3N_{atom} - 6$ internal vibrations for an isolated molecule and they are spread across the entire energy range from 0 to 4400 cm^{-1} (the upper bound is the stretching mode in dihydrogen). Fundamental transitions are, therefore, relatively rare and the density of states is zero at most energies but rises abruptly at those energies where a vibration occurs. We normalise the integrated $g(\omega)$ of each component in every mode to unity, as was done in (§2.5.3). When taken with our normalisation, terms like $g(\omega)_{int}$ are absent from the scattering law equations. This stems from the common practice of using commercial programs to calculate the dynamics of molecules and taking their output as input data in the calculation of $S(Q, \omega)$. The local density of states in the C–H stretching region of the benzene example (§2.6.1.1) is six, $g(\nu_{C-H})_{int} = 6$.

Table 2.6 Vibrational degeneracy of modes.

Character of the mode	Degeneracy
A and B	1
E	2
F or T	3
G	4
H	5

Our normalisation of the internal modes requires the external modes to have a total $g(\omega)_{ext}$ of six, three librations and three translations. This is in contradistinction to the solid state physics convention that normalises the whole density of states (internal plus external) to unity. scattering law equations based on this convention need $g(\omega)$ terms to balance the distribution of intensities between the external and internal modes.

2.6.2.1 The Einstein approximation

The $g(\omega)$ can be used to provide a simple approximation to the lattice vibrations, the Einstein approximation. Let us begin by agreeing that single characteristic frequencies, $(\omega_E)_j$, could be chosen to individually represent each of the six types (three translational and three rotational) of external mode. The $(\omega_E)_j$ value is the density of states weighted mean value of all of the frequencies over which that external mode, j, was dispersed. Normalising as discussed above:

The weighted mean value required is, at low temperature

$$\int_{(\omega_j)_{\text{min}}}^{(\omega_j)_{\text{max}}} g(\omega_j) d\omega = 1 \qquad (2.55)$$

$$\frac{1}{(\omega_{\text{E}})_j} = \int_{(\omega_j)_{\text{min}}}^{(\omega_j)_{\text{max}}} \frac{g(\omega_j)}{\omega_j} d\omega \qquad (2.56)$$

Where the integral covers the range from the minimum energy of that type of vibration, $(\omega_j)_{\text{min}}$, to the maximum, $(\omega_j)_{\text{max}}$. Einstein was the first to approximate the vibrational frequencies of a monatomic crystal to a single frequency. We introduce a rather more severe form of this approximation; a single weighted mean frequency, ϖ_{E}, is chosen to be representative of all six external vibrations of a molecule in its crystal.

$$\frac{1}{\varpi_{\text{E}}} = \sum_{j=1}^{6} \frac{6}{(\omega_{\text{E}})_j} \qquad (2.57)$$

2.6.2.2 Band shapes and van Hove singularities

One particular aspect of the density of states is the local shape it gives to spectral bands. Where there are many states piled up over a small range of frequencies the intensity will be stronger than if few states are spread thinly. Strong sharp features, or singularities (named after van Hove), commonly appear in the external mode region of spectra. There are several types of singularity but two are worth outlining, the acoustic and optic types, the relationship of the density of states to these typical dispersion curves is shown in Fig. 2.6. The acoustic type arises from a dispersion curve (or branch) that starts steeply at low energy (about zero) but soon modifies to a lesser slope and finally the dispersion flattens completely. The respective shape of the spectrum, given by this density distribution of the external states, is; weak at low energy becoming stronger as the slope is reduced and maximising at the highest energy, where the dispersion is flat.

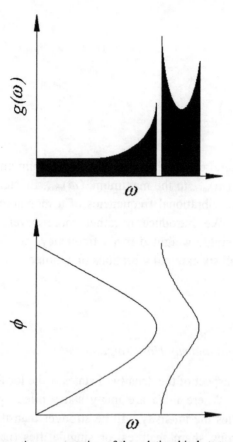

Fig. 2.6 A diagrammatic representation of the relationship between, below, the acoustic (left) and the optic (right) types of dispersion and, above, their respective density of vibrational states, $g(\omega)$, for the external modes of a solid. The symbols are defined in the text. Notice, that we have inverted the conventional representation of dispersion to maintain a conventional display of the spectrum.

On the high energy side of this maximum the intensity falls abruptly since there are no states of this vibrational mode in that energy region. The optic type starts at typical reststrale energies, say 100 cm^{-1}, with a sharp strong onset, typically weakening quickly, to regain intensity before falling again, abruptly, to zero intensity. Several overlapping singularities may be seen in complex spectra so it is worth noting that sharp phonon (i.e. external mode) features only occur for single quantum excitations.

2.6.3 Phonon wings

In condensed matter texts it is conventional to discuss the vibrations of crystals in terms of a pseudo particle, the phonon (which, like the neutron, can have a range of quantised energies and wavelengths) [1, 2]. Strictly, all vibrations (both internal and external) in molecular crystals are thus phonons. We shall avoid this pedantry and reserve phonon to indicate that a vibration has a mostly external character. Other relevant aspects of the external dynamics of molecules are given in §4.3 and §10.1. Above (§2.5) we made the choice to treat the internal vibrations separately from the phonons and to deal with the molecule in isolation. This is only acceptable because the two sets of vibrations are naturally separated, either harmonically or temporally [9]. However, the external vibrations cannot be ignored since, just as combination modes are generated from the ranks of the internal fundamentals, so the external modes are a source for combination with any internal vibration involving common atoms. This is familiar from the ultraviolet spectroscopy of molecules where weak combinations with the internal vibrations (also called phonon side bands) are seen alongside the intense electronic transitions.

The total scattering law is the result of the convolution of the internal and external dynamics, which stems from Eq. (2.29).

$$S^{\bullet}_{total}(\boldsymbol{Q},\omega)_l = S(\boldsymbol{Q},\omega_{int})_l \otimes S(\boldsymbol{Q},\omega_{ext})_l \ \sigma_l = S^{\bullet}(\boldsymbol{Q},\omega_{int})_l \otimes S(\boldsymbol{Q},\omega_{ext})_l \ (2.58)$$

The external vibrations can be expanded along similar lines to our treatment of the internal dynamics, except that the vibrating mass is now the effective molecular mass. The resulting spectrum of the external modes is somewhat similar to that arising from the internal modes.

No exchange of energy $\quad [\delta(\omega)$

1 phonon events $\qquad ....\delta(\omega - p_1) + \delta(\omega - p_2) + \delta(\omega - p_3).......$

2 phonon events $\qquad ....\delta(\omega - 2p_1).... + \delta(\omega - p_1 - p_2) + ...$

In a crystal there are very many more vibrations, of the order N_{mol}, and they are indistinguishable as individual features. The total molecular spectrum includes combinations between internals and externals and takes the following form:

$$\delta(\omega)\left\{\ \left[\delta(\omega)+\delta(\omega-p_1)\cdots+\delta(\omega-n\,p_j)+\cdots\right]\right.$$

$$+\delta(\omega-v_1)\left[\delta(\omega)+\delta(\omega-p_1)\cdots+\delta(\omega-n\,p_j)+\cdots\right]$$

$$\left.+\delta(\omega-v_2)\left[\delta(\omega)+\delta(\omega-p_1)\cdots+\delta(\omega-n\,p_j)+\cdots\right]\cdots\ \right\}$$

The elastic line is dressed with the full spectrum of phonons in all their orders, as indeed is each individual internal transition but it remains to be determined what form, or shape, this dressing takes.

To proceed further we require the form of the argument in the Debye-Waller factor (§2.5.1.2) this is given in Appendix 2, Eq. (A2.83), which is simply recast in terms of the external vibrations.

$$2W_{ext}=\left(Q^2\cdot\sum_j\left\{{}^j U_l^2\coth\left(\frac{\hbar\omega_j}{2k_B T}\right)\right\}\right) \tag{2.59}$$

The ${}^j U_l^2$ are the mean square displacements of the scattering atom, still labelled l, caused by the *external* vibrations of individual molecules and we have used j to label the $6N_{mol}$ different phonons. Again we can define a total displacement tensor for the atom but now related to the external vibrations.

$$(A_l)_{ext}=\sum_j {}^j U_l^2 \tag{2.60}$$

We can also define a value for $(\alpha_l^v)_{ext}$; this is related to A_{ext}, but measured in the direction of the *internal* vibrational displacement under consideration, see Eq. (2.41)

$$(\alpha_l^v)_{ext}=\frac{1}{5}\left\{\left({}^{Tr}A_l\right)_{ext}+2\left(\frac{{}^v B_l:(A_l)_{ext}}{{}^{Tr}({}^v B_l)}\right)\right\} \tag{2.61}$$

We invoke the Einstein approximation, see Eq. (2.57), and assume that $(\alpha_l^v)_{ext}$ is the displacement that stems from ϖ_E. Thus there is only a single external vibrational mode and so A_{ext} is isotropic. In a manner analogous to Eq. (2.44) we write:

$$S(Q,\varpi_E)_l^n = \frac{\left(Q^2 \left(\alpha_l^v\right)_{\text{ext}}\right)^n}{n!} \exp\left(-Q^2 \left(\alpha_l^v\right)_{\text{ext}}\right) \tag{2.62}$$

The strength of the different phonons is again given by the orders, $n = 0, 1, 2 \ldots$, exactly as shown for the internal modes, (§2.5.1.3). The first order term is the relative strength of the one-phonon spectrum; its shape is the observed density of external states, $g(\omega)_{\text{ext}}$ (§2.6.2). The $n = 2$ term gives the strength of the two-phonon spectrum. In vibrational spectroscopic terms this would be expressed as a first overtone or combination. However, this language is inappropriate since there are no sharp features in the two-phonon spectrum that can be traced back to their one phonon origins.

The shape of the two-phonon spectrum is given by the convolution of the one-phonon spectrum with itself, $g(\omega)_{\text{ext}} \otimes g(\omega)e_{\text{xt}}$. The shape of the three-phonon spectrum, $n = 3$, is obtained by convolving the one- and two-phonon spectra, and so on. The weight of each contribution is given by

$$S(Q,\varpi_E) = \sum_n \frac{\left(Q^2 \left(\alpha_l^v\right)_{\text{ext}}\right)^n}{n!} \exp\left(-Q^2 \left(\alpha_l^v\right)_{\text{ext}}\right) = 1 \tag{2.63}$$

(The scattering cross section is omitted from Eq. (2.63) since this function will be convolved with $S^{\bullet}(Q,\omega_{\text{int}})$, Eq. (2.58).) Notice that the *shapes* of the phonon wings are given by the $g(\omega)$ext convolutes but that they have a total strength of unity, not the strength of the phonons themselves, which is normalised to six (five for linear molecules). The convolution of the wing shapes with the internal transition is trivial, since each internal transition is so sharp that any other function convolved with it remains unchanged in shape. We see that, whilst the total intensity of an internal transition is unchanged by convolution with the external dynamics, it no longer falls exclusively at the unique energy transfer value associated with a given vibration, ω_0, the band origin. Rather it is distributed into stronger or weaker combinations with the spectrum of the external modes and spreads away from ω_0, into the phonon wings. The shape of an observed band in the neutron spectrum is thus the same as the shape of the elastic line plus the lattice mode region of the spectrum, provided that the two spectral regions were measured at the same Q value. (Note, this is difficult to achieve in practice.)

Phonon wings are probably the most important band shaping processes in inelastic neutron scattering spectroscopy and this theme is developed in later chapters. The intensity arising from the νth internal transition and remaining at the band origin, ω_0, is termed the zero-phonon-band intensity, often found in the literature as S_0. From Eq. (2.62), for $n = 0$

$$S_0(Q,\omega_v) = S(Q,\omega_v)_{\text{int}} \exp(-Q^2 (\alpha_i^y)_{\text{ext}}) \tag{2.64}$$

That intensity arising from the νth internal transition but not remaining at ω_0 is the intensity in the phonon wing, S_W.

$$S_W(Q,\omega_v) = S(Q,\omega_v)_{\text{int}} - S_0(Q,\omega_v) \tag{2.65}$$

We shall calculate the redistribution of the intensity for our simple example, the bifluoride ion.

2.6.4 Worked example—phonon wings of the bifluoride ion

We remain with our earlier example of the [FHF]$^-$ ion and since this is centrosymmetric only translational external vibrations will result in the displacement of the hydrogen atom. The dynamics of the low cross section fluorine atoms can be ignored. The ionic mass is 39 amu and the total hydrogen displacements (Å) of the internal modes are (§2.5.3):

$$(A_{\text{H}})_{\text{int}} = \begin{bmatrix} 0.0132 & & \\ & 0.0132 & \\ & & 0.0119 \end{bmatrix}_{\text{H}}$$

$$^{\text{Tr}}(A_{\text{H}})_{\text{int}} = 0.0383 \text{ Å}^2 \tag{2.66}$$

The anisotropic displacement parameters of the hydrogen atom, as determined from diffraction studies [10], are:

$$(A_{\text{H}})_{\text{total}} = \begin{bmatrix} 0.0195 & & \\ & 0.0195 & \\ & & 0.0165 \end{bmatrix}$$

$$^{\text{Tr}}\left(A_{\text{H}}\right)_{\text{total}} = 0.0555 \text{ Å}^2 \tag{2.67}$$

These represent the total vibrational displacements, internal and external, of the hydrogen atom (§2.4), so using:

$$\left(A_{\text{H}}\right)_{\text{total}} = \left(A_{\text{H}}\right)_{\text{int}} + \left(A_{\text{H}}\right)_{\text{ext}} \tag{2.68}$$

We may extract the external A tensor of the hydrogen. Furthermore, we can determine the external α values that control the distribution of intensities arising from internal vibrations.

$$\left(A_{\text{H}}\right)_{\text{ext}} = \begin{bmatrix} 0.00630 & & \\ & 0.00630 & \\ & & 0.00460 \end{bmatrix}$$

$$^{\text{Tr}}\left(A_{\text{H}}\right)_{\text{ext}} = 0.0172 \text{ Å}^2 \tag{2.69}$$

The antisymmetric stretch, at 1370 cm^{-1}, will be used as the example, and from Eq. (2.61):

$$\left(\alpha_{\text{H}}^{1370}\right)_{\text{ext}} = \frac{1}{5}\left\{ ^{\text{Tr}}\left(A_{\text{H}}\right)_{\text{ext}} + 2\frac{^{1370}B_{\text{H}} : \left(A_{\text{H}}\right)_{\text{ext}}}{^{\text{Tr}}\left(^{1370}B_{\text{H}}\right)} \right\} =$$

$$\left(\alpha_{\text{H}}^{1370}\right)_{\text{ext}} = \frac{1}{5}\left(0.055\text{Å}^2 + 2\times 0.0046\text{Å}^2\right) = 0.0128 \text{ Å}^2 \tag{2.70}$$

Table 2.7 The weights of the phonon wing orders, n, calculated for the antisymmetric stretching mode of the [FHF]⁻ ion, at a Q of 7 Å^{-1}.

Order n	Pre-exponential factor $(Q^2.\alpha_{1370})$ $(n/n!)$	Phonon order weights S_n
0	1.000	0.534
1	0.627	0.335
2	0.197	0.105
3	0.038	0.022
4	0.006	0.004
	Total	1.000

Again we chose $Q = 7$ Å^{-1}, and proceed to calculate the strengths of the different orders, n. The argument, $Q^2\alpha$, gives an external Debye-Waller factor of:

$$\exp\left(-Q^2 \left(\alpha_H^{1370}\right)_{ext}\right) = \exp(-49 \times 0.0128) = 0.534 \qquad (2.71)$$

We apply Eq. (2.72) and obtain the weights of the phonon orders given in Table 2.7. Approximately 98% of the total intensity is found distributed across only the first three phonon orders. The zero-phonon line, or band origin, remains the strongest feature but, at this value of Q^2, the first order event is quite strong. If we had chosen $Q = 9$ Å^{-1}, the first order contribution would have been the stronger.

We complete the exercise by calculating the Einstein frequency (§2.6.2.1) that represents the lattice modes. The low temperature expectation value for the atomic displacement in any mode is, Eq. (A2.82), for u^2 (in Å^2), a reduced oscillator mass, $^v\mu_l$ (in amu), and a vibrational frequency, ϖ_E (in cm^{-1}).

$$\left|u^2\right| = \left(16.9 \text{ Å}^2 \text{ amu cm}^{-1}\right)\frac{1}{^v\mu_l \, \varpi_E} \qquad (2.72)$$

Substituting the molecular mass for the reduced mass and α ($= 0.0128$ Å2) for u^2 implies a putative Einstein frequency of about 34 cm^{-1}. For the purpose of this exercise the one phonon lattice mode spectrum will be represented by a single, sharp transition line at this energy. (This is mathematically convenient but unrealistic, see Chapter 5.) The phonon wings appear to higher frequencies of the band origin, as a series of lines spaced at 34 cm^{-1} intervals, see Table 2.8.

Table 2.8 The distribution of the intensity of the antisymmteric stretching mode at 1370 cm^{-1}, amongst the band origin, $n = 0$, the wing orders $n = 1, 2, 3$, see Fig. 2.7.

Order, n	0	1	2	3	Total intensity
ω / cm^{-1}	1370	1404	1438	1472	
Relative intensity	0.534	0.335	0.105	0.022	0.996
Calculated intensity	4.66	2.89	0.93	0.20	8.7

In Fig. 2.7 we show how the original band intensity, 8.7 (from Table 2.5), would be redistributed among the different wing orders. Band shaping processes, similar to that calculated for the bifluoride ion, will apply to all vibrational transitions, fundamentals, overtones or combinations. The spectral shape involving the redistribution of intensity away from the band origin develops as the momentum transfer changes and is shown, for the example of the antisymmetric stretching mode in the bifluoride ion, in Fig. 2.7. Here we see that at low to moderate values of Q, most of the intensity remains in the band origin. At intermediate values of Q, intensity has been distributed away from the band origin to higher frequencies. Although the origin may retain sufficient peak intensity to remain individually identifiable it has lost much of its intensity. At high values of Q, the band origin has been reduced to insignificance and the intensity resides in the wings. Notice how, at high Q, the envelope of the individual wings approximates to a broadly Gaussian shape. This form can be fitted to a true Gaussian shape and well-defined values for its centre, and width can be extracted (§2.6.5.1).

As the value of Q continues to increase the centre of the Gaussian envelope moves out to higher frequencies and its width expands. The envelope's central intensity maximum decreases dramatically and the total intensity falls, since the Debye-Waller factor is smaller. Eventually the envelope will broaden and weaken to such an extent that it disappears into the experimental background. This simple picture nicely summarises the effects of phonon wings but it will be considerably modified by the introduction of more realistic treatments of the external vibrations of molecular crystals, see Chapter 5. However, the model remains sufficiently robust to provide an introduction to the effects of molecular recoil.

2.6.5 Molecular recoil

As given above, the overall shape of an internal vibrational transition critically depends on the value of Q at which it was measured. At low Q the band should stand sharp and, possibly, intense (with its origin, ω_0, probably close to any optical counterpart). At high Q most of the intensity appears at higher frequencies, broadened into a Gaussian envelope, and weakened. The impact that Q makes on an INS spectrum is seen in Fig. 3.31. It would be as well then to understand what characteristics of a sample determine if an experimentally chosen Q value is high or low.

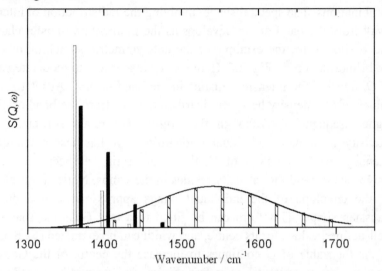

Fig. 2.7 How phonon wings develop as the momentum transfer increases. The wings on the stretching mode of the bifluoride ion are used, the band origin is at 1370 cm^{-1}, see §2.6.4. The wings were calculated at a low Q value, 4 Å^{-1} (on the left of each group, unfilled), an intermediate Q value, 7 Å^{-1} (centre of each group, as given in Table 2.8, filled), and a high Q value, 20 Å^{-1} (on the right of each group, hatched). The Gaussian envelope is drawn to best fit the high Q results of each group and shows the effects of recoil scattering as discussed in (§2.6.5).

Whenever the external vibrations produce large anisotropic displacement parameter values for the scattering atoms it will exaggerate the impact of any given value of Q. The phonon wing envelope will move to even higher frequencies and the response will broaden. Only two characteristics of a sample bear on its anisotropic displacement parameter (with samples at low temperatures), the effective molecular mass, μ_{eff}, and the Einstein frequency, ϖ_E, see (§2.6.2.1). The lighter the effective mass and the lower the Einstein frequency the greater will be the anisotropic displacement parameter. To proceed further we must consider what is meant by the effective mass of a molecule.

2.6.5.1 The Sachs-Teller mass tensor

Except for unusual cases, like the bifluoride ion, most hydrogenous molecules carry their hydrogen atoms in terminal positions, typically somewhat distant from the molecular centre of mass. When a neutron strikes such a hydrogen atom it is the molecule as a whole that responds, either through the excitation of an internal normal mode (involving only molecular deformation), or through the excitation of the lattice modes (involving no molecular deformation). The effective mass of the scattering atom during internal mode excitation is its reduced mass, $\nu\mu_l$, obtained by the normal coordinate analysis procedures covered in Chapter 4. If, however, the energy of the neutron is small in comparison with the internal modes but large compared with the free molecular rotational levels then the system can be treated as scattering from a classical rigid rotor.

Consider the case of methane where the incident neutron is travelling along any one of the four equivalent C–H bond directions towards the carbon atom. Then, at the moment of impact, the hydrogen atom appears with the full molecular mass (or translational inertia), 16 amu. If, however, the neutron approaches the hydrogen atom in a direction perpendicular to the C–H bond then rotational inertia is more important. A scattering atom at some distance from the centre of mass of a molecule with a high rotational inertia will have a considerable effective mass, m_{eff}. The mass tensor, $M_{\text{S-T}}$, concept of Sachs and Teller provides a method of calculating the effective mass of an atom, at position r_x, r_y, r_z, [11]. When the Cartesian molecular coordinate system, (x, y, z), is centred on the molecular centre of mass and aligned with the principal axes of rotational inertia, the moments of inertia are I_x, I_y, I_z. The effective mass, m_{eff}, of the atom is the scalar mass of its tensor $|M_{\text{S-T}}|$. In terms of the librational and translational inertia:

$$\frac{1}{M_{\text{S-T}}} = \frac{1}{M_{\text{Lib}}} + \frac{1}{M_{\text{Trans}}}$$

$$\frac{1}{M_{\text{lib}}} = \begin{bmatrix} \left(\dfrac{r_y^2}{I_z} + \dfrac{r_z^2}{I_y}\right) & \dfrac{-r_x \cdot r_y}{I_z} & \dfrac{-r_x \cdot r_y}{I_y} \\[2ex] \dfrac{-r_y \cdot r_x}{I_z} & \left(\dfrac{r_z^2}{I_x} + \dfrac{r_x^2}{I_z}\right) & \dfrac{-r_y \cdot r_z}{I_x} \\[2ex] \dfrac{-r_z \cdot r_x}{I_y} & \dfrac{-r_z \cdot r_y}{I_x} & \left(\dfrac{r_x^2}{I_y} + \dfrac{r_y^2}{I_x}\right) \end{bmatrix} \tag{2.73}$$

$$\frac{1}{M_{\text{trans}}} = \begin{bmatrix} \dfrac{1}{m_{\text{mol}}} & & \\[2ex] & \dfrac{1}{m_{\text{mol}}} & \\[2ex] & & \dfrac{1}{m_{\text{mol}}} \end{bmatrix}$$

As a straightforward example we can calculate the $(m_{\text{eff}})_{\text{H}}$ of any hydrogen atom of methane. The axis system is most conveniently oriented through the carbon atom along the three S_4 axes of improper rotation that characterise the point group symmetry of this T_{d} molecule. In this axis system each hydrogen atom is equidistant from each axis, r, and the rotational inertia about each axis is equivalent, $I = 4m_{\text{H}} r^2$. Then

$$\frac{1}{M_{\text{S-T}}} = 3\left[\left(\frac{r^2/2}{4m_{\text{H}} r^2} + \frac{r^2/2}{4m_{\text{H}} r^2}\right)\right] + 3\left[\frac{1}{m_{CH_4}}\right] = 3\left[\left(\frac{2}{16} + \frac{2}{16}\right)\right] + 3\left[\frac{1}{16}\right]$$

$$\left|M_{\text{S-T}}\right|_{\text{H}} = \frac{^{\text{Tr}}(M_{\text{S-T}})}{3} = (m_{\text{H}})_{\text{eff}} = \frac{16}{5} = 3.2 \text{ amu} \tag{2.74}$$

The effective mass of each hydrogen atom in methane is thus 3.2 amu. (The effective mass of the carbon, $(m_C)_{\text{eff}}$, will be the molecular mass, 16 amu, since there can be no rotational contributions.) When struck by a neutron the simple condition of momentum and energy conservation requires the more energetic neutron to loose energy to the methane.

$$m_n v_i + m_{\text{eff}} v_{\text{mol,i}} = m_n v_f + m_{\text{eff}} v_{\text{mol,f}}$$

$$k_i + p_{\text{mol,i}} = k_f + p_{\text{mol,f}} \tag{2.75}$$

But the molecular velocity, v_{mol}, is zero along any direction, including that of the neutron momentum vector, when averaged over every direction that the molecular motions take in the sample. Therefore,

$$k_i - k_f = Q = p_{mol,f} = m_{eff} v_{mol,f} \qquad (2.76)$$

The energy lost by the neutron was absorbed in making the molecule recoil and is called the recoil energy, E_r. In the case of the methane molecule, measured at a momentum transfer of 9 Å^{-1}, the recoil energy would be (the conversion factor from Å^{-2} to cm^{-1}, 16.7, was introduced in Eq. (2.21)).

$$E_r = \frac{m_{eff}(v_{mol})^2}{2} = \frac{Q^2}{m_{eff}} = \frac{81 \times 16.7}{3.2} = 422 \text{ cm}^{-1} \qquad (2.77)$$

The elastic line, which appears at an energy transfer of zero for zero momentum transfer, is now displaced by 422 cm^{-1} to higher energies. A similar effect is seen on all the internal transitions. Thus, for example, the doubly degenerate deformation mode of methane, v_4, appearing at 1534 cm^{-1} in the infrared, would be centred at 1956 cm^{-1} (=1534 + 422) in the INS (if measured at the momentum transfer value of 9 Å^{-1}). The standard deviation of the Gaussian, Γ (cm^{-1}), is given by [12]:

$$\Gamma^2 = \omega E_r = \frac{4 E_k E_r}{3} = \frac{16.9}{4U^2} = \left(\frac{\Delta}{2.36}\right)^2 \qquad (2.78)$$

Here ω is equated with the characteristic, Einstein, frequency of the lattice, ϖ_E, and U^2 is the mean square displacement of the scattering atom. Their relationship to the average kinetic energy of the scattering atom, E_k (assuming harmonic potentials) and the full width at half height (FWHH), Δ, of the Gaussian are also given.

As an exercise the reader may wish to recalculate the distribution of phonon wing intensities for the earlier example of the bifluoride (§2.6.4) at a value of 20 Å^{-1}. The results of such a calculation are shown in Fig. 2.7, where the individual components are enveloped in a broad Gaussian. The envelope is centred at, E_r =153 cm^{-1} away from the band origin, at 1370 cm^{-1}, and it has a width (FWHH) Δ = 180 cm^{-1}. We proceed to calculate both the m_{eff} and ϖ_E from these parameters using Eqs. (2.77) and (2.78)

$$\left(\frac{180\,\text{cm}^{-1}}{2.36}\right)^2 = 5840 = \varpi_E\,E_r = 153\,\varpi_E \qquad \text{thus} \qquad \varpi_E = 38\,\text{cm}^{-1}$$

(2.79)

$$E_r = 153 = \frac{Q^2\,16.7}{m_{\text{eff}}} = \frac{400 \times 16.7}{m_{\text{eff}}} \qquad \text{thus} \qquad m_{\text{eff}} = 44\,\text{amu}$$

The characteristic frequency, previously determined (§2.6.4) as 34 cm^{-1} is here found to be 38 cm^{-1}, whilst the calculated effective mass of the bifluoride is 44 amu, close to the molecular mass of 39 amu. The two calculations are clearly self-consistent, as they must be since the phonon wings are simply the start of the molecular recoil in the lattice. However, the extreme naivety of the Einstein model is not usually successful at modelling the lattice dynamics of even simple systems.

Generally it is difficult to achieve the neutron momentum transfers needed to recoil heavy effective masses. As the lattice forces stiffen so the effective mass will increase beyond that calculated for the molecule in the gas phase. Significant recoil effects are only found for the lightest effective masses, those of methane, water and methyl groups being cases in point. It is also often impossible to reconcile the simple characteristic frequencies extracted from recoil broadening effects with the sophisticated external dynamics of even the simplest of samples. Other complications include the possibility that different parts of the same molecule may recoil independently and, ultimately, that atomic recoil effects will be observed [12].

2.7 Conclusion

This chapter has introduced the theoretical foundations used to understand the INS intensities experimentally observed in neutron scattering from hydrogenous molecular systems. The important role of the large incoherent cross section of hydrogen was seen to be the main reason why good quality spectra are obtained from hydrogenous compounds. The concept of neutron momentum transfer was explored and the ease of calculating intensities through the scattering law was underlined.

The modern methods of treating INS spectra obtained on powders were introduced and the simple example of the bifluoride ion was treated in detail. The most important band shaping processes were introduced and a full treatment of phonon wings was given. With this as a foundation we may proceed in the following chapter to apply the theory to some more realistic yet still straightforward examples.

2.8 References

1   G.L Squires (1996). *Introduction to the Theory of Thermal Neutron Scattering*, Dover Publications Inc., New York.

2   M.T. Dove (1993). *Introduction to Lattice Dynamics*, Cambridge University Press, Cambridge, London.

3   H.H. Chen-Mayer, D.F.R. Mildner, G.P. Lamaze & R.M. Lindstrom (2002). J. Appl. Phys., 91, 3669–3674. Imaging of neutron incoherent scattering from hydrogen in metals.

4   J.L. Finney & A.K. Soper (1994). Chem. Soc. Rev.,23, 1–10. Solvent structure and perturbations in solutions of chemical and biological importance.

5   M. Plazanet, N. Fukushima, M.R. Johnson, A.J. Horsewill, & H.P. Trommsdorff (2001). J. Chem. Phys., 115, 3241–3248 The vibrational spectrum of crystalline benzoic acid: Inelastic neutron scattering and density functional theory calculations.

6   H.B. Burgi & S.C. Capelli (2000). Acta Cryst., A56 403–412. Dynamics of molecules in crystals from multi-temperature anisotropic displacement parameters. I Theory.
    R. J. Nelmes, (1980). Acta Cryst., A36, 641–653. Anisotropy in thermal motion and in secondary extinction.

7   M. Bée (1988). *Quasielastic Neutron Scattering*, Adam Hilger, Bristol.

8   F. Fillaux, N. Leygue, J Tomkinson, A. Cousson, & W. Paulus (1999). Chem. Phys., 244, 387–403. Structure and dynamics of the symmetric hydrogen bond in potassium hydrogen maleate: a neutron scattering study.

9   M. Warner, S.W. Lovesey & J. Smith (1983). Z. Phys. B-Condens. Matter, 51, 109–126, The theory of neutron scattering from mixed harmonic solids.

10 T.C. Waddington, J. Howard, K.P. Brierley & J Tomkinson (1982). Chem. Phys. 64, 193–201. Inelastic neutron scattering spectra of the alkali metal (Na,K) bifluorides: the harmonic overtone of v3

11 V.F. Turchin (1965). *Slow Neutrons*. Israel Program for Scientific Translations Ltd., Jerusalem.

12 J. Mayers, ISIS, CCLRC, UK, Private communication.

3

Instrumentation and Experimental Methods

In order safely and successfully to carry out an inelastic neutron scattering (INS) experiment, there are five essential components:
- a source of neutrons (§3.1)
- a means to transport the neutrons (§3.2)
- a neutron detection system (§3.3)
- a method to analyse the energy of the neutrons (§3.4).
- the sample (§3.5).

In this chapter we will describe how these are accomplished and how the sample may be introduced so as to optimise the measurement. Some safety aspects of neutron experimentation will also be presented (§3.6). More detailed discussion of neutron sources and instrumentation can be found in [1–3].

3.1 Neutron sources

The neutron was discovered by Chadwick [4] in 1932, who bombarded beryllium with α-particles (^{4}He) from a ^{210}Po source.

$$^{210}_{84}\text{Po} \rightarrow {}^{206}_{82}\text{Pb} + {}^4_2\text{He}$$

$$^4_2\text{He} + {}^9_4\text{Be} \rightarrow {}^{12}_6\text{C} + {}^1_0\text{n} + 5.7\,\text{MeV}$$

(3.1)

The experiment was successful because the nucleons (protons and neutrons) are arranged in shells in the core, analogous to electron shells, and in ^{9}Be the highest energy neutron is the only nucleon in the outermost shell. Thus exactly as sodium has a low first ionisation energy because the 3s electron is outside a filled shell, the highest energy neutron in ^{9}Be is comparatively weakly bound and so is expelled when the compound nucleus formed from the α-particle and the ^{9}Be decays. Nuclear reactions are weak neutron sources with an energy distribution similar to that from a reactor. Such sources are useful for testing neutron detectors but are too weak for scattering experiments.

A number of methods have been used to generate sufficiently strong neutron sources, but all current facilities are either fission reactors or spallation sources, so we will consider only these two methods.

3.1.1 Reactor sources

Research nuclear reactors use thermal neutrons to induce fission in a critical mass of ^{235}U to produce high-energy (fast, or hot) neutrons:

$$_0^1 n_{thermal} + {}_{92}^{235}U \rightarrow \sim 2.5 \, _0^1 n_{fast} + \text{fission products} + 180 \, \text{MeV} \qquad (3.2)$$

(1 MeV = 10^6 eV = 8.07×10^9 cm^{-1} = 9.65×10^7 kJ mol^{-1}). Of the ~ 2.5 neutrons produced per fission event, one is required to maintain the nuclear reaction, ~ 0.5 neutrons are lost to absorption and one is available to leave the core and be used experimentally. Since ^{235}U occurs naturally at only 0.7% abundance, the use of enriched (>90% ^{235}U) uranium is required. This has lead to concerns about nuclear weapons proliferation and there is a drive to use lower levels of enrichment in research reactors.

A research reactor differs fundamentally from a nuclear power reactor where the desired product is heat. The heat produced by a research reactor is an undesirable by-product and its removal limits the maximum size (and hence neutron flux) of a research reactor core. The latest and most advanced research reactor is the 20 MW FRM-II reactor (Munich, Germany) shown in Fig. 3.1 [5]. It has an unperturbed maximum neutron flux of about 8×10^{14} cm^{-2} s^{-1}. The most powerful research reactor is the Institut Laue Langevin (ILL, Grenoble, France), Fig. 3.2, at 57 megawatts (MW) [6]; 20 MW research reactors are more common.

Research reactors are based on the swimming pool concept: the reactor core sits in the middle of a large tank of water that acts as both coolant and moderator and allows sufficient space for specialised moderators and guide tubes. FRM-II is designed to provide the maximum possible thermal neutron flux from a small core cooled with light water surrounded by a large moderator tank filled with heavy water (D_2O). Because the thermal power is limited to 20 MW, the reactor core consists of only one cylindrical fuel element that contains about 8.1 kg of uranium as U_3Si_2 (93% ^{235}U) dispersed in an aluminium matrix.

An unusual exception to the swimming pool reactor design is the IBR-2 reactor (Dubna, Russia) [7], shown in Fig. 3.3. The core of this

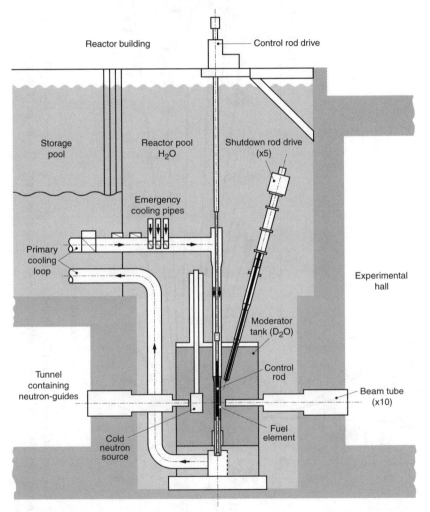

Fig. 3.1 Vertical cross section through the FRM-II reactor. Reproduced from [5] with permission from Technische Universität München.

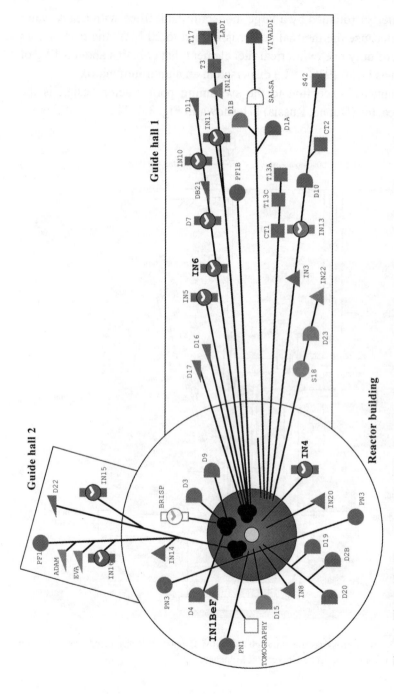

Fig. 3.2 The Institut Laue Langevin (ILL, Grenoble, France) showing the instrument layout. Those used for vibrational spectroscopy (IN1BeF, IN4, IN6) are shown in bold. Reproduced from [6] with permission from the Institut Laue-Langevin.

reactor is an irregular hexagonal tube composed of fuel element sub-assemblies. There are seven fuel elements (plutonium dioxide) in each double-walled steel vessel and it is surrounded by a stationary reflector with control and safety rods. Around the reactor, water moderators are viewed by beam-tubes. Power pulses with a frequency of 5 Hz are generated by two moveable reflectors. When they both approach the core, neutrons are directed back into the core increasing the rate of fission and a power pulse develops. The advantages of the system are that the average power is low, 2 MW, but the peak power during the 320 μs pulse is very high, 1500 MW.

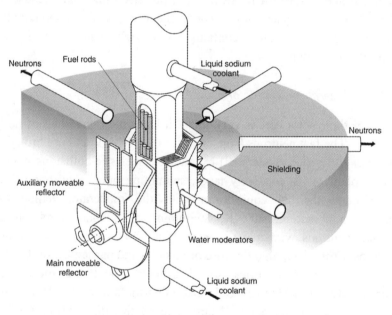

Fig. 3.3 Schematic of the IBR-2 reactor [7].

Safety is clearly a major consideration and research reactors are designed to fail-safe to prevent fission product release. Reactors operate under a triple containment philosophy. The first container is the cladding of the fuel itself, the second is the swimming pool; which is made from heavy, 1.5 m thick, concrete lined with stainless steel. Finally the whole reactor is housed inside a reinforced building that is kept at a slightly sub-ambient pressure and is accessed by an air-lock.

3.1.1.1 Biological shielding

Neutrons are highly penetrating and only weakly interact with matter, so it is reasonable to ask why they are dangerous. The human body, composed of hydrogenous material, is an excellent moderator of high energy (MeV) neutrons. The inelastic collisions result in hydrogen atoms being ripped out of tissue resulting in the formation of high energy charged particles and ions that cause massive damage. The suppression of high energy neutrons dominates the shielding requirements immediately around sources and necessitates materials that can both moderate and absorb them. Steel is the preferred material because it is affordable in the quantities needed. Steel is usually augmented by concrete because concrete contains up to 20% water; thus it is an effective moderator.

3.1.1.2 Moderators on reactor sources

Since the reactor core is very small, it leaks more than half of its fast neutrons. The fast neutrons have energies of ~ 1 MeV and are brought to useful energies (i.e. moderated) by multiple inelastic collisions with the D_2O coolant of its, relatively large, swimming pool. Neutrons achieve approximate thermal equilibrium with the moderator temperature; ambient temperature, 300 K, swimming pool moderators produce *thermal* neutrons with a mean energy of ~208 cm^{-1}. (Some of these neutrons diffuse back into the core of reactors and participate in the chain reaction.) Neutron energies are distributed about the mean value approximately as a Maxwell-Boltzmann distribution (§3.1.2.2). Neutrons with energies that are not well described by this distribution are 'non-Maxwellian' or 'epithermal'.

The distribution of thermal neutrons is not optimal for many experiments for which higher or lower energies are desirable. This is achieved by installing small, specialised moderators within the ambient moderator pool. These moderators operate at different temperatures from the pool and produce their peak neutron fluxes at different neutron energies. Most commonly, long wavelength, low energy (slow, or cold) neutrons are desired. A typical cold-source contains ~ 2.5 kg of liquid deuterium at 25 K and provides its peak flux at an energy of ~ 40 cm^{-1}. Less frequently installed on reactors is a hot source, typically an insulated 14 kg graphite block heated to ~ 2400 K by absorption of γ-radiation from the reactor.

3.1.2 Spallation sources

A diagrammatic representation of the spallation process is shown in Fig. 3.4. A heavy metal target is bombarded with pulses of high-energy protons from a powerful accelerator. Protons hitting nuclei in the target material trigger an intranuclear cascade, exciting individual nuclei into a highly energetic state. The nuclei then release energy by evaporating nucleons (mainly neutrons), some of which will leave the target, while others go on to trigger further reactions. Each high-energy proton delivered to the target results in the production of approximately 15 neutrons. This results in an extremely intense neutron pulse and only

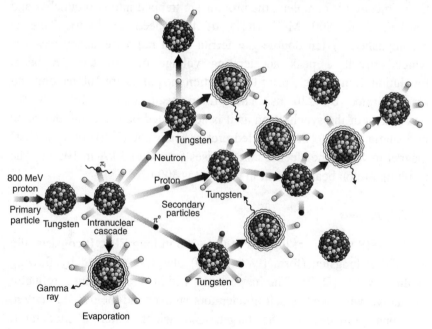

Fig. 3.4 Diagrammatic representation of the spallation process.

modest, compared to a reactor, heat production in the neutron target. At ISIS (Chilton, UK) [8], the time averaged heat production in the target is 160 kW, but in the pulse, the neutron brightness exceeds that of the most advanced steady state sources. The equivalent reactor power would be 16000 MW!

The ISIS Facility is shown in Fig.3.5. The proton beam production begins in the ion source where H^- ions are generated by dissociation of H_2 gas followed by electron transfer from the caesium-coated cathode. The H^- ions are extracted by an electric field into a radio frequency quadrupole accelerator that injects the ions into the linear accelerator (linac) with an energy of 665 keV. The linac accelerates the beam to 70 MeV (37% of the speed of light) providing a 200 μs long, 22 mA H^- pulse. On entry to the synchrotron, the H^- ion beam is passed through a 0.3 μm thick aluminium oxide stripping foil that removes both electrons from the H^- ions converting them to protons. The proton beam is injected during approximately 130 cycles of the synchrotron to allow accumulation of 4.2×10^{13} protons with minimal space-charge effects. Once injection is complete, the protons are trapped into two bunches and accelerated to 800 MeV (84% of the speed of light). This is accomplished by ten double-gap ferrite-tuned radio frequency cavities, which provide a peak accelerating voltage of 140 kV per beam revolution. Immediately prior to extraction the pulses are 100 ns long and are separated by 230 ns. The proton beam makes about 10,000 revolutions of the synchrotron as it is accelerated before being deflected in a single turn into the extracted proton beam line (EPB) by a 'kicker' magnet in which the electric current rises from 0 to 5 kA in 100 ns. The resulting proton beam current in the EPB is 300 μA.

3.1.2.1 Targets

The ISIS target is at the end of the ~200 m long EPB. It consists of a set of thin tungsten (formerly tantalum) plates surrounded by flowing cooling water (D_2O). The target sits within its beryllium reflector assembly, with very small moderators above and below the target. Neutrons produced in the target, and which enter a moderator, continually interact losing energy at every collision. Those that escape

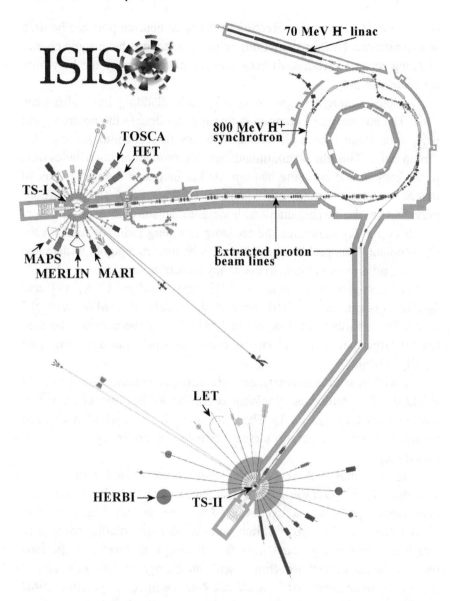

Fig. 3.5 Plan of the ISIS Facility (Chilton, UK) [8]. The major components are shown, INS instruments are indicated. TS-I is the high power 50 Hz target station and TS-II is the long wavelength, cold neutron 10 Hz target station. TS-II is scheduled to be operational in 2008.

from the moderator in the direction of the open neutron port can be used in experiments. Neutrons emerging in other directions have some chance of being scattered (reflected) back into the moderators by the beryllium reflector assembly

Depleted uranium targets with a suitable cladding have also been used. The advantage of uranium is that it gives double the neutron yield per proton since fission also occurs but this fission continues after the proton pulse. This almost continuous neutron production contributes only to the background on some instruments but increases the useful flux of others. A more serious consideration is that under the conditions prevalent in the target, uranium undergoes anisotropic crystallisation, which eventually punctures the cladding releasing radionuclide into the D_2O coolant. This problem has proven to be insurmountable to date and there are no uranium targets in use at any spallation source.

The next-generation sources, SNS (Oak Ridge, USA) [9] and J-PARC (Tokai, Japan) [10], have design goals of 1 MW, with the possibility of further development to 3 MW. Liquid mercury will be used for the target because solid targets cannot dissipate the heat generated rapidly enough.

As with reactor sources targets are strongly contained and heavily shielded. Their biological shielding is similar to, but thicker than, that found on reactors (§3.1.1.1). The extra shielding is required since most neutrons from spallation targets remain unmoderated and very penetrating.

ISIS and most other spallation sources are pulsed sources. The exception is SINQ (Villigen, Switzerland) [11], which is the only continuous spallation source, Fig. 3.6. The proton beam is pre-accelerated in a Cockcroft-Walton column—a high-voltage machine in which rectifiers charge capacitors that discharge to accelerate the ions into the linear accelerator (linac) with an energy of 870 keV and is brought up to an energy of 72 MeV in a 4-sector injector cyclotron. Final acceleration to 590 MeV occurs in the 8-sector main ring cyclotron, from which the beam is transported to the experimental hall in a shielded tunnel. The beam is bent downwards through a sloping drift tube for onward transport to the SINQ target station. Underneath the target

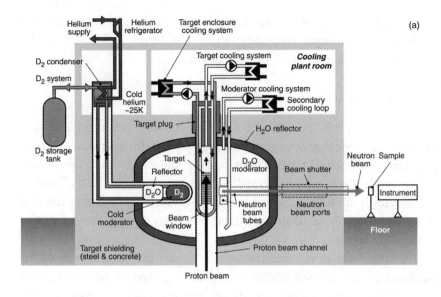

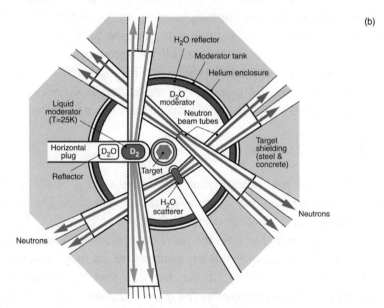

Fig. 3.6 Schematic of the SINQ target station (a) vertical section, (b) horizontal section. Reproduced from [11] with permission from SINQ.

station, the proton beam is diverted vertically by bending magnets into the heavy metal target (presently lead). The target is situated in the central tube of the heavy water moderator tank.

3.1.2.2 Moderators on spallation sources

As with reactor sources, the neutrons initially produced are very energetic ca 2 MeV and must be moderated to useful energies. A major difference between spallation and reactor sources is that the pulsed source moderators are very small, about one litre or less. Thermal equilibrium is not fully achieved in this volume and a significant fraction of the neutrons, those retaining a relatively high energy, are present as epithermal neutrons.

The small size of the moderators also means that all neutrons of the same energy are created within a very short period of time, the pulse width. Short pulse widths, ca 10 μs, are essential for good energy resolution (§3.4). Targets are often surrounded by several moderators, these run at temperatures optimised to produce peak neutron flux at different energies. Water moderators (300 K) produce peak fluxes at ca 200 cm^{-1}, methane (112 K) ca 70 cm^{-1} and dihydrogen (20 K) ca 40 cm^{-1}. The flux distribution, $J(E_i)$, of the ISIS moderators is shown in Fig. 3.7. In the Maxwellian region (§3.1.1.2), the distribution is described by:

$$J(E_i) = \Xi_n \frac{E_i}{\left(k_B T_{eff}\right)^2} \exp\left(\frac{-E_i}{k_B T_{eff}}\right) \tag{3.3}$$

where Ξ_n is the integrated Maxwellian intensity, T_{eff} is the effective moderator temperature and k_B is the Boltzmann constant. The values of T_{eff} for the different moderators are: H_2O, 390 K; CH_4, 128 K; H_2, 32 K. These are slightly higher than the measured moderator temperatures of 300, 112 and 20 K, showing that the neutrons are not fully moderated.

The exception is SINQ. It can be seen that the layout closely resembles that of a reactor (compare Figs. 3.1, 3.2 and 3.6), consequently the neutrons are heavily moderated as in a reactor. Rather than a spallation source, SINQ is best regarded as a medium power reactor operating without the disadvantages of using fissile material.

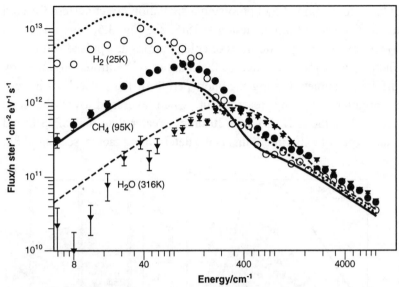

Fig. 3.7 Flux distribution from the ISIS moderators in the Maxwellian region, the points are experimental measurements, the lines are from Eq. (3.3).

3.1.3 Which to use—reactor or spallation source?

Reactor sources are much more common than spallation sources; there are around 20 reactors that produce core fluxes $>10^{14}$ cm^{-2} s^{-1}. To generate the proton beam needed for a spallation source requires considerable infrastructure and by 2004 there were only five spallation sources world-wide: ISIS [8], IPNS (Argonne, USA) [12], LANSCE (Los Alamos, USA) [13], KENS (Tsukuba, Japan) [14] and SINQ (Villigen, Switzerland) [11] with two more under construction, SNS (Oak Ridge, USA) [9] and J-PARC (Tokai, Japan) [10]. Reactor sources are also much more developed, the first neutron experiments were carried out in the 1950s and the ILL opened in 1975. In contrast the first spallation user facility, opened only in 1980, with ISIS in 1985.

The energy spectrum of the neutrons produced at the two types of source is distinctly different as shown in Fig. 3.8a. This provides a degree of complementarity: whereas reactors produce large numbers of cold and thermal neutrons, spallation sources produce many more high-energy neutrons. However, it has become clear that there are significant advantages to be gained by optimising a spallation target system for cold neutrons. The second target station at ISIS, see Fig. 3.5, (due on-line in 2008) will exploit the possibilities of cold neutrons at a pulsed source

There is a major difference in the number of neutrons produced by the different sources. During the short period of the pulse ISIS is very much brighter than any reactor but, when averaged over the time between pulses the ILL flux is ~ 30 times greater than that of ISIS and for many types of experiment this is the determining factor, see Fig. 3.8b.

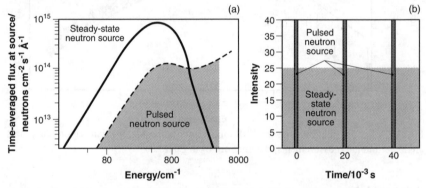

Fig. 3.8 Comparison of (a) the energy spectrum and (b) the flux distribution in time produced by a reactor and a spallation source.

Vibrational spectra obtained over the whole range requires relatively high neutron energies in the epithermal region and, as will be seen from later chapters in this book, spallation sources are pre-eminent in this field. However, where a limited energy range is acceptable reactor sources can be very powerful.

As can be seen from Fig. 3.9, the future clearly lies with spallation sources. This is driven by engineering considerations, compared to a reactor, the heat-load per spallation neutron is much smaller. There are also broader political and environmental reasons; fissile material is not required in spallation sources and less active waste is produced.

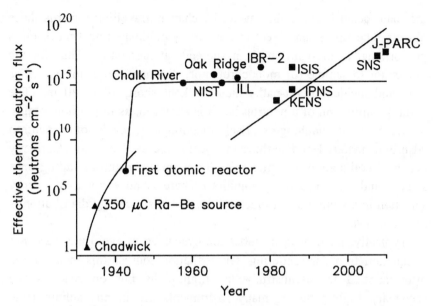

Fig. 3.9 The evolution of neutron flux since the discovery of the neutron. ▲ charged particle sources, ● reactor sources, ■ spallation sources.

3.2 Neutron transport

Neutrons are lead from the source along beam-tubes. These are usually arranged tangentially in order to avoid a direct view of the source's (either the reactor core or the spallation target) primary radiation (fast neutrons and gamma rays). The primary radiation dominates the safety requirements of the instruments (and hence cost) and also contributes much of the instrumental background. As can be seen from Figs. 3.1, 3.2, 3.3, 3.5 and 3.6, the beam-tubes are buried in, or arranged as close to, the moderator as possible, to capture the maximum possible flux.

3.2.1 Neutron beam-tubes

Neutrons are neutral and little can be done to influence their trajectories, beam-tubes serve only to remove the moderator from the neutron's path. Indeed, neutrons hit monochromators, or samples, held in the main beam because they were, by chance, travelling towards those objects. The flux at an object in a naturally collimated neutron beam is proportional to the solid angle it subtends at the source. Thus for the moderator to sample (or monochromator) distance, d_i the flux varies as $1/d_i^2$ and incident flight-paths must be kept short, 10 to 20 m. (The natural collimation of a neutron beam is achieved, as in pin-hole optics, by a series of diaphragms. Each diaphragm is made of a neutron adsorbing material and defines the beam size at that point. Correctly designed collimation will maximise the umbra of neutrons falling on a sample and minimise the penumbra.) There is no element of energy selection in a simple beam-tube and the exit beam energy distribution is said to be *white*.

Typically, incident beam-tubes are evacuated to avoid the 0.2% m^{-1} attenuation from air scattering. On spallation sources this technique is often applied to all instrumental flight-paths but on reactors the secondary flight-paths of many instruments are in air, helium (low absorption, some scattering), or argon (low scattering, some absorption).

3.2.2 Neutron guides

The need to reduce d_i can be obviated, at least for instruments requiring long wavelength (low energy) neutrons by the use of neutron guides. These are square or rectangular section tubes made from optically flat glass that has been metal coated. They work because long wavelength neutrons undergo total external reflection from the metal surface and are retained within the guide. The, very small, critical angle of reflection, γ_c, is given by:

$$\sin\gamma_c \approx \gamma_c = \lambda\sqrt{\frac{\langle\rho_N\rangle\langle b_{coh}\rangle}{\pi}} \tag{3.4}$$

where λ is the neutron wavelength (Å), $\langle\rho_N\rangle$ is the mean atom-number density and $\langle b_{coh}\rangle$ is the mean coherent scattering length (§2.1).

The preferred metal is ^{58}Ni since this has a large value of b_{coh} and hence $\gamma_c/\lambda \approx 0.1$ Å^{-1}. Although a perfect guide transports all of the flux that enters close to the moderator to its exit many of these neutrons travel paths that diverge considerably from the central axis of the beam.

A modest degree of energy selection can be introduced through the use of bent neutron guides. These guides follow the circumferences of circles struck with very long radii, *ca* 1 km. Energetic neutrons fail to negotiate these gentle curves and only long wavelengths are transmitted. Unfortunately the wavelength dependence of Eq. (3.4) means that, for the epithermal energies needed in vibrational spectroscopy, nickel guides are ineffective.

A recent development has been the use of supermirrors, which comprise alternating layers of different scattering length densities. This increases the critical angle by a factor m, where m is the ratio of the critical angle of the supermirror to that of nickel. The value of m has steadily improved such that the state-of-the-art supermirrors have m ~ 4. Such devices have considerable promise for neutron vibrational spectroscopy. Table 3.1 shows the potential gains in flux for the instrument TOSCA (§3.4.2.3) by the use of a m = 4 supermirror. Even at 2500 cm^{-1} there is a gain of more than a factor of three and the performance improves dramatically as the neutron energy falls. It is very likely that future vibrational spectroscopy instruments will use supermirror guides rather than simple beam tubes.

Table 3.1 Potential gain in flux (relative to a simple beam-tube) as a function of energy by the use of an m = 4 supermirror guide on TOSCA.

Energy / cm^{-1}	Gain
100	37
290	19
660	10
1350	6
2640	3.7

3.3 Neutron detection and instrument shielding

A major advantage of neutrons is that they are uncharged and penetrate deeply into matter. The concomitant disadvantage is that neutrons are difficult to stop, either to detect or eliminate.

3.3.1 Detection

A critical requirement of a detector of any particle is that it should detect only the particles of interest and an ideal detector would discriminate between particles of different energy. This is not possible in neutron detectors because the energy of the particle is insignificant in comparison with the energy of the nuclear reaction used in its detection. All detection methods rely on nuclear reactions to generate the charged particles that ionise the surrounding medium. The most useful reactions are:

$$_0^1n + _2^3He \rightarrow \ _1^3H + _1^1H + 0.77\,MeV \tag{3.5}$$

$$_0^1n + _3^6Li \rightarrow \ _1^3H + _2^4He + 4.79\,MeV \tag{3.6}$$

$$_0^1n + _5^{10}B \rightarrow \ _2^4He + _3^7Li + \gamma\,(0.48\,MeV) + 2.3\,MeV \quad (93\%)$$
$$\rightarrow \ _2^4He + _3^7Li + 2.79\,MeV \quad (7\%) \tag{3.7}$$

^{3}He and ^{10}B (as BF_3) are used in gas tubes and ^{6}Li is used in scintillator detectors. For vibrational spectroscopy gas tubes are most commonly used. At research facilities, the use of $^{10}BF_3$ is disfavoured on grounds of low detection efficiency and safety.

Electrical pulses from the detectors are sorted by analogue electronics to reduce the occurrence of false detection. The most common approach is to discriminate against signals that are too weak or too strong. Electrical pulses stronger than a minimum value, the lower level of discrimination, are accepted and transmitted, digitally, as a count. Pulses that are stronger than the upper discrimination level are also rejected.

3.3.1.1 Helium detectors

A helium gas tube is shown schematically in Fig. 3.10a. It consists of an earthed steel tube filled with ^{3}He; the pressure is usually around 10 bar. A high voltage, ~ 1800 V, anode runs down the length of the tube. The charged ionisation products caused by the proton and triton ($^3H^+$), Eq. (3.5), are accelerated towards the anode, causing further ionisation and an avalanche effect. This results in a gain of up to a factor of 10^5 and single neutron detection is readily achievable.

Fig. 3.10b shows an idealised output from a helium tube. There are a large number of low (MeV) energy events from gamma ray ionisation. There are steps at 0.19 MeV and 0.58 MeV, and a large peak with a sharp cut-off at 0.77 MeV. Conservation of momentum requires that the

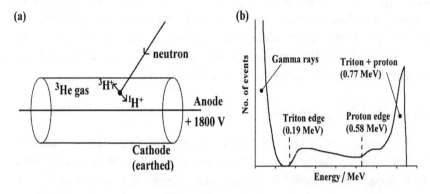

Fig. 3.10 (a) Schematic of a ³He gas tube and (b) idealised pulse height spectrum from a ³He gas tube.

proton and triton are emitted in opposite directions, with the velocity of the proton three times that of the triton. Thus the total energy released, 0.77 MeV, is partitioned as 0.58 MeV to the proton and 0.19 MeV to the triton. Below the step at 0.19 MeV (0.58 MeV) the proton (triton) has travelled no significant distance in the helium but became trapped by the detector wall. The large peak at 0.77 MeV is where both particles contribute to the ionisation and the electric pulse that passes down the anode.

Helium tubes can also be used as position sensitive detectors. The simplest method is to have an array of tubes and note which tube detects the neutron. To measure the position along a single tube, the central anode is replaced by a resistive wire. The charge produced by the ionisation, travels to both ends of the wire and is collected. If the charge collected is c_1 and c_2, then the position along the wire (of length l) where the charge originated is given by $lc_1/(c_1 + c_2)$. The spatial resolution is typically 10—25 mm.

The advantages of ³He tubes are that they are insensitive to γ–rays and magnetic fields, they are quiet (a 300 mm by 25 mm diameter tube has a background of *ca* 4 counts per hour) and they are a well-proven, mature technology. Their disadvantages are that a spatial resolution of less than 10 mm is difficult to achieve and their electrical connections leave a few centimetres of dead space at each end of the detector. Also, since the drift velocity of the ions is slow, the detector's response time suffers, and they are not suitable for high count-rate applications. For vibrational spectroscopy this is not a problem because count-rates are generally low and ^{3}He tubes will probably remain the detectors of choice for vibrational spectroscopy in the next few decades.

3.3.1.2 Scintillator detectors

For cases where the limitations of helium tubes are a problem, scintillator detectors are used. These are solid state detectors where the ^{6}Li is doped into a phosphor, usually ZnS, which emits a flash of light as the ions from Eq. (3.6) pass through it. The flash is detected by a photomultiplier tube (PMT). The PMT can be directly attached to the scintillator or can be remotely coupled to it by a light pipe, lens or fibre optic system. The use of fibre optics alleviates space constraints and allows 'odd-shaped' detectors to be constructed. The scintillator material is much denser than a gas so they can be much thinner (1 mm is typical), hence their spatial resolution can be much better; detectors with 4096 pixels of area 3×3 mm^2 are available. Since the use of one PMT per pixel would result in bulky and prohibitively expensive detectors, a coincidence counting scheme is used. Each pixel is viewed by several fibre optics and pixel 1 is coupled to PMTs 1, 2, 3 and 4. Pixel 2 is connected to PMTs 1, 2, 3 and 5 and so on. A count is only registered for pixel 1 if light is detected in PMTs 1, 2, 3 and 4 *simultaneously*. Each PMT is connected to about 100 pixels, in this way 32 PMTs can code for 4096 pixels. Unfortunately scintillators are sensitive to light and γ–rays, their PMTs also require shielding against magnetic fields.

3.3.1.3 Beam monitors

All neutron instruments require beam monitors to measure the incident flux. There may also be a transmission monitor after the sample.

Since the incident flux should remain as high as possible, monitors are necessarily insensitive. Monitors can be either a low pressure ^{3}He tube or a scintillator. In the latter case, beads of scintillator glass are placed on wires arranged like a ladder and attached to a PMT.

3.3.2 Instrumental shielding

All neutrons that fall on a detector will be detected, more or less efficiently. Ideally, detectors would only be exposed to neutrons with the correct characteristics; since other neutrons simply contribute to the instrument background they must be reduced to a minimum.

This need is met by appropriate shielding. The same process of moderation found at the heart of the source is again exploited but this is now followed by adsorption. Most materials have a 1/(neutron velocity) dependence to their absorption cross section and slower neutrons are more readily absorbed. (The $1/v$ dependence simply comes from the fact that the slower the neutron, the longer it spends near an absorbing nucleus).

The most commonly used absorbing materials on neutron instruments are boron, cadmium and gadolinium. The three elements have very different absorption characteristics, as can be seen from their absorption cross sections shown as a function of energy in Fig. 3.11.

Boron is a typical $1/v$ absorber, so is used in intimate contact with an hydrogenous compound. For bulk shielding, wax tanks are used. These consist of a braced metal skin that is filled with wax that has had sufficient of a boron containing compound, usually sodium tetraborate decahydrate, $Na_2B_4O_7.10H_2O$, suspended in it to give 5 wt % boron in the product. When this solidifies, it forms a material that has sufficient mechanical strength to be used as a structural element in the beamline, while having excellent shielding properties. For smaller items, borated polyethylene is used, which is similar but can be cut and shaped exactly as can polyethylene.

Boron is also used in 'crispy mix', which is a suspension of B_4C particles in resin. The resin content varies depending on the application. The high resin content (30 wt %) mix is easy to cast, has good mechanical strength and moderating properties. It is used to hold and shield detectors and also around the sample environment tank. Low resin content (4-10 wt %) is used where it is important to absorb rather than scatter the neutrons and is used to reduce albedo from the collimation pieces used to define the incident beam (§3.2.1).

Metal shielding materials often depend upon neutron-capture resonances. As with the infrared adsorption of molecules, materials have specific energies at which adsorption occurs, resulting in an excited system. Most metals have a wealth of resonances, usually at MeV energies. Cadmium is unusual in that it has a low energy resonance (*ca* 800 cm^{-1}). This makes it extremely effective at removing low energy neutrons. Sheets of cadmium metal (0.5–2 mm thick) are easily bent and shaped making it a convenient material with which to work (if it is being sawn or welded stringent precautions are necessary to prevent inhaling the toxic dust or vapour). Its main disadvantage is that on neutron capture, it decays emitting a γ-ray that requires further lead shielding.

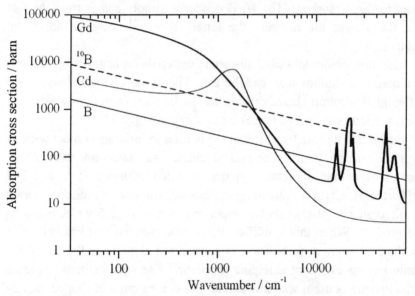

Fig. 3.11 Absorption cross sections of boron, cadmium and gadolinium as a function of energy. Reproduced from [2] with permission from Taylor and Francis.

Gadolinium is also very effective at low energies and has nuclear resonances that improve its performance at high energies. However, it is expensive and is most often found as gadolinium paint giving a layer of shielding to fine surfaces.

3.4 Neutron spectrometers

The aim of an INS experiment is to measure the scattering intensity $S(Q, \omega)$ as a function of (Q, ω). This requires knowledge of both the energy transfer E_t and the momentum transfer Q:

$$E_t = \hbar\omega = E_i - E_f \tag{3.8}$$

$$Q = k_i - k_f \qquad |k| = k = 2\pi/\lambda \tag{3.9}$$

For INS spectroscopy there are three main types of spectrometer in use: triple axis (§3.4.1), which is rarely used to study hydrogenous materials; more relevant are instruments that fix the final energy which are known as *indirect geometry instruments* and those that fix the incident energy which are known as *direct geometry instruments*. Examples of indirect geometry (§3.4.2, filter and analyser) spectrometers and direct geometry (§3.4.3, chopper) instruments are discussed in turn.

3.4.1 Triple axis spectrometers

The original INS spectrometer invented by Brockhouse and co-workers [15] at Chalk River, Canada, was the triple axis spectrometer. A modern example, IN1 [16] at the ILL, is shown in Fig. 3.12. Triple axis spectrometers are optimised for coherent INS spectroscopy and are rarely used for incoherent INS work. These instruments were designed to exploit the constant flux of reactor sources. A detailed description of the theory and practice of triple axis spectrometry is given in [17].

In their mode of operation a monochromatic neutron beam, of wavevector k_i, is selected from the white incident beam, using Bragg reflection from a single crystal, the monochromator. This beam illuminates the sample. In a selected direction from the sample a second single crystal, the analyser, is oriented such that only a given final wavevector, k_f, is reflected onto a single detector. The choice of monochromator and analyser crystals depend on the energy transfer range of interest, typically Cu(200) is used for 200–150 cm^{-1} and Cu(220) for 1000–2000 cm^{-1}.

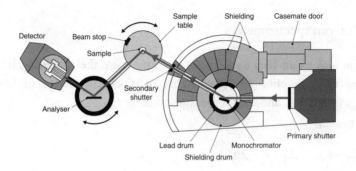

Fig. 3.12 Schematic of the triple axis spectrometer IN1 at the ILL. Reproduced from [16] with permission from the Institut Laue-Langevin.

If neutrons are observed in the detector, then some process in the sample changed k_i to k_f. The important point is that a given configuration of the spectrometer corresponds to a single point in (Q,ω) space. Either k_i or k_f, can be changed stepwise and the scattering angle varied so as to scan (Q,ω) space and detect all possible neutron scattering processes. By virtue of the three axes of rotation (monochromator, sample and analyser), the instrument is enormously flexible and, in principle, any point in (Q,ω) space is accessible. The great strength of triple axis spectroscopy is simultaneously its biggest disadvantage; it is a point-by-point method and data accumulation is very slow. Further, the use of Bragg reflection for both the monochromator and the analyser reduces the final flux, which is very weak, and introduces a particular structure to the background. Any process that simultaneously fulfils both Bragg conditions will be detected. One such occurs when a Bragg harmonic from the monochromator has the same wavelength as a different harmonic of the analyser, i.e. when $n\ k_i = n'\ k_f$ (where, n and n' are integers, $n \neq n'$). Triple axis spectrometers viewing hot sources are most suited to vibrational spectroscopic studies but suffer badly from this background problem.

3.4.2 Indirect geometry instruments

Instruments that work with any fixed final energy are conventionally known as *indirect geometry instruments*. However, we shall limit our consideration to those spectrometers with low final energies. They are remarkably simple to design, relatively cheap to build, easy to operate and have an output, which is both similar to optical spectra and easily compared to calculation. They are ideally suited to exploitation by the chemical and biological scientific communities.

Indirect geometry spectrometers have no requirement (within the limitations implied by the use of $S^\bullet(Q, \omega)$, §2.5.1) to calibrate detector efficiencies, on either continuous or pulsed sources (compare §3.4.3). Since the final energy of the neutrons never varies the detection efficiency is constant. Variations arising from differing discrimination levels (§3.3.2) could play a significant role, except that (on low final energy instruments) all detectors follow almost the same path in (Q,ω) space (§3.4.2.3). Occasionally there is a need to calibrate the detected intensity in respect of the sample mass and standard analytical chemical techniques can be readily adapted to this circumstance.

3.4.2.1 Filter instruments at reactor sources

The flux detected on triple axis spectrometers (§3.4.1) can be greatly increased by replacing the analyser crystal with a beryllium filter, since a much larger detector area can be used. Beryllium is transparent to neutrons with energies less than *ca* 40 cm^{-1}. Higher energy neutrons are Bragg scattered out of the beryllium and are not transmitted. The beryllium edge is very discriminating, a 15 cm thickness of beryllium transmits only 1 in 10^4 neutrons of the wrong energy. It acts as a band-pass filter with a more or less sharp cut-off. The sharpness of the cut-off is improved by cooling the filter below 100K, see Fig. 3.13, and such filters are routinely operated at liquid nitrogen temperature. (Any neutron scattering in the beryllium, below 40 cm^{-1}, is mainly due to its phonons. Cooling the beryllium reduces their number, decreases the scattering and improves the transmission. If the transmission is 100% at 5 K this degrades to about 75% at 100K, and about 30% at room temperature.)

Graphite filters work on exactly the same principles but their cut-off is at even lower energy, *ca* 12 cm^{-1}, but is less discriminating and should be operated in conjunction with beryllium. The filter materials are compared in Fig. 3.13.

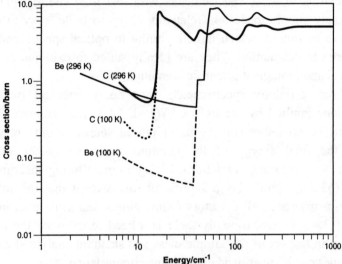

Fig. 3.13 Total scattering cross section of beryllium (black line) and graphite (grey line) at room temperature (solid line) and at 100K (dashed line). Reproduced from [19] with permission from the National Institute of Standards and Technology.

The incident energy is selected stepwise across the spectrum and, since the final energy is fixed below 40 cm^{-1}, the energy transfer is obtained. This is the working principle of the spectrometer IN1BeF [16] at the ILL, which was, for many years, the best spectrometer for neutron vibrational spectroscopy.

The resolution is improved by using a graphite filter but at the cost of a large decrease in the detected flux [18]. Increasing the total detector area can compensate for this loss of detected flux and is the guiding philosophy behind the instrument FANS (Filter Analyzer Neutron Spectrometer) at NIST (Gaithersburg, USA), Fig. 3.13 [19]. The sample table and the pie-shaped filter-detector assembly pivot about the monochromator axis as the incident neutron energy is scanned between 40 and 2000 cm^{-1}. These energy limits are set by the largest and smallest Bragg angle respectively that the monochromator can achieve.

The filter-detector assembly consists of two wedges covering a total scattering angle range of ±150° (1.13 steradians), Figs. 3.14 and 3.15. They consist of blocks of polycrystalline beryllium and graphite, at 77 K. A significant decrease in background is achieved by placing a second Be filter after the graphite. Collimators capture any neutrons not travelling radially from the sample and different filter sections are also separated by absorbing material, again intended to capture any stray Bragg diffracted neutrons.

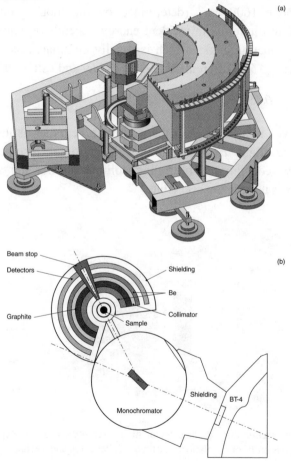

Fig. 3.14 The beryllium filter instrument FANS at NIST (Gaithersburg, USA) (a) exploded view, (b) schematic. Reproduced from [19] with permission from the National Institute of Standards and Technology.

An increase in sensitivity ($\times$ 4) is planned by the use of a double focussing monochromator, where many small crystals will be arranged on a curved surface, analogous to a parabolic mirror. This will provide an incident flux of almost 10^7 neutrons cm^{-2} s^{-1}.

The resolution of filter instruments is determined by; either, the bandpass of the filter and the detector response (at low energy transfers); or, the monochromator characteristics (at higher energy transfers). The resolution of FANS is ~ 10 cm^{-1} at low energies and at high energies is ~ 50 cm^{-1} above 1600 cm^{-1} (determined by the monochromator). The variation of resolution with incident energy depends on uncertainties in the monochromating angle, which is usually dominated by the crystal's mosaic spread, $\Delta\theta_B$. Differentiating Bragg's law and using Eq. (2.21)

$$\frac{\Delta E}{E} = \frac{\Delta\lambda}{\lambda} = \Delta\theta_B \cot\theta_B \qquad (3.10)$$

For a given monochromator the resolution degrades as the cotangent of the Bragg angle, θ_B. Some improvement to the resolution can be imposed by reducing the beam-line collimation but the sacrifice in neutron flux is usually unacceptable.

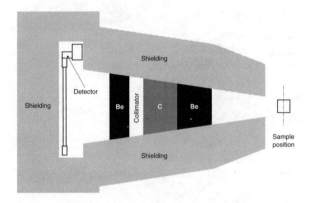

Fig. 3.15 Cross section through the detector-filter assembly on FANS. Reproduced from [19] with permission from the National Institute of Standards and Technology.

To record a spectrum on these instruments, the energy range to be scanned is selected. Since choices can be made about the energy range to be covered, interest can be focussed on a particular energy transfer region, or experimental time saved by avoiding unprofitable parts of the spectrum.

The monochromator is rotated to the starting energy and data are collected for a fixed number of incident neutrons, as detected by the monitor. This eliminates variations in the incident flux due to the distribution of neutron energies in the source. The monochromator is then rotated to provide the next incident energy and the process repeated. The energy step is open to choice but it seems unwise to have fewer than five discrete data points in a resolution element. Such that if the instrumental resolution is 50 cm^{-1} the energy step should be $\leq$ 10 cm^{-1}. Intensities collected in this way are immediately proportional to $S^{*}(Q, \omega)$ without further treatment and, to within the resolution of the spectrometer, the energy transfer is the incident energy minus half the energy of the filter edge.

The fixed monitor count used in the data collection is chosen to optimise its statistical errors. Spectra are accumulated over several hours or overnight (or longer) for weak samples. On FANS a good signal to noise spectrum from *ca* 8 g of a typical organic compound can be recorded in about five hours (excluding the time to cool the sample (§3.5.2)). The errors in INS spectroscopy are, as in infrared, Raman and NMR, governed by Poisson statistics, so the error on a data point of n counts, is $\pm \sqrt{n}$ This 'tyranny of the square root' requires the measurement time to double before the statistics improve by only $\sqrt{2}$.

Although the statistical errors of a spectrum are reduced by counting at each point for longer, a better appreciation of all the contributing errors is obtained by repeating a scan. Repeated scans are not always reproducible to within their statistical errors, the differences represent contributions from systematic errors. Data collected at the beginning of a scan were not collected under precisely the same conditions (nor indeed with the detector in the same location) as that at the end. Backgrounds emanating from experiments on neighbouring spectrometers can make a significant contribution to these systematics. Also, the fast neutron background increases significantly at high energy transfers, here the Bragg angle at the monochromator is small and the sample views the reactor core more directly. This contribution often depends somewhat on the status of the reactor fuel element in its burn cycle.

A subtler problem arises from contamination of the incident neutron beam. From Eqs. (3.11) and (2.21) we see that the energies of the Bragg orders reflected from the monochromator are related as the squares of the natural numbers (1, 4, 9..). An incident beam of nominal energy 250 cm^{-1} also has flux at 1000 cm^{-1} and 2250 cm^{-1}. Strong transitions at high energy will scatter these contaminating neutrons into the detector. Under these circumstances, spectra with few low energy features may show extra weak bands. This is of greater concern for spectrometers that view hot sources. Finally, the choice of materials from which to make a versatile monochromator is restricted and copper is almost universally favoured despite its significant incoherent cross section. The background introduced by incoherent contamination of the incident beam is estimated by off-setting the monochromator by a few degrees every, say, tenth energy step. This background data is then accumulated for a fixed time, not monitor count.

These instruments are energy calibrated by using a material with strong Bragg reflections and a well-known structure, such as nickel. For a given monochromator setting, a known Bragg reflection e.g. Ni(111) appears in a detector at a particular scattering angle, θ (measured with respect to the straight-through beam). Then, from Bragg's law:

$$\theta = 2\theta_B$$
$$n\lambda = 2d_B \sin\theta_B$$

(3.11)

hence λ but $E = 660/\lambda^2$, Eq. (2.21), and the corresponding incident energy is determined. When more than one monochromator is available the process is repeated for each.

3.4.2.2 Filter instruments at spallation sources

Filter instruments are also used at pulsed sources, but they operate differently from those at continuous sources. At a pulsed source, each neutron is 'time-stamped' at its moment of creation and this makes time-of-flight techniques the method of choice. The aim of the design is to exploit energy dispersion during neutron flight time. Although neutrons of all incident energies are produced almost simultaneously in the moderator, because of their different velocities they become dispersed along the beam-tube as they travel away from the moderator.

Since the mass of the neutron (m_n, kg) is known, as are the neutron flight distances (d, m), for an elastic scattering (diffraction) process the flight time (t, μs) determines the neutron velocity (v_n) and hence its energy (E, cm^{-1}) since (cf Eq. 2.21)

$$E = \frac{m_n v_n^2}{2} = \frac{m_n}{2}\left(\frac{d}{t}\right)^2 = 42.2 \times 10^6 \; \frac{\text{cm}^{-1}\,\mu\text{s}^2}{\text{m}^2}\left(\frac{d}{t}\right)^2 \tag{3.12}$$

For an inelastic process, the total flight time, t_t, is given by:

$$t_t = t_i + t_f$$
$$= \frac{d_i}{v_i} + \frac{d_f}{v_f}$$
$$= \left(d_i \Big/ \sqrt{\frac{m_n}{2E_i}}\right) + \left(d_f \Big/ \sqrt{\frac{m_n}{2E_f}}\right) \tag{3.13}$$
$$= \left(\sqrt{\frac{m_n d_i^2}{2(E_t - E_f)}}\right) + \left(\sqrt{\frac{m_n d_f^2}{2E_f}}\right)$$

where t_t is the sum of the incident (before scattering) t_i and final t_f (after scattering) flight times Thus it is necessary to know the distance from the source to the sample, d_i, the sample to detector distance, d_f, and either the incident, E_i, or final, E_f, energy as well as the total flight time, t_t.

The FDS (Filter Difference Spectrometer) [20] at LANSCE (Los Alamos, USA) uses a combination of beryllium and BeO filters to define the final energy. The instrument is shown schematically in Fig. 3.16. It sits at 13 m from a 283 K water moderator, the neutrons scattered by the sample pass through either a beryllium or beryllium oxide filter to the detectors. The Be and BeO segments are symmetrically arranged about the straight through beam such that they mirror each other at the same scattering angle. The difference spectrum (Be minus BeO) is then generated and converted to energy transfer.

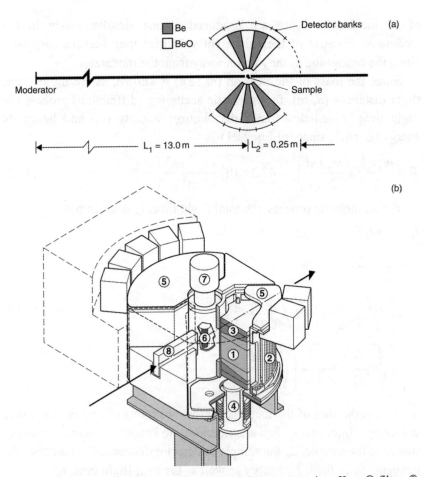

Fig. 3.16 The FDS at LANSCE (a) schematic and (b) cut-away view. Key: ① filter, ② detectors, ③ shielding, ④ closed cycle refrigerator to cool filter, ⑤ shielding, ⑥ sample (usually annular or cylindrical), ⑦ cryostat for sample, ⑧ incident beam tube. Reproduced from [20] with permission from Elsevier.

An example of FDS output, for potassium hydrogen maleate at 15 K is shown in Fig. 3.17. The improved resolution of the (Be—BeO) spectrum, over the simple Be filter method, is marked. This results from two factors; the narrower bandpass used in the difference spectrum and the removal of the long-time tails of the incident pulse. The long-time tails are seen illustrated in the insets in Fig. 3.17, which show the shapes of the elastic lines. The energy transfer range is 50—5000 cm^{-1}, the lower limit is set by the bandpass of the Be filter and the low flux at low energies

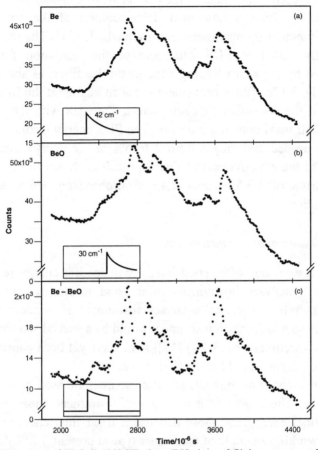

Fig. 3.17 Treatment of FDS (LANSCE) data: INS time-of-flight spectrum of potassium hydrogen maleate recorded with (a) Be filter, (b) BeO filter and (c) the difference (Be–BeO). The inset in each spectrum shows the lineshape of the elastic line associated with each measurement. Reproduced from [20] with permission from Elsevier.

Detailed considerations of resolution, operation and calibration are similar to those for crystal analyser spectrometers (§3.4.2.3) see also Appendix 3.

The longer the flight-path, the more dispersed are the neutrons and the better the incident resolution. The filters determine the resolution characteristics of the final flight path and the best flux is obtained when the incident and final resolutions are matched.

A more recent refinement of the FDS analysis has been to enhance its resolution by using a numerical deconvolution of the spectrometer response function by maximum entropy methods [21]. The technique is particularly suited to FDS data because the resolution function is determined by the sharp leading edge of the Be filter; as shown in the inset to Fig. 3.17c. The improvement is not so marked if the BeO edge is used, or if the resolution function were a Gaussian with the same full width at half maximum and intensity. The FDS energy resolution can be varied as needed and ranges from 2 to 5% of the energy transfer. An example of the effectiveness of this approach is shown in Fig. 3.18 for the spectrum of 1,3,5-triamino-2,4,6-trinitrobenzene before and after deconvolution.

3.4.2.3 Crystal analyser instruments

Filter instruments offer good intensity at modest energy resolutions and crystal analyser instruments offer good resolution with modest intensities. While a crystal analyser instrument is conceivable at a continuous source (such an instrument would be a specialised form of the triple axis spectrometer (§3.4.1) [17]) none have yet been constructed to exploit the advantages of low final energies.

Crystal analyser instruments at pulsed sources are used in the time-of-flight mode. There are two main types of instrument: those that use a variable final energy and those that use a fixed final energy. There is only one working example of the former type at present, PRISMA [22] at ISIS. This is used exclusively for coherent inelastic neutron scattering and so will not be considered further here.

Of the low final energy crystal analyser instruments, there were several early spectrometers that incorporated some of the elements found in current machines. The first instrument to include all the elements was installed at KENS (Tsukuba, Japan) in the early 1980's [18]. Similar instruments were commissioned over the next decade: NERA-PR at IBR-2 (Dubna, Russia) [24], CHEX at IPNS (Argonne, USA) [25] and TFXA at ISIS (Chilton, UK) [26]. The operating principle of all these instruments is the same and it will be described for TFXA, which was the most successful of these first generation instruments. A cut-away drawing of the instrument is shown in Fig. 3.19.

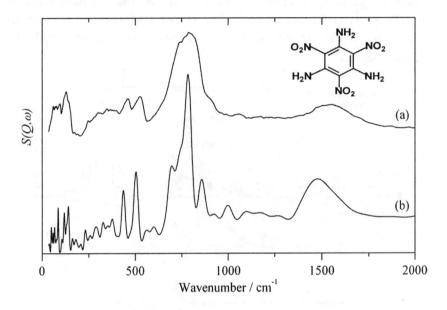

Fig. 3.18 Example of the improvement in resolution obtainable by the use of the maximum entropy method on FDS (LANSCE) data. INS spectrum of 1,3,5-triamino-2,4,6 trinitrobenzene before (a) and after (b) deconvolution.

TFXA stood 12.1 m from an ambient water moderator, every pulse illuminated the sample with a white spectrum of incident neutrons, the fast (energetic) neutrons arrived early and the slow neutrons arrived last. Some of those neutrons with sufficient energy to promote an internal transition in the sample were scattered towards the secondary spectrometer that viewed the sample at a scattering angle of 135°. Planes in the analyser crystal, highly oriented pyrolytic graphite (002), were set to select the final neutron energy of ~32 cm^{-1}. These neutrons were Bragg diffracted onto the detector assembly, passing through a beryllium filter on the way. Neutrons of all energies, except those Bragg reflected, passed through the graphite and were absorbed in the shielding. Each detector assembly consisted of 13 ^{3}He detectors, that had been 'squashed' to an oval shape giving a maximum thickness of ~6 mm. These were specially chosen to be reasonably *inefficient* and so discriminate disproportionately against the high energy background neutrons (§3.3.1.1). A great advantage of TFXA was that there were no moving parts in the spectrometer. The use of a white beam also means that it is possible to obtain diffraction data simultaneously with the inelastic data.

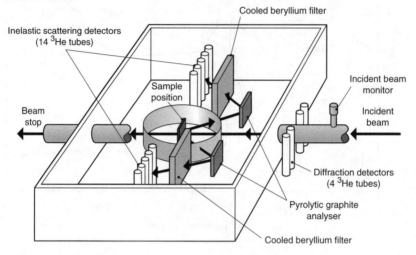

Fig. 3.19 Cut-away drawing of TFXA at ISIS (Chilton, UK) [26].

In crystal analyser instruments the final energy is determined by the analyser, the role of beryllium is restricted to the removal of high order reflections. These have energies 4, 9, 16… times that of the first order energy. The effectiveness of beryllium is illustrated in Fig. 3.20 [23]. In its absence, Fig. 3.20a, very strong higher order reflections are present, which totally swamp any inelastic features. In Fig. 3.20b a 20 cm Be filter was installed and the higher order reflections are completely suppressed allowing the inelastic features to be seen. The combination of Be and pyrolitic graphite is used on all crystal analyser instruments. Graphite crystals are chosen for their very high Bragg reflectivity and small incoherent cross section.

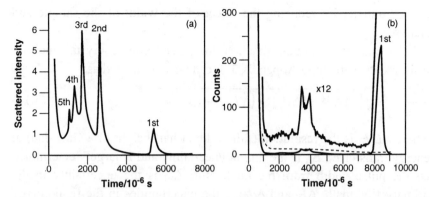

Fig. 3.20 Time-of-flight spectrum of vanadium (LAM-D, KENS) (a) without and (b) with a 20 cm Be filter. Note that the higher order reflections present in (a) are completely suppressed in (b). (The 1st order reflection occurs at a different time because the detector was in a slightly different position after removal of the filter). Reproduced from [23] with permission from Elsevier.

Although the beryllium filter may be placed either before or after the graphite analyser; positioning it after the analyser is much the better choice. This is because the beryllium scatters the neutrons it rejects, over 4π steradians, but does not absorb them. As such it acts as a weak neutron source and contributes to the background. By positioning the beryllium after the analyser, only those few neutrons from the higher order reflections remain to be scattered and cause background.

3.4.2.3.1 The resolution of crystal analyser instruments

Crystal analyser instruments on pulsed sources have two subtle features that improve their sensitivity and resolution, time focussing and energy focussing [23]. These are explained in more detail in Appendix 3 but both rely on having the planes of the sample, the analyser and the detectors parallel. Time focussing improves the resolution at high energy transfers and applies when, irrespective of their final energy, all neutrons take the same time-of-flight from sample to detector. This removes the uncertainty in the final energy. Energy focussing applies when, irrespective of their divergence, all neutrons with the same final energy are reflected to the same detector. This improves the resolution at low energy transfer by reducing the effect of the width of the sample and analyser.

The energy resolution function for pulsed source crystal analyser instruments is given in Appendix 3. It can be summarised as:

$$\frac{\Delta E}{E_t} = \frac{1}{E_t}\sqrt{\left\{(\Delta t_t)^2 + (\Delta d_i)^2 + (\Delta E_f)^2 + (\Delta d_f)^2\right\}} \qquad (3.14)$$

where $\Delta E/E_t$ is the resolution (full width at half height of a Gaussian) at energy transfer E_t, Δt_t is the timing uncertainty and is largely determined by the pulse width from the moderator; Δd_i is the uncertainty in the primary flightpath, ΔE_f and Δd_f are the uncertainties in the final energy and the final flight-path respectively. Both these quantities depend on the thickness of the sample (d_s), the graphite analyser (d_g) and detector (d_d).

There are various techniques for measuring the flight-path lengths but none are entirely satisfactory and there always remains an uncertainty, ca $\pm$ 1 cm. (Since the neutron can start anywhere in the moderator, scatter from any point in the sample and be detected anywhere in the detector.) However, the flight-paths in neutron instruments are so long that the relative error is small and the contribution from Δd_i is negligible.

The other three factors are all significant although the dominant factor is ΔE_f. The individual contributions to the resolution of TFXA are shown in Fig. 3.21a; (d_i = 12.1 m, d_f = 0.70 m, E_f = 28 cm^{-1}, d_g = 0.002 m, d_d = 0.006 m and d_s = 0.002 m). Now:

$$\left(\frac{\Delta E_f}{E_t}\right)^2 \propto E_i \left(\frac{d_f}{d_i}\right)\left\{(d_s)^2 + (2d_g)^2 + (d_d)^2\right\} \qquad (3.15)$$

Improvements in resolution can be obtained by decreasing the thickness' d_s, d_g, d_d and increasing d_i. The greatest thickness is d_g (thick crystals reflect more neutrons) and, since the terms are added in quadrature, this term will dominate. (Resolution is, conventionally on indirect geometry instruments with low final energy, given as a fraction of *energy transfer*. Care must be taken when making comparisons with direct geometry spectrometers where resolution is given as a fraction of the *incident energy* (§3.4.4).) TFXA was upgraded to TOSCA, see Fig. 2.21, in two stages, in the first the detector thickness was decreased from

6 mm to 2.5 mm [27] and the instrument was then moved from 12.13 m to 17 m [28]. The effect of the changes can be seen in Fig. 3.21b.

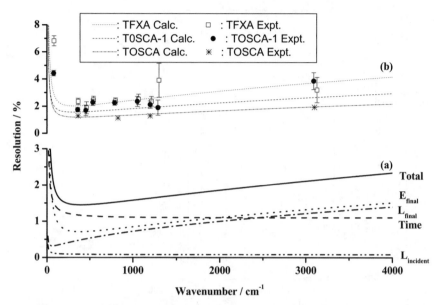

Fig. 3.21(a) The resolution function and its components for TFXA, (b) the evolution of the resolution on TFXA, TOSCA-1 and TOSCA.

In Fig. 3.22 we show a time-of-flight spectrum obtained on TFXA and its corresponding energy transfer spectrum. It can be seen that the spectrum is contained in the interval 1000–13000 μs. Qualitatively, the improvement in resolution resulting from the increased primary flight-path arises because the neutrons become more spread out in time which increases the useful time interval to 1000—18000 μs.

Increasing the primary flight-path improves the resolution but decreases the incident flux by half ($\sim (12.1/17.0)^2$) because of the reduced solid angle. Increasing the sample area, or detector area (or by using the guide described in §3.2.2) can compensate for this loss. On TOSCA the sample size and detector areas were enlarged. To accommodate more detectors, TOSCA has analyser-detector modules in both backward and forward scattering as seen in Fig. 3.23. The result is that TOSCA is presently the world's best crystal analyser spectrometer for both flux and resolution and is likely to remain so until instruments are built at SNS or J-PARC.

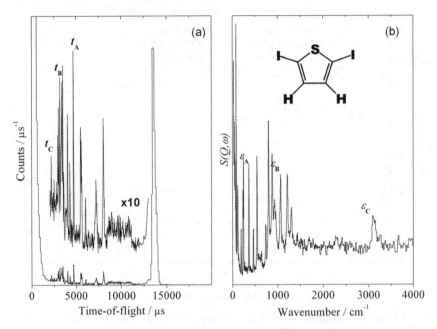

Fig. 3.22 (a) The raw time-of-flight data of 2,5-diiodothiophene (notice the strong elastic line at 13600 μs), (b) the resulting INS spectrum.

3.4.2.3.2 Frame overlap suppression

The increased incident flight-path has a second consequence. The intense feature at 13600 μs, in Fig. 3.22a, is the elastic line for TFXA, d_i = 12.1 m. As the flight path is increased, the elastically scattered neutrons reach the sample later and the elastic line occurs at later times. On TOSCA where d_i = 17.5 m, it occurs at 22600 μs. However, ISIS operates at 50 Hz and each time frame is only 20000 μs long. Thus the elastic line from one pulse of neutrons would occur in the next time frame; frame overlap. An alternative viewpoint is that fast neutrons from the current frame have overtaken the slow neutrons from the previous frame. Frame overlap is catastrophic because the uniqueness of the time-stamp, so crucial to the analysis of the spectrum, is lost.

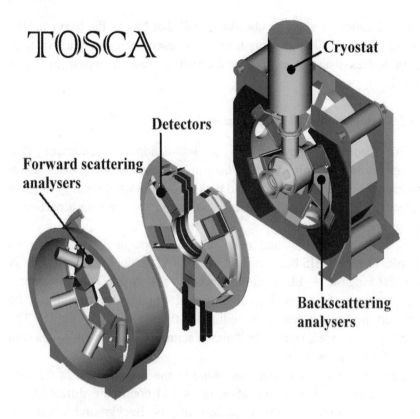

Fig. 3.23 The INS spectrometer TOSCA at ISIS.

In Fig. 3.22a there is also an intense feature at very short time (<1000 µs) which is due to the prompt flash caused mostly by γ–rays and unmoderated, very high energy neutrons. These are undesirable because the prompt γ–rays can saturate some detectors and the very high energy neutrons can become thermalised in the shielding and leak out to contribute to the instrument background. Both problems are tackled by removing prompt-flash products from the beam, with a Nimonic chopper. This is a rotating arm with a block of strongly scattering metal at its end (300 mm long and 1.5 mm wider than the beam).

The chopper is phased to the pulsed source such that the beam is blocked for the first 500 µs of the frame. On the leading edge of the arm is a 'tail-cutter', a sheet of absorbing B_4C that blocks the beam for the period 17000—20000 μs. This timing is chosen such that neutrons with energies less than 40 cm^{-1} are adsorbed so cannot appear in frame overlap.

3.4.2.3.3 Operation and calibration

To record a spectrum on a low-bandpass instrument data accumulation is simply started. There are no choices to be made about the energy range to be covered and data across the whole energy range is collected for every sample. Interest cannot be focussed on a particular energy transfer region.

Spectra are accumulated over several hours, or longer for weak samples (with an ISIS beam current of 180 μA, the count-rate on TOSCA is in the range 0.1—1.0 neutrons detector^{-1} s^{-1}). A good signal-to-noise ratio spectrum from *ca* 3 g of a typical organic compound can be recorded in about six hours (excluding the time to cool the sample). The counting errors are governed by Poisson statistics, and the error on a data point of n counts is $\pm \sqrt{n}$.

The systematic errors that contribute to the uncertainties in the data from reactor based instrumentation (§3.4.2.1) contribute differently to the data from time-of-flight based instruments. Background levels will still vary dependant on the operations of neighbouring spectrometers but since the spectra are accumulated pulse by pulse no particular region of the spectrum is affected by this variation. Spectra are reproducible, within their statistical errors, even when measured years apart.

The data are collected as detected counts *versus* time-of-flight and must be divided by the monitor spectrum to remove the variation of the incident flux with energy. Time-of-flight and energy transfer are not linearly related, see Eq. (3.12) and data are collected in short (>3 μs^{-1}), equal time-periods, or bins, and these correspond to broader and broader energy steps across the spectrum. After conversion into an energy transfer *versus* $S(Q, \omega)$ spectrum, the spectra in individual detectors are added. The conversion to energy makes use of the parameters extracted from the calibration procedure.

The method of calibration [29] uses a material with strong, sharp transitions whose energies are accurately known from infrared and Raman spectra (thus can be traced back to certified standards) and which can be identified in the time-of-flight spectrum. The best calibrant known to date is 2,5-diiodothiophene (see Fig. 3.19 for the structure). There are three unknown values for the parameters, d_i, d_f and E_f that must be determined. Three transitions that span the energy range of the spectrometer are selected, these have known transition energies ε_A, ε_B and ε_C and are observed (in an individual detector) at times t_A, t_B, t_C. Substituting in Eq. 3.13:

$$t_A = \left(d_i \middle/ \sqrt{\frac{m_n}{2(E_f + \varepsilon_A)}} \right) + \left(d_f \middle/ \sqrt{\frac{m_n}{2E_f}} \right)$$

$$t_B = \left(d_i \middle/ \sqrt{\frac{m_n}{2(E_f + \varepsilon_B)}} \right) + \left(d_f \middle/ \sqrt{\frac{m_n}{2E_f}} \right) \qquad (3.16)$$

$$t_C = \left(d_i \middle/ \sqrt{\frac{m_n}{2(E_f + \varepsilon_C)}} \right) + \left(d_f \middle/ \sqrt{\frac{m_n}{2E_f}} \right)$$

The three non-linear simultaneous equations, Eq. (3.16), in three unknowns are solved to give the desired parameter values for that detector. Fig. 3.22 shows (a) the raw flight time data for a single detector with the three transitions used for calibration indicated and (b) the resulting INS spectrum of 2,5-diiodothiophene. The flight time *versus* counts spectrum of each detector is also corrected for the incident flux distribution (as measured by the incident beam monitor). All calibration procedures are specific to a given sample position. If the position of the calibrant differs from that of the sample the conversion to energy transfer will be wrong, less so at low energy transfers than high.

The great advantage of crystal analyser spectrometers is that there are no moving parts in the spectrometer. Even physically moving the instrument is not problematic because TOSCA uses a kinematic mount system and the whole spectrometer can be removed and replaced within a day but the calibration remains unaltered. An advantage of this feature is that the conversion to $S(Q,\omega)$ can be initiated automatically when the current spectrum has finished accumulating, with full confidence in the results. This highlights one of the great strengths of crystal analyser spectrometers on pulsed sources: they are very simple to operate. Further, the spectrum that is obtained is similar to the more familiar infrared and Raman spectra.

3.4.2.3.4 Momentum transfer aspects of low final energy instruments

The price that is paid for the good resolution, high flux and simplicity of operation of the low final energy instruments is that they have a fixed trajectory through (Q,ω) space, as shown in Fig. 3.24. The reason for this is shown in the inset in Fig. 3.24.

For most energy transfers the incident energy (and hence k_i) is much larger than the final energy (and hence k_f), so the momentum transfer Q is almost equal to k_i irrespective of the scattering angle. From the Cosine Rule for the solution of triangles, Eq. (2.23):

$$Q^2 = k_i^2 + k_f^2 - 2k_i k_f \cos\theta \tag{3.17}$$

but with $k_i \gg k_f$ this reduces to:

$$Q^2 \approx k_i^2 \tag{3.18}$$

and from Eq. (2.21), energy in cm^{-1} and Q in Å^{-1},

$$E = 16.7 Q^2 \quad \text{then} \quad Q^2 \approx E/16.7 \tag{3.19}$$

As shown in Fig. 3.24, and implied by Eq. (3.18), the trajectory through (Q,ω) space is almost independent of the scattering angle. This is the trajectory taken by all indirect geometry instruments that have fixed low final energies, say less than 40 cm^{-1}, and applies regardless of the techniques used to define the final energy

Further, the direction of momentum transfer is almost parallel to the incoming neutron beam, again irrespective of the energy transfer. The condition to observe an INS transition is that Q is parallel to the direction of atomic displacement (§2.5.2). Thus experiments analogous to optical polarisation measurements are possible, for suitably aligned samples. An example using polyethylene is shown in Fig. 10.6.

The low final energy has another consequence. In comparison with direct geometry instruments (§3.4.3), the data from low final energy

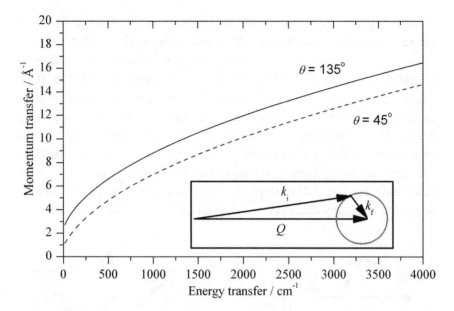

Fig. 3.24 The trajectory through *(Q, ω)* space of an INS spectrometer with a low fixed final energy for two scattering angles: 45° (forward) and 135° (backscattering). The inset shows the scattering triangle for the case where $k_i \gg k_f$ hence $Q \approx k_i$ irrespective of the scattering angle.

spectrometers is collected at relatively high Q. As such the intensities suffer from disadvantageous Debye-Waller factors, especially if the sample is not cooled to <30 K (§2.5.1.2, and §A2.4)

3.4.3 Direct geometry instruments

Direct geometry instruments use choppers or crystal monochromators to fix the incident energy and they are found on both continuous and pulsed sources. To compensate for the low incident flux resulting from the monochromation process, direct geometry instruments have a large detector area. This makes the instruments expensive, they are generally twice the price of a crystal analyser instrument. At present, they are used infrequently for the study of hydrogenous materials, so we will limit our discussion to a chopper spectrometer at a pulsed source and a crystal monochromator at a continuous source.

Direct geometry instruments are fundamentally different from the indirect geometry instruments with a low final energy in that they are able to measure both Q and ω independently. Consider the scattering triangles shown in Fig. 3.25a. Neutrons are scattered with the same energy transfer (and hence k_f) but in different directions, or scattering angle θ. The momentum transfer, Q, is different for each scattering angle, thus detectors positioned at different angles allow the measurement of Q at a given energy transfer. Conversely, neutrons that are scattered through the same angle but with different energy transfer (and hence k_f), Fig. 3.25b, allow the measurement of the energy transfer at a given Q for any detector, applying Eqs. (3.17), (3.8) and (2.21):

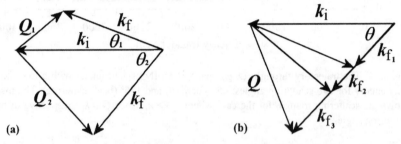

(a)   (b)

Fig. 3.25 Scattering triangles for a direct geometry instrument. (a) Detectors at different angles give different Q at constant energy transfer and (b) an individual detector measures energy transfer at constant Q.

$$Q^2 = k_i^2 + k_f^2 - 2k_i k_f \cos \theta$$

$$\frac{\hbar^2 Q^2}{2m_n} = E_i + E_f - 2\{E_i E_f\}^{1/2} \cos \theta \qquad (3.20)$$

$$\frac{\hbar^2 Q^2}{2m_n} = 16.7 Q^2 = 2E_i - \hbar\omega - 2\{E_i(E_i - \hbar\omega)\}^{1/2} \cos \theta$$

Thus a detector positioned at a scattering angle θ will perform a scan in time whose locus is a parabola in (Q,ω) space. Fig. 3.26 shows the trajectories for a range of angles in the interval 3–135°. This gives rise to a characteristic 'bishop's mitre' type of plot. As the incident energy increases, the Q-range also increases: e.g. $0.3 \leq Q \leq 9.1$ Å^{-1} with $E_i = 400$ cm^{-1} and $0.6 \leq Q \leq 20.3$ Å^{-1} with $E_i = 2000$ cm^{-1}.

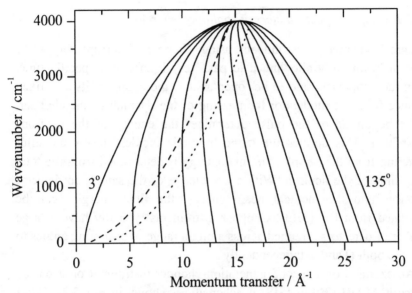

Fig. 3.26 Trajectories in (Q, ω) space for a direct geometry spectrometer with detectors at angles 3, 20, 30, 40, 50, 60, 70, 80, 90, 100, 110, 120 and 135° and with an incident energy of 4000 cm^{-1}. The dashed lines are the trajectories of an indirect geometry instrument (low-bandpass) using scattering angles of 45 (long dashes, forward scattering)) and 135° (short dashes, backscattering) and a final energy of 28 cm^{-1}.

Direct geometry spectrometers have no fixed final energy and the detectors will capture slow neutrons more efficiently than fast neutrons, correcting for this relative efficiency within a given detector follows the (1/neutron velocity) rule (§3.3.2). However, although detectors are very similar one to another, the precise settings of the discriminator levels (§3.3.1) often vary and the relative detection efficiencies are affected. This efficiency variation must be eliminated since each detector takes a significantly different trajectory through (Q, ω) space and the data from all detectors are used to produce the final $S(Q, \omega)$ map (§5.4). This normalisation is achieved by recording the spectrum of a vanadium sample, this metal is an almost purely incoherent scatterer and has a low absorption cross section. The elastic intensities recorded in all detectors are scaled to be equal, as they should be to first order (although detailed corrections are required in practice).

3.4.3.1 Chopper spectrometers at spallation sources

Direct geometry instruments using choppers interrupt the white incident beam, only allowing it to pass very briefly at a specific time, when the chopper is said to be 'open'. Since the chopper sits at a fixed distance from the moderator the phasing of the opening of the chopper, to the beginning of the pulse determines the energy of the neutrons transmitted. This defines the monochromatic incident beam and after scattering from the sample, the final energy is determined from the total flight time of the neutron. With a knowledge of the sample to detector distances and the incident beam energy, the final energies can be calculated from Eq. (3.13). Chopper instruments typically have a large number of detectors arranged across a broad range of scattering angles to measure both Q and ω independently.

An example of a second generation chopper instrument on a pulsed source is MARI [30] at ISIS. A schematic is shown in Fig. 3.27. After the moderator, a Nimonic chopper is used to remove the gamma flash and improve the background. (As used on TOSCA (§3.4.2.3.2) but without the tail-cutter). The beam is then monochromated using a Fermi chopper, see Fig 3.28. This is a metal drum (rotor) approximately 15 cm in diameter pierced by a beam sized hole. The hole is filled with a slit package, thin sheets of highly absorbing material, such as boron, interleaved with sheets of the neutron transparent aluminium. These are all arranged vertically and when open the slits lie along the beam direction. Since it takes a finite time for neutrons to traverse the drum, the slits are curved in opposition to the direction of rotation to increase the transmission. By changing the slit package, the instrument can be optimised for either flux or resolution, Fig. 3.29.

The rotor is suspended vertically, by magnets, in a vacuum and rotates at speeds up to 600 Hz (36000 rpm), the fastest speed is determined by the mechanical properties of the chopper. The incident neutron energy, E_i, $(70 \leq E_i \leq 8000\ cm^{-1})$ is selected by phasing the opening time of the chopper to the neutron pulse from the target.

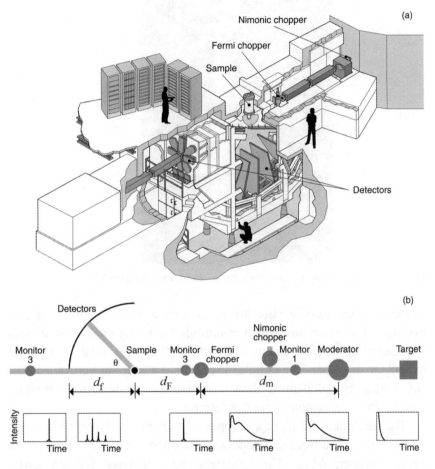

Fig. 3.27 (a) Cut-away drawing of MARI. (b) Components of a chopper instrument.

The monochromated neutrons are scattered by the sample and recorded by ^{3}He detectors (§3.3.1.1). The azimuthal low angle bank consists of eight radial arms of detectors ranging from 3° to 12°. The rest are arranged in a vertical scattering plane and cover from 12° to 135°.

The detectors between 12° and 30° are arranged on a Debye-Scherrer cone to improve the Q-resolution. To reduce the background all internal

Fig. 3.28 A Fermi chopper. The rotor with the slit package is on the left.

surfaces in the sample tank are lined with a neutron absorbing low hydrogen B_4C resin mix which minimises the background arising from the scattering of high energy neutrons. B_4C is also used as a shield behind the detectors. The sample tank also has a series of vertical slats of B_4C acting as collimators, these ensure that only neutrons travelling radially from the sample strike the detector.

The resolution, $\Delta E/E_i$, of a chopper instrument largely depends on the width of the pulse from the moderator, Δt_m, and the opening time of the Fermi chopper, Δt_{chop}. The moderator term is fixed for any given moderator but the second term depends on the rotation speed of the chopper and the slit width. The resolution is given by

$$\left(\frac{\Delta E}{E_i} \right)^2 = \left[\left(\frac{\Delta E_m}{E_i} \right)^2 + \left(\frac{\Delta E_{chop}}{E_i} \right)^2 \right] \tag{3.21}$$

where, using the definitions shown in Fig. 3.26.

$$\frac{\Delta E_{\mathrm{m}}}{E_{\mathrm{i}}} = 2\left(\frac{\Delta t_{\mathrm{m}}}{t_{\mathrm{chop}}}\right)\left[1 + \frac{d_{\mathrm{F}}}{d_{\mathrm{f}}}\left(1 - \frac{E_{\mathrm{t}}}{E_{\mathrm{i}}}\right)^{3/2}\right] \tag{3.22}$$

and

$$\frac{\Delta E_{\mathrm{chop}}}{E_{\mathrm{i}}} = 2\left(\frac{\Delta t_{\mathrm{chop}}}{t_{\mathrm{chop}}}\right)\left[1 + \frac{d_{\mathrm{m}} + d_{\mathrm{F}}}{d_{\mathrm{f}}}\left(1 - \frac{E_{\mathrm{t}}}{E_{\mathrm{i}}}\right)^{3/2}\right] \tag{3.23}$$

Here t_{chop} is the neutron's flight time from the moderator to the chopper and the lengths, d, are defined in Fig. 3.26. On MARI since all the detectors are at the same distance from the sample, the resolution is the same for all detectors and is ~1% E_{i}. (On direct geometry spectrometers, resolution is given as a fraction of the incident energy. Note that for indirect geometry instruments with low final energy the resolution is given as a fraction of energy transfer (§3.4.2.3.1)).

In Fig. 3.29 we show the resolution and flux of MARI for different slit packages, rotation rates and incident energies. The S slit package has relatively wide slits so has high flux and modest resolution, the A, B and C slit packages have narrower slits (hence better resolution but less flux) and are optimised for incident energies of 4000, 1600 and 800 cm^{-1} respectively. Note that the rotation rate has a marked influence on the resolution, since it modifies the chopper 'open' time. It can be seen that the resolution is best at maximum energy transfer i.e. when $E_{\mathrm{t}} \approx E_{\mathrm{i}}$ and degrades by about a factor of 2 to 3 (depending on the choice of slit package) as the elastic line is approached. This is the reverse of the indirect geometry machines where the resolution is best at low energy transfer and degrades as it increases. Thus the two types of instrument are highly complementary.

3.4.3.1.1 Calibration

To obtain $S(Q, \omega)$ from the measured data it is necessary to know the instrumental flight paths, the energy of the incident neutrons and the angles of the detectors. The incident energy is found from the beam monitors. The first monitor is normally placed before the Fermi chopper to monitor the incident flux for the purposes of normalisation. The second and third are placed just after the Fermi chopper and beyond the sample at the end of the spectrometer respectively. The distance between the second and the third monitors is determined by direct measurement. The times that the peak intensity of the monochromatic neutron beam appears in the monitors are measured. Knowing the distance between the two monitors, $(d_f + d_F)$ Fig 3.26, the difference in the monitor times, $(t_F - t_f)$ gives the incident energy, from Eq. (2.21).

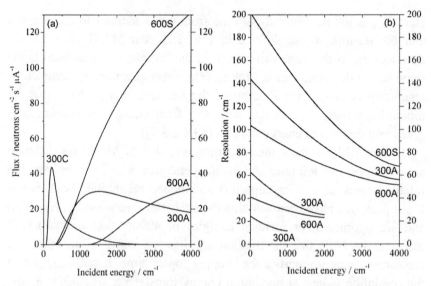

Fig. 3.29   The effect on (a) the flux at the sample and (b) the resolution of MARI resulting from different slit packages (S, A, C), rotation rates (300, 600 Hz) and incident energies.

$$E_i = 42.2 \times 10^6 \left(\frac{d_f + d_F}{t_F - t_f} \right)^2 = \frac{660}{\lambda^2} \tag{3.24}$$

The detector angles are found using Bragg scattering from a well-defined sample such as nickel (§3.4.2.1.1). Substituting, into Eq. (3.24), the total flight path from moderator to detector, d_t, and t the time of arrival of the, say the Ni(111) Bragg peak in the detector. We then substitute the result into Bragg's law (and knowing the Ni(111) spacing, d_B) obtain the scattering angle, $\theta = 2\theta_B$.

$$\lambda = 1.56 \times 10^{-4} (t/d_t) = 2d_B \sin \theta_B \tag{3.25}$$

3.4.3.2 *Crystal monochromator instrumentation on a continuous source*

An example of crystal monochromator instrument on a continuous source is IN4 [31] at the ILL, shown in Fig. 3.30. IN4 sits at the end of a short thermal neutron guide, thus the lack of hot neutrons effectively restricts the energy transfer range to less than $\sim$800 cm^{-1}.

On a continuous reactor source, it is necessary first to generate a pulsed beam. On IN4 the beam is pulsed and fast neutrons and γ-rays that would give background are largely eliminated by two background choppers. These are rapidly rotating choppers, which act as a low-pass filter. Energy selection is in two parts: *via* diffraction from pyrolytic graphite (PG); or copper: PG(002) for 46–165 cm^{-1}, PG(004) for 182–660 cm^{-1}, Cu(111) for 115–422 cm^{-1} and Cu(220) for 293–1000 cm^{-1}. Each face of the monochromator is an assembly of 55 small crystals arranged to focus the divergent incident beam onto a small area at the sample position. Supplementary energy selection is by a Fermi chopper that rotates at speeds of up to 650 Hz. The neutrons scattered by the sample are detected by a large area array of ^{3}He detectors (§3.3.1.1). The operation, calibration and resolution of these instruments is similar

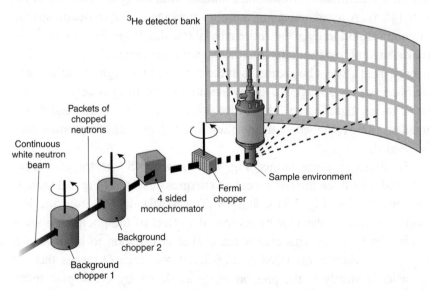

Fig. 3.30 Schematic of the direct geometry spectrometer IN4 at the ILL. Reproduced from [32] with permission from the Institut Laue Langevin.

to those on pulsed sources (§3.4.3.1.1) except that their resolution also involves consideration of mosaic spread as with other monochromator instruments (§3.4.2.3.1).

3.4.4 Choosing the optimal technique

At a spallation source both types of instrument are usually present, see Table 3.2, and both have advantages and disadvantages. The crystal analyser instruments offer excellent resolution (particularly below 1500 cm^{-1}), a wide energy transfer range and simplicity of operation but the information is obtained at relatively high Q. They are optimised for hydrogenous samples and generally give excellent results for these types of sample.

Fig. 3.29 illustrates one of the great strengths of the direct geometry instruments: the incident flux and the resolution are directly under the experimentalist's control in a way that is not possible on a crystal analyser or beryllium filter instrument. This makes these instruments extremely versatile but correspondingly more difficult to operate. There is also the additional information available from the Q-dependence of the data (§5.4). A disadvantage of these instruments is that to obtain spectra at good resolution over the entire energy transfer range it is necessary to make several separate measurements with different E_i. The incident energy must always be greater than the energy of the highest transition to be observed and, if access to any sizeable range in Q is required, may need to be significantly above that energy. However, the higher the incident energy, the worse the resolution so a degree of compromise may be needed.

To illustrate these points, Fig. 3.31 shows INS spectra of toluene obtained with three different incident energies on MARI, Fig. 3.31a, b, c, and on TOSCA, Fig. 3.31d, at ISIS. The MARI data is from the low angle, i.e. low Q, detector banks thus the effect of the phonon wings is minimised (§2.6.3). This allows the C-H stretch region to be observed, Fig. 3.31a, whereas on TOSCA, Fig. 3.31d, the large Q means that the intensity is mostly in the phonon wings as shown by the displacement, broadening and weakening of the peak relative to the MARI data. Note that access to a Q–range that is lower than that on TOSCA, is only possible if the incident neutron energy is higher than the transition studied. This is clearly seen in Fig. 3.26, where the top of the mitre ($= E_i$) is mid-way between the Q-values of the TOSCA forward and backscattering banks. However, because the degree of intensity transfer into the phonon wings and the magnitude of the argument of the Debye-Waller factor both depend on Q^2 (§2.6.3), even a modest reduction in Q may allow the fundamentals to be observable. Typically, an incident energy 10 to 20% higher than the feature of interest is enough to allow its observation.

The resolution ($\sim$1—2% E_i) is relatively poor because the incident energy is large. Note that the difference between the spectra is the result of the different Q-values that are used, it is *not* the difference in resolution of the instruments, it is the access to lower Q that is crucial. In fact, across most of the energy transfer range the resolution of TOSCA is comparable or superior to that of MARI but the trajectory in (Q,ω) means that above 1500 cm^{-1} it is rare that it can be exploited.

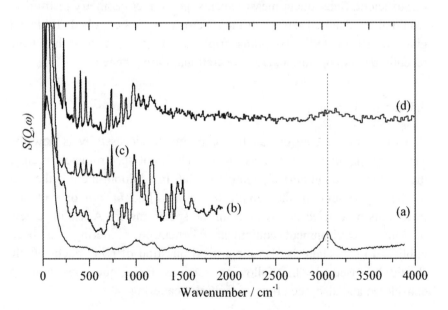

Fig. 3.31 INS spectra of toluene at 20K obtained on TOSCA and at low Q on MARI. MARI spectra at (a) 4000, (b) 2000 and (c) 800 cm^{-1} incident energy respectively, the TOSCA spectrum is shown in (d).

Decreasing E_i to 2000 cm^{-1} improves the resolution considerably but loses the C—H stretch region, it is also possible to observe the C—C stretching modes at 1200—1600 cm^{-1} that are scarcely seen on TOSCA. Similarly decreasing E_i to 800 cm^{-1} gives good resolution but most of the spectrum is lost. In this energy region the momentum transferred during scattering, be it on TOSCA or MARI, is so low that the differences between the two spectra, Fig. 3.31c and d, are determined mostly by the resolution of the individual instrument. The TOSCA spectrum has better resolution over much of the energy transfer range and was measured in the same time it took to record one of the MARI spectra but suffered from phonon wing effects above about 1500 cm^{-1}. In this example, the Q-dependence of the data has not been used.

Clearly both types of instruments are highly complementary and both have strengths and weaknesses. Ideally, the same sample would be run first on an indirect geometry instrument which would provide a rapid, but still fairly detailed overview of the subject. In many instances this would be sufficient. Subsequent measurements on a direct geometry instrument would allow detailed aspects of the spectroscopy to be probed. Table 3.2 gives a list of INS (excluding triple axis) spectrometers that have recently been in operation, are in operation, or are planned.

3.5 Sample handling

Crucial to carrying out an INS experiment successfully is how the sample is mounted in the spectrometer. Different spectrometers require the sample to be cylindrical, annular or flat. In addition, there is the aim of the experiment: is the sample to be modified *in situ* or will any treatments be carried out away from the spectrometer? A useful review of sample environment equipment for neutron scattering has been published [42]. There is also the critical question of how much sample should be used. In the following sections these questions will be considered and also, the effect of multiple scattering.

3.5.1 Sample quantity and multiple scattering

Inelastic neutron scattering is a rare process in part because neutron sources are so weak. Thus one might be inclined to load as much sample as possible into the beam. Within the limits set by multiple scattering, this is an excellent idea. Multiple scattering arises when a neutron is scattered twice (or more) from separate atoms within the body of the sample (compare §2.5.1.3, Table 2.1). The most likely process is two elastic scattering events but elastic scattering is of no consequence for the energy transferred in an INS experiment.

An elastic scattering event and an inelastic event (in whatever order) is potentially more troublesome, because of the extra distance travelled by the neutron. Fortunately, compared to the total flightpath, the extra distance travelled is negligible (since the scattering occurs within the sample, whose size is small compared to the instrument flightpaths) and the energy transfer remains substantially correct. However, the direction of the neutron's travel has changed more than once thus the apparent momentum transfer Q (in both magnitude and direction) will be incorrect, since to calculate Q single scattering is assumed.

The chance of two inelastic events occurring is low, except for the very strongest scattering samples. In this case the flight time would give an apparent peak at an energy transfer equal to the sum of the two losses. (Multiple inelastic events are exploited in moderators (§3.1.1.2) to bring very energetic neutrons into a more useful energy range. In a sense samples are simply very ineffective moderators.)

This simple analysis suggests that the double inelastic event is detrimental to the spectra collected on all instruments, whereas the (elastic + inelastic) case is detrimental only for direct geometry instruments. This is true even for powder samples, since the magnitude of the momentum transfer is also lost. This contamination is most problematic for data obtained from the low scattering angle detectors. The data in these detectors are nominally obtained at low Q but multiple scattering injects high Q information into their data.

This is a much less serious problem for indirect geometry instruments with low final energies. These spectrometers work close to the maximum

Table 3.2 INS instruments. (Triple axis spectrometers are excluded, since they are mainly used for coherent INS. A list can be found in [17]).

Instrument	Location	Type	Energy range/ cm^{-1}	Resolution/(%$\Delta E/E_T$)	Status	Ref.
IN1BeF	ILL (France)	Be filter	300 - 3600	8 - 12	Operational	[16]
BT4	NIST (USA)	Be filter	80 - 1600	3	Superseded by FANS	[19]
FANS	NIST (USA)	Be filter	80 - 1600	3	Operational	[19]
FDS	LANSCE (USA)	Be filter	50 - 5000	5 (2 with MaxEnt)	Operational	[20]
Filter Detector	BARC (India)	Be filter	80 - 2000	20	Operational	[32]
TFXA	ISIS (UK)	Crystal analyser	-16 - 8000	2.5	Superseded by TOSCA	[26]
TOSCA	ISIS (UK)	Crystal analyser	25 - 8000	1.5	Operational	[28]
LAM-D	KENS (Japan)	Crystal analyser	-16 - 8000	2.5	Operational	[33]
NERA-PR	IBR-2 (Russia)	Crystal analyser	16 - 4000	2.5	Operational	[24]
KDSOG-M	IBR-2 (Russia)	Crystal analyser	40 - 3000	15	Operational	[34]
CAS	IPNS (USA)	Crystal analyser	-16 - 8000	2.5	Superseded by CHEX	[25]
CHEX	IPNS (USA)	Crystal analyser	-16 - 8000	2.5	Not operational	
VISION	SNS (USA)	Crystal analyser	25 - 8000	1.5	Proposed for 2006 start of SNS	
HET	ISIS (UK)	Direct geometry	0 - 8000	2 - 4	Operational	[35]
MARI	ISIS (UK)	Direct geometry	0 - 4000	1 - 2	Operational	[30]
MAPS	ISIS (UK)	Direct geometry	0 - 8000	1.5 - 3	Operational	[36]
MERLIN	ISIS (UK)	Direct geometry	0 - 4000	2 - 4	Successor to HET. Operational 2005	[37]
IN4	ILL (France)	Direct geometry	0 - 800	2 - 5	Operational	[31]
LRMECS	IPNS (USA)	Direct geometry	0 - 5000	7-14	Operational	[38]
HRMECS	IPNS (USA)	Direct geometry	0- 8000	3 - 6	Operational	[39]
PHAROS	LANSCE (USA)	Direct geometry	0 - 8000	0.5 - 1	Operational	[40]
INC	KEK (Japan)	Direct geometry	0 - 8000	5 - 10	Operational	[41]
	SNS (USA)	Direct geometry			Proposed for 2006 start of SNS	
	J-PARC (Japan)	Direct geometry			Proposed for 2007 start of J-PARC	

value of Q, for a given energy transfer (§3.4.2.3) and no multiple scattering can much influence this value. Analysis of spectra recorded on TFXA and TOSCA suggests that up to ~ 25% of the recorded spectrum can result from multiple scattering without seriously degrading the spectral quality. (Note, however, that strongly scattering samples are usually thick and the sample thickness enters directly into the resolution characteristics of TOSCA, Eq. (3.14).)

Correcting spectra for multiple scattering contamination is non-trivial (and we do not discuss such correction procedures) but fortunately, in the INS of hydrogenous systems, it is easier to avoid the problem. Samples should not scatter too much and experience over many years suggests that a sample that scatters 10% (or less) of the incident neutrons is certainly acceptable on all spectrometers. The mass of sample that scatters this proportion of neutrons can be calculated from an equation that is directly analogous to the Beer-Lambert law of optical absorption spectroscopy (§2.0) Eq.(2.1).

$$\frac{J_f}{J_i} = \exp(-d_s\,C\,\sigma_{\text{total}}) \tag{3.26}$$

where J_i and J_f are the incident and final, or transmitted, fluxes respectively, d_s is the sample thickness and $C\sigma_{\text{total}}$ is the total scattering cross section summed over all atoms, per unit area. For a multiple scattering calculation the absorption cross section is irrelevant since it is only the scattered neutrons that are important. So:

$$C\sigma_{\text{tot}} = \frac{\rho N_A}{M}\sum_l \sigma_l \tag{3.27}$$

ρ is the density (kg m^{-3}), N_A is the Avogadro number, M is the molar mass (kg), σ is the total scattering cross section of each atom, l, in the molecule in barns (10^{-28} m^2) (§2.1.3). As an example, for toluene, C_7H_8 ($\rho = 865$ kg m^{-3}, $M = 0.092$ kg mole^{-1}):

$$C\sigma_{\text{tot}} = \frac{\rho N_A}{m_{\text{mol}}}\sum_l \sigma_{\text{inc}} = \frac{\rho N_A}{M}\{7\sigma_C + 8\sigma_H\} \tag{3.28}$$

$$= \frac{865\,(6.02\times10^{23})}{0.092}\{7(5.55\times10^{-28}) + 8(82.0\times10^{-28})\} = 393\text{ m}^{-1}$$

The quantity required for a 10% scatterer (i.e. 90% transmission) is:

$$0.9 = \exp(-d_s \, C \, \sigma_{tot}) = \exp(-d_s 393) \quad \text{so} \quad d_s = 3 \times 10^{-4} \text{ m} \qquad (3.29)$$

Assuming a beam size of 5×5 cm^2, as used on MARI, gives a sample volume of 0.7 cm^3 and a sample mass of 0.6 g. On TOSCA (beam size 4 × 4 cm^2), a 1 mm pathlength cell containing 1.4 g would be a 32% scatterer but would still give an excellent spectrum, almost untroubled by any adverse multiple scattering effects. Good rule-of-thumb measures of the amount of sample required are given by Eqs. (3.26) and (3.27).

3.5.2 Cryogenics

The Debye-Waller factor, that suppresses all neutron intensities (§2.5.1.2), is strongly temperature dependent and INS measurements are generally carried out at 30K, or less, this is almost mandatory on indirect-geometry low final energy instruments. As an example, Fig. 3.32 shows the spectrum of perdeuteropolyethylene at room temperature and

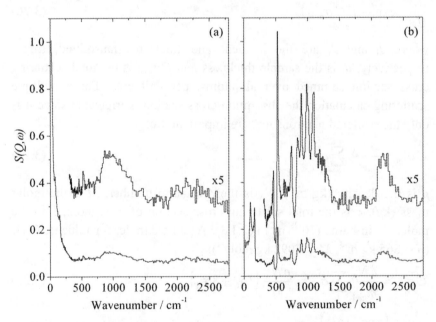

Fig. 3.32 INS spectrum of perdeuteropolyethylene at (a) room temperature and (b) at 6K.

at 6K recorded on TFXA. The difference is marked. The room temperature spectrum is almost universal, in that *any* organic compound will give a very similar spectrum. Liquid helium cryostats or closed cycle refrigerators are used to cool the samples.

Standard liquid helium cryostats can be used at any temperature above the lambda point of helium, ~1.5K, and specialised types of cryostat routinely operate as low as 0.05K. For most samples of chemical interest, there is no practical difference between spectra recorded at 4.2K and at 20K (unless phase transitions occur), thus the additional expense and complexity of very low temperatures cannot be justified.

Closed cycle refrigerators are simple, cheap to operate and very reliable. For the majority of samples, the temperatures in the range 10–20K that are achieved by closed cycle refrigerators are entirely satisfactory. Closed cycle refrigerators that operate to 5K have been commercially available for sometime and their cooling power is increasing. They are already starting to supplant liquid helium cryostats in some circumstances. This trend is certain to continue, particularly given the increasing cost and reducing availability of liquid helium.

The sample can be cooled either by directly attaching it to the cold finger of the cryostat or, indirectly, by helium exchange gas. With the sample directly attached, it and the cryostat must be warmed to room temperature each time a sample is changed, then evacuated to ~10^{-6} mbar and cooled again. This is a complex and time consuming process. Indirect attachment uses helium exchange gas for cooling. The cryostat cools a central chimney, at the bottom of which the sample is mounted on the end of a special pole, a centrestick. (Conventionally the neutron beam passes through the lower part of a cryostat.) The sample is cooled by conduction across a low pressure, 15 mbar, of helium exchange gas. To change sample, the chimney is filled with helium gas to atmospheric pressure, the centrestick withdrawn and exchanged for another with the next sample attached. The chimney is then evacuated to its operational pressure and the sample allowed to cool. The sample changeover can easily be accomplished in less than half-an-hour. The centrestick system is so successful and so flexible that it is the preferred method at virtually all neutron scattering establishments.

3.5.3 Conventional samples

Conventional samples are pure compounds, or mixtures of such, where the only requirement is to measure their INS spectrum at low temperature, irrespective of how exotic they may be. Each INS spectrometer has different requirements in terms of sample shape: FANS at NIST uses cylindrical samples, TOSCA at ISIS uses a flat plate and MARI at ISIS uses an annular shape.

Sample containers are thin-walled cells of either aluminium or vanadium. Vanadium is required if diffraction measurements are to be included. Solids can be simply wrapped in aluminium foil. Liquids, however, must be held in sealed cans. Indium wire provides a good seal, because it is ductile, easily created and, more importantly, it has a very low coefficient of thermal expansion. Thus a can that is sealed at room temperature will remain sealed after having been cycled to 20K. If the compounds are air or moisture sensitive (solid or liquid) they are loaded into the cans in a glove-box. Gases are more problematic, one successful method is have a suitable volume of the gas attached above the cell. The whole is lowered into the cryostat where the gas is first liquefied and allowed to fill the cell, subsequently the temperature is lowered to freeze the liquid.

With modern instrumentation, it is possible to measure a spectrum in a few hours, which is comparable to the time needed to cool typical samples to 30K and might leave the spectrometer standing idle for long periods. There are a number of possible approaches to reducing the time wasted by waiting for samples to cool down. For FANS at NIST (§3.4.2.1.1), the system is to use two cryostats, while the first sample is being measured, a second sample is cooling off-line in another cryostat. The cryostats are then interchanged and the second sample measured. For TOSCA at ISIS (§3.4.2.2.2), a 24 position automatic sample changer is used. While the sample changer takes 6–12 hours to cool, averaging across the samples means that the cooling time per sample is 30 minutes or less.

3.5.4 Temperature, pressure and magnetic field

Neutrons are highly penetrating so common engineering materials such as aluminium and steel can be used for equipment. The design and manufacture of complex sample environment is, therefore, simplified but this benefit is somewhat offset by the need for relatively large sample sizes. Nonetheless, neutron scattering experiments can be carried out at temperatures from 0.05 to 2000K, at pressures from 10^{-9} to 10^5 bar and in magnetic fields up to 7 Tesla. The current trend is to combine two of these variables. Thus cryomagnets to measure below 4.2K in a magnetic field are commercially available and low temperature high-pressure experiments are becoming routine. Most of these types of experiments are the province of condensed matter physics and beyond the scope of this book.

Temperature is a variable that is sometimes used, although as explained earlier there are limits on the range, see Fig. 3.32. The exception is when a direct geometry spectrometer is used. The Debye-Waller factor can be reduced by collecting data at small Q as shown in Fig. 3.31. Chemists have been slow to exploit the possibilities of direct geometry instruments, a situation that is ripe for change. One method of changing temperature is to warm the cryostat, which is generally slow and inefficient. It is better to heat the sample directly by having cartridge heaters placed directly on the sample holder.

Low temperature, high pressure experiments have sometimes been carried out, but the large sample size needed restricts the available pressure range. For experiments up to 6 kbar a 'helium intensifier' can be used. This consists of a pressure can mounted on the end of centrestick that has high pressure gas lines that run to outside the cryostat. The pressure is generated by a two-stage compressor starting from helium gas at cylinder pressure. With this system it is possible to change the pressure *in situ*. High pressures, up to 25 kbar, can be attained using a McWhan cell. The sample is held between two ceramic anvils and the load is applied to a piston by a hydraulic press. When the desired pressure is reached, the piston is locked in place to maintain the pressure. Pressurisation can be done only *ex situ* at room temperature so the method is laborious if several pressures are required. The cell itself weighs ~ 10 kg so cooling it to 20K is also very slow.

3.5.5 Catalysts and in situ experiments

INS studies of catalysts are described in Chapter 7. These experiments present a particular set of problems that need to be addressed. Foremost is the fact that INS is not an intrinsically surface sensitive technique. This is overcome by using large samples to maximise the number of surface sites and hydrogenous adsorbates to give the highest possible contrast. Supported metal catalysts, zeolites and oxides often have low densities, while metal powders have high densities but low surface areas. Both situations require a large volume cell to place sufficient sample in the beam.

The experiments can be conducted in a number of ways. The samples, surface plus adsorbate, can be prepared completely off-line and loaded into a suitable sealed cell for the measurement. This is generally less favoured because of concerns about the catalyst during its transfer into a sample cell. A more common approach is to load only the surface sample into the cell and all the operations are carried out on the sample in the cell but off-line from the spectrometer. These operations usually involve an activation cycle, followed by an adsorption stage and possibly a desorption stage. INS spectra are recorded after each stage and the spectrum of the adsorbed species is the difference spectrum: INS spectrum after the adsorption stage minus the INS after the activation stage. The repositioning of samples after off-line treatment requires care. This is especially true if the INS spectrum of the bare catalyst is structured since a valid difference spectrum critically depends on accurate relocation. However, careful experimental design can largely eliminate this problem.

Catalysts with a heavy-metal content are an exception to the rule that chemical samples do not usually become radioactive. This radioactivity must be allowed to decay to safe levels before it can be handled without very special apparatus or precautions. To avoid delays whilst waiting for this decay, several cells are run in parallel: while one is being measured, another is being treated and a third is being left to decay. This method ensures that the spectrometer is in use at all times. It is productive but very demanding on the experimental team.

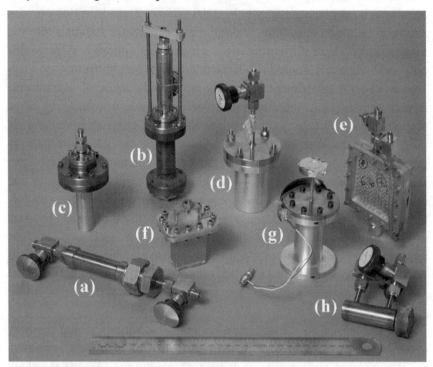

Fig. 3.33 Catalyst cells: (a) high temperature (> 600 K), flow-through, Swagelok™ sealed stainless steel cell; and (b) high temperature gold wire sealed zirconium cell. O-ring sealed aluminium cells: (c) low volume, (d) high volume and (e) flow-through. (f) Aluminium cell for off-line prepared samples; (g) low temperature (<350 K) indium wire sealed aluminium cell. (h) High temperature, flow-through, Conflat ™ sealed stainless steel cell for use on MARI, all the others are for TOSCA. The scale in the foreground is 0.3 m.

A further option is to carry out all the operations on the sample in the cell while it is in the cryostat, at the correct temperature, in the beam. This has considerable advantages: only one cell is needed and there are no concerns about sample relocation. The experiment becomes much more flexible and readily accommodates changes in the original plan, e.g. increasing the amount of adsorbate. The disadvantages are that the experimental arrangements are more complex: the transfer lines transporting the adsorbate to the sample must be kept warm to prevent the adsorbate freezing and blocking the line. Moreover, chemical reactions are virtually non-existent at 20K, generally only physisorption occurs, so it may be necessary to heat the sample. The adsorption process may also be time consuming, leaving the spectrometer standing idle for long periods. These difficulties are such that this approach is generally only used when it is desired to adsorb dihydrogen on a substrate.

For catalyst experiments the cells are usually made of aluminium, steel or zirconium. Aluminium and zirconium have lower scattering cross sections than steel (also with very low absorption cross sections) and so produce less background than steel. Zirconium has good mechanical properties though it is more difficult to machine and more expensive than steel. Aluminium is more easily workable than steel or zirconium but more difficult to seal. Rubber or polymer O-ring seals can be used but these severely limit the maximum temperature (< 250°C) that can be used in the activation cycle and the O-rings do not seal well at low temperature. The integrity of O-ring seals is also a concern as the sample warms to room temperature after the measurement. Metal-metal seals do not suffer from these problems and are easy to make with zirconium or steel.

The experience of the authors after several iterations is that the best system is to use steel cells with the minimum wall thickness permitted for the required temperature and pressure conditions. An off-the-shelf sealing system, such as Conflat™ flanges with copper gaskets or metal tubes with Swagelok™ fittings work well. There is also the need to allow the adsorbate access to the sample and all-metal bellows type valves are the best choice. These survive temperature cycling well. A variety of cells that have been used at ISIS are shown in Fig. 3.33. Fig. 3.34 shows a centrestick equipped for gas handling.

3.5.6 Safety

In describing the common types of experiment some hazards have already become apparent. There are three types of hazard to consider: radiological, chemical and environmental. At all neutron scattering laboratories, a safety assessment of the experiment that covers all of these categories is carried out before the sample even approaches the beamline so that a safe working protocol is established

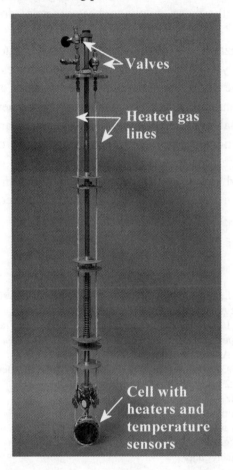

Fig. 3.34 Catalyst cell mounted on a gas-handling centrestick, length 1.40 m.

The radiological hazard attracts the most attention but is probably the easiest to assess. From knowledge of the incident neutron spectrum and the elemental composition of the sample (and its cell) it is possible to predict the degree of activation and the decay time needed. As can be seen from Appendix 1, absorption cross sections for light elements aregenerally small, but increase with increasing atomic number. Since much of chemistry deals with light elements this is fortunate and a typical organic material will never become highly radioactive. Catalyst systems, especially those involving heavy-metals, do become active (§3.5.5).

Chemical hazards are dealt with by good laboratory practice and appropriate safety equipment; fume hoods, glove boxes, gloves, glasses and clothing. The environmental hazard arises when an experimentalist is working in an unfamiliar area. Neighbouring experiments may well be operating unusual equipment and often several hazards are in close proximity. Additionally, the size of the equipment is often larger than most chemists are accustomed to in the laboratory. All of these risks are manageable, and provided the procedures laid out in the established working protocol are followed, the risks are reduced to better than statutory levels.

3.6 References

1   K. Skold & D.L. Price (1986). *Methods of Experimental Physics*, Vol 23 Neutron Scattering, Part A, Academic Press, London.
2   C.G. Windsor (1981). *Pulsed Neutron Scattering*, Taylor and Francis, London.
3   Eds. R.J. Newport, B.D. Rainford & R. Cywinski (1988). *Neutron Scattering at a Pulsed Source*, Adam Hilger, Bristol.
4   J. Chadwick (1932). Nature, 120, 312. Possible existence of a neutron.
5   http://www.frm2.tu-muenchen.de/frm2/index_en.html
6   http://www.ill.fr
7   http://nfdfn.jinr.ru/userguide-97/ibr-2/
8   http://www.isis.ac.uk
9   http://www.sns.gov
10   http://jkj.tokai.jaeri.go.jp/
11   http://sinq.web.psi.ch/
12   http://www.pns.anl.gov/
13   http://www.lansce.lanl.gov/index_ext.htm
14   http://neutron-www.kek.jp/
15   B.N. Brockhouse (1955). Can. J. Phys., 33, 889—891. Neutron scattering & the frequency distribution of the normal modes of vanadium metal. B.N. Brockhouse (1955). Phys. Rev., 98, 1171. Slow neutron spectrometry— a new tool for the study of energy levels in condensed systems. B.N. Brockhouse (1995). Rev. Mod. Phys., 67, 735—751. Slow neutron spectroscopy & the grand atlas of the physical world. (This paper is from Brockhouse's Nobel prize address and gives the history of the early years of triple-

axis spectroscopy. The earliest references to the technique are the 1955 papers above. The latter paper is an abstract of a paper presented at the 1955 American Physical Society meeting in New York).

16 http://www.ill.fr/YellowBook/IN1

17 G. Shirane, S.M. Shapiro & J.M. Tranquada (2002). *Neutron Scattering with a Triple-Axis Spectrometer*, Cambridge University Press, Cambridge.

18 H. Jobic & H.J. Lauter (1988). J. Chem. Phys. 88, 5450—5456. Calculation of the effect of the Debye-Waller factor on the intensities of molecular modes measured by neutron inelastic scattering. Application to hexamethylenetetramine.

19 http://rrdjazz.nist.gov/instruments/fans/

20 A.D. Taylor, E.J. Wood, J.A. Goldstone & J. Eckert (1984). Nucl. Inst. Meth. Phys. Res., 221, 408—418. Lineshape analysis & filter difference method for a high intensity time-of-flight inelastic neutron scattering spectrometer.

21 D.S. Sivia, P. Vorderwisch & R.N. Silver (1990). Nucl. Insts. Meth. Phys. Res., A290, 492—498. Deconvolution of data from the filter difference spectrometer - from hardware to maximum-entropy.

22 U. Steigenberger, M. Hagen, R. Caciuffo, C. Petrillo, F. Cilloco & F. Sacchetti (1991). Nucl. Insts. Meth. Phys. Res., B53, 87—96. The development of the PRISMA spectrometer at ISIS.

23 S. Ikeda & N. Watanabe (1984). Nucl. Insts. Meth.A (Phys. Res.), 221, 571—576. High resolution tof crystal analyzer spectrometer for large energy transfer incoherent neutron scattering.

24 I. Natkaniec, S.I. Bragin, J. Branowski & J. Mayer (1994). ICANS XII Proceedings of the Twelth International Collaboration on Advanced Neutron Sources 24—28 May 1993, Rutherford Appleton Laboratory Report RAL-94-025, 8996. Multicrystal inverted geometry spectrometer NERA-PR at the IBR-2 pulsed reactor.

25 http://www.pns.anl.gov/instruments/chex/

26 J. Penfold & J. Tomkinson (1986). *Rutherford Appleton Laboratory Report RAL-86-019.* The ISIS time focused crystal spectrometer, TFXA.

27 Z.A. Dowden, M. Celli, F. Cilloco, D. Colognesi, R.J. Newport, S.F. Parker, F.P. Ricci, V. Rossi-Albertini, F. Sacchetti, J. Tomkinson & M. Zoppi (2000). Physica B 276,, 98—99. The TOSCA incoherent inelastic neutron spectrometer: progress and results.

28 D. Colognesi, M. Celli, F. Cilloco, R.J. Newport, S.F. Parker, V. Rossi-Albertini, F. Sacchetti, J. Tomkinson & M. Zoppi (2002). Appl. Phys. A 74 [Suppl.], S64S66. TOSCA neutron spectrometer; the final configuration.

29 V. Rossi-Albertini, D. Colognesi & J. Tomkinson (2001). J. Neutron Res., 8, 245—259. A study on the calibration of a time focused inelastic neutron scattering spectrometer.

30 M. Arai, A.D. Taylor, S.M. Bennington & Z.A. Bowden (1992). In Recent Developments in the Physics of Fluids, eds. W.S. Howells & A.K. Soper, F321-F328, Adam Hilger, Bristol. MARI—a new spectrometer for liquid & amorphous materials.

31 http://whisky.ill.fr/YellowBook/IN4/

32   V.C. Rakhecha (2002). Neutron News, 13,11–15. Neutron beam facilities at Bhabha
      atomic research centre.
33   K. Inoue, T. Kanaya, Y. Kiyanagi, K. Shibata, K. Kaji, S. Ikeda, H. Iwasa & Y.
      Izumi (1993). Nucl. Insts. Meth. A (Phys. Res.), 327, 433—440. A crystal analyzer
      type inelastic spectrometer using the pulsed thermal neutron source.
34   http://nfdfn.jinr.ru/userguide-97/kdsog.html
35   A.D. Taylor, B.C. Boland, Z.A. Dowden & T.J.L. Jones (1987). *Rutherford
      Appleton Laboratory Report RAL-87-012*. HET. The high energy inelastic
      spectrometer at ISIS.
36   http://www.isis.rl.ac.uk/excitations/maps/
37   http://www.isis.rl.ac.uk/excitations/merlin/index.htm
38   http://www.pns.anl.gov/instruments/lrmecs/
39   http://www.pns.anl.gov/instruments/hrmecs/
40   http://www.lansce.lanl.gov/research/Pharos.html
41   M. Arai, M. Kohgi, M. Itoh, H. Iwasa, N. Watanabe, S. Ikeda & Y. Endoh (1989).
      ICANS X Proceedings of the Tenth International Collaboration On Advanced
      Neutron Sources Held at Los Alamos 3–7 October 1988, Institute of Physics
      Conference Series, Institute of Physics, Bristol, 97, 297–308. Development of a
      chopper spectrometer at KENS
      S. Itoh, M. Arai & M. Kawai (2002). Appl. Phys. A-Mater, 74, Part 1 Suppl. S,
      S198—S200. Improvement of the performance of the chopper spectrometer, INC.
42   I.F. Bailey (2003). Z. Kristallogr., 218, 84—95. A review of sample environments in
      neutron scattering.

4

Interpretation and Analysis of Spectra using Molecular Modelling

In this chapter we discuss the theoretical modelling of inelastic neutron scattering (INS) spectra. We shall see that INS spectra can be calculated accurately, both their intensities and frequencies. The modelling enables us to assign the spectra, the prerequisite for chemical and structural interpretations.

4.1 Modelling—the classical and *ab initio* approaches

The calculation of molecular structures and dynamics has been an area of knowledge that has advanced dramatically in the last twenty years. With the advent of more powerful computers, and their increased affordability, it is now possible to perform calculations on small-to-medium size molecules on a personal computer using *ab initio* methods. Our ability to do this has changed the way we interpret and analyse INS spectra. In particular, we take advantage of the fact that the intensity of the spectral lines in INS spectra is not subject to the photon selection rules, unlike infrared and Raman spectroscopy. This particular characteristic gives an increased certainty to the spectral assignments when aided by theoretical calculations.

Although, strictly speaking, the description *ab initio* implies the solution of the quantum mechanical problem without parameterization, quantum mechanics is generally employed only in the calculation of the energy hypersurface that the nuclei feel and move over. The minimization of structures and calculation of their vibrations is most often a classical mechanics problem.

Embedded in the theoretical framework of the study of molecules and solids, there is an approximation, the Born-Oppenheimer approximation, which is justified in most cases and allows us to use classical mechanics to calculate the structure and dynamics of molecules and solids.

4.1.1 The Born-Oppenheimer approximation

Since we shall use the Born-Oppenheimer approximation [1] throughout this chapter it is worth describing in some detail. The adiabatic Born-Oppenheimer approximation states that the mass difference between the electron and a nucleus is so big that the separation of the Schrödinger equation into terms describing the nuclear and electronic motion is allowed. During nuclear motion, the electrons move as if the nuclei are fixed in their *instantaneous* positions. The Hamiltonian is expanded in powers of a parameter κ. For a molecule of N_{atom} atoms, labelled l, each of mass m_l

$$\kappa = \left(\frac{m_e}{M_0} \right)^{\frac{1}{4}} \quad \text{where} \quad M_0 = \frac{1}{N_{atom}} \sum_{l=1}^{N_{atom}} m_l \tag{4.1}$$

and m_e is the electron mass and M_0 is the mean atomic mass. In the expansion of the Hamiltonian, the zero order term is the electronic energy, the second order term is the vibrational energy and the fourth term is the rotational energy; the first and third order terms vanish. The validity of the Born-Oppenheimer treatment depends on the following assumptions:

(a) The vibrational energy separation is about two orders of magnitude smaller than the electronic energy separation

$$\Delta E_{vib} \approx \kappa^2 \Delta E_{elec} \tag{4.2}$$

(b) The rotational energy levels separation is about two orders of magnitude smaller than the separation between vibrational energy levels:

$$\Delta E_{rot} \approx \kappa^2 \Delta E_{vib} \tag{4.3}$$

(c) Vibrational displacement coordinates are *ca* κ times the size of the bond lengths. With these conditions generally satisfied, the separation of the electronic and nuclear motions is valid [2].

As we shall see later in this chapter, the classical mechanics formulation that the energy of the nuclei can be written without making any explicit reference to the electronic contribution is a statement that is only valid because of the Born-Oppenheimer approximation. Also, in *ab initio* methods, we assume that the electronic motion is not coupled with the motions of the nuclei; therefore the ground state electronic density can be calculated for a fixed configuration of the atomic nuclei. The calculation of the forces on the nuclei is performed regarding the nuclei as classical particles whose interaction is described by the corresponding electron density. In fact the minimization of the structures is achieved within classical mechanics.

In molecular calculations, the minimum energy is usually reached when the atoms are in their equilibrium positions. Care must be taken in calculations on extended solids, since the relaxation of the interatomic distances in most codes usually does not include zero-point effects. We shall see later that these effects can have appreciable consequences when dealing with light elements, especially hydrogen.

4.2 Normal mode analysis of molecular vibrations

Molecules consist of atoms held together, bonded, by interatomic forces. These interatomic forces are electromagnetic interactions arising from the motion of electrons and determine the shape and size of molecules. At absolute zero, in the classical interpretation, the molecules have an equilibrium configuration that is determined by the energy minimum of these interatomic forces. We shall analyse how the motions of atoms in a molecule can be calculated and describe how the interactions between atoms in molecules are calculated.

The atoms in a molecule undergo vibrations around their equilibrium configuration within the quantum mechanical picture, even at zero temperature. The application of elementary dynamical principles to these small amplitude vibrations leads to normal mode analysis. Crystalline solids can naively be thought of as big molecules; but solving the equations becomes impossible unless the periodicity of the unit cell is included; whereupon major simplifications of the algebra are introduced.

4.2.1 Vibrations in molecules

In order to introduce the theory of vibrational spectroscopy in inelastic neutron scattering, we make some simplifications that will help us to understand the concepts. First we shall deal with the vibrational modes of molecules in a vacuum or in a dilute gas phase. Note, however, that in INS experiments the sample is cooled to *ca* 20 K, therefore the molecules are part of an extended solid. However, because the forces that keep the atoms in the molecule are often larger than the forces that molecules experience from other molecules in the condensed phase, isolated molecule calculations can be good models.

The N_{atom} atoms that constitute a polyatomic molecule can, before they are bonded, each move in three perpendicular, energy independent directions. The number of degrees of freedom is therefore $3N_{atom}$. (We choose to represent these directions as the Cartesian frame x, y, z, where e_x, e_y, e_z are the unit vectors.) When the atoms bind into a molecule the translations along, and the rotations around, the x, y, z, axes are the only energy independent motions of the molecule. The remaining $3N_{atom} - 6$ motions, or degrees of freedom, are not translations or rotations of the molecule; they are the internal vibrations of the molecule. In a linear molecule there are only two rotations since the third axis coincides with the molecular axis itself; therefore the remaining degrees of freedom are $3N_{atom} - 5$. These vibrations define the total number of vibrational modes (the 'normal modes') and, by their superposition, any possible internal motion of the molecule can be composed.

A normal mode or normal vibration of a polyatomic system is defined as a vibrational state in which each atom moves in simple harmonic motion about its equilibrium position, each atom having the same frequency of oscillation with, generally, all atoms moving in phase.

There is no unique coordinate system in which to perform the vibrational analysis. We have chosen the Cartesian system but in the early literature it was common to use complicated internal coordinate systems to facilitate the computations and exploit molecular symmetry. However, easy access to robust and well proven programs has removed much of the necessity of such complexities. Moreover, the modern computer resources now available enable us to limit our analysis to the simpler Cartesian coordinate system.

4.2.2 Calculation of vibrational frequencies and displacements

In order to find the equation of motion of the atoms in a molecule we need to express the kinetic and potential energies as a function of the atomic coordinates. The coordinates that we shall use describe the displacements of the atoms from their equilibrium positions, u. Here $(u_l)_x$, $(u_l)_y$, $(u_l)_z$ are the magnitudes of displacements of the atom l in a molecule from its equilibrium position, referred to the Cartesian frame. Using the time derivatives of these coordinates, we can write the kinetic energy of a molecule, E_k^{mol}, containing N_{atom} atoms with masses m_l:

$$2E_k^{mol} = \sum_{l=1}^{N_{atom}} m_l \left[\left(\frac{d(u_l)_x}{dt} \right)^2 + \left(\frac{d(u_l)_y}{dt} \right)^2 + \left(\frac{d(u_l)_z}{dt} \right)^2 \right] \tag{4.4}$$

where $\quad u_l = (u_l)_x e_x + (u_l)_y e_y + (u_l)_z e_z$

We replace the $(u_l)_x$ terms with mass-weighted coordinates, q, where, for example:

$$q_1 = \sqrt{m_1}(u_1)_x, q_2 = \sqrt{m_1}(u_1)_y, q_3 = \sqrt{m_1}(u_1)_z$$

$$q_4 = \sqrt{m_2}(u_2)_x, q_5 = \sqrt{m_2}(u_2)_y, q_6 = \sqrt{m_2}(u_2)_z \tag{4.5}$$

$$q_i = \sqrt{m_i}(u_i) \qquad ; m_i = m_{l+1} \quad \text{for} \quad i = 3l+n$$
$$\text{where} \quad n = 1,2,3 \text{ and } l = 0,1,2,\cdots(N_{atom}-1)$$

Double indices, like (l, x) are thus replaced by the expanded single index, i, the summation of Eq. (4.4) now runs up to $3N_{atom}$, but is simplified to:

$$2E_k^{mol} = \sum_{i=1}^{3N_{atom}} \dot{q}_i^2 \qquad \text{where} \quad \dot{q} = \frac{dq}{dt} \tag{4.6}$$

With the exception of diatomic molecules, the potential energy surface that each atom experiences is very complicated. We approximate the molecular potential energy, V^{mol}, in the neighborhood of the equilibrium positions of the atoms with a Taylor series expansion:

$$2V^{mol} = 2V_0^{mol} + 2\sum_{i=1}^{3N_{atom}} \left(\frac{\partial V^{mol}}{\partial q_i}\right)_0 q_i + \sum_{ij=1}^{3N_{atom}} \left(\frac{\partial^2 V^{mol}}{\partial q_i \partial q_j}\right)_0 q_i q_j + \dots \qquad (4.7)$$

where V_0^{mol} is the value of the energy minimum at molecular equilibrium and i and j are the expanded indices of different atoms. Since the atoms are positioned at their equilibrium positions, the derivative of the potential energy at the equilibrium positions is zero:

$$\left(\frac{\partial V^{mol}}{\partial q_i}\right)_0 = 0 \qquad (4.8)$$

Also, since we are considering infinitesimal vibrational amplitudes, the terms higher than quadratic can be neglected because $q_i q_j \gg q_i q_j q_k$. (This approximation will be inadequate when strong anharmonicities are present.) If we choose the minimum of energy, V_0^{mol}, as the arbitrary zero of our energy scale, the potential energy, within our approximation is:

$$2V^{mol} = \sum_{ij=1}^{3N_{atom}} \left(\frac{\partial^2 V^{mol}}{\partial q_i \partial q_j}\right)_0 q_i q_j \qquad (4.9)$$

We write:

$$\left(\frac{\partial^2 V^{mol}}{\partial q_i \partial q_j}\right) = f_{ij} \qquad (4.10)$$

Then f_{ij} is a force constant that represents the change in potential energy caused by the combined displacements of coordinates q_i and q_j. The matrix formed by all such force constants is the dynamical matrix; then:

$$2V^{mol} = \sum_{ij=1}^{3N_{atom}} f_{ij} q_i q_j \qquad (4.11)$$

We now invoke the Lagrange equation:

$$\frac{d}{dt}\left(\frac{\partial L}{\partial \dot{q}_j}\right) - \frac{\partial L}{\partial q_j} = 0, \quad j = 1, 2, 3 \cdots \qquad (4.12)$$

It has the vibrational frequencies and amplitudes as its solutions. The Lagrangian, L, is a function of the kinetic and potential energy:

$$L^{mol}(q,\dot{q}) = E_k^{mol}(q,\dot{q}) - V^{mol}(q,\dot{q}) \tag{4.13}$$

and in our case, a harmonic system, the kinetic energy depends only on the velocities and the potential energy depends only on the coordinates. Then Eq. (4.13) simplifies to:

$$\frac{d}{dt}\left(\frac{\partial E_k^{mol}}{\partial \dot{q}_j}\right) + \frac{\partial V^{mol}}{\partial q_j} = 0, \quad j = 1, 2, \cdots, 3N_{atom} \tag{4.14}$$

We now replace E_k^{mol} and V^{mol} by their expressions in Eqs. (4.6) and (4.11) and recognize that the factor of two in Eq. (4.11) accounts for the double summation over the i, j indices. The following set of $3N_{atom}$ equations appears:

$$\ddot{q}_j + \sum_{i=1}^{3N_{atom}} f_{ij} q_i = 0, \quad j = 1, 2, \cdots, 3N_{atom} \quad \text{where} \quad \ddot{q}_j = \frac{d^2 q_j}{dt^2} \tag{4.15}$$

The molecular vibrations take the form of $3N_{atom}$ independent simple harmonic oscillators, one on each coordinate q_i, of general form:

$$q_i = \sqrt{m_i}\left(u_i^{mol}\right)\cos\left(\varepsilon + \sqrt{\bar{\lambda}_v}\ t\right) \tag{4.16}$$

where u_i^{mol} (the vibrational eigenvector) is the amplitude of the molecular deformation in terms of the mass weighted coordinate q_i, ε is the phase of the vibration and t is time. The eigenvalue, which is related to the frequency ω_v of the vibrational mode v, is given by $\bar{\lambda}_v$ (written with a bar to distinguish it from the symbol λ for wavelength).

$$4\pi^2 \omega_v^2 = \bar{\lambda}_v \tag{4.17}$$

With all atoms moving in phase and at the same frequency for each particular mode, we can write the following set of algebraic equations using Eq. (4.16).

$$\sum_{i=1}^{3N_{atom}} \left(f_{ij} - \bar{\lambda}_v \delta_{ij}\right)\sqrt{m_i}\left(u_i^{mol}\right) = 0, \quad j = 1, 2, \cdots, 3N_{atom} \tag{4.18}$$

where δ_{ij} is the Kronecker delta function that is unity when $i = j$ and zero otherwise.

Besides the trivial solution, $u_i^{mol} = 0$ that corresponds to no vibration, there is a specific set of displacement values for which the solutions are non-trivial, the eigenvectors. These values are those where the eigenvalues, λ, satisfy the secular determinant equation:

$$\begin{vmatrix} (f_{11} - \lambda_1) & f_{12} & \cdots & f_{1,3N_{atom}} \\ f_{21} & (f_{22} - \lambda_2) & \cdots & f_{2,3N_{atom}} \\ \vdots & \vdots & \ddots & \vdots \\ f_{3N_{atom},1} & f_{3N_{atom},2} & \cdots & (f_{3N_{atom},3N_{atom}} - \lambda_{3N_{atom}}) \end{vmatrix} = 0 \qquad (4.19)$$

Since the solutions to all eigenvalue problems appear as ratios the eigenvectors are unitless; conventionally they are normalised by:

$$^{v}L_i^{mol} = \frac{\sqrt{m_i}\left(^{v}u_i^{mol}\right)}{\sqrt{\sum_{j=1}^{3N_{atom}} m_j \left(^{v}u_j^{mol}\right)^2}} \quad \text{or} \quad ^{v}L_{l'}^{mol} = \frac{\sqrt{m_{l'}}\left(^{v}u_{l'}^{mol}\right)}{\sqrt{\sum_{l=1}^{N_{atom}} m_l \left(^{v}u_l^{mol}\right)^2}} \qquad (4.20)$$

where $^{v}L_i^{mol}$ is the normalised vibrational amplitude of coordinate i in mode v. Note the equivalence of both expressions in Eq. (4.20), since the index running over i expresses the individual Cartesian components of the l' index. There is one 'normal mode coordinate' associated with each normal mode frequency. The normal mode coordinates are defined using the mass weighted Cartesian displacements according to the following linear equations:

$$^{v}Q = \sum_{i=1}^{3N_{atom}} {}^{v}L_i^{mol} q_i \qquad \text{for} \qquad v = 1, 2, \cdots, 3N_{atom} \qquad (4.21)$$

(Whilst retaining the conventional symbol for a normal coordinate, Q, we have distinguished it from the momentum transfer vector, Q, by its font.) The normal coordinate system is important since it represents the displacement vectors in Cartesian coordinates and simplifies the expressions for the kinetic and potential energy. This allows us to rewrite Eqs. (4.6) and (4.11) as:

$$2E_k^{mol} = \sum_{v}^{3N_{atom}} {}^{v}\dot{Q}^2 \qquad \text{and} \qquad 2V^{mol} = \sum_{v}^{3N_{atom}} \lambda_v {}^{v}Q^2 \qquad (4.22)$$

Note that the expression of the potential involves no cross products and is quadratic in Q; it also leaves the kinetic energy in its original form. Six of the $3N_{atom}$ roots of the secular equation have the value zero, these modes correspond to the 3 translational and 3 rotational modes, leaving $3N_{atom} - 6$ normal modes. For a proof the reader should refer to [3].

4.2.3 The quantum problem

The advantage of using the normal coordinate system becomes apparent when we analyse the Schrödinger equation:

$$-\frac{h^2}{8\pi^2 m}\nabla^2\psi + V\psi = E\psi \tag{4.23}$$

In classical mechanics the vibrations and rotations of a molecule are strictly separable. In quantum mechanics, however, the rotations and vibrations are only so approximately [3]. If we ignore the complications that can arise from the coupling, the total wave function for rotations and vibrations, ψ, can be written:

$$\psi \approx \psi_{vib}\,\psi_{rot} \tag{4.24}$$

The function ψ_{rot} is the solution of the rotational problem. The vibrational part, ψ_{vib}, is a function of the normal coordinates and is the vibrational wave function. Substituting Eq. (4.24) in Eq. (4.23) and ignoring the rotational and translational contributions, the Schrödinger equation for the vibrational wave function will be:

$$-\frac{h^2}{8\pi^2}\sum_{v=1}^{3N_{atom}-6}\frac{\partial^2\psi_{vib}}{\partial^v Q^2} + \frac{1}{2}\sum_{v=1}^{3N_{atom}-6}\lambda_v\,{}^v Q^2\,\psi_{vib} = E_{vib}^{mol}\,\psi_{vib} \tag{4.25}$$

where E_{vib}^{mol} is the vibrational energy. It is evident that the new wave equation, written in normal coordinates, is in the form of a function that can be separated into $3N_{atom} - 6$ equations, one for each normal coordinate. The energy and wave function are written:

$$E_{vib}^{mol} = \sum_v E_v \quad \text{and} \quad \psi_{vib} = \prod_v \psi\left({}^v Q\right) = \psi\left({}^1 Q\right)\psi\left({}^2 Q\right)\cdots\psi\left({}^{3N_{atom}-6} Q\right) \tag{4.26}$$

The wave equation is satisfied if each function $\psi\left({}^v Q\right)$ and energy E_v satisfy equations of the form:

$$-\frac{h^2}{8\pi^2}\frac{d^2\psi\left(^vQ\right)}{d^vQ^2}+\frac{1}{2}\lambda_v\,^vQ^2\,\psi\left(^vQ\right)=E_v\,\psi\left(^vQ\right) \tag{4.27}$$

Each equation is now a total differential equation in one variable, ^vQ. This is the linear harmonic oscillator equation in terms of the normal coordinate ^vQ. The solution is then expandable as the product of harmonic oscillator functions, one for each normal mode, and the total energy corresponds to the sum of the energies of the $3N_{atom}-6$ oscillators.

4.2.4 The energy levels of the harmonic oscillator

The energy levels of the vth quantum harmonic oscillator are:

$$E_v=\left(n_v+\tfrac{1}{2}\right)\hbar\omega_v,\quad n_v=0,1,\dots \tag{4.28}$$

where n_v is the vibrational quantum number and ω_v is the classical frequency. Therefore, from Eq. (4.26), the total vibrational energy is:

$$E_{vib}^{mol}=\sum_{v=1}^{3N_{atom}-6}E_v=\sum_{v=1}^{3N_{atom}-6}\left(n_v+\tfrac{1}{2}\right)\hbar\omega_v \tag{4.29}$$

The summation is for every normal mode coordinate with its associated quantum number. The lowest energy level is the energy level where all quantum numbers are zero, the ground state. The energy of the ground state is the zero-point energy and is a feature of quantum systems.

When a single quantum number is unity (all others are zero) the molecule is at a fundamental level, in a first excited state. Transitions to this state are the 'fundamental transitions'. When only one of the quantum numbers is greater than unity (all others are zero) the corresponding energy levels are overtone levels, the 'overtone transitions'. When two or more quantum numbers have non-zero values, the levels are combination levels, the 'combination transitions'[4].

Since transitions can occur only between levels of different energy, the energy that can be transferred, for a given normal mode v, to and from a molecule is:

$$h\omega_v^{nn'}=h\omega_v^{n'n}=E_v^{n'}-E_v^{n}\qquad n\neq n' \tag{4.30}$$

If $E_v^{n'}>E_v^{n}$ the molecule gains energy from an incident neutron (analogous to the photon in Raman spectroscopy); this is a Stokes process and the scattered particles constitute Stokes radiation. If, on the other hand, $E_v^{n'} < E_v^{n}$ the molecule transfers energy to the neutron, this is an anti-Stokes process. Only Stokes processes from the ground state are of relevance to the vast majority of INS vibrational spectra (i.e. neutron energy loss). The sample temperatures prohibit anti-Stokes events and the complications that occasionally arise in the infrared from hot bands is avoided. Since only Stokes transitions are permitted the vibrational quantum number n_v is equivalent to the order n in the Scattering Law (§2.5.1.3).

The amplitude of the motion of the harmonic oscillator is given by:

$$<u^2> = \frac{\hbar}{2\mu\omega};$$

using eq 4.30, for the transition, n=1←0

$$<u^2> = \frac{\hbar^2}{2\mu E_\omega} = \left(16.9\text{Å}^2 \text{ amu cm}^{-1}\right)\frac{1}{\mu E_\omega} \tag{4.31}$$

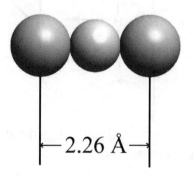

Fig. 4.1 The [FHF]⁻ ion.

4.2.5 Worked example—vibrational frequencies of the bifluoride ion

Vibrational frequencies and eigenvectors are part of the output of most computational chemistry packages. It is instructive to work through the calculation of the vibrational modes of the linear bifluoride [FHF]⁻

ion, Fig. 4.1. Because it is linear it will have $3N_{atom}-5 = 4$ normal modes. The vibrational bands of this molecular ion have been observed, by diode laser spectroscopy: $v_1=583.0539(13)$, $v_2=1286.0284(22)$, and $v_3=1331.1502(7)$ cm^{-1} [5,6]. From the rotational constant, the equilibrium F–F internuclear distance was $2.27771(7)$ Å.

Because of symmetry, the cross terms in the dynamical matrix that contain the derivatives in perpendicular directions vanish and

$$\frac{\partial^2 V^{mol}}{\partial x_i \partial y_j} = \frac{\partial^2 V^{mol}}{\partial x_i \partial z_j} = \frac{\partial^2 V^{mol}}{\partial y_i \partial z_j} = 0$$

The normal modes are represented in Fig. 4.2.

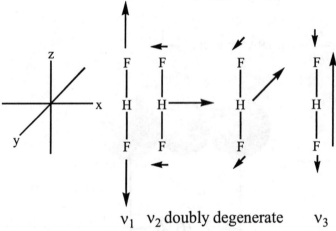

v_1 v_2 doubly degenerate v_3

Fig. 4.2 The normal modes of the [FHF]⁻ ion.

All the atoms move along the same direction for a given mode. Normal vibrations are along the axis of the molecule (along z) or perpendicular to it (xy-plane). The modes perpendicular to the molecular axis are degenerate because of the indistiguishability of x and y. We then have a simple matrix for the eigenvalue problem. We rearrange the reduced coordinates so that all the x coordinates, all the y coordinates and all the z coordinates are in separate groups. The dynamical matrix is then blocked as:

$$
\begin{pmatrix}
f_{11} & f_{12} & f_{13} & & & & & & \\
f_{12} & f_{22} & f_{23} & & 0 & & & 0 & \\
f_{13} & f_{23} & f_{33} & & & & & & \\
& & & f_{44} & f_{45} & f_{46} & & & \\
& 0 & & f_{45} & f_{55} & f_{56} & & 0 & \\
& & & f_{46} & f_{56} & f_{66} & & & \\
& & & & & & f_{77} & f_{78} & f_{79} \\
& 0 & & & 0 & & f_{78} & f_{88} & f_{89} \\
& & & & & & f_{79} & f_{89} & f_{99}
\end{pmatrix}
\tag{4.32}
$$

In this representation of the matrix we have ordered the terms as follows:

- $f_{11} = f_{33} = f_{44} = f_{66}$ is the force constant on a F atom when displaced perpendicularly to the FHF axis.
- $f_{22} = f_{55}$ is the force constant on the H atom when displaced perpendicularly to the FHF axis.
- $f_{77} = f_{99}$ is the FF stretching force constant.
- f_{88} is the force constant on the H atom when moved along the FHF axis.

The solution of the eigenvalue problem is separated for each submatrix. The first two submatrices are identical since we cannot distinguish between x and y, and the third submatrix has the displacements in the z direction. Using data from a simple calculation from GAUSSIAN98 (§4.5) on the [FHF]⁻ ion the three matrices are written below. Note that the dynamical matrix most commercial programs save on file is not given in reduced coordinates. The transformation from the Cartesian into the reduced coordinates is, recalling the results of Eq. (4.5), for an example along x:

$$
f_{ij}^{\text{reduced}} = \frac{\partial^2 V^{\text{mol}}}{\partial q_i \partial q_j} = \frac{1}{\sqrt{m_i m_j}} \frac{\partial^2 V^{\text{mol}}}{\partial x_i \partial x_j} = \frac{f_{ij}^{\text{Cartesian}}}{\sqrt{m_i m_j}}
\tag{4.33}
$$

Knowing this result we can calculate the frequencies and displacements for different masses (as for isotopic substitution). The force constants depend *inter alia* on the atom type, but not on their masses. Thus if we have the dynamical matrix for the [FHF]⁻ ion we can use it to calculate the frequencies and displacements of the [FDF]⁻ ion. The eigenvalues are given in atomic units, hartree (bohr² amu)⁻¹ and can be converted to our conventional units of energy, cm⁻¹, by:

$$\lambda_v = 4\pi\omega_v^2 \qquad \omega_v = \left(5140\ \text{cm}^{-1}\sqrt{\text{bohr}^2\text{amu/hartree}}\right)\sqrt{\lambda_v} \qquad (4.34)$$

We have broken the problem into three separate parts, for convenience, and we solve each part separately. Usually, however, the full matrix is solved directly and its roots (λ_v, $v = 1,..9$ in this case) appear in frequency order. Below we shall number the roots as they would appear from a full calculation. The dynamical matrix for the motions along x (or y) in reduced coordinates is:

$$\begin{pmatrix} f_{11} & f_{12} & f_{13} \\ f_{12} & f_{22} & f_{23} \\ f_{13} & f_{23} & f_{33} \end{pmatrix} = \begin{pmatrix} f_{44} & f_{45} & f_{46} \\ f_{45} & f_{55} & f_{56} \\ f_{46} & f_{56} & f_{66} \end{pmatrix} =$$

$$(4.35)$$

$$= \begin{pmatrix} 0.0141 & -0.0283 & 0.0141 \\ -0.0283 & 0.0567 & -0.0283 \\ 0.0141 & -0.0283 & 0.0141 \end{pmatrix}$$

Solving the secular determinant of this matrix, Eq. (4.19), we obtain $\lambda_1^x = 0$, $\lambda_4^x = -2.8 \times 10^{-6}$, $\lambda_7^x = 0\ 0.0577$ (hartree (bohr² amu)⁻¹). The second eigenvalue is non-zero because of numerical errors and the only real vibration is $\omega_7 = 1235$ cm⁻¹. The corresponding displacements (the eigenvectors) are given in the mass weighted coordinates of Eq. (4.5):

$$^1Q_x = \begin{pmatrix} 0.698 \\ 0.161 \\ 0.698 \end{pmatrix}, \quad ^4Q_x = \begin{pmatrix} 0.707 \\ 0.0 \\ -0.707 \end{pmatrix}, \quad ^7Q_x = \begin{pmatrix} -0.114 \\ 0.987 \\ -0.114 \end{pmatrix} \qquad (4.36)$$

These are the x components of each Q vector for the displacements of the atoms, F, H and F, (in order from top down) for a given eigenvalue, see Fig. 4.3. As a consequence of their normalisation, Eq. (4.20), each of the three eigenvectors of Eq. (4.36) sums as squares to unity (e.g. $0.698^2 +0.161^2 +0.698^2 = 1$). Removing the mass weighting transforms the vectors into Cartesian coordinates, $\tilde{u}$, after Eq. (4.5), (e.g. $0.114/\sqrt{19} = 0.026$, n.b. $m_H = 1.0073$ amu).

$$
{}^{1}\widetilde{u}_x = \begin{pmatrix} 0.16 \\ 0.16 \\ 0.16 \end{pmatrix}, \quad {}^{4}\widetilde{u}_x = \begin{pmatrix} 0.162 \\ 0.0 \\ -0.162 \end{pmatrix}, \quad {}^{7}\widetilde{u}_x = \begin{pmatrix} -0.026 \\ 0.983 \\ -0.026 \end{pmatrix} \tag{4.37}
$$

The first eigenvector is a whole body translation along the x axis, Fig. 4.3, the second a rotation of the molecule in the x,z plane. The third corresponds to a bending mode component, motion of the hydrogen atom on the x axis, perpendicular to the axis of the molecule, the F atoms move in antiphase and keep the centre of mass of the molecule stationary. The solutions to the y submatrix are the same, yielding λ_2^y, λ_5^y and λ_8^y.

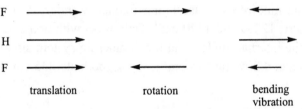

| | translation | rotation | bending vibration |

Fig. 4.3 Representation of the displacement vectors of Eq. (4.37). Not to scale.

The Cartesian displacements are rendered in conventional units, Å^2, by scaling the sum of the squared displacements of each eigenvector, in Eq. (4.36), to the total displacement calculated for a classical oscillator at that frequency, Eq. (A2.81) (see also §4.3.7). For ω_7 (1235 cm^{-1}) as our example, at low temperatures:

$$
u_i^2 = \widetilde{u}_i^2 \frac{16.9}{\omega_i}, \quad u_7^2 = \frac{16.9}{\omega_7} \begin{pmatrix} \dfrac{\left(Q_7^x\right)_F^2}{m_F} \\[2mm] \dfrac{\left(Q_7^x\right)_H^2}{m_H} \\[2mm] \dfrac{\left(Q_7^x\right)_F^2}{m_F} \end{pmatrix}; \quad \begin{aligned} \left(u_7^2\right)_F &= 0.0137 \times 0.00068 \\ \left(u_7^2\right)_H &= 0.0137 \times 0.966 \\ \left(u_7^2\right)_F &= 0.0137 \times 0.00068 \end{aligned} \tag{4.38}
$$

For the submatrix along z

$$\begin{pmatrix} f_{77} & f_{78} & f_{79} \\ f_{78} & f_{88} & f_{89} \\ f_{79} & f_{89} & f_{99} \end{pmatrix} = \begin{pmatrix} 0.136 & -0.0349 & -0.101 \\ -0.0349 & 0.0697 & -0.0349 \\ -0.101 & -0.0349 & 0.136 \end{pmatrix} \qquad (4.39)$$

The eigenvalues for this matrix are: $\lambda_3^z = 0$, $\lambda_6^z = 0.0699$ and $\lambda_9^z = 0.0125$, (hartree (bohr^2 amu)$^{-1}$), which correspond to the real frequencies $\omega_6 = 575$ cm^{-1} and $\omega_9 = 1370$ cm^{-1}. The eigenvector displacements of the z submatrix are, Å^2,

$$^3u_z^2 = \begin{pmatrix} 0.0008 \\ 0.0008 \\ 0.0008 \end{pmatrix}, \quad {}^6u_z^2 = \begin{pmatrix} 0.00078 \\ 0.0 \\ -0.00078 \end{pmatrix}, \quad {}^9u_z^2 = \begin{pmatrix} -9 \times 10^{-6} \\ 0.0119 \\ -9 \times 10^{-6} \end{pmatrix} \qquad (4.40)$$

We arrange the nine roots in ascending frequency order; 0, 0, 0, 10^{-6}, 10^{-6}, 575, 1235x, 1235y and 1370 cm^{-1}. Only roots with non-zero values are retained, the internal modes, and the numbering system simplifies to that shown in Fig. 4.2, ν_1 (at 575 cm^{-1}), $\nu_{2a\,(2b)}$ (both at 1235 cm^{-1}) and ν_3 (at 1370 cm^{-1}).

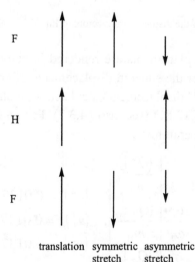

Fig. 4.4 Representation of the displacement vectors along the z axis. Not to scale.

A comparison of ^7Q with ^9Q shows that the mass weighted displacement vectors are equal in value but along different axes. This is the consequence of the symmetry of the force constants in the dynamical matrix and the relative atomic masses. It demonstrates that, whilst the frequencies depend strongly on the force constant values, the displacement vectors are much less sensitive, in this case insensitive. The symmetry of the force constants is the most important factor in determining the magnitude and direction of the eigenvectors and they are not strongly dependent on the frequency values. Different *ab initio* methods often calculate significantly different values for the vibrational frequencies. However, they will all generate very similar mass weighted atomic displacement vectors. The influence of poorly calculated *ab initio* frequencies on INS intensities is introduced through the transformation to conventional units. As can be seen from Eq. (4.38) the transformation involves the, possibly poorly, calculated frequencies as denominators. This has a bearing on the analysis procedures (§5.3.1).

Before leaving this topic we should write a few words about 'scaling' corrections. In GAUSSIAN, raw frequency values computed at the Hartree-Fock level contain known systematic errors due to the neglect of electron correlation, resulting in an overestimation of about 10%. Therefore, it is usual to scale the predicted Hartree-Fock frequencies by an empirically determined factor, 0.893. Frequencies calculated with methods other than Hartree-Fock are also often similarly scaled to compensate for known systematic errors in calculated frequencies, e.g. in DFT calculations the scaling factor is typically 0.97.

4.2.6 Comparison with experiment—sodium bifluoride

Although it is not possible to measure INS on gases, isolated molecule calculations, which strictly refer to the gas phase, are used successfully to interpret and analyse INS spectra. As an exercise we shall see how good, or not, a gaseous bifluoride [FHF]⁻ model is for the solid Na[FHF]. A detailed analysis of the calculation of the INS intensities is presented in §2.5.3 and §2.6.4.

The INS spectrum of Na[FHF] is compared with that calculated for [FHF]⁻ using the atomic displacements given by GAUSSIAN98, in Fig. 4.5. The quality of the calculations is good enough to immediately assign the main features of the spectrum. The first three frequencies are close to the experimental values, but some differences are apparent. The large feature at 1200 cm^{-1} in the experimental spectrum is a split peak. The splitting is an LO-TO effect, which is a characteristic of long range ionic interactions in crystal lattices (§2.6.1.3). This feature, therefore, cannot be calculated in the isolated molecule approximation, but see (§4.3.4).

The antisymmetric stretch near 1400 cm^{-1} is calculated as a sharp band; its frequency is underestimated by the calculation, but there is also another difference. The mode is actually dispersed across a broad energy range and the motion has an amplitude dependency on the frequency. There is also a contribution to the intensity in this area from the phonon wings of the 1200 cm^{-1} band (§2.6.4). This modest level of calculation is often sufficient for molecular solids that have no intermolecular interactions and we are not interested in studying the external modes. However, when the nature of the interactions is long range or we are looking at the low energy range of the spectrum, the effects of dispersion and the strength of intermolecular forces become apparent and this simplistic approach is inappropriate.

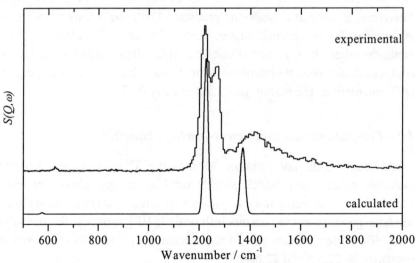

Fig. 4.5 INS spectrum of Na[FHF] (above) [7], calculated from GAUSSIAN98 (below).

The very weak feature observed at 604 cm^{-1} and calculated at 575 cm^{-1} corresponds to the ν_1, symmetric stretch. The weakness of the band is directly due to the complete absence of any hydrogen displacement in this mode and the very small total cross section of fluorine.

4.2.7 A molecular modelling example—adamantane

The calculation of the vibrational properties of hydrocarbons using density functional theory is usually in very good agreement with the INS experimental spectra. The treatment of the data has to take into consideration not only the contribution of the fundamentals but also the combinations and overtones. The effect of the lattice vibrations is included in an *ad hoc* manner (§5.3). As an example of the usefulness of the isolated molecule approach, in Fig. 4.6 we show the experimental and calculated spectra of adamantane (tricyclo[3.3.1.1^{3,7}]decane, $C_{10}H_{16}$).

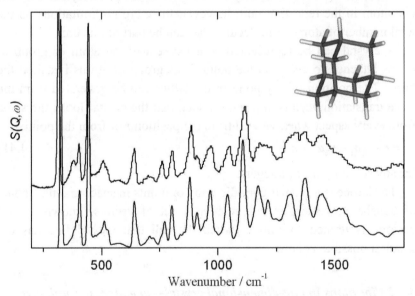

Fig. 4.6 The calculated (bottom) and experimental (top) INS spectrum of adamantane.

This is an especially favourable example since the globular nature of the molecule results in only weak intermolecular interactions. Nonetheless the agreement between observation and calculation is very impressive, particularly as no scaling of the calculated frequencies was applied. This example demonstrates the adequacy of the isolated molecule approximation when used appropriately.

4.3 The vibrational problem in the solid state

A more realistic analysis of the neutron vibrational spectra must include the intermolecular interactions present in the solid state. We now proceed to introduce these forces in a systematic manner.

4.3.1 The solid state—crystals

An ideal crystal is a periodic array of atoms with a certain pattern (unit) repeating itself infinitely in space. The simplest crystal contains one atom in the repeating unit; however, there are no limitations on the total number of atoms or molecules that can be part of this unit.

A crystal can be represented as a lattice with an atom or group of atoms attached to every lattice point. This group of atoms is called the basis. The position of any point in the lattice can be generated from the three translation vectors, a_1, a_2, a_3 such that the crystal looks the same from every aspect when viewed from the position r or from the point:

$$r' = r + n_1 a_1 + n_2 a_2 + n_3 a_3 \qquad (4.41)$$

for any set of n_1, n_2, n_3 integers.

The lattice is a regular periodic array of points in space and the choice of translation vectors is not unique. A set of vectors, known as cell vectors, is termed 'primitive' when the cell that they define has the smallest possible volume.

4.3.2 Vibrations in one-dimensional crystal—one atom per unit cell

We consider a simple case of an infinite one-dimensional chain of atoms, with only one atom, mass m, per unit cell. They are connected by springs, of force constant f, and interact only with their nearest neighbours. The atom l is displaced, u_l, from its equilibrium position. Then u_{l+1} and u_{l-1} are the (different) displacements of the neighbouring atoms. The potential energy arises from the relative atomic displacement, through the extension or compression of their shared spring.

$$V = \sum_{l=1}^{N_{atoms}} \frac{1}{2} f(u_l - u_{l+1})^2 \quad \text{where} \quad N_{atoms} \to \infty \qquad (4.42)$$

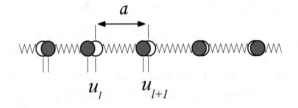

Fig. 4.7 A section of an infinite chain of atoms connected with springs separated at their equilibrium positions by a distance a and subject to a displacement, u_l, from equilibrium.

The equation of motion is:

$$m\frac{d^2 u_l}{dt^2} = m\ddot{u}_l = f(u_{l-1} - 2u_l + u_{l+1}) \qquad (4.43)$$

Using Bloch's theorem and cyclic boundary conditions [8] the atomic displacements can be expressed as a product of a maximum amplitude, $(^j u_l)_k$, and a phase factor, the wave vector k. Solutions to Eq. (4.43) are:

$$^j u_l (t) = (^j u_l)_k \exp(i(kla - \omega t)) \qquad (4.44)$$

where $|k| = k$. Here j-superscript, the external mode index, specifies the type of vibration, it is suppressed in subsequent equations to avoid confusion with the coordinate index j-subscript. In this particular case only one type of vibration can be supported by the chain. Eq. (4.44) can be compared to Eqs. (4.16), (A2.60) and (A2.61), they all represent a sinusoidal displacement of the atom but here the phase factor is given in terms of a wavelength, λ. This has the value $2\pi/ka$ (or $2\pi/k$, if λ is expressed in terms of the lattice spacing, a). Thus Eq. (4.44) corresponds to a simple harmonic oscillator with a frequency, compare Eq. (4.17):

$$\omega_v(k) = 2\sin\left(\frac{ka}{2}\right)\sqrt{\frac{f}{m}} \quad \Leftrightarrow \quad k(\omega_v) = \frac{2}{a}\arcsin\left(\frac{\omega_v}{2}\sqrt{\frac{m}{f}}\right) \qquad (4.45)$$

This classic result exemplifies many aspects of the dynamics of lattices. The eigenvalue, ω_v, is a periodic function of k (conventionally written $\omega_v(k)$ we shall find it convenient to use the $k(\omega_v)$ form, §10.1.2). There is no longer a single valued vibrational frequency, as found for the isolated molecule. This dependence of vibrational frequency on k, in Eq. (4.45), is called the vibration dispersion relation or more simply its dispersion; that of Eq. (4.45) is shown in Fig. 4.8.

$$k\left(\omega_v\right)=k'\left(\omega_v\right) \quad \text{where} \quad k'=\left(k+\frac{2\pi}{a}\right) \quad \text{so} \quad -\frac{\pi}{a}\leq k \leq \frac{\pi}{a} \qquad (4.46)$$

These values define the Brillouin zone. Any value, k', that falls outside this zone corresponds to a motion that already exists for a k value within the zone, and any wavevector outside of the first Brillouin zone, k' can be written as:

$$k' = n\frac{\pi}{a}+k , \qquad (4.47)$$

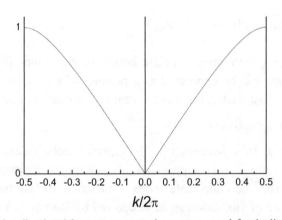

Fig. 4.8 Plot of the vibrational frequency versus the wave vector k for the linear chain.

In Eq. (4.47) k lies in the range defined in Eq. (4.46), n is an integer. A phase difference of greater than 2π between atoms has no physical meaning. At the centre of the zone (the 'gamma point', Γ) k is zero. From Eq. (4.45), the atoms cannot move in phase and at the same time maintain the centre of mass of the chain stationary, so at the zone centre the frequency goes to zero.

The reader should be very much aware of the possibilities for confusion that might arise between the neutron wavevector (a property of the neutron) and the lattice wavevector (a property of the lattice vibrations). In the absence of specific indications context should be sufficient to avoid such confusion.

4.3.3 Vibrations in one-dimensional crystal—two atoms per unit cell

The next level of complexity is the one dimensional chain, but now having two atoms per unit cell, the same spacing and for simplicity the same force constants on the springs. The atoms have different masses, m_1 and m_2. (A system of the same atoms, A, but with alternating force constants, ...A-A...A-A..., gives similar results.)

The equations of motion, not presented here, are more complicated and their solutions are given in Eq. (4.48).

$$\omega_\pm^2 = f\left(\frac{1}{m_1}+\frac{1}{m_2}\right) \pm f\sqrt{\left(\frac{1}{m_1}+\frac{1}{m_2}\right)^2 - 4\frac{\sin^2(ka)}{m_1 m_2}} \qquad (4.48)$$

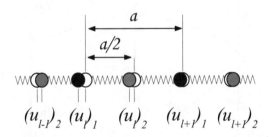

Fig. 4.9 A section of an infinite chain of atoms with two atoms, 1 and 2, per unit cell.

When plotted against k, these roots comprise two modes, or branches, (i.e. j-superscript now has two values, 1 and 2) see Fig. 4.10. The ω_- root is proportional to k as $k \rightarrow 0$ (similar to the solutions for the monoatomic chain). This is the acoustic mode (or branch) since it is analogous to the long wavelength vibrations of an elastic continuum (sound waves). For the second mode, the optic mode, or branch:

$$\omega_+ = 2f\left(\frac{1}{m_1} + \frac{1}{m_2}\right) \qquad \text{as } k \to 0 \qquad\qquad (4.49)$$

The optic branch is separated from the acoustic branch, but as the wavevector approaches the zone boundary, $k\to\pi/a$, the two modes approach in frequency. For the optical branch, at the limit $k\to 0$, both light and heavy atom sub lattices are vibrating against each other as rigid units, while keeping the centre of mass stationary. (In terms of the alternating system of forces the diatomic molecule, A-A, of each cell is vibrating in phase with its neighbours.) If the two atoms carry opposite charges, this gives an oscillating dipole moment that is optically active; hence its name. Note that there are frequency values for which there are no solutions. This feature is characteristic of elastic waves and there is a frequency gap at the boundary of the Brillouin zone. In understanding INS intensities the amplitude of motion for each atom, that is its maximum displacement, is important and the amplitude depends on the lattice wave vector [9].

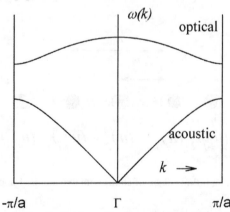

Fig. 4.10 Optic (ω_+) and acoustic (ω_-) branches for a diatomic linear chain.

We summarise, without proof, relevant properties of the amplitude by taking as an example a linear chain with two atoms of masses $m_2/m_1=3$ and solving the corresponding equations of motion. The acoustic modes, Fig. 4.11, near to the centre of the Brillouin zone are the oscillations of the unit as a whole and the amplitudes of motion must be the same for all the atoms in the unit cell. At shorter wavelengths, larger k, the amplitude of the oscillation of the heavier atoms becomes larger while the lighter atom has smaller amplitude.

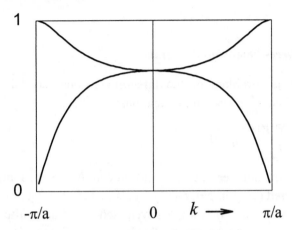

Fig. 4.11 Amplitude in the acoustic mode for the heavy (line above) and light (line below) atoms in a diatomic chain, with a mass ratio of 1:3.

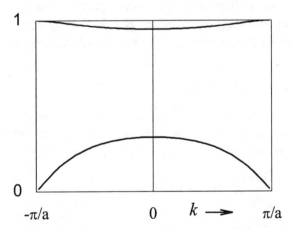

Fig. 4.12 Amplitude in the optic mode for the light (above) and heavy (below) atoms in a diatomic chain, with a mass ratio of 1:3.

At the zone boundary, largest k, the atomic motion in the acoustic mode involves only the heavy atoms while the light atom remains stationary. For the optical modes at the gamma point, both atoms move against each other while keeping the centre of mass stationary; the lighter atom has the larger amplitude of motion, Fig. 4.12. As the zone boundary is approached only the lighter atom moves while the heavier atom is stationary.

4.3.4 The three-dimensional crystal

The general problem of the three-dimensional crystal consists in solving the following system of equations:

$$\ddot{q}_j + \sum_{i=1}^{3N_{atom}N_{cell}} \frac{\partial^2 V(r)}{\partial q_j \partial q_i} q_i = 0 \tag{4.50}$$

Here N_{atom} is the number of atoms per unit cell, N_{cell} is the number of unit cells considered (very large), i and j labels include the x, y, z components of different atoms and q is a mass weighted coordinate, after Eq. (4.5). The interaction cannot depend on the absolute positions of the atoms but their relative positions. Again a major simplification to this equation comes from the application of the Bloch theorem and cyclic boundary conditions [8]. We shall represent the positions of all atoms in the solid by reference to a chosen unit cell, the zeroth cell (superscript 0), then for any k:

$$q_i = q_i^0 \exp(i k.r) \text{ and } q_j = q_j^0 \tag{4.51}$$

The summation over the $3 N_{atom}N_{cell}$ atomic coordinates is written as a double summation, for the atoms in the unit cell with an index i and over r eventually extended all over space. Eq. (4.50) can now be written:

$$\ddot{q}_j^0 + \sum_r \sum_{i=1}^{3N_{atom}} \frac{\partial^2 V(r)}{\partial q_i \partial q_j} q_i^0 \exp(i k.r) = 0$$

$$\ddot{q}_j^0 + \sum_{i=1}^{3N_{atom}} \left(\sum_r \frac{\partial^2 V(r)}{\partial q_i \partial q_j} \exp(i k.r) \right) q_i^0 = 0 \tag{4.52}$$

We define force constants:

$$f_{ij}(k) = \sum_r \frac{\partial^2 V(r)}{\partial q_i \partial q_j} \exp(i k.r) \tag{4.53}$$

The force constants now have an explicit dependence on the vector k. The periodicity of the crystal simplified the problem and we have a set of $3N_{atom} \times 3N_{atom}$ equations for every value of k:

$$\sum_i^{3N_{atom}} (f_{ij}(k) - \lambda \delta_{ij}) u_i = 0 \qquad j = 1, 2, \cdots, 3N_{atom} \tag{4.54}$$

The process of finding the frequencies and the amplitudes of the displacement is mathematically the same as in the molecular case (§4.2.2), with the exception of the k dependence. Therefore we must calculate the frequencies and amplitudes across the whole range of k values. Normalising the vibrational amplitude, after Eq (4.20), we have:

$$^v L_l^{crys}(k) = \frac{\sqrt{m_l} \, ^v u_l^{crys}(k)}{\sqrt{\sum_{l'=1}^{N_{atom}} m_{l'} (^v u_{l'}^{crys}(k))^2}} \tag{4.55}$$

where $^v L_l^{crys}(k)$ is the normalized vibrational amplitude of the lth atom in mode v at the point k in the Brilluoin zone. In a three dimensional system there are three acoustic modes and $3N_{atom} - 3$ optical modes. Thus the mode index v now includes both the internal and external modes j-superscript is suppressed. The changed nature of the vibrational displacements is indicated by specific reference to the crystal $^v L^{crys}(k)$ and its dependence on the wavevector k (§4.3.6).

For an elastic wave in a medium that is both isotropic and continuous the acoustic phonons have three different wave-polarisations, one longitudinal, with the atomic displacements in the direction of the wave propagation and two with transverse polarisation, the atomic displacements perpendicular to the propagation vector. In a crystal, the transverse modes are not necessarily degenerate except in specific symmetry directions and, because the atoms are located in discrete positions, the velocity of propagation will depend on its direction.

4.3.5 Example of a simple system—lithium hydride

Lithium hydride, LiH, is a rather simple solid; there are two atoms in the primitive cell and four electrons. It has the rock salt structure and its primitive cell contains one lithium and one hydrogen atom, Fig. 4.14. There are six dispersion curves of which three are acoustic and three optical. We can compare our calculations with the measured dispersion curves from LiD (rather than LiH because the deuterium coherent scattering cross section is required to measure the relative displacements of pairs of atoms (§2.1).

A triple axis spectrometer (§3.4.1) was employed in these measurements [10,11,12,13]. The vibrational calculations were performed on the primitive cell once its calculated Helmholtz free energy, F, had been minimised using the quasi-harmonic approximation. Quantum effects are present due to the lightness of both atoms [14,15].

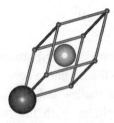

Fig. 4.13 The primitive cell of LiH. The small sphere is Li^+ and the large sphere H^-.

The dynamical matrix was calculated and the solution of the eigenvalue and eigenvector problem obtained. The dispersion curves of LiD are given in Fig. 4.14, where they have been plotted along the conventional high symmetry directions of the crystal. If we analyse the vibrations at the gamma point in LiD we have one optical mode at around 800 cm^{-1}, and another at 425 cm^{-1}, the three acoustic modes have to be zero at this point. In the higher energy mode the atom with greatest displacement is oscillating in the direction of propagation and corresponds to a longitudinal mode. The mode at lower energy is actually degenerate and corresponds to the two transverse optical modes: the atoms are moving perpendicular to the direction of propagation.

This is an example of LO-TO splitting of the vibrational modes according to their direction of motion at the gamma point (§2.6.1.3, §4.2.6). For any general direction the polarisation of the waves is not strictly longitudinal or transverse and the polarisation vectors are dependent on k as well as k. Similarly, the acoustic modes split into a longitudinal and two transverse modes. These modes vary in frequency along different directions in the unit cell [16].

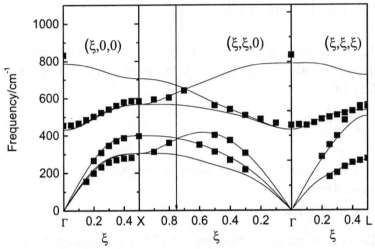

Fig. 4.14 Density of vibrational states calculated for LiD along some high symmetry directions. The squares represent the experimental data [10,17] and the lines are the calculated dispersion curves. ξ is the reduced wavevector component.

4.3.6 Calculation of the scattering law

The INS intensity, $S^\bullet(Q,\omega)$, as calculated from the Scattering Law, Eq. (2.41), is related to the mean square atomic displacements, weighted by the incoherent scattering cross sections. What is required to calculate this quantity is the mean square atomic displacement tensor, ${}^v\boldsymbol{B}_l$, and this can be obtained from the crystalline equivalent of ${}^v\boldsymbol{L}_l{}^{mol}$ (§A2.3), the normalised atomic displacements in a single molecule Eq. (4.20). This is ${}^v\boldsymbol{L}_l{}^{crys}(\boldsymbol{k})$ and was introduced above, in Eq. (4.55). We have seen how dispersion requires characteristic vibrations in crystals to be a function of k such that the branches appear across a, usually narrow, frequency band ω_ν (but not the single frequency that is found for isolated molecules). Values of L can be obtained at any (k, ω_ν) using the dynamical matrix but, because we are using the incoherent neutron scattering cross section (§2.1), the k space information is inaccessible. We thus abridge $^\nu L$ to $^\nu Ł$ by integrating over k at various frequencies (say at an interval of every wavenumber) across the band ω_ν, one for each mode ν.

$$^\nu Ł_l^{crys}(\omega_\nu) = \int_{k=-\pi/a}^{\pi/a} {}^\nu L_l^{crys}(k, \omega_\nu)\, dk \tag{4.56}$$

This is entirely equivalent to integrating over both k and ω_ν but expressing the distribution in ω_ν by the local density of states $g(\omega_\nu)$ (§2.6.2). If:

$$^\nu \overline{Ł}_l^{crys} = \int_{(\omega_{min})_\nu}^{(\omega_{max})_\nu} {}^\nu Ł_l^{crys}\, d\omega \tag{4.57}$$

then the local distribution of intensity of any one component of the νth mode could be written:

$$S(Q, \omega)_1^\nu \propto g(\omega_\nu)\, {}^\nu \overline{Ł}_l^{crys} \tag{4.58}$$

However, in our convention, see Eq. (2.55), $g(\omega_\nu)$ integrates to unity across its local spectral range, we thus avoid the extra step of Eq. (4.58) and work directly with $^\nu Ł_{l,}^{crys}(\omega_\nu)$. First we simplify Eq. (4.56), replacing the integral by a sum limited to a selection of discrete k points. For example in the case of Na[FHF], as we shall see below, we construct a three-dimensional grid of $16 \times 16 \times 16$ points evenly distributed across the k space of the first Brillouin zone. Individual $^\nu Ł_{l,}^{crys}(\omega_\nu)$ values (at each frequency across the band, ω_ν) now relate to individual $^\nu B_l$ values, yielding $^\nu B_l(\omega_\nu)$, and by extending Eq. (A2.82).

$$^\nu B_l(\omega_\nu) = {}^\nu Ł_l(\omega_\nu) \left(\frac{\hbar}{2 m_l \omega_\nu} \right) \coth\left(\frac{\hbar \omega_\nu}{2 k_B T} \right) \tag{4.59}$$

The $^\nu B_l(\omega_\nu)$ are manipulated appropriately (§A2.5) to generate the required $S^\bullet(Q, \omega)$ values (again $g(\omega_\nu)$ remains absent).

4.3.7 The-k-space grid—computational and instrumental aspects

Although the preferred method of calculating a spectrum is to perform an *ab initio* calculation on an extended solid, extracting frequencies and displacements across the Brillouin zone, on a fine k-grid, this approach can be computationally very expensive. In plane wave codes like 'CASTEP'[18], 'CPMD' [19], 'PWSCF'[20], 'VASP'[21], 'ABINIT' [22], and some others, the number of plane waves that are taken into consideration, the selected correlation functional and the choice of pseudopotential will all have an impact on the quality of the calculations. Some codes (e.g. ABINIT) alleviate the problem by permitting frozen phonon calculations at the symmetry zone boundary, i.e. (0,0,0), (1/2,0,0), (1/2,1/2,0) and (1/2,1/2,1/2) and so determine the dynamical matrix at these points. The code then interpolates values of the dynamical matrix for all the points within the Brillouin zone and uses these to calculate the solution to the vibrational problem inside the zone.

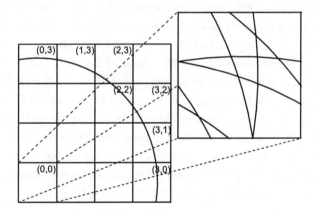

Fig. 4.15 A two-dimensional representation of the momentum transfer in a polycrystalline system; the circle represents a Q value of 2.1 zone widths, which for adamantane corresponds to about 2 Å^{-1}.

We must now show that INS techniques used to measure $S^{*}(Q,\omega)$ evaluate the same object as that calculated. What is calculated is an approximation to the k integral across the Brillouin zone. It is important to know that the neutron spectrometer used also measures a suitable selection of the whole zone. This can be seen by reference to Fig. 4.15.

In the figure, one corner of the single plane of neighbouring zones of an arbitrary crystal is shown diagrammatically, also shown on the figure is the sector of a circle. The circle is struck from the centre of the first zone and has, in this example, a radius of about 3.1 zone widths. The radius represents some specific value of the neutron momentum transfer, Q, used in the INS measurement (experimental uncertainties will give the circular line some width). Only its magnitude is important since it refers to powdered samples (if the measurement were done on a single crystal then the value of Q would be given by a small point on the circle).

It can be seen that the circle sweeps through each zone it crosses along a different trajectory such that it samples the dynamics in each zone differently. Given that the dynamics of every zone is replicated in that of the first zone then the INS measurement does indeed sample across the whole of the first zone, Fig. 4.15. This radius of the circle must be large in comparison to the zone width, since a small circle will hardly sample more than an unrepresentative part of the zone. The larger the volume of the crystallographic unit cell the smaller is that of its Brillouin zone. From Eq. (4.46), the width of the zone is $k = 2\pi/a$ and, for the lattice parameter of adamantine, $a = 9.36$ Å, $k = 0.671$ Å^{-1}. As can be seen from Fig. 4.15 the circle radius should be at least a factor two or three greater than the reciprocal lattice parameter, in our example of adamantane the minimum radius, i.e. minimum Q value, would need to be about 2.0 Å^{-1}. This condition is satisfied on low-bandpass spectrometers, which are commonly used in the topics covered in this book, for energy transfers above about 50 cm^{-1} (§3.4.2.3).

4.3.8 Comparison with experiment—sodium bifluoride

As an example of the connection between the dispersion relationship, the calculated spectrum and the observed $S^*(Q,\omega)$ we shall revisit Na[FHF]. The calculations were performed with a density functional theory plane wave code using pseudopotentials [22]. After optimising the geometry, a normal mode analysis was performed on a 16×16×16 grid in the first Brillouin zone. The dispersion curves were also calculated along the high symmetry directions in the crystal, Fig. 4.16a. The modes that involve mainly the motion of the hydrogen are dispersed according to the high energy curves of Fig. 4.16a.

Using the displacements calculated in the grid we obtain the INS spectrum of the [FHF]⁻ ion that was calculated earlier using the isolated molecule approximation (§4.2.6). We see that the degenerate frequencies at 1225 cm⁻¹ now demonstrate the observed LO-TO splitting, these correspond to the hydrogen atom moving in the plane perpendicular to the FHF axis. The mode that appears above 1300 cm⁻¹, due to the hydrogen atom moving along the FHF axis, is very dispersed and $S^\bullet(Q,\omega)$ extends from 1400 cm⁻¹ to 2100 cm⁻¹. We also find the fluorine atoms moving against each other at about 575 cm⁻¹. The external modes that could not be calculated in the isolated molecule approach (§4.2.5) appear here as a natural consequence of treating the full crystal. A final picture of the calculated spectra compared with the experiment is presented in Fig. 4.17.

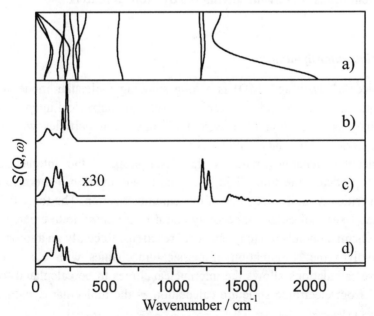

Fig. 4.16 (a) The dispersion curves along one high symmetry direction of Na[FHF]. The individual densities of states of: (b) Na, (c) H, (d) F from the ABINIT program [22].

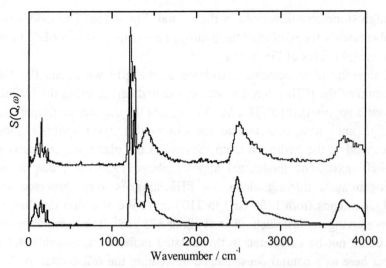

Fig. 4.17 INS spectrum of Na[FHF]. Experimental (top) and calculated (bottom) [7].

4.4 Calculations that avoid solving the dynamical matrix

4.4.1 Molecular dynamics

Molecular dynamics (MD) is a long standing molecular modelling technique [23]. In molecular dynamics we solve by numerical integration the equations of motion of a system of interacting particles with or without periodic boundary conditions.

Molecular dynamics requires the description of the interactions between particles, the force field, the validity and quality of the results depend critically on the accuracy of this parameterization. The force field approach fixes molecular connectivity and it is not possible to create (or break) chemical bonds or study chemical reactions. Recently an approach that involves the combination of classical mechanics with electronic structure calculations allows the internuclear forces to be calculated 'on the fly' from electronic structure calculation as the molecular dynamics develops [19].

We consider a system of N_{atom} interacting particles moving under the influence of the forces between them. Where the position of the lth atom, $r_l(t)$, and its velocity, $v_l(t)$, evolve over time, t. Since, from Newton's equations of motion, where m_l is the atomic mass:

$$m_l \ddot{r}_l = F_l \tag{4.60}$$

We know the force, F_l, on this particle and:

$$F_l = -\frac{\partial V}{\partial r_l} = F_l\left(r_1, r_2, \cdots, r_{N_{atom}}\right) \tag{4.61}$$

In principle the force on each particle depends on the coordinates of all the N atom positions, $F_l = F_l(r_1, r_2, \ldots, r_N)$, and this makes Eq. (4.61) a set of $3N_{atom}$ coupled second order differential equations. A unique solution is obtained for a given set of initial conditions $\{r_l(0), v_l(0)\}$, for all l. Newton's equations determine the full set of coordinates and velocities as a function of time, therefore, specifying the classical state of the system at any time t. Since the analytical solution is usually not available a numerical time integration scheme must be applied. The accuracy of the numerical integration depends on how short the period is between consecutive moments in time, the time step, Δt. If the forces in the system are highly nonlinear the divergence of the trajectories from two very similar initial conditions becomes exponential. However, it has been proved that the numerical solution is statistically equivalent to the true solution, within a bounded error, provided that the same average physical observables are calculated.

4.4.2 The velocity autocorrelation function

If we take the scalar product of an atom's velocity with its velocity a short time later, t, and take an average over all the atoms of the same type then we may represent it as a convolution $v(0) \otimes v(t)$, also commonly written $\langle v(0).v(t) \rangle$. The Fourier transform of a convolution is equivalent to product of the Fourier transforms of the two functions taken separately. Self convolution, autocorrelation, yields the square of the functions under Fourier transformation and only the real part of the transformation is available, the power spectrum, $\wp(\omega)$.

$$\wp(\omega) = \int_{-\infty}^{\infty} v(0) \otimes v(t) \exp(-i\omega t)\,dt \quad \text{and} \quad \int_{0}^{\infty} \wp(\omega)\,d\omega = 1 \qquad (4.62)$$

An atom's average position, squared, $r(t)^2$ in a solid, measured after a long time, t, tends to a constant, this is the total vibrational mean squared displacement, A. The contribution to A arising from a specific vibrational mode, v, at ω_v, is the strength of the power spectrum at ω_v.

$$A = A\int_{0}^{\infty} \wp(\omega)\,d\omega = \sum_{v=1}^{3N-3} A \int_{(\omega_v)_{min}}^{(\omega_v)_{max}} \wp(\omega_v)\,d\omega_v = \sum_{v=1}^{3N-3} {}^{v}u^2$$

$$\text{so} \quad A \int_{(\omega_v)_{min}}^{(\omega_v)_{max}} \wp(\omega_v)\,d\omega_v = {}^{v}u^2 \qquad (4.63)$$

4.4.3 Computational considerations

The normal modes analysis finds the eigenvectors and corresponding eigenvalues of the dynamical matrix. This matrix is a $3N \times 3N$ matrix (where N is the number of atoms); the computational effort scales as N^3. If N is ca 100 the dynamical matrix diagonalisation is easily performed. However, for $N > 1000$ this method becomes difficult to implement. For biomolecules N can exceed 10000 [24]. Also the time scales involved in the different vibrations are very different: the characteristic time for a C—H stretch is about 100 times smaller than the time scale associated with the motions of the lattice. Therefore if the calculation aims to reproduce the complete velocity autocorrelation function it must run with small time steps for a period of time long enough to have good statistics for the low frequency motions. Imposing rigidity on the molecule removes the internal vibrations but allows the simulation to run using a longer time step, one that is in line with the characteristic times of intermolecular motions [25,26,27,28,29]. Recently a 'driven molecular dynamics' approach has been introduced [30].

4.5 *Ab initio* methods

4.5.1 Hartree-Fock method

The original *ab initio* approach to calculating electronic properties of molecules was the Hartree-Fock method [31,32,33,34]. Its appeal is that it preserves the concept of atomic orbitals, one-electron functions, describing the movement of the electron in the mean field of all other electrons. Although there are some inherent deficiencies in the method, especially those referred to the absence of correlation effects. Improvements have included the introduction of many-body perturbation theory by Mollet and Plesset (MP) [35] (MP2 to second-order; MP4 to fourth order). The computer power required for Hartree-Fock methods makes their use prohibitive for molecules containing more than very few atoms.

4.5.2 Density functional theory

Currently the most popular approach to carrying out electronic structure calculations is density functional theory (DFT). The central concept is the electron probability density, $\rho(r)$. This density is a function of the three spatial coordinates only, instead of the $3N$ coordinates that are necessary to describe the Schrödinger wave function. The potential and kinetic energies of the molecule are expressed in terms of the electron density such that the total energy contains an unknown, but universal functional of $\rho(r)$. Note that the original wave function possesses all the coordinates for every electron in the system. In practice it is convenient to retain the wavefunction of individual electrons to calculate the electron density. The theoretical basis of DFT can be found in [36,37]. It is necessary to make some approximation to the functional and this is sometimes referred to as the 'level of theory' used in the calculation.

For non-periodic systems it is standard practice to use localised basis functions to describe the individual electron wavefunctions. In the case of periodic systems, the use of plane waves is the most favoured method, because of the simplicity of the Fourier analysis of plane waves. There are many excellent accounts of the principles and applications of DFT and we refer the reader to these [38].

As we have seen before, the vibrational spectra of molecules require the calculation of the second derivatives of the energy with respect to the atom coordinates about their equilibrium positions. Many DFT program codes can perform the calculation of the dynamical matrix either using analytical second derivatives or numerical differences. As a result their output includes a set of eigenvectors and vibrational frequencies which are the input to programs that calculate INS spectra (§5.3).

4.6 Use of force fields derived from classical mechanics

Many systems of interest are too large to be tackled using *ab initio* methods and here force field methods can be useful. Force field methods do not explicitly include the electrons, rather the energy of a system is a function only of the nuclear coordinates. The main application of molecular mechanics modelling is in the area of big systems (thousands of atoms are not uncommon). The calculations can be performed in a fraction of the computer time that would be required for an *ab initio* calculation. Their accuracy is determined by the quality of the parameterization of the force field.

The model of the interactions in molecular mechanics is simple and the Born-Oppenheimer approximation must be applicable. The most common molecular modelling force fields are based on the very simple picture of intra- and intermolecular forces. The energy is expressed as a function of a series of reference values for the bond lengths and angles. The electrostatic interactions can also be considered in this framework.

The most common force fields include two-, three- and four-body potential energy functions. It is possible to use more complicated force fields but it is usually impractical. The two-body terms usually describe bond lengths and non bonded interactions, the three-body terms describe the angle bending and the four-body term describes bond rotations or torsions.

The force fields have to be defined through their functional form as well as the parameters. Different force fields can use the same functionality with very different parameters (or force fields with different functional forms) and still provide results of comparable accuracy. This highlights the problem of transferability of force fields and parameters. Force fields are often developed to study certain characteristics and properties of materials and that information forms part of the force field itself. Force fields that are derived to reproduce the structural properties of solids and molecules do not always provide good results when calculating vibrational spectra. This does not mean that the force field fails; it is simply being used outside its range of applicability.

Force fields are empirical and never unique such that the choice of the 'best' function reduces to a choice of which field performs best in the context of the work in hand. In the force field picture there is the concept of atom type. If we use *ab initio* methods, we define only the atom number (and possibly its mass for an isotopic substitution calculation). In molecular mechanics, however, we need to be much more specific. Sometimes the hybridization of an atom in a given configuration is needed, carbon atoms can be sp^3 (tetrahedral) or sp^2 (trigonal planar) or sp (linear). For each hybridization type the angle between the bonds is different.

The description of the electrostatic interactions can be important. Force fields developed for organic molecules often ignore electrostatic interactions. In ionic systems, on the other hand, the description of the electrostatic interaction is crucial for any sensible reproduction of material properties. The most basic approach is to use point charges located at the atomic position. Unfortunately, the question of what value to allocate to a charge has no simple answer. Even in ionic materials the allocated charge on an atom will not be its formal ionic charge, but some 'effective' charge. The effective charge is calculated using a variety of methods and its value affects the transferability of force fields.

Electron distribution is affected by the presence of external electric fields and the polarisation interaction that arises in the presence of other atoms. One of the most common models to take into account polarisation is the 'shell' model. In the shell model the total charge of the atom is distributed between the nucleus and a mass-less spherical shell. This shell is connected to the nucleus by a spring of a given force constant.

The force field specifies an analytical expression for the interaction potentials, it is thus possible to calculate the minimum energy configuration and solve the eigenvalue problem described earlier.

4.7 The ACLIMAX program

Solving the vibrational problem, as presented in this chapter, provides the eigenvalues (vibrational frequencies) and the eigenvectors (atomic displacements) that are needed to calculate $S^a(Q,\omega)$. Recently we have developed and released ACLIMAX [39] as 'freeware'. This takes *ab initio* outputs to generate a spectrum from a low-bandpass spectrometer (§3.4.2.3). The program can be downloaded from:

http://www.isis.rl.ac.uk/ molecularspectroscopy/tosca/.

ACLIMAX is a stand alone program written in Visual Basic but using Fortran for the mathematical calculations, see Fig. 4.18. It can be used to interpret the INS spectra taken on any low-bandpass spectrometer (§3.4.2.3). Currently the program accepts input from the following *ab initio* calculations: GAUSSIAN98 [40], DMOL3 [41], CASTEP [18] and ADF [42] and, in principle, any eigenvalue and eigenvector output in the correct units is usable.

We prefer to support the output from standard packages that are continually developed. They are fast, accurate, widely supported, robust, simple to use and can be mastered quickly. They are widely exploited by optical spectroscopists and, especially important, these programs are familiar to the wider chemical community. Moreover, they automatically incorporate much of the theory of molecular vibrations [43].

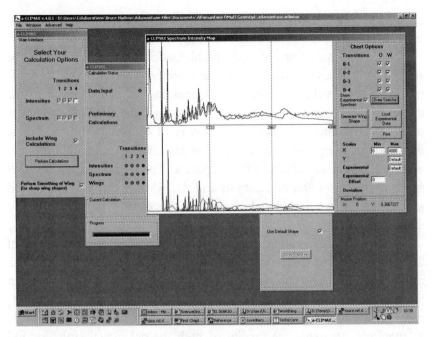

Fig. 4.18 Snapshot of the ACLIMAX graphical user interface.

4.8 Conclusion

In order to test the quality and validity of the theoretical models we need to compare with experiments. Nowadays it is possible to calculate the position and vibrations of atoms in solids and molecules, the comparison with experiment, however, is usually limited to only the description of the atomic positions, structure. Comparisons with the dynamical data are not routinely made and when made it is most generally restricted to infrared and Raman data, which are limited to information at the gamma point and it also presents difficulties on the theoretical calculation of spectral intensities.

The calculated infrared and Raman spectra are obtained from the responses of the electronic cloud to the nuclear displacements. Discrepancies in the calculation of the electron density, which should have little impact on the structure, might exacerbate the resulting optical intensities.

INS spectroscopy avoids this problematic feedback effect, in the sense that it measures nuclear motion directly. This provides a simple relation between the amplitude of the motion, the cross section of the nuclei and the vibrational frequency to determine the intensity and shape of the spectral lines. The electron density is, of course, used to determine the potential energy surface that controls the nuclear motion but it is not involved in the calculation of the spectra. The combined use of INS spectroscopy, *ab initio* calculations and ACLIMAX can provide a good test of the molecular model, assessing the quality of structure and dynamics.

Furthermore, INS can be used to compare the relative effectiveness of the many different theoretical tools used in *ab initio* calculations. The results from *ab initio* calculations depend on the level of theory used in the computation. The combined use of INS results and ACLIMAX, provides the most stringent test of the level of theory used in *ab initio* calculations, assessing the quality of both the calculated structure and dynamics.

4.9 References

1   M. Born & J.R. Oppenheimer (1927). Ann. Phys. (Leipzig), 74, 1–31. Zur Quantumtheorie der Molekulen.
2   P.R. Bunker & P. Jensen (2000). *Computational Molecular Spectroscopy*, John Wiley, New York.
3   E.B Wilson, J.C. Decius & P.C. Cross (1955). *Molecular Vibrations, the Theory of Infrared and Raman Vibrational Spectroscopy*, McGraw-Hill, New York.
4   I. Mills, T. Cvitaš, K. Homann, N. Kallay & K. Kuchitsu (1993). *Quantities, Units and Symbols in Physical Chemistry*, 2nd Edition, International Union of Pure and Applied Chemistry (IUPAC), Blackwell Scientific Publications, Oxford.
5   S. Kawahara, T. Uchimaru, & K Taira (2001). Chem. Phys., 273, 207–216. Electron correlation and basis set effects on strong hydrogen bond behavior: a case study of the hydrogen difluoride anion.
6   K Kawaguchi & E Hirota (1987). J. Chem. Phys., 87, 6838–6841. Diode-laser spectroscopy of the v_3 and v_2 bands of FHF$^-$ in 1300 cm^{-1} region.

7   G.J. Kearley, J. Tomkinson & J. Penfold (1987). Z. Phys. B, 69, 63–65. New constraints for normal-mode analysis of inelastic neutron-scattering spectra: application to the HF_2^- ion.

8   G.L Squires (1996). *Introduction to the Theory of Thermal Neutron Scattering*, Dover Publications Inc., New York.

9   S.R. Elliott (1998). *The Physics and Chemistry of Solids*, John Wiley & Sons, Chichester.

10   J.L. Varble, J.L. Warren and J.L. Yarnell (1968). Phys. Rev., 168, 980–989. Lattice dynamics of lithium hydride.

11   R. Vacher, M. Boissier & D. Laplaze (1981). Solid State Commun., 37, 533–536. Brillouin-scattering study of anharmonicity of lithium hydride at the origin of acoustic branches

12   G. Roma, C.M. Bertoni & S. Baroni (1996). Solid State Commun., 98, 203–207. The phonon spectra of LiH and LiD from density-functional perturbation theory.

13   W. Dyck & H. Jex (1981). J. Phys. C Solid State Phys., 14, 4193–4215. Lattice-dynamics of alkali hydrides and deuterides with the NaCl type-structure,

14   A.A. Maradudin (1971). *The Theory of Lattice Dynamics in the Harmonic Approximation*, 2nd Edition, Solid State Physics, Suppl. 3, Academic Press, New York, London.

15   N.L. Allan, G.D. Barrera, J.A. Purton, C.E. Sims & M.B. Taylor (2000). Phys. Chem. Chem. Phys., 2, 1099–1111. Ionic solids at elevated temperatures and/or high pressures: lattice dynamics, molecular dynamics, Monte Carlo and *ab initio* studies.

16   G. Auffermann, G. Barrera, D. Colognesi, G. Corradi, A.J. Ramirez-Cuesta & M. Zoppi (2004). J. Phys. C: Condens. Matter, 16, 5731–5743. Hydrogen dynamics in heavy alkali metal hydrides obtained through inelastic neutron scattering.

17   A.C. Ho, R.C. Hanson & A. Chizmeshya (1997). Phys. Rev. B, 55, 14818–14829. Second-order Raman spectroscopic study of lithium hydride and lithium deuteride at high pressure.

18   M.D. Segall, P.J.D. Lindan, M.J. Probert, C.J. Pickard, P.J. Hasnip, S.J. Clark & M.C. Payne (2002). J. Phys.: Condens. Matter, 14, 2717–2744. First-principles simulation: ideas, illustrations and the CASTEP code.

19   R. Car & M. Parrinello (1985). Phys. Rev. Lett., 55, 2471–2474. Unified approach for molecular dynamics and density functional theory. (Available at http://www.CPMD.org/)

20   PWscf (Plane-Wave Self-Consistent Field), http://www.pwscf.org/

21   G. Kresse & J. Furthmüller (1996). Phys. Rev. B, 54, 11169–11186. Efficient iterative schemes for *ab initio* total-energy calculations using a plane-wave basis set. (http://cms.mpi.univie.ac.at/vasp/).

22   X. Gonze, J.-M. Beuken, R. Caracas, F. Detraux, M. Fuchs, G.-M. Rignanese, L. Sindic, M. Verstraete, G. Zerah, F. Jollet, M. Torrent, A. Roy, M. Mikami, Ph. Ghosez, J.-Y. Raty & D.C. Allan (2002). Computational Materials Science, 25, 478–492. First-principles computation of material properties: the ABINIT software project. (http://www.abinit.org/)

23   M.P. Allen & D.J. Tildesley (1987). *Computer Simulation of Liquids*, Oxford Science Publications, Clarendon Press, Oxford.

24   G.H. Li & Q. Cui (2002). Biophys. J., 83, 2457–2474. A coarse-grained normal mode approach for macromolecules: An efficient implementation and application to Ca^{2+}-ATPase.

25   G.R. Kneller (2000), Chem. Phys., 261, 1–24. Inelastic neutron scattering from damped collective vibrations of macromolecules.

26   T. Rog, K. Murzyn, K. Hinsen & G.R. Kneller (2003). J. Comp. Chem., 24, 657–667. nMoldyn: A program package for a neutron scattering oriented analysis of molecular dynamics simulations.

27   A.J. Ramirez-Cuesta, P.C.H. Mitchell, S.F. Parker & P.M Rodger (1999). Phys. Chem. Chem, Phys., 1, 5711–5715. Dynamics of water and template molecules in the interlayer space of a layered aluminophosphate. Experimental inelastic neutron scattering spectra and molecular dynamics simulated spectra.

28   A.J. Ramirez-Cuesta, P.C.H. Mitchell, A.P. Wilkinson, S.F. Parker & P.M Rodger (1998). Chem. Comm., 23, 2653–2654. Dynamics of water molecules in a templated aluminophosphate: molecular dynamics simulation of inelastic neutron scattering spectra.

29   A.J. Ramirez-Cuesta, P.C.H. Mitchell, & P.M Rodger (1998). J. Chem. Soc. Faraday Trans., 94, 2249–2255. Template-framework interactions in chiral AlPOs.

30   J.M. Bowman, X.B. Zhang & A. Brown (2003). J. Chem. Phys., 119, 646–650. A normal-mode analysis without the Hessian: A driven molecular dynamics approach.

31   D.R. Hartree (1928). Proc. Cambridge Phil. Soc., 24, 89–122. The wave mechanics of an atom with a non-coulomb central field.

32   J.C. Slater (1929). Phys. Rev., 34, 1293–1322. The theory of complex spectra.

33   J.C. Slater (1930). Phys. Rev., 35, 210–211. Note on Hartree's method.

34   V. Fock, (1930). Z. Phys., 61, 126–148. Naherungsmethode zur Losung des quantenmechanischen mehrkorperproblems.

35   C. Møller & M.S. Plesset (1934). Phys. Rev., 46, 618–622. Note on an approximation treatment for many-electron systems.

36   V. Kohn & L. J. Sham (1965). Phys. Rev., 140, A1133–A1138. Self-consistent equations including exchange and correlation effects.

37   R. G. Parr & W. Yang (1989). *Density Functional Theory of Atoms and Molecules*, Oxford University Press, New York.

38   W. Koch & M.C. Holthausen, (2001). *A Chemist's Guide to Density Functional theory*, 2nd Edition, Wiley-VCH; Weinheim.

39   A.J. Ramirez-Cuesta (2004). Comput. Phys. Comm., 157, 226–238. ACLIMAX 4.0.1, The new version of the software for analyzing and interpreting INS spectra.

40   M.J. Frisch, G.W. Trucks, H.B. Schlegel, G.E. Scuseria, M.A. Robb, J.R. Cheeseman, V.G. Zakrzewski, J.A. Montgomery, Jr., R.E. Stratmann, J.C. Burant, S. Dapprich, J.M. Millam, A.D. Daniels, K.N. Kudin, M.C. Strain, O. Farkas, J. Tomasi, V. Barone, M. Cossi, R. Cammi, B. Mennucci, C. Pomelli, C. Adamo, S. Clifford, J. Ochterski, G.A. Petersson, P.Y. Ayala, Q. Cui, K. Morokuma, D.K.

Malick, A.D. Rabuck, K. Raghavachari, J.B. Foresman, J. Cioslowski, J.V. Ortiz, B.B. Stefanov, G. Liu, A. Liashenko, P. Piskorz, I. Komaromi, R. Gomperts, R.L. Martin, D.J. Fox, T. Keith, M.A. Al-Laham, C.Y. Peng, A. Nanayakkara, C. Gonzalez, M. Challacombe, P.M.W. Gill, B. Johnson, W. Chen, M.W. Wong, J.L. Andres, C. Gonzalez, M. Head-Gordon, E.S. Replogle & J.A. Pople (1998). *GAUSSIAN98,* Revision A.3, Gaussian, Inc., Pittsburgh.

41   B. Delley (2000). J. Chem. Phys., 113, 7756–7764. From molecular to solid with the DMOL3 approach.

42   ADF Program System, Scientific Computing & Modelling Nv Vrije Universiteit; Theoretical Chemistry De Boelelaan 1083; 1081 HV Amsterdam; The Netherlands

43   J.B. Foresman & A. Frisch (1996). *Exploring Chemistry with Electronic Structure Methods*, 2nd Edition, Gaussian Inc., Pittsburgh.

5

Analysis of INS spectra

Modern data treatment emphasises a synthetic approach to the interpretation of INS spectra, where the observed spectrum is compared with one obtained from calculations based on a putative *ab initio* structure and force field for the system. This is a powerful method and certain of success when the nature of the sample is not in question. However, our understanding of the most interesting samples is perhaps more tenuous than we should like, as will be the case for the adsorbed state of molecules on surfaces or for very recently discovered materials. It is when the synthetic approach fails badly (as when the calculated spectrum bears no relationship to that observed) that resort must be had to the analytical techniques of spectral interpretation.

In this chapter it is our objective to provide an understanding of some of the simple tools of INS spectral analysis. After a general consideration of the use of model compounds (§5.1) three examples of representative spectra are worked through in detail: a simple inorganic hydrogenous system, (§5.2); a typical organic molecule, benzene (§5.3), and a molecular metal hydride, rubidium hexahydridoplatinate(IV) (§5.4). The data from the first two compounds were obtained on an indirect geometry, high resolution low-bandpass spectrometer, TOSCA, and that of the third was taken on a direct geometry spectrometer, MARI.

Before obtaining an INS spectrum all relevant information on the sample should have been studied and an idea formulated as to the likely bands in the spectrum and the broad distribution of intensities. Often the best estimate of the likely spectrum will result from *ab initio* calculations and this partly synthetic approach will be used to inform our interpretation of the benzene spectrum.

The analysis of the INS spectrum of ammonium bromide is treated in the absence of any prior calculation. Here the likely spectrum is a composite drawn from the types of eclectic information available for molecular compounds and shows how these might be pieced together.

5.1 General considerations—model compounds and the INS database

The spectroscopy of molecular materials increases in intricacy as more oscillators interact (§2.6.1). This leads to an increase in the number of transitions if the number of coupled oscillators remains small and their local geometry defined, as with the discrete dispersion of the internal modes of molecules like benzene. Alternatively it leads to continuous dispersion, when the number of coupled oscillators is large and their crystalline geometry defined, as with external modes. When the geometry of oscillators is not well defined, heterogeneous band broadening effects are observed. Some aspects of all these responses are present in most INS spectra and, unfortunately, this spectral richness is very challenging. Whilst the examples chosen to illustrate this chapter have relatively simple chemical structures of high symmetry, it is not always possible, or desirable, to restrict INS studies to such straightforward systems. This leaves us with the need to understand what may be rather complex spectra. One approach is to search for some aspects of a measured spectrum that may be recognisable in the spectra of structurally related systems; the so-called 'model compounds'. Because of the broad frequency range and good resolution of modern low-bandpass spectrometers, their spectra are uniquely appropriate to this approach. A database of the spectra of compounds recorded on TOSCA is available at

http://www.isis.rl.ac.uk/insdatabase/

The complete list of all published high-resolution work to date (2004) is given in Appendix 4.

Data accumulated on all low band pass spectrometers are almost always obtained under the same instrumental conditions (§3.4.2.2.2), irrespective of any subsequent data treatment. Spectra obtained from the same system but measured years apart should thus be equal to within errors and even old INS spectra of model compounds continue to remain relevant.

Some thought must, therefore, be given to how INS spectra of simple model compounds from the database can be combined to inform the understanding of INS spectra from more complex systems. Generally, common structural motifs found in different molecules do not necessarily produce common spectral patterns. This is a feature of molecular vibrational spectroscopy, irrespective of technique, but is especially limiting for INS.

Optical techniques, because they involve insignificant momentum transfer (§2.3), avoid the intensity loss caused by the Debye-Waller factor. They readily obtain data from the high-frequency vibrations that are more localised on groups of a few atoms within the molecule. These characteristic frequencies change only subtly from sample to sample and are familiar as 'group frequencies', like the C=O stretch about 1700 cm^{-1} or C–H about 3000 cm^{-1}. High resolution INS spectroscopy yields its most exploitable information at lower momentum transfers, which on low-bandpass spectrometers correspond to low energy transfers, say less than 1500 cm^{-1}. At these frequencies the vibrational modes are not localised but involve atomic displacements over the molecule as a whole. These modes tend to undergo significant change from system to system. The use of model compounds to understand INS spectra is a powerful technique but not one that can be applied thoughtlessly (see also §8.1).

5.2 Ammonium bromide

In common with many small ions and molecules in lattices at room temperature the ammonium ions are more or less freely rotating, as can be determined by NMR [1]. This rotation smoothes the tetrahedral structure of the ammonium ion and allows it to behave as a pseudo-spherical ion. Hence, at ambient temperature ammonium bromide adopts the CsCl crystal structure but this is replaced by other structures at low temperatures as thermal energy is drained from the rotational motion. Site group and factor group considerations [2] are absent and the internal vibrations of the ammonium ion reflect the tetrahedral symmetry of the free ion, see Table 5.1 [3]. Apart from the internal vibrations there are some early optical results showing a strong *restrahl* response about 150 cm^{-1} [4]. A full set of dispersion curves (§2.6.1, 2.6.2) exists but only for ND$_4$Cl in a different structural phase [5]. The initial requirement is to produce a concordance between the dispersion curve of Fig. 5.1 and the optical observations. Since only the orientation of the ammonium ions is affected by the crystal structure it is reasonable to use the dispersion data to produce spectral estimates, whilst acknowledging that the atomic masses (Cl for Br and ND$_4$ for NH$_4$) are wrong.

Table 5.1 The internal vibrations of the ammonium ion [3].

Description	Mode	Character	ω / cm^{-1}
Antisymmetric N—H stretch	ν_3	F_2	3145
Symmetric N—H stretch	ν_1	A_1	3040
Symmetric HNH deformation	ν_2	E	1680
Antisymmetric HNH deformation	ν_4	F_2	1400

The dispersion curves are conveniently labelled in Fig. 5.1, the transverse acoustic (TA) and longitudinal acoustic (LA) branches are seen rising from the Brillouin zone centre at zero energy transfer. The optical branches (TO, LO) lie fairly flat across the zone in the energy range about 150 to 300 cm^{-1}.

The optical branches in ND$_4$Cl are thus dispersed over the range 170 to 285 cm^{-1} (§2.6.1.3) with the TO grouped mostly around the 160 to 200 cm^{-1} region. This is where most of the optic branch intensity will reside, in agreement with the analysis of the *restrahl* data from the bromide [2]. The TO will be about 150 cm^{-1} and the LO about 220 cm^{-1}, when some allowance is made for the difference between the masses of the bromide and chloride ions. The acoustic branch intensity peaks around 130 cm^{-1}, where all three branches flatten. The distribution of intensities between the optic and the acoustic depends on how the masses move in the different vibrations (§4.3.4). Crudely, the acoustic modes move both the ammonium and halide ions together but the optic vibrations have them beating against one another [6].

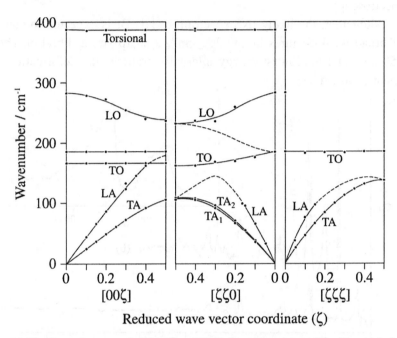

Fig. 5.1 The dispersion curves of ND_4Cl, at 295 K. Reproduced from [5] with permission from Springer Verlag.

Simple reduced mass considerations imply that the optical modes, involving a smaller reduced mass, will have the larger vibrational displacement and give the stronger bands. Any difference introduced by deuteration will have had only a modest impact on these vibrations since they are translational in character and the ammonium mass only increases from 18 to 22 amu.

However, the torsional vibration will be very strongly affected by this increased mass. Torsional vibrations involve motions around the centre of mass and the mass of the nitrogen atom plays no role in this displacement. The moment of inertia of ND_4 is twice that of NH_4 and the ratio of their torsional frequencies is, therefore, $\sqrt{2}$, Eq. (5.1). Thus the perhydro ammonium ion should have its torsion at about 550 cm⁻¹. The band should be sharp since dispersion is small and the branch is seen to be quite flat in Fig. 5.1. It is because of their low oscillator mass and non-dispersed nature that torsions almost always appear strongly in INS spectroscopy.

Altogether, the likely INS spectrum of NH_4Br is expected to consist of broad weak features below 200 cm^{-1}, a sharp strong libration about 550 cm^{-1} and the lowest energy internal vibrations, the deformations at about 1400 cm^{-1}.

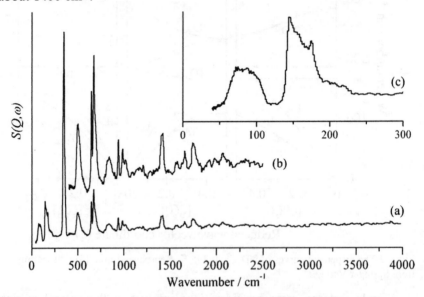

Fig. 5.2 The INS spectrum of NH_4Br [7]; (a) the whole vibrational range; (b) the same spectrum from 350 cm^{-1} with intensities ×3; (c) the low frequency region.

5.2.1 Observed INS spectrum of ammonium bromide

The observed INS spectrum of NH_4Br is given in Fig. 5.2. The spectrum was taken on the indirect geometry low-bandpass spectrometer TOSCA at ISIS (§3.4.2.2.2). The spectrum consists of a series of features across the whole spectrum, which sits on a gently rising background. However, bands appear between 500 and 1400 cm^{-1}, where none were expected and there are no bands about 3000 cm^{-1}, where the stretches should occur. The following analysis proceeds by discussing the main spectral regions individually.

5.2.1.1 The low energy region

This region covers up to *ca* 300 cm^{-1} in energy transfer, see Fig. 5.2(c). There are two features, an ill-shaped band at about 80 cm^{-1} and a finely structured band centred at 160 cm^{-1}. These bands appear reasonably close to the values suggested from the likely spectrum. The acoustic modes were expected at 130 cm^{-1} (for ND$_4$Cl) but observed at 80 cm^{-1} for NH$_4$Br.

We can ascribe these frequency differences to the effect of the different reduced masses, μ, and force constants, f. From considerations of simple harmonic motion [3].

$$\hbar\omega \propto \sqrt{\frac{f}{\mu}} \tag{5.1}$$

The optic modes were expected to start about 160 and end about 200 cm^{-1}, an average of 180 cm^{-1}. In Fig. 5.2 it starts dramatically, with a very clear van Hove singularity (§2.6.2.2), at 140 cm^{-1} and stops, almost as cleanly with a second singularity at 180 cm^{-1}, an average of 160 cm^{-1}. Here the ions are moving in antiphase and the reduced mass μ is,

$$\frac{1}{\mu} = \frac{1}{m_{NH_4}} + \frac{1}{m_{Br}} \quad \text{or} \quad \frac{1}{\mu} = \frac{1}{m_{ND_4}} + \frac{1}{m_{Cl}} \tag{5.2, 5.3}$$

We estimate f from the bulk moduli, $f = (C_{11} + 2(C_{12}))/3$ [8], and substitute for f and μ into Eq. (5.1)

$$\left(\frac{\omega_{Br}}{\omega_{Cl}}\right)^2_{TO} = \left(\frac{160}{180}\right)^2 = 0.790 \qquad \frac{f_{Br}\,\mu_{Cl}}{f_{Cl}\,\mu_{Br}} = \frac{0.166 \times 13.6}{0.185 \times 14.7} = 0.829 \tag{5.4}$$

Thus the difference in forces between the chloride and bromide is as important as the changed masses. The intensity in the LO branches extends the optic band out to about 220 cm^{-1}.

5.2.1.2 The librational mode region

The librations or torsions of the ammonium ion on its site give rise to the very intense and sharp band at 345 cm^{-1}, see Fig. 5.2(a) and shown truncated in Fig. 5.2 (b). The truncation is needed to clearly expose the sequence of bands starting at about 500 cm^{-1} and reaching across the region up to the lowest internal mode. These bands were not predicted when we formulated the likely spectrum and they require a convincing explanation. In INS spectroscopy the most usual place to begin would be by considering the possibility of combination bands. (Readers more familiar with optical techniques will find that the strength of overtones and combinations is one of the aspects of INS that marks it as different from infrared or Raman.) The only possibilities for combination with the libration are the translational phonons and the TO frequency is a good candidate at 160 cm^{-1} (= 505–345). Considering the impact of phonon wings raises a question regarding the applicability of the theory as previously developed (§2.6.3).

Phonon wings were introduced in the context of their impact on the internal vibrational spectrum of molecules but the librational mode is an external mode, it is itself a phonon. However, this is only a question of classification and semantics. The phonon wing treatment is simply one approach to calculating the intensities of combination bands. It is the method of choice when detailed information on the external mode atomic displacements is absent. We now proceed to apply the phonon wing treatment of §2.6.3, from which we shall obtain the value of the mean square displacement of the ammonium ion due to the translational vibrations of the lattice, α_{tran} (which is but one of the contributions to the full α_{ext}.)

If the phonon wing model is reasonable the value obtained will be realistic. In preparation for this we calculate the relevant Q values (§2.3) for this spectrometer at the librational transition, Eq. (5.5).

The Q^2 value at the centre of the first order phonon wing, with an energy of 505 cm^{-1} is different, 23.9 Å^{-2}. The intensities of the zero phonon band and the first order phonon wing are found by integrating the observed spectrum across the individual transitions. In this case from 310 to 380 cm^{-1} for the librational mode's band origin intensity, S_O, and from

380 to 600 cm^{-1} for the first wing, see Fig. 5.2(b). Their ratio, S_1/S_0, is 0.368 (§2.6.4), Eq. (5.6).

$$E_f = 28 \times cm^{-1} = k_f^2 \times 16.7 \times \frac{cm^{-1}}{\text{Å}^{-2}}$$

$$k_f^2 = 1.68 \times \text{Å}^{-2}$$

$$E_i = E_f + \hbar\omega = 28 \times cm^{-1} + 345 \times cm^{-1} = 373 \times cm^{-1} = k_i^2 \times 16.7 \times \frac{cm^{-1}}{\text{Å}^{-2}}$$

$$k_i^2 = 22.3 \times \text{Å}^{-2}$$

$$Q^2 = k_f^2 + k_i^2 - 2 \times k_i \times k_f \times \cos\theta$$
$$= 1.68 \times \text{Å}^{-2} + 22.3 \times \text{Å}^{-2} - 2 \times 1.30 \times \text{Å}^{-1} \times 4.72 \times \text{Å}^{-1} \times \cos 45°$$
$$= 15.4 \times \text{Å}^{-2}$$

$$(5.5)$$

$$\frac{S_1}{S_0} = \frac{Q_1^2 \alpha_{tran} \exp(-Q_1^2 \alpha_{tran})}{1 \exp(-Q_0^2 \alpha_{tran})} = \frac{23.9 \alpha_{tran} \exp(-23.9 \alpha_{tran})}{\exp(-14.4 \alpha_{tran})} = 0.368 \quad (5.6)$$

The value of α_{tran} is 0.0184 Å^2, measured at 20K. The nitrogen suffers the same translations as the hydrogen atoms but no displacement in the librational, and little in the internal, vibrations. Therefore, we can compare our result with diffraction results for the isotropic thermal parameter B (= $8\pi^2\alpha$) of the nitrogen atom. First we adjust our value to the temperature of the diffraction measurements, 300 K, from Eq. (2.59):

$$\alpha_T = \alpha_0 \coth\left(\frac{\hbar\varpi_E}{2k_B T}\right) \quad (5.7)$$

where α_0 is the zero-point value and α_T is the value at temperature T. We shall assume that our low temperature α value is a good approximation to α_0 and that the average TO energy represents the lattice vibrations, ϖ_E = 160 cm^{-1}. At room temperature, α_T = 0.0281 Å^2, and so B = 2.22 Å^2. This is clearly in good agreement with the diffraction results, B_N = 2.31 [9]. Our α value is indeed reasonable and confirms the assignment of the phonon wings.

We now calculate the strength of the other phonon wings. The relevant Q^2 value, 32.2 Å^{-2}, corresponds to an energy transfer of 665 cm^{-1} ($= 345 + 2 \times 160$). From Eq. (2.63):

$$\frac{S_2}{S_0} = \frac{\left(Q_2^2 \alpha\right)^2 \exp(-Q_2^2 \alpha)}{2\exp(-Q_0^2 \alpha)} = \frac{(32.2 \times 0.019)^2 \times \exp(-0.612)}{2\exp(-14.4 \times 0.019)} = 0.133 \quad (5.8)$$

Based on the intensity at the band origin, the second wing intensity can be calculated to be about 22% of the intensity found between 600 and 800 cm^{-1}. (There is also a contribution, of about 5%, from the third wing stretching to almost 1000 cm^{-1}.)

The remaining sharp features between 600 and 800 cm^{-1} are not phonon wings and require a different explanation. After phonon wings other combinations and overtones are the prime candidates for sources of INS intensity (§2.5.2.3). The calculation of overtones is the most straightforward and we proceed to determine if they could provide an explanation for the bands. In the harmonic approximation, overtones fall at simple multiples of their fundamental and the strongest bands in Fig. 5.2(b) are at about 670 cm^{-1}, or twice the value of the strongest band in the spectrum. This reasonable assignment can be confirmed by demonstrating that the oscillator mass of the band at 345 cm^{-1} is related to that of the ammonium ion. Here we shall use the special characteristics of indirect geometry spectrometers with low final energy (§3.4.2.3.4). The momentum transferred on these spectrometers is almost all of that available from the incident neutron. We use this feature, recall Eqs. (3.18), (2.21) and (2.72).

$$Q^2 \approx \left|k_i^2\right| = E_i/16.7 \approx E_{tr}/16.7 = \omega_v/16.7$$

$$^v u^2 = \frac{16.9\text{Å}^2 \text{ amu cm}^{-1}}{\mu \omega_v} \approx \frac{\text{amu}}{\mu} \frac{16.7 \dfrac{\text{cm}^{-1}}{\text{Å}^{-2}}}{\omega_v} = \frac{\text{amu}}{\mu} \frac{1}{Q^2}; Q^2 u^2 \approx \frac{\text{amu}}{\mu} \quad (5.9)$$

(Notice that, $Q^2\mu^2$ is unitless; μ is given in atomic mass units, 1 amu.) If the atomic displacement of the scattering atom in the molecule, u^2, is exclusively derived from one vibration then the mass, μ, given by Eq. (5.9) is the oscillator mass for that unique mode, μ_v. Otherwise, μ, represents only a parameterisation in terms of an effective mass, m_{eff}, with no obvious interpretation (for an interesting comment on a related issue see [10]). However, where the displacements are dominated by a single mode then we may approximate, $m_{eff} \approx \mu_v$ (as the approximation falters then $m_{eff} > \mu_v$).

On low-bandpass spectrometers, when the energy transferred is doubled then the momentum is also approximately doubled such that the momentum transferred to an overtone, $Q_{02}{}^2$, is twice that transferred at its fundamental, $Q_{01}{}^2$. Substituting this into the scattering law expressions for overtone intensities, Eq (2.36) gives:

$$\frac{S_{0-2}}{S_{0-1}} = 2\left(Q_{01}^2 u^2\right)\exp\left(Q_{01}^2 u^2\right) = 2\left(\frac{1}{m_{eff}}\right)\exp\left(-\frac{1}{m_{eff}}\right) \qquad (5.10)$$

$$\frac{1}{m_{eff}} = \frac{1}{\mu_v + \mu_0}$$

The ratio of the intensity of an overtone to its fundamental is given by the effective mass, m_{eff}, which increases as extra vibrational contributions increase the total displacements of the scattering atoms. Ultimately, the effective mass, m_{eff}, may be quite different from the oscillator mass, μ_v. This is related to, but subtly different from, effects observed on direct geometry spectrometers (§5.3.2).

The Eq. (5.10) applies to the total intensity from the vibration and band origin intensities, $S_{2 \leftarrow 0}/S_{1 \leftarrow 0} = 0.352$, must be corrected for any intensity loss, due to phonon wing effects. We extracted a value for α, above, and can use this directly:

$$\frac{\exp(-Q_{01}^2\alpha)}{\exp(-2Q_{01}^2\alpha)}\left(\frac{S_{2 \leftarrow 0}}{S_{1 \leftarrow 0}}\right)_{obs} = \exp(Q_{01}^2\alpha)\left(\frac{S_{2 \leftarrow 0}}{S_{1 \leftarrow 0}}\right)_{obs} = 1.3 \times 0.352 = 0.459 \ (5.11)$$

However, this method is not always possible and in those cases some approximation must be used. A common and successful approach is to assume that the value of Q^2 changes little across the region of a transition and its observed wing intensity. The integrated intensity across this region is then taken to be the total intensity, as corrected for wing effects. In the case of ammonium bromide the application of Eq. (5.10) is reasonable and $m_{eff} \approx \mu_v$. The intensities are integrated over 310 to 380 cm^{-1} for the fundamental intensity and the corrected intensity, see above, from 620 to 750 cm^{-1} for the overtone, from Fig. 5.2, gives:

$$0.459 = 2\left(\frac{1}{\mu_v}\right)\exp\left(-\frac{1}{\mu_v}\right) \tag{5.12}$$

The oscillator mass μ_v is 3.2 amu. This must be compared to the mass of the hydrogen atoms in the librational motion of the ammonium ion, 4 amu. Of importance here is not the precise value of the oscillator mass but that it is light, no assignment of this band to heavy atom motions is tenable.

Our earlier suggestion that the extra bands in this spectral region were overtones is thus confirmed and this allows us to assign the intensity about 1000 cm^{-1} to the second overtone, $(3\leftarrow0)$ at 1035 $(=3\times345)$.

Finally we turn to the structured nature of the overtone bands. First we note that the structure in bands at low energies, due to dispersion effects, becomes washed out in their overtone spectra. In this respect the bands about 650 and 944 cm^{-1}, assigned to components in the librational $(2\leftarrow0)$ and $(3\leftarrow0)$ manifolds, are unusually sharp. They are sharper than the fundamental from which they derive and their widths are limited only by the spectrometer's resolution. These effects have been assigned to the impact of anharmonicity, since the librational well is more sinusoidal than quadratic in shape. Further analysis would address, mostly, band positions and this type of treatment can be followed in several texts [11]. We shall not pursue this avenue further except to draw attention to the distribution of intensities between the bands within any one overtone manifold. The intensities should be distributed according to the local density of states (§2.6.2) of the symmetry species of the individual components in each manifold; tables of these species have been published [12]. In this case the librational ground state species for a T_d ion, under T symmetry, is F_2 and this gives rise to the $(2\leftarrow0)$ overtone species A_1, E and F_2. The distribution of intensity in the $(3\leftarrow0)$ band will be according to A_1, F_1, $2F_2$.

5.2.1.3 The internal mode region

The internal modes can be seen in Fig. 5.2(b) above 1200 cm^{-1}. This region of the spectrum is very rich and again there are too many features to be explained by the two fundamentals alone. The external mean square displacement value is so big that most of the intensity arising from the internal vibrations is actually found in the phonon wings (§2.6.5). Here the full wing is operative and includes translational and librational contributions. Indeed few of the sharp features present in this part of the spectrum are due to zero phonon transitions of the fundamentals. This is but a slightly more severe case of the analysis that will be covered in detail for benzene and we end our analysis of the pure ammonium bromide salt here.

However, other ammonium halides enable us to explore the spectral impact of the most extreme phonon wing effects, molecular recoil.

5.2.2 Molecular recoil

When the pseudo-spherical ammonium ion is mostly replaced by a truly spherical ion the complex sequence of phase changes found in the pure ammonium halides is suppressed. The mixed potassium ammonium halide salts retain their NaCl cubic structure down to the lowest temperatures. The alkali metal ions support the structure leaving the ammonium ions as free to rotate at 1 K as at 300K [13]. The INS spectrum of this system is quite different from the pure salt and there are no sharp features in any region of the spectrum. We shall analyse the impact that this freedom has on the internal modes about 1400 cm^{-1}.

5.2.2.1 The internal modes under recoil.

The INS spectrum of the $(NH_4)_{0.5}K_{0.5}I$ mixed salt is compared with that of pure NH_4I in Fig. 5.3. The sharp phonon wing features seen in the INS of NH_4I have been washed out into a broad smooth band. A schematic diagram of how different systems respond under recoil is given in Fig. 5.4. The intensity appears as a broadened feature at higher energy transfer (§2.6.5) and in this case the broad band is centred at 2130 cm^{-1}. When a molecule recoils, all of the momentum transferred from the neutron is taken up by the molecule acting as a single particle of effective mass m_{eff}.

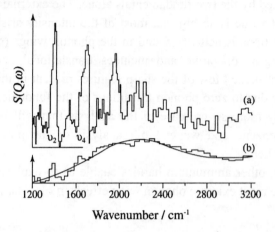

Fig. 5.3 Comparison of the INS of the internal modes of NH_4^+ in the pure iodide (a) and in the mixed salt (b), $(NH_4)_{0.5}K_{0.5}I$. Reproduced from [14] with permission from Elsevier.

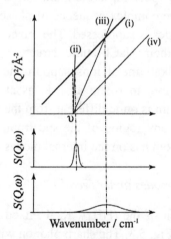

Fig. 5.4 A schematic diagram showing how different systems respond on a low-bandpass spectrometer under recoil. The trajectories are shown in the uppermost diagram. The spectrometer's trajectory (i), the recoil of a heavy mass (ii), the recoil of an intermediate mass (iii) and the recoil of a unit mass (iv). Below are shown the corresponding INS spectra associated with the heavy mass, above, and intermediate mass, below. Reproduced from [14] with permission from Elsevier.

From Eqs. (5.9) and (2.77):

$$E_{\text{total}} \approx \frac{Q^2}{m_n} = \omega_v + E_r = \omega_v + \frac{Q^2}{m_{\text{eff}}} \tag{5.13}$$

Dividing through by the recoil energy, E_r, we obtain, allowing $m_n = 1$amu.

$$m_{\text{eff}} = \left(\omega_v / E_r\right) + 1 \tag{5.14}$$

All of the intensity at 2130 cm^{-1} came from internal transitions at 1400 and 1720 cm^{-1}, these were the F mode, ν_2, and the E mode, ν_4. Their density of states weighted average energy is, $\omega_v = 1528$ cm^{-1} (= $(3 \times 1400 + 2 \times 1720)/5$). Then $E_r = 602$ (= $2130 - 1528$) cm^{-1} and so $m_{\text{eff}} = 3.5$ amu. We can compare this effective mass to the ammonium ion's Sachs-Teller mass (§2.6.5.1) is, from Eq. (2.74), $|M_{\text{S-T}}|_H = 3.2$ amu. The difference between the calculated and observed masses is probably due to the restricted translational space available to the ion in the lattice. The kinetic energy from Eq. (2.78) is 90 cm^{-1} [14].

5.3 Benzene

Benzene is used here to represent the archetypal organic molecular fragment. Its INS spectrum is based on *ab initio* calculation. Since very few organic systems have had their external modes studied in detail this part of the spectrum will be more or less unknown but expected below about 200 cm^{-1}. In the absence of an accepted assignment scheme an *ab initio* calculation of the vibrational spectrum is a very good place to start since this will eventually be used to analyse the spectrum. Isolated molecule calculations (§4.5) cannot prove useful in this regard if there are strong intermolecular effects, like intermolecular hydrogen bonding.

5.3.1 The internal modes

In Table 5.2 we draw together the optical results and the generally accepted assignment scheme for benzene alongside the results of an *ab initio* calculation. The likely spectrum will then be that produced by programs like ACLIMAX [16]. Immediately we scale the calculated frequencies by about 0.95, this is a crude but effective correction for known errors at the B3LYP level of *ab initio* calculation (§4.2.5).

Since we have no information on the relative strengths of the internal and external vibrations the simplest isotropic representation is used and its initial strength is estimated (§2.6.3). The observed INS spectrum is shown in Fig. 5.5(upper), where it is compared with the calculated spectrum, Fig. 5.5(lower), using program default values.

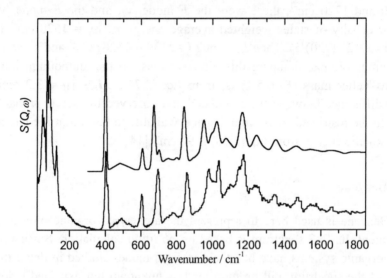

Fig. 5.5 Comparison of the observed, lower trace, and the calculated, upper trace, INS spectra of benzene. This represents the beginning of the analysis process using the scaled (×0.95) positions of the bands and a naïve, isotropic phonon wing.

The first thing to note is that the overall pattern is good, we have obviously measured a benzene spectrum very similar to that calculated. We can immediately conclude that there are no strong intermolecular interactions in the crystal.

However, the strength of the external modes has been seriously underestimated, the internal mode band origins are too strong and the phonon wings too weak. The external mode contribution parameter is adjusted by eye so that the observed and calculated distribution of intensities more closely resembles one another in the internal mode region. Most of the sharp peaks observed in the INS correspond to internal vibrations, the phonon wing effects are significant but do not dominate the spectrum and there is no evidence of molecular recoil. At this point some workers will prefer to stop, return to the *ab initio* calculations and attempt to improve the quality of those results.

Alternatively we may continue in our attempts to match the individual *ab initio* peaks more nearly with the INS data. This process is exactly analogous to the scaling factor used above except that here each peak has its own scaling factor. The calculated frequencies are simply changed to the observed values, taking care to respect the symmetry character of the modes. The intensity observed in INS, *S*, is controlled more by the mass of the individual atoms involved in a given vibration than the forces holding them together, (§4.2.5) and see Eq. (2.36).

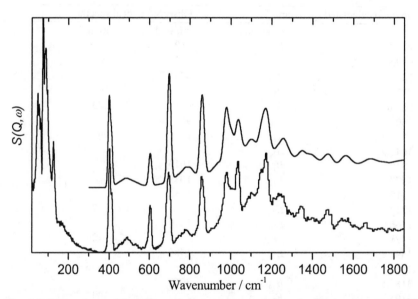

Fig. 5.6 Comparison of the observed, lower trace, and the calculated, upper trace, INS spectra of benzene. This represents the intermediate stage of the analysis process using adjusted band positions, still with the simple, isotropic phonon wing, now better adjusted to the distribution of intensities.

Recalling Eq. (5.9), on low-bandpass spectrometers the intensity is a slowly varying function of the energy transferred ($\propto Q^2$). Therefore, <u>small</u> but seemingly arbitrary changes to the *ab initio* frequencies will have little impact on the calculated *ab initio* intensities.

$$\frac{dS}{dQ^2} \approx \frac{dS}{dE} \approx 0 \tag{5.15}$$

This is equivalent to accepting that off-diagonal terms in the force constant matrix remain unchanged if the principal force constants do not vary greatly. Of course, large changes to the frequencies cannot be justified under this approximation since this is equivalent to significant changes in the principal force constants. Indeed, the largest error introduced into the determination of INS intensities from *ab initio* calculations comes from poorly calculated frequencies, through Eq. (4.38) (§4.2.5).

Table 5.2 The calculated and observed transitions, $\omega/$ cm^{-1}, of benzene [15].

ab initio calculated spectrum		Observed optical spectra	
	Character (a)	infrared	Raman
422.6	E_{2u}	410	
639.9	E_{2g}		606
707.5	A_{2u}	673	
737.9	B_{2g}		
885.6	E_{1g}		849
1001.7	E_{2u}	957	
1021.7	A_{1g}		992
1045.9	B_{2g}		
1049.0	B_{1u}	1010	
1077.8	E_{1u}	1038	
1218.7	B_{2u}	1150	
1232.5	E_{2g}		1178
1366.0	B_{2u}	1310	
1421.1	A_{2g}		1326
1541.8	E_{1u}	1486	
1653.9	E_{2g}		1596
3181.8	B_{1u}	3067	
3191.7	E_{2g}		3047
3209.9	E_{1u}	3080	
3224.2	A_{1g}		3062

(a) under the point group D_{6h}

This is not an exercise in triviality intended to best fit the observed spectra by haphazardly moving calculated intensities to arbitrary frequencies. The results from infrared and Raman spectroscopy fixes the symmetry character of the vibrational mode; this information is also available from the *ab initio* calculation and the two should match one another. This matching imposes severe constraints on the process. The most common problem is that the optical assignments are incomplete or in doubt (typically because some optical features are weak or inactive). If after this manipulation the calculated and observed spectra fail to reasonably match it raises significant doubts about the whole scheme and possibly about the molecular structure.

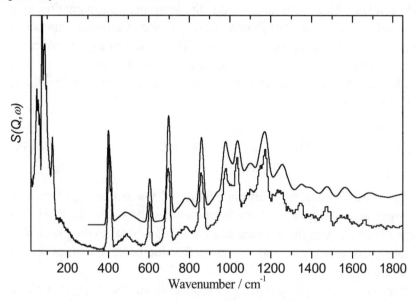

Fig. 5.7 Comparison of the observed, lower trace, and the calculated, upper trace, INS spectra of benzene (with the vertical offset removed the two spectra overlay each other convincingly). This represents the final stage of the analysis process using a specifically shaped, anisotropic phonon wing.

5.3.2 Impact of the external modes

First we introduce a more realistic approximation to the phonon spectrum. The phonon spectrum is generally taken to be the sharply structured low-energy part of the observed INS spectrum. In the case of benzene the selected spectral range is between 20 and 150 cm^{-1}. Using this shape the phonon wings can be further refined within the isotropic approximation, though it is appropriate to go further.

Anisotropy can be introduced through a simple approximation: the forces within the crystal are assumed to be isotropic and the molecule responds as a non-deforming unit. Each hydrogen has its own, generally different (but in the case of benzene not), effective mass. The anisotropy is introduced through the Sachs-Teller procedure for determining the effective mass of each hydrogen atom (§2.6.5.1). The strength of the Debye-Waller factor estimated under the isotropic approximation as a spherical volume is distributed amongst the Cartesian directions inversely as the components of the Sachs-Teller mass of a given atom.

Following a number of manual iterations in the program to best match the intensities the external modes displacement tensor A_{ext} is extracted, see Fig. 5.6.

$$A_{ext} = \begin{bmatrix} 0.0083 & & \\ & 0.0023 & \\ & & 0.0144 \end{bmatrix} \qquad {}^{Tr}A_{ext} = 0.025 \qquad (5.16)$$

In this specific case the Cartesian frame of the tensor is referred to hydrogen in a C–H bond with its direction along the y axis, the x axis lies perpendicular to y in the benzene plane and z is normal to the plane. The anisotropy is very clear, with the greatest value being six times that of the smallest.

Because of this anisotropy, two modes of similar energy at about 1000 cm^{-1} will suffer different degrees of attenuation from phonon wing effects. A mostly radial motion, like the ring-breathing mode (displacements along y in this example), will have an external Debye-Waller factor of 0.69. However, a mostly tangential motion, like an in plane C–H deformation (along x), will have a Debye-Waller factor of 0.55. This explains why the C–H stretching modes of benzene, which have only radial displacements, are so strong.

5.3.2.1 Environmental effects and static disorder

The tensor of Eq. (5.16) can be compared with the crystallographic results provided they are referred to the same frame and obtained at the same temperature (from perdeuterobenzene at 15 K [18])

$$A_{ext} = \begin{bmatrix} 0.0108 & & \\ & 0.0061 & \\ & & 0.0101 \end{bmatrix} \qquad {}^{Tr}A_{ext} = 0.027 \qquad (5.17)$$

Clearly the two techniques, diffraction and INS, are in good agreement, given that Eqs. (5.16) and (5.17) are the result of approximations.

The value of A_{ext} is determined by the local crystal structure and it changes as the structure evolves. The INS spectrum of benzene adsorbed on zeolite is shown in Fig. 5.8 (§7.5.4.1) [17]. Here the benzene sits on an uncongested surface and intermolecular forces restrain the molecule only weakly. The basic spectral features of the benzene molecule remain visible in Fig. 5.8. The benzene is thus physisorbed and has not reacted with the surface.

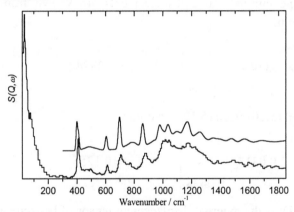

Fig. 5.8 Comparison of the calculated, upper trace, INS spectrum of benzene (as shown in Fig. 5.6) compared with the observed spectrum, lower trace, of benzene adsorbed on zeolite. Reproduced from [17] with permission from The Royal Society of Chemistry.

Using the same assignment scheme as for crystalline benzene and the shape of the external phonon spectrum taken from the low frequency INS spectrum shown in Fig. 5.8 (this region of the INS spectrum is stronger than in that of crystalline benzene). The best value for A_{ext} in this system is

$$A_{ext} = \begin{bmatrix} 0.0237 & & \\ & 0.0064 & \\ & & 0.0409 \end{bmatrix} \qquad {}^{Tr}A_{ext} = 0.071 \qquad (5.18)$$

The external displacements of benzene are much greater in the zeolite than in the crystal, hence the more intense external mode region. We may compare this value with that extracted from diffraction data taken on the same system [19]. The diffraction data were analysed under the constraints of the rigid molecule approximation, the reported anisotropic displacement parameter tensor must, therefore, include both internal and external modes

$$A_t = \begin{bmatrix} 0.064 & & \\ & 0.038 & \\ & & 0.063 \end{bmatrix} \qquad {}^{Tr}A_t = 0.165 \qquad (5.19)$$

Fortunately, the internal mode displacements are known from the *ab initio* calculation.

$$A_{int} = \begin{bmatrix} 0.0131 & & \\ & 0.0064 & \\ & & 0.0189 \end{bmatrix} \qquad {}^{Tr}A_{int} = 0.0384 \qquad (5.20)$$

Subtracting Eq. (5.20) from (5.19) we obtain.

$$A_{ext} = \begin{bmatrix} 0.0504 & & \\ & 0.0313 & \\ & & 0.0438 \end{bmatrix} \qquad {}^{Tr}A_{ext} = 0.1255 \qquad (5.21)$$

This, Eq. (5.21), is the external displacement tensor of benzene adsorbed on zeolite, as measured by diffraction. It is about twice that measured by INS, Eq. (5.18), which is a difference too great to be explained by simple experimental errors.

The explanation for the difference lies in the fact that the diffraction experiment is measuring both the dynamic, or thermal, and the static disorder (§2.4). Benzene is not organised across the surface of the zeolite in a very precise pattern but is disordered, each molecule happily sitting slightly misaligned with its neighbours. The difference between the results for the two techniques is a measure of this disorder, subtracting Eq. (5.18) from Eq. (5.21).

$$A_{\text{disorder}} = \begin{bmatrix} 0.0504 & & \\ & 0.0313 & \\ & & 0.0438 \end{bmatrix} - \begin{bmatrix} 0.0237 & & \\ & 0.0064 & \\ & & 0.0409 \end{bmatrix}$$

$$= \begin{bmatrix} 0.027 & & \\ & 0.025 & \\ & & 0.003 \end{bmatrix} \quad {}^{\text{Tr}}A_{\text{disorder}} = 0.055 \tag{5.22}$$

The value for the out-of-plane, z, disorder 0.003 Å^2 is probably at the limit of experimental errors and should be ignored. The in-plane disorder is equal in both x and y directions and represents a positional uncertainty of about ±0.1 Å and an angular uncertainty of about $\pm4°$.

5.4 Molecular systems using a direct geometry spectrometer

Direct geometry spectrometers differ considerably from their indirect counterparts. Naturally each technique offers certain advantages and suffers from distinct disadvantages. The most significant advantage offered by direct geometry spectrometers is access to the variation of intensity with Q (§3.4.3.1). However, optimising a measurement to cover an adequate range in Q is complicated by the need to achieve enough spectral resolution to separate out the transitions of interest, free from other contributions (§3.4.4). These requirements are contradictory, since broader ranges in Q occur for smaller differences between E_i and E_f but better energy resolution is achieved at greater energy differences.

It is difficult to demonstrate this on the observed spectra of typical organic systems because of their large number of transitions. Therefore, we have chosen to discuss the spectrum of a simple molecular hydride, rubidium hexahydridoplatinate(IV), $Rb_2[PtH_6]$ [20]. Every hydrogen is equivalent within this octahedral, O_h, ion and the ions are sufficiently separated that they do not interact. The high mass of platinum completely suppresses whole body translational modes and the libration occurs at high enough energy to be treated as an internal mode. Altogether, this has the advantage of reducing the number of transitions and simplifies the instrumental optimisation process discussed above. This will enable the practical aspects of the spectroscopy to be underlined. Note that, despite the high molecular symmetry the local atomic symmetry of each hydrogen atom is only D_4. Before we turn to the spectrum of the hydride it is instructive to consider the special case of helium.

5.4.1 A special case—liquid helium

The special case of liquid helium allows us to present one aspect of direct geometry spectroscopy in particular. Since helium atoms interact so weakly, there are almost no restoring forces in the liquid and only the conservation of momentum plays a significant role in its INS spectrum. The spectrum of a mixture of liquids ^{3}He and ^{4}He is shown in Fig. 5.9 [21], as can be seen from the figure it consists of two continuous traces. The first response, with a slope of 5.6, is from the light isotope of helium, ^{3}He. The second response, of slope 4.2, is from the common isotope of helium, mass four. (The ratio of the slopes is 4.2/5.6 = 3/4.) There are no excitations in the spectrum and the observed response is the result of atomic recoil.

The conservation of momentum restricts the scattering intensity to lie on a line with a slope related to the mass of the scattering atom. The lines in Fig. 5.9 trace the maximum scattering strength from each isotope and at a given value of energy transfer the response is broad in Q. Substituting the atomic mass for the effective mass in Eq. (2.77):

$$E_r = \frac{16.7Q^2}{m} \quad \text{and} \quad \frac{dE}{dQ^2} = \frac{16.7}{m} \tag{5.23}$$

The slope is given in the unconventional units, $cm^{-1}(\text{Å}^{-2})^{-1}$. (Conventionally the momentum transfer in Fig. 5.9 would be linear in Q and the lines of equivalent mass become quadratic curves (see Fig. 9.18).)

Conservation of momentum is also found in the oscillator responses from molecules but here the *simultaneous* conservation of energy as well as momentum dramatically restricts the scattering. It must still reach its maximum strength on specific mass lines but only at specific energy values.

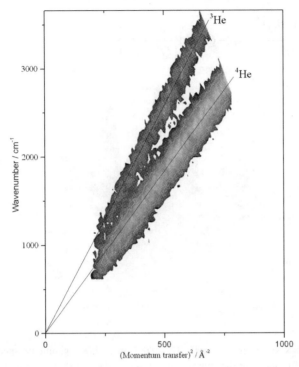

Fig. 5.9 The recoil of helium isotopes in a mixture of ^{3}He and ^{4}He (MARI, ISIS) [21]. The lines trace the maximum scattering of each isotope. Note that the momentum transfer is unconventionally given in Å^{-2}.

5.4.2 Rubidium hexahydridoplatinate(IV)

The INS spectrum of rubidium hexahydridoplatinate(IV), $Rb_2[PtH_6]$, obtained on the direct geometry spectrometer MARI, is given in Fig. 5.10. As anticipated from the considerations presented above the spectrum consists of a series of ridges, all relatively well-defined in energy but broad as a function of Q. The point where the scattering reaches a maximum (as a function of Q) defines the mass-line that joins that point to the origin, for an effective mass, m_{eff}.

From the data of Fig. 5.10 we can determine the point at which the intensity reaches a maximum for a given vibration, see also Fig. 2.4. Further, differentiating Eq. (2.41) with respect to Q^2.

$$\frac{dS}{dQ^2} = \frac{\left(Q^{2^{Tr}}\left(^{v}B_1\right)\right)^{n}}{3\,n!}\exp\left(-Q^2\alpha_1^{v}\right)\left\{\frac{n}{Q^2}-\alpha_1^{v}\right\} \qquad (5.24)$$

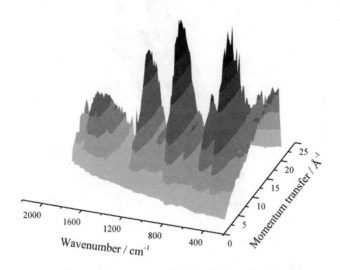

Fig. 5.10 The INS spectrum of $Rb_2[PtH]_6$, taken on a direct geometry spectrometer (MARI, ISIS) [20]. Showing the ridges that are due to the variation of band intensities with momentum and energy transfer.

Eq. (5.24) has a maximum at

$$\left\{\frac{n}{Q^2}-\alpha_1^{v}\right\}=0 \quad \text{and so} \quad Q^2\,\alpha_1^{v}=n \qquad (5.25)$$

All single quantum events, $n = 1$, involving the same α_1^{v} reach their maximum intensity at the same value of Q^2 (and so of Q) and all higher quantum events, $n > 1$, reach their maximum at a higher fixed value of Q. Combining Eqs. (5.25) and (2.72) yields

$$Q^2\alpha_1^{v}=Q^2\left(^{v}u_1^2+{}^{0}u_1^2\right)=Q^2 16.9\left\{\frac{1}{\mu_v\,\omega_v}+\frac{1}{\mu_0\,\omega_v}\right\}=1 \qquad (5.26)$$

where the difference $(\alpha_1^{v}-{}^{v}u_1^2)$ is represented by ${}^{0}u_1^2$, the mean square atomic displacement of the atom due to all the other vibrations, except v, of the system. It is convenient to represent this displacement through a dummy vibration, which is supposed to exist at the same energy as the transition of interest, ω_v. The oscillator mass, μ_0, of the dummy vibration is adjusted to give the correct ${}^0u_1{}^2$ for an oscillation at ω_v. Then

$$\omega_v = Q^2 16.9 \left\{ \frac{1}{\mu_v} + \frac{1}{\mu_0} \right\}$$

$$\frac{d\omega_v}{dQ^2} = \frac{dE}{dQ^2} = 16.9 \left\{ \frac{1}{\mu_v} + \frac{1}{\mu_0} \right\}$$

(5.27)

Comparing Eqs. (5.23) and (5.27) we see that the effective mass, m_{eff}, is:

$$\frac{1}{m_{eff}} \approx \frac{1}{\mu_v} + \frac{1}{\mu_0}$$

(5.28)

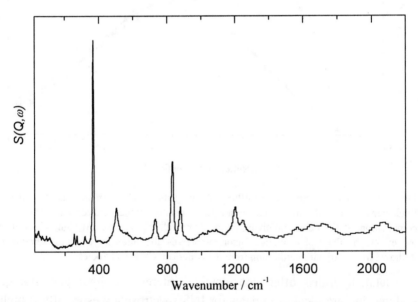

Fig. 5.11 The INS spectrum of $Rb_2[PtH_6]$ taken on a low band pass spectrometer (TOSCA). Reproduced from [20] with permission from the American Chemical Society.

In the limit that ${}^vu_1{}^2 = \alpha^v_1$ then $m_{eff} = \mu_v$. The scattering from the atoms in a sample approaches this limit when the great majority of their displacement arises from a single vibrational mode. However, unless this approximation is valid the oscillator mass and the effective mass diverge (for an interesting comment on a related issue see [10]). The mass μ_0 falls as the difference, $(\alpha^v_1 - {}^vu_1{}^2)$, increases and the observed effective mass, m_{eff}, may be quite different from the calculated oscillator mass, μ_v (as the approximation falters then $m_{eff} < \mu_v$).

This is related to similar but subtly different effects observed on indirect geometry spectrometers (§5.2.1.2). The INS spectrum of $Rb_2[PtH_6]$, obtained on the low-bandpass spectrometer TOSCA, is given in Fig. 5.11.

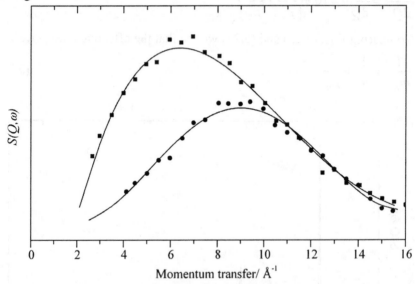

Fig 5.12 The INS spectrum of $Rb_2[PtH_6]$ showing the variation of the intensity of the fundamental, at 820 cm^{-1}, with momentum transfer, upper curve, and that of a combination, at 1200 cm^{-1}, involving the fundamental, lower curve. At high values of Q the effects of the Debye-Waller factor suppresses the intensity of both vibrations. Redrawn from [20] with permission from the American Chemical Society.

Metal hydrides offer considerable interest as hydrogen storage systems and are ideal candidates for INS spectroscopy especially if high point molecular symmetry is limiting the effectiveness of optical techniques, as in the case of the hexahydrides [22].

Apart from the optical data and the crystal structure there are only the results of *ab initio* calibrations to guide our expectations of the INS spectrum. The INS spectrum of $Rb_2[PtH_6]$ is given in Fig. 5.10. The profiles of two transitions in the $Rb_2[PtH_6]$ spectrum are compared in Fig. 5.12. These are obtained by integrating the intensity observed within an energy band, $\Delta\omega_{integral}$, including the transition energy, $\omega_v \pm 15$ cm^{-1}. Both curves show the intensity rising smoothly as Q increases. This smooth increase is a feature of the INS from incoherent scattering samples. A leading edge structured in Q is clear evidence of significant coherent scattering [23].

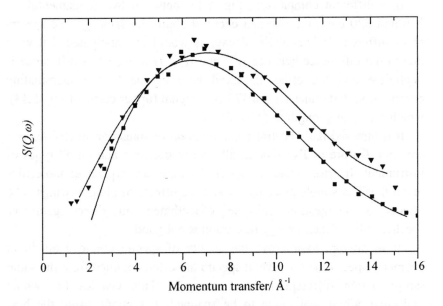

Fig. 5.13 The INS spectrum of $Rb_2[PtH_6]$ showing the variation of the intensity of one fundamental, at 370 cm^{-1}, with momentum transfer, full triangles, and comparing it with that of another fundamental, at 820 cm^{-1}, full squares. The effects of spectral contamination are discussed in the text. Redrawn from [20] with permission from the American Chemical Society.

The stronger feature, at 820 cm^{-1}, is the fundamental and it reaches its maximum strength at the Q value of 6.5 Å^{-1} ($Q^2 = 42.25$ Å^{-2}). The weaker feature, at 1200 cm^{-1}, has its maximum at a higher Q ($Q^2 = 81$ Å^{-2}). From Eq. (5.25) the ratio of the Q values indicates how many quanta were exchanged, in this case two (1.92 = 81/42.3).

Subsequently both curves decrease in intensity at high Q, due to the Debye-Waller factor and from Eq. (5.25) $\alpha_1^v = 0.0247$ Å^{-2} (= 2/81). Because of this unusual sample the determination of α_1^v is unclouded by concerns over the impact of phonon wings (§5.3.3).

Alternatively, the variation of the logarithm of the observed intensity, ln$S$, with ln$Q$ can be used to determine the nature of a given transition. This function is linear and has a slope of about two for one-quantum transitions and about four for two-quantum transitions. This method has been used on data from triple-axis spectrometers.

In a different comparison, Fig. 5.13, between two fundamentals at 370 and 820 cm^{-1}, the maxima occur at about the same Q. The spectra show different Debye-Waller decay, as would be anticipated if it were due to the difference between α^{370} and α^{820}. However, this difference is negligible and the effect observed here is due to a contaminating contribution that underlies the 370 cm^{-1} signal (upper curve of Fig. 5.13), which grows in strength as Q increases.

It is stressed that spectral contamination of transition intensities is a common feature of INS data at all energy transfers and on all types of instrument. It arises from the spectral congestion typical of molecules (even those as simple as benzene) and the effects of phonon wings. The effects of congestion can be exacerbated on direct geometry spectrometers if their energy resolution is not good.

In an attempt to improve the quality of spectra obtained on direct geometry spectrometers several spectra are often obtained from the same sample using different incident energies. This enables the whole molecular vibrational range to be spanned in sections using the best resolution available. Subsequent amalgamation of these partial spectra into a consistent picture that explains the whole range can be troublesome (§3.4.4). Clean spectra as straightforward as Rb$_2$PtH$_6$ are uncommon.

5.4.3 Phonon wings

The impact of phonon wings on the spectra observed on direct geometry spectrometers can be as taxing as their effects on indirect geometry spectrometers at the same Q values. The spectrum of Rb$_2$[PtH$_6$]

was specially chosen with an eye to avoid phonon wings so we shall use data from a more typical molecule to explore their effects.

A simplified spectrum of v_{30} in adamantane is shown in Fig. 5.14 [24], notice the unusual perspective of the figure. In this example the spectrum is relatively sharp and it is possible to clearly distinguish the one phonon contribution.

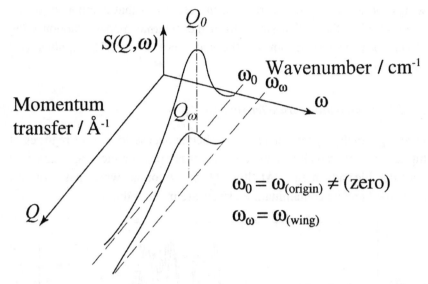

Fig. 5.14 The development of a phonon wing shown simplified for the v_{30} mode of adamantane.

The phonon wing orders develop as a series of ridges running parallel to the parent transition, the zero phonon band at ω_0. (Only the first order wing is shown in Fig.5.14.) This ridge, at ω_w, mimics the rise and fall of the one-phonon spectra observed on the high energy side of the elastic line. The order reaches its maximum intensity at higher Q than the parent transition. It is clear that, even in a modestly congested spectrum, the wings from lower transitions may fall accidentally degenerate with higher parent transitions. Combined intensities are not easily separated and the parent bands can not, necessarily, be analysed in isolation. In many ways this points up the strength of the synthetic approach to understanding INS spectra.

When analysing INS spectra from direct geometry spectrometers, where the samples have significant phonon wings, the choice of the width of the energy integration, $\Delta\omega_{\text{integral}}$, plays an important role. By suitably choosing this width the intensities associated with the wings might be excluded, or specifically included. The results of the two choices are different. Excluding the phonon wing intensity (narrow $\Delta\omega_{\text{integral}}$) will give information on S_O, Eq. (2.64). Including the phonon wings (broad $\Delta\omega_{\text{integral}}$) will remove the effects of that component of the Debye-Waller factor that stems from the external modes. Phonon wing effects are integrated out of the data and Eq. (2.41) applies (for fundamentals).

5.4.4 Low momentum transfer spectra

Direct geometry spectrometers have better access to low Q regimes at higher energy transfers than low-bandpass spectrometers and this is a considerable advantage. At low Q the phonon wing has not yet developed and its contaminating effects are much reduced.

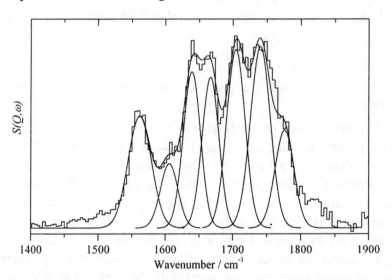

Fig 5.15 The INS spectrum of $Rb_2[PtH_6]$ obtained at low momentum transfer (MARI, ISIS), showing the effective suppression of the phonon wing contributions. Reproduced from [20] with permission from the American Chemical Society.

Unfortunately at low Q the strength of the observed bands is also underdeveloped and rather weak, long counting times are needed to obtain the good statistics needed to carry the analysis further. An example of the quality that is available from low Q data is given in Fig. 5.15. The problems that can arise from a too strongly scattering sample, measured at low Q, have already been discussed (§3.5.1).

5.5 Conclusion

In this chapter we have presented and discussed the use of the basic tools for spectral analysis. However, it should always be remembered that synthetic methods are more powerful, more likely to achieve success and, probably, quicker. Analytical methods should only be used to treat spectra that are either, particularly straightforward, or, paradoxically, particularly difficult. The first category represents simple problems that can be dealt with more efficiently by a rapid analysis than by a drawn-out *ab initio* calculation. The second category represents a failing synthetic approach and under these circumstances analysis offers, possibly, the only way forward.

5.6 References

1   D.E. Woessner & B.S. Snowden, Jr. (1967). J. Chem. Phys., 47, 378–381. Proton spin-lattice relaxation temperature dependence in ammonium bromide.
2   W.G. Fateley, F.R. Dollish, N.T. McDevitt & F.F. Bentley (1972). *Infrared and Raman Selection Rules for Molecular and Lattice Vibrations: The Correlation Method*, Wiley-Interscience, New York.
3   K. Nakamoto (1997). *Infrared and Raman Spectra of Inorganic and Coordination Compounds*. Part A: Theory and Applications in Inorganic Chemistry. 5th Ed., John Wiley, New York.
4   C.H. Perry & R.P. Lowndes (1969). J. Chem. Phys., 51, 3648–3654. Optical phonons and phase transitions in the ammonium halides.
5   H.G. Smith in: G. Venkataraman & V.C. Sahni (1970). Rev. Mod. Phys., 42, 409–470. External vibrations in complex crystals; also in: H. Biltz & W. Kress (1979). *Phonon Dispersion Relations in Insulators*. Springer tracts in modern physics, 10, Springer-Verlag, Berlin.
6   M.T. Dove (1993). *Introduction to Lattice Dynamics*. Cambridge University Press, , Cambridge, London.

7   P.S. Goyal, J. Penfold & J. Tomkinson (1986). Chem. Phys. Lett., 127, 483–486. Observation by neutron incoherent inelastic spectroscopy of split higher harmonics of the ammonium ion librational mode in the cubic phase of NH_4Br.

8   C.W. Garland & C.F. Yarnell (1966). J. Chem. Phys., 44, 1112–1120. Temperature and pressure dependence of the elastic constants of ammonium bromide, (table 5).

9   R.S. Seymour & A.W. Pryor (1970). Acta Cryst., B 26, 1487–1491. Neutron diffraction study of NH_4Br and NH_4I.

10 D.G. Truhlar & B.C. Garrett (2003). J. Phys. Chem., A 107, 4006–4007. Reduced mass in the one-dimensional treatment of tunnelling.

11 D. Smith (1988). Chem. Phys. Lett., 145, 371–373. The anharmonic librational energy levels of the ammonium ion in the low-temperature phase of ammonium chloride.

12 G. Herzberg (1945). Molecular Spectra and Molecular Structure: Infrared and Raman of Polyatomic Molecules, Vol. 2. Van Nostrand, New York.

13 R. Mukhopadhyay, B. A. Dasannacharya, J. Tomkinson, C. J. Carlile & J. Gilchrist (1994). J. Chem. Soc. Faraday Trans., 90, 1149–1152. Rotational excitations of NH_4^+ ions in dilute solutions in alkali-metal halide lattices.

14 J. Tomkinson (1988). Chem. Phys., 127, 445–449. The effect of recoil in the inelastic neutron scattering spectra of molecular vibrations.

15 G. Varsanyi (1969). Vibrational Spectra of Benzene Derivatives, Academic Press, New York.

16 D.J. Champion, J. Tomkinson & G.J. Kearley (2002). Appl. Phys., A 74, S1302-S1304. a-CLIMAX: a new INS analysis tool.

17 H. Jobic, A. Renouprez, A.N. Fitch & H.J. Lauter (1987). J. Chem. Soc. Faraday Trans. I, 83, 3199- 3205. Neutron spectroscopic study of polycrystalline benzene and of benzene adsorbed in Na-Y zeolite.

18 G.A. Jeffrey, J.R. Ruble, R.K. McMullan & J.A. Pople (1987). Proc. Roy. Soc. Lond., A 414, 47–57. The crystal structure of deuterated benzene.
    G. Filippini & C.M. Gramaccioli (1989). Acta Cryst., A 45, 261–263. Benzene crystals at low temperature: a harmonic lattice-dynamical calculation.

19 A.N. Fitch, H. Jobic & A. Renouprez (1986). J. Phys. Chem., 90, 1311–1318. Localisation of benzene in sodium Y zeolite by powder neutron diffraction. (Somewhat less detailed is: R. Goyal, A.N. Fitch, H. Jobic (2000). J. Phys. Chem., B104, 2878-2884. Powder neutron and X-ray diffraction studies of benzene adsorbed in zeolite ZSM-5.)

20 S.F. Parker, S.M. Bennington, A.J. Ramirez-Cuesta, G. Auffermann, W. Bronger, H. Herman, K.P.J. Williams & T. Smith (2003). J. Am. Chem. Soc., 125, 11656–11661. INS, Raman spectroscopy and periodic DFT studies of Rb_2PtH_6 and Rb_2PtD_6.

21 W.G. Stirling, Physics Department, Liverpool University, UK, Personal communication, see also R.T. Azuah, W.G. Stirling, J. Mayers, I.F. Bailey & P.E. Sokol (1995). Phys. Rev., B 51, 6780–6783. Concentration-dependence of the kinetic-energy in the He_3He_3 mixtures.

22 D. Bublitz, G. Peters, W. Preetz, G. Auffermann & W. Bronger (1997). Z. Anorg. Allg. Chem., 623, 184–190. Darstellung, 195 Pt-NMR-, IR- und Raman-spectren sowie normalkoordiatenanalyse der komplexionen [Rb_2PtH_n D_{6-n}], n = 1–6.

23 A.C. Hannon, M. Arai & R.G. Delaplane (1995). Nuc. Instrum. and Meth., A 354, 96–103. A dynamic correlation function from inelastic neutron scattering data.

24 S. F. Parker, unpublished work.

6

Dihydrogen and Hydrides

In this Chapter we present those aspects of dihydrogen rotational spectroscopy that are significant in INS spectroscopy, (§6.1). We then discuss the dynamics of dihydrogen weakly and strongly perturbed by surface interactions (§6.2) and sorbed by carbons (§6.3) and by microporous oxides including zeolites (§6.4). We describe INS studies of dihydrogen complexes (§6.5), hydrogen in metals (§6.6) and metal hydrides (§6.7). For INS studies of dihydrogen on catalysts and its dissociation to hydrogen atoms see Chapter 7: although dihydrogen can remain molecular on Ru/C (§7.3.1.10), and molybdenum disulfide (§7.6.5).

Pure rotational transitions of symmetrical diatomic molecules like dihydrogen are forbidden in infrared spectroscopy by the dipole selection rule but are active in Raman spectroscopy because they are anisotropically polarisable. They are in principle observable in INS although the scattering is weak except for dihydrogen. These rotational transitions offer the prospect of probing the local environment of the dihydrogen molecule, as we shall see in this chapter.

Interest in the rotational spectroscopy of dihydrogen has been stimulated by the search for hydrogen storage materials. The interest in nanomaterials for hydrogen storage is that the nanopores might afford strong binding forces for H_2 molecules. Because of the simplicity of the measurements described below, the strength of the rotational transition observed and its narrow width, the results obtained from similar samples on different instruments will often be thrown into sharp contrast. Here we report the published results somewhat uncritically but the informed reader will do well to appreciate the experimental error that accompanies the result from any given instrument. At the time of writing (2004), the most accurate energy transfer value, at hydrogen rotor energies, was $\pm 1.5\%$ (§3.4.2.3.1 and Appendix 3). The position of the dihydrogen rotor peak is thus not determined by INS to better than, 118 ± 2 cm^{-1}.

In the solid state at low temperatures, small molecules like methane and dihydrogen are able to find or create enough space within their local lattice environment to rotate. Unrestricted rotations usually have a marked impact on INS spectra: the effects of molecular recoil are seen as a broad wing on the high energy side of a vibrational or rotational transition (§2.6.5, §5.2.2). However, the typical energy range of rotational spectroscopy, a few wavenumbers or less, cannot be accessed by neutron spectrometers designed to study the usual energy range of molecular vibrations. The exception is dihydrogen with its first rotational transition at 118 cm^{-1}. The transition occurs at a higher energy than other rotational transitions as a consequence of the low mass of the H_2 molecule (2 amu) and its short bond (0.746 Å), which give it a uniquely high rotational constant, $B_{rot}^{HH} = 59.3$ cm^{-1} [1]. (B_{rot}^{HH} refers the molecule to its ground vibrational state. The spectroscopic value, $(B_{rot}^{HH})_e = 60.9$ cm^{-1}, is different since it is referred to a putative state involving no vibration.) Direct excitation of the rotational spectrum of dihydrogen will therefore produce transitions that intrude into the lower vibrational region. The only intramolecular vibration of hydrogen has its harmonic fundamental near 4400 cm^{-1}, observed at 4160 cm^{-1} [2].

6.1 The rotational motion of diatomic molecules

The Schrödinger equation for a system of two particles, can be separated [3]. We define the reduced mass, μ,

$$\mu = \frac{m_1 m_2}{m_1 + m_2}; \text{if } m_1 = m_2 = m_H \text{ then } \mu = \frac{m_H}{2} \qquad (6.1)$$

Then in spherical coordinates, Fig. 6.1, we have (where r, θ and ϕ are the polar coordinates of a nucleus)

$$\frac{1}{r^2}\frac{\partial}{\partial r}\left(r^2\frac{\partial\psi(r,\theta,\phi)}{\partial r}\right)+\frac{1}{r^2\sin\theta}\frac{\partial}{\partial\theta}\left(\sin\theta\frac{\partial\psi(r,\theta,\phi)}{\partial\theta}\right)+$$

$$\frac{1}{r^2\sin^2\theta}\left(\frac{\partial^2\psi(r,\theta,\phi)}{\partial\phi^2}\right)+\frac{2\mu}{\hbar}(E-V(r))\psi(r,\theta,\phi)=0 \tag{6.2}$$

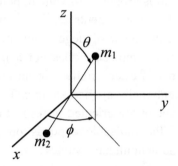

Fig. 6.1 Spherical coordinates, the angle θ is the polar angle and ϕ the azimuth.

Under the Born-Oppenheimer approximation Eq. (6.2) separates into three equations of the single variables r, θ and ϕ. The solutions for the angular functions $\Theta(\theta)$ and $\Phi(\phi)$ are given by:

$$\Phi_M=(\phi)=\frac{1}{\sqrt{2\pi}}\exp(iM\phi) \tag{6.3}$$

and

$$\Theta_{JM}=\sqrt{\frac{(2J+1)(J-|M|)!}{2(J+|M|)!}}\ P_J^{|M|}\cos\theta \tag{6.4}$$

The total angular wavefunction is the product of both angular terms, Eq. (6.3) and Eq. (6.4)

$$\psi(r,\theta,\phi)=\Psi(r)\Phi_M(\phi)\Theta_{JM}(\theta)=\Psi(r)Y_{JM}(\theta,\phi) \tag{6.5}$$

Where the simultaneous wavefunctions of the angular momentum operators, $\hat{J}^2$ and $\hat{J}_z^2$ are expressed in the spherical harmonics, Y_{JM}. The total angular momentum due to the rotation of the molecule is

$$\hat{J}\psi=\sqrt{\hat{J}_x^2+\hat{J}_y^2+\hat{J}_z^2}\ \psi=\hbar\sqrt{J(J+1)}\psi \tag{6.6}$$

The integers J and M are the angular momentum quantum numbers, with allowed values

$$J = 0, 1, 2, \cdots \quad ; \quad M = 0, \pm 1, \pm 2, \cdots \pm J \tag{6.7}$$

The magnetic quantum number, M, reflects the quantised nature of the z-components of the angular rotational momentum vector. In the gas phase the molecule is in an isotropic, zero valued, potential field and the relative orientations of the z-components are irrelevant. The number of orientations only represents the total degeneracy of the Jth rotational state. Here the energy of the molecule does not depend on M and the magnetic quantum number is often omitted. In an anisotropic field the unique direction is conventionally aligned along z. The z-components of the angular momentum vector are oriented in respect of this, laboratory, axis and the energy of the molecule depends on J and M. The z-components of angular momentum have values

$$\hat{J}_z \psi = M \hbar \psi \tag{6.8}$$

The rotational energy of the molecule in free space is [4]

$$E_{JM} = E_J = J(J+1)\frac{\hbar^2}{2I} = J(J+1)B_{rot} \tag{6.9}$$

where I, the moment of inertia of the molecule, is

$$I = 2\left(\frac{r}{2}\right)^2 m \; ; \text{ for dihydrogen } I_{HH} = \mu r_{HH}^2 \tag{6.10}$$

and B_{rot} is the rotational constant:

$$B_{rot} = \frac{\hbar^2}{2I} \; ; \text{ for dihydrogen } B_{rot}^{HH} = \frac{\hbar^2}{2I_{HH}} \tag{6.11}$$

It is useful to have a pictorial representation of these rotational states and the angular probability density function, $P(\theta, \phi)$, offers some advantages in this respect. We shall expand on this later but first some of the specifics of the dihydrogen rotational states must be given.

6.1.1 The rotational spectroscopy of dihydrogen

In the solid and liquid state, dihydrogen has many unusual features that mark it as different from other molecular solids and liquids. The most remarkable and important difference is that, even in solid dihydrogen, the free rotor states that describe the rotational motion of an isolated molecule are almost unperturbed by interactions between the neighbouring dihydrogen molecules [1] Solid dihydrogen is a 'quantum crystal'; dihydrogen molecules with their centres of mass localised at lattice positions rotate freely even at the lowest temperatures. This remarkable behaviour is due to the weak intermolecular interactions (nearest neighbour distances of ~3.8 Å) and the small moment of inertia of the dihydrogen molecule.

6.1.2 Ortho- and para-hydrogen

Quantum mechanical restrictions on the symmetry of the rotational wave function produce two molecular species for dihydrogen: ortho- and para-hydrogen. The hydrogen nucleus, a particle with half-integer spin, is a fermion. Its quantum mechanical description must obey Fermi statistics, as expressed through the Pauli principle. This imposes a symmetry requirement on molecular wavefunctions and forbids the occupation of certain states. Identical nuclei are indistinguishable; their exchange can change the sign of the total wavefunction. The relevant wavefunctions are those for molecular rotation, $Y_{JM}(\theta,\phi)$, and nuclear spin, ψ_{spin}. We have

$$\hat{P}_{12}\psi_{total} = (\pm 1)Y_{JM}(\theta,\phi)\,\psi_{spin} \tag{6.12}$$

Here the $\hat{P}_{12}$ operator exchanges identical nuclei, $r \rightarrow -r$. For a fermion the negative sign (-1) applies in Eq. (6.12) and the total wave function has to be antisymmetric under the exchange of indistinguishable particles. Such exchange occurs during a rotation about a C_2 axis [5]The symmetry of the rotational wavefunctions under exchange is

$$Y_{JM}(\pi - \theta, \phi + \pi) = (-1)^{J} Y_{JM}(\theta,\phi) \tag{6.13}$$

Also, we may represent the spin states by $\uparrow$, spin 'up', and $\downarrow$, spin 'down', if we represent the wavefunction in the Dirac notation $|J,M\rangle$. The spin wavefunctions are thus:

$$\psi_{spin}^{sym} = \begin{cases} \uparrow_1\uparrow_2 & |+1,+1\rangle \\[2mm] \dfrac{\uparrow_1\downarrow_2 + \downarrow_1\uparrow_2}{\sqrt{2}} & |+1,\ 0\rangle \\[2mm] \downarrow_1\downarrow_2 & |+1,-1\rangle \end{cases} \qquad (6.14a)$$

$$\psi_{spin}^{asym} = \frac{\uparrow_1\downarrow_2 - \downarrow_1\uparrow_2}{\sqrt{2}} \qquad |\ 0,\ 0\rangle \qquad (6.14b)$$

The even values of J correspond to antisymmetric wavefunctions and must be combined with symmetric spin wavefunctions. The only allowed rotational states of spin paired, or antiparallel, ($\uparrow\downarrow$) dihydrogen are thus those with, $J = 0, 2, 4...$ This defines *para-hydrogen*. There are $2J+1$ possible spin states for each acceptable J value and it follows from Eq. (6.7) that there is only one spin state for the para-hydrogen ground state, $J = 0$.

The odd rotational values of J correspond to symmetric rotational wavefunctions and, in order to have an antisymmetric total wavefunction, we must combine these with antisymmetric nuclear spin wavefunctions. The odd rotational states $J = 1, 3, 5...$ are combined with a symmetric nuclear spin wavefunction (where nuclear spins are parallel, or unpaired, $\uparrow\uparrow$). This defines *ortho-hydrogen* and again its degeneracy, in the gas phase, is the $2J+1$ possible spin states.

In the absence of a catalyst (§6.1.6) transitions between the spin paired and spin unpaired states are rare, the species are 'spin-trapped'. Dihydrogen is effectively a mixture of two stable species, para-hydrogen and ortho-hydrogen, where the molecules mix freely and there is little exchange between the two populations. The two fluids have slightly different physical properties. Whilst the ground state rotational energy of dihydrogen, zero, is the minimum rotational energy for para-hydrogen molecules, the minimum energy of ortho-hydrogen molecules is $2B_{rot}$. In an equilibrium mixture at room temperature only the $J = 0$ and $J = 1$ states are significantly occupied and in proportion to their degeneracies, namely 1 and 3. Dihydrogen gas is thus a mixture of 25% para-hydrogen and 75% ortho-hydrogen

We can evaluate the energies of the first three rotational states, from Eq. (6.9) and the B_{rot} value given earlier:

$$E_{0,M} = 0; \quad E_{1,M} = 2B_{rot} = 118 \, cm^{-1}; \quad E_{0,M} = 6B_{rot} = 354 \, cm^{-1} \tag{6.15}$$

Excitations *within* the para-hydrogen population start from the ground state and the first transition is:

$$\Delta E_{J-J'} = E_2 - E_0 = 6B_{rot} = 354 \, cm^{-1} \tag{6.16}$$

Excitations *within* the ortho-hydrogen population start from the $J = 1$ state and the first transition is:

$$\Delta E_{J-J'} = E_3 - E_1 = (12 - 2)B_{rot} = 593 \, cm^{-1} \tag{6.17}$$

6.1.2.1 The molecular isotopes of dihydrogen

Deuterium ($^2H_2$) nuclei have $I_D = 1$; the $J = 0, 2, 4..$ states are ortho-deuterium and the $J = 1, 3, 5..$ states are para-deuterium. Here the rotational constant has a value, $B_{rot}^{DD} = 29.9 \, cm^{-1}$. The total neutron scattering cross section of deuterium (7.64 barn) is too low to enable its common use in INS experiments but HD , can be readily measured.

In HD the nuclei are different and so distinguishable, the Pauli principle no longer operates and transitions between any two quantum states, J, are allowed without restriction to odd or even number and no spin trapping occurs. The rotational energies are still given by Eq. (6.9) and the rotational constant for HD is, $B_{rot}^{HD} = 44.7 \, cm^{-1}$ [1] We shall see that the use of HD in INS spectroscopy has benefits.

6.1.3 The angular probability density function, P(θ, φ)

The angular probability density function, $P(\theta, \phi)$, provides the basis for a pictorial description of the rotational states of dihydrogen. The function is a surface represented in polar coordinates. Any point on the surface can be joined to the origin by a line. The (θ, ϕ) coordinates of the line correspond to the orientation of the dihydrogen molecular axis and the probability of any particular orientation occurring is given by the length of the line, $P(\theta, \phi)$. Only two rotational states are of significance for our discussion, $J = 0$ and $J = 1$. In the ground state, $J = 0$, $P(\theta, \phi)$ is spherical, see §6.2.1, and all molecular orientations are equally probable.

The first rotational state, $J = 1$, see §6.2.1, has a more complex shape imposed by the magnetic quantum numbers, M. The probability distribution function can be discussed in terms of two component shapes, which can be represented as a toroidal or 'ring-doughnut' shape with its maximum probability in the x,y plane (horizontal) and, an 'hour-glass' shape, with its maximum probability along the z axis, (vertical). (Recall that θ is the angle between the normal to the x,y-plane (the z-axis) and the H—H axis.)

The ring-doughnut represents the $M = \pm1$ states; a dihydrogen molecule in a $M = \pm1$ state is most likely (>85% probability) to be found with its molecular axis within the range $45° < \theta < 135°$. The direction of its horizontal component is equally probable in ϕ.

The hour-glass represents the $M = 0$ state; a dihydrogen molecule in this state is most likely (>64% probability) to be found within the ranges, $\theta < 45°$ or $\theta > 135°$. The direction of its horizontal component is equally probable in ϕ. Notice that both the ring-doughnut and the hour-glass shapes occupy large volumes of space and considerable latitude is available in the values of θ.

6.1.4 The scattering law for dihydrogen rotations

In principle all the rotational transitions of dihydrogen are observable by INS spectroscopy but some transitions are rather weak since the incoherent neutron scattering cross section used generally in this book is inapplicable (§2.1). This derives from the strict nuclear spin correlation found in dihydrogen. Transitions within the separate para- or ortho-hydrogen manifolds are controlled by the coherent cross section of hydrogen, 1.76 barn, and these transitions are too weak to be observed in its INS spectrum.

A fortunate consequence of the weakness of these transitions is uniquely beneficial for INS spectroscopy since all elastic scattering from dihydrogen is effectively suppressed. Elastic scattering is observed when no exchange of energy occurs, usually $J(0\leftarrow0)$; the transition is thus restricted to a single manifold and the cross section is low.

Spin-flip transitions, between the manifolds, are controlled by the incoherent cross section of hydrogen, 80.3 barn, and are enabled by spin exchange with the neutron, $I_n = 1/2$. Thus, for the $J(1\leftarrow0)$ transition

$$\left(\uparrow\downarrow\right)^{J=0}_{H_2} + \left(\uparrow\right)^{E_i}_{n} \xrightarrow{\sigma_{inc}} \left(\uparrow\uparrow\right)^{J=1}_{H_2} + \left(\downarrow\right)^{E_f}_{n}; \qquad \Delta E_{0-1} = E_i - E_f \qquad (6.18)$$

The reason for the strength of the $J(1\leftarrow0)$ feature in the INS spectrum of dihydrogen is its access to the incoherent scattering cross section. The strong features of the dihydrogen INS spectrum are, therefore, the $J(1\leftarrow0)$, $J(3\leftarrow0)$,... $J(odd\leftarrow0)$; $J(3\leftarrow2)$,... $J(odd\leftarrow2)$ etc and the $J((2\leftarrow1)$,... $J(even\leftarrow1)$; $J(4\leftarrow3)$,... $J(even\leftarrow3)$ etc transitions. However, since most INS spectroscopy involves samples equilibrated to their ground state (§6.1.6), we usually have only $J = 0$ as the initial state. A full treatment of the INS spectrum of dihydrogen might take as its starting point the Young-Koppel model [6] which has seen some recent refinement [7,8]. However, this is rather more mathematically sophisticated than is required for our purposes. The rotational INS spectrum of dihydrogen can be well enough interpreted by a simplified approach since there are, usually, only few transitions. There is no need to be concerned about the distribution of intensity between many different transitions, as is required for the vibrational problems that are covered elsewhere in this book.

6.1.5 An outline of the INS spectrum of dihydrogen

The INS spectrum of solid dihydrogen, as obtained on a low-bandpass spectrometer, is shown in Fig. 6.2, with the sample equilibrated to 13 K. It consists of a sharp, intense transition centred at 118 cm^{-1}; to higher energies there is a strong and broad shoulder (with some weak overlying features). The shoulder, centred at *ca* 300 cm^{-1}, is smooth except for a sharp curtailment at energies below the rotational transition. The intensity in this shoulder comes from rotational transitions displaced by the translational recoil of the dihydrogen molecule. Applying Eq.

(5.13), for the recoil energy, we have $E_r = (300 - 118) = 182$ cm^{-1}, yielding a very light effective mass, *ca* 2.5 amu.

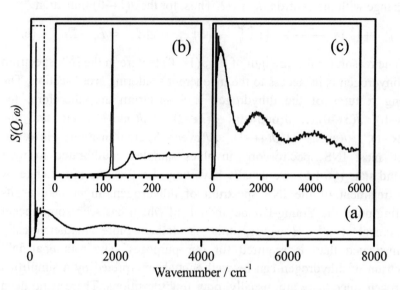

Fig. 6.2 The INS spectrum of solid dihydrogen, 13 K. (a) The full spectrum over the maximum available range (0—8000 cm^{-1}), (b) the area around the rotational line at 118 cm^{-1} and (c) expanded vertical axis to show how recoil impacts the other rotational transitions (see text).

As solid dihydrogen is warmed the effects of recoil become more marked; near the melting point, 15 K, the transition line is reduced to an edge, Fig 6.3. In the liquid even the edge disappears and the onset of recoil occurs below the rotational transition. In the solid the effects of recoil are complete at higher energy transfers (which on low-bandpass spectrometers have higher Q values) and we then see broad bands centred about 1850 and 4000 cm^{-1}. These bands correspond to the recoiled $J(3\leftarrow0)$, 1428 cm^{-1}, and $J(5\leftarrow0)$, 3570 cm^{-1}, rotational transitions, see Fig. 6.2(c).

When dihydrogen interacts more or less strongly with its environment the spectra depend on the form of the interaction potential. If the hindrance to translation is increased, the contribution of recoil to the spectrum is reduced and the relative strength of the rotational line, compared to that of the shoulder, is greater. If different molecules experience different isotropic potential field strengths the rotational transition could appear broadened. The effects of weak anisotropic potentials also lead to broadened transitions. Strong anisotropy leads to much more dramatic effects, since under the influence of an anisotropic field the *M* degeneracy is lifted.

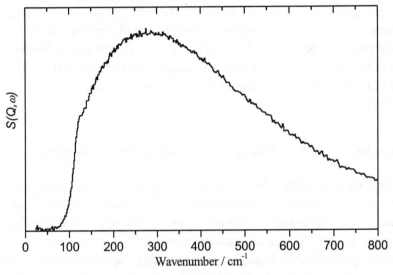

Fig. 6.3 The backscattering spectrum of liquid hydrogen at 14.1K. Note that the rotational line has disappeared and a sharp edge is now seen at 118 cm^{-1}.

6.1.6 Experimental considerations—the conversion to para-hydrogen

In the case of isolated or non-interacting molecules spin conversion is forbidden. However, a magnetic field gradient will catalyse the conversion of hydrogen and either a magnetic or electric field will catalyse conversion of deuterium. Temperature variations in dihydrogen can also trigger conversion. Dihydrogen molecules convert slowly (about 2% per hour in the solid) through their interaction with each other's magnetic dipole, homogeneous conversion. In low density gas phases this conversion rate is very slow and high purity samples, once prepared, can be kept for up to a year with little change. There are techniques to prepare both ortho- and para-hydrogen of better than 99% purity [1]

Liquid dihydrogen, when kept below its boiling point, will convert to para-hydrogen over a period of several days. Bringing the liquid into contact with a paramagnetic catalyst accelerates this process, with conversion time constants of seconds to minutes. One way of efficiently producing millilitres of liquid para-hydrogen is to liquefy the hydrogen in a cell containing a nickel catalyst. The cell should have a slight temperature gradient; the resulting convection will remove the already formed para-hydrogen from the catalyst leaving the catalyst exposed to other ortho- molecules. It is possible to achieve equilibrium at the boiling point (20.4 K) within 20 h, when the concentration of para-hydrogen will be 97.8%.

6.2 The INS spectrum of dihydrogen in an anisotropic potential

The unique direction of the anisotropic potential is taken to lie along the z-axis. A common situation occurs for dihydrogen adsorbed onto a surface, where the x- and y- axes are defined in the plane of the surface. Below we present the general case in terms of this typical example. As the out-of-plane forces, those along z, securing the dihydrogen to the surface increase in strength, the rotational states probing the near surface will become perturbed [9]. The total energy of the system, in polar coordinates, $V(z, \theta, \phi)$, can be separated

$$V(z,\theta,\phi) = V_0 + f_{ads}(z-z_0)^2 + V(\theta,\phi) \qquad (6.19)$$

where, V_0, represents some minimum isotropic potential and depends on the distance of the molecule to the surface, z. If we consider just the orientational terms, a commonly used expression for the potential that governs the rotation of the molecule is [10]

$$V(\theta,\phi) = \left[a + \frac{b}{2}\cos 2\phi\right]\sin^2\theta \qquad (6.20)$$

where θ is the polar angle (the angle between the H–H bond and the surface normal, z), ϕ is the azimuthal angle (between the x axis, defined in the plane of the surface, and the projection of the H–H bond onto the surface plane). The values of a and b give the relative weights of the potential and the factor two in 2ϕ represents how the symmetry of the field matches the C_2 molecular symmetry. Treating the orientational potential as a perturbation and expanding it in spherical harmonics we can solve the Schrödinger equation for the system. The relevant Hamiltonian is

$$\hat{H}_{J'M'JM} = J(J+1)B_{rot}\,\delta_{J'J}\,\delta_{M'M} + \left\langle Y_{J'M'}\left|V(\theta,\phi)\right|Y_{JM}\right\rangle \tag{6.21}$$

The resultant matrix is diagonalised numerically to determine the energy states of the perturbed rotor (see §6.4.1 below). A minimum of 64 rotational levels must be calculated to obtain an error smaller than $B_{rot}\times10^{-5}$ in the calculated energies. The energy levels are shown in Fig. 6.4. Since most experiments involve para-hydrogen in its ground state Fig. 6.4 refers all energies to the energy of that state. Note that the absolute energy of the ground state actually rises as the potential deepens and the convention of Fig. 6.4 is not universally followed in the literature. The labels of the states refer to the unperturbed case since, strictly, J and M are no longer good quantum numbers in the presence of an external field.

We shall now develop a simplified description of the rotational states, which will rely on 'model' conditions that can be applied to observations. However, the literature is quite inconsistent in the titles that are given to these models and this provides a field rich in opportunity for confusion arising from misinterpreted nomenclature.

6.2.1 Planar rotor in an attractive field—the 2-D type

When $b \approx 0$ and $a < 0$, then the potential in Eq. 6.20 is most attractive when the molecule lies *parallel* to the surface, see Fig. 6.5. Those rotational states represented by the, planar, ring-doughnut ($J = 1$, $M = \pm1$, $45° < \theta < 135°$) are stabilized by this potential. The relative energies of the $M = \pm1$ states fall as the potential deepens but remain degenerate. The molecular axis is unrestricted in ϕ if $b = 0$, but it will become more restricted as b increases, see below. As a becomes more attractive (more negative), the potential securing the molecule to the plane increases and the energy of the state represented by the hour-glass ($J = 1$, $M = 0$, $\theta < 45°$ or $\theta > 135°$) increases, Fig. 6.6., becoming more energetic as the well deepens.

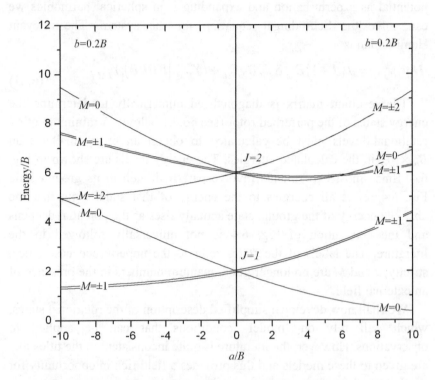

Fig. 6.4 The energies of the rotational states of dihydrogen in the field of Eq. (6.20), relative to the $J = 0$, ground state. Given as a function of the parameter a/B. A nominal value of $b = 0.2 B$ is used to remove the degeneracy in M.

6.2.1.1 Perturbed planar rotor in a deep attractive field—the 2-DP type

In the limit of very large, negative, a values the ring-doughnut collapses to a circle, θ is now $90°$, and $V(\theta, \phi)$ (Eq. 6.20) becomes $(b/2)\cos(2\phi)$. The energy levels are given by:

$$E_{J'} = M^2 B_{rot} \qquad M = 0, \pm 1, \pm 2 \cdots \pm J \qquad (6.22)$$

Sometimes J' is found in the literature instead of M. The E_J are the energies of a rotor confined to rotate in a plane, the axis of rotation being the normal to that surface, the z-axis. The energy of the first transition, $M(1\leftarrow0)$, is now B_{rot}.

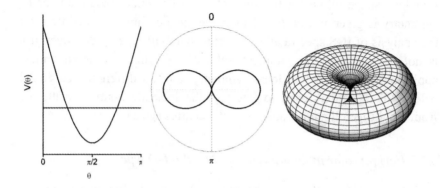

Fig 6.5 Representation of the 2D type constraining potential energy: left, the potential energy as a function of the polar angle θ; centre, a representation of the potential energy function as a cut in the xz-plane; right, the three-dimensional representation of the potential in polar coordinates.

For these large negative a values we have the limiting case of the 2-D type model. The degeneracy of the $M = \pm1$ states is removed as the value of the b parameter increases.

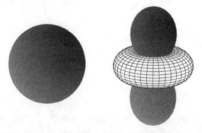

Fig. 6.6 Diagramatic representation of the probability distribution function of the rotor. On the left, $J = 0$. On the right, $J = 1$, the solid lobe (the hour-glass shape) represents $M = 0$ and the wireframe (ring-doughnut) represents $M = \pm1$.

The probability density surface, already collapsed to a circle by the strong a field, now becomes deformed in a manner dependent on the local symmetry. This is the perturbed, 2-DP type, spectroscopic model.

Local C_2 symmetry will introduce a single waist into the probability density circle converting it to a double lobe shape (almost a figure of eight). Local C_4 symmetry will introduce a double waist generating a four lobe shape (almost a four-leaf clover). The energy diagram for C_2 symmetry is given in Fig. 6.5. This is the most commonly used field for the analysis of INS spectra and conforms to a 2DP type model (even if it is not obvious why the simplest field, of C_2 symmetry, is so widely applicable even when the local symmetry appears different, see §6.2.3 below). This field is used to interpret the inelastic neutron scattering from dihydrogen ligands bound to metal centres (§6.5).

6.2.2 Upright rotor in an attractive field—the 1-D type

When $b \approx 0$ and $a > 0$, the potential is most attractive when the molecular axis is *perpendicular* to the surface.

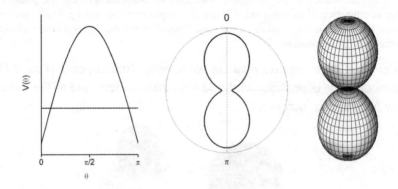

Fig. 6.7 Representation of the potential energy for a molecule that is aligned perpendicular to a plane. Left, the potential energy as a function of the azimuthal angle θ; centre a representation of the potential energy function as a cut in the *xz*-plane; right, the three- dimensional representation of the potential in polar coordinates.

The state represented by the, upright, hour-glass ($J = 1$, $M = 0$, $\theta < 45°$ or $\theta > 135°$) is stabilized by this field, see Fig. 6.7. The molecular axis is unrestricted in ϕ if $b = 0$, but becomes more restricted as b

increases. It is the ring-doughnut ($J = 1$, $M = \pm 1$) states that are now of higher energy.

In the limit of very large, positive, a values the potential constrains the molecule into the z-axis, the energy difference between the ground state ($J=0$, $M=0$) and the first rotational level, ($J=1$, $M=0$) becomes very small, and negligible for values of $a > 30$ B_{rot}. The transition energy drops below the operational range of energies available to the spectrometers included in this book.

It is possible to find strongly perturbed 1-D type systems [10] where the value of b is significant. As in the 2-DP case this raises the degeneracy of the $M = \pm 1$ states but the splitting occurs in a region that is experimentally problematical, it also raises the energy of the $M = 0$ state. Taken together these effects make any definitive assignment of a spectrum to a perturbed 1-D type difficult.

6.2.3 Experimental consequences

We have used limiting cases to describe the lowest energy excited states of dihydrogen in the different anisotropic attractive fields, $a < 0$ gives a planar, or '2-D type' state, ($J = 1$, $M = \pm 1$), and $a > 0$ gives an upright, or '1-D type' ($J = 1$, $M = 0$). The perturbed deep potential '2-DP type' was also presented.

Fortunately, in principle, it is simple to distinguish between the two main types of spectral distribution, at least for modest values of a and b. In the 2-D type spectrum the strongest transition should appear at energies above B_{rot} but below $2B_{rot}$, its relative strength being about twice that of the higher transition. For the 1-D type spectrum the strongest transition should be that at the higher energy. In both cases the intensity weighted mean frequency should be about $2B_{rot}$ except where a significant b value has displaced it. Beyond moderate values for the parameters a and b discussion of INS spectra in terms of rotational states may be inappropriate, especially for the higher energy transitions. A more adapted language might be in terms of the two librations of the dihydrogen molecule oscillating in a two-fold deep well. In a sufficiently deep well the transitions will appear to be approximately harmonic and possibly structured. This structure has its origins in the rotational nature of the states but in the language of librations is described as 'tunnel' splitting and can give rise to very low energy transitions.

Low energy transitions, ca 10 cm^{-1} or less, are observed (but not on the spectrometers which are the subject of this Book) when: very deep wells give a low energy 1-D type first state; a deep 2-D type becomes perturbed to 2-DP; or, unusual spin traps have been created in high energy 'librational' states. These transitions are collectively called 'tunnelling' transitions and they have been successfully exploited, especially in the study of dihydrogen ligands (§6.5).

The assignment of appropriate models to a given system has not always been strongly founded. The number of observable rotational transitions is few and the presence of any translational vibrations merely complicates the picture. Often several similar solutions can be found for widely different a and b values [11]. Perhaps the greatest possibility for confusion occurs between 1-D and 2-DP type spectra. Early INS work avoided a full description of the potential in terms of spherical harmonics and usually worked within simplifying approximations specific to individual cases. The relationship between those models and the forms we develop here is not necessarily straightforward.

In what follows, we shall discuss the use of INS in the study of dihydrogen as a molecular probe of the interaction between the molecule and its environment.

6.3 Dihydrogen on graphite and carbons

Carbon is commonly used as a catalyst support and in fuel cell electrodes. A strong interest in single wall carbon nanotubes (SWNT) developed in 1997 when it was reported that they were efficient hydrogen storage systems, up to 10 wt% of hydrogen being claimed [12]. Subsequently the hydrogen-storage abilities of graphitic nanofibers and alkali doped multiwalled nanotubes were reported [13,14]. It appears now that the claimed storage capacities of carbon materials are closer to 2.5 wt% [15].

This work stimulated a number of neutron scattering studies of dihydrogen in carbonaceous materials aimed at elucidating the orientation and strength of interaction of the confined H_2 molecule within a nanopore. Observation of the dihydrogen rotational transition by INS is the ideal method—with the proviso that well resolved spectra are essential to observe any splittings and spectral shifts.

6.3.1 Graphite and other carbons and carbon nanostructures

Interaction of dihydrogen with carbons causes shifts or splittings of only a few wavenumbers away from the gas values but often accompanied by changes of band width. Peak positions and assignments are given in Table 6.1. A thorough account of computed and experimental spectra is provided in [16]. The dynamics of dihydrogen molecules inside a carbon nanotube was studied as a function of tube radius. In a cylindrical confining potential the coupled rotation-vibration levels of the confined H_2 molecule were calculated. Inclusion of coupling between rotations and translations was necessary to obtain the correct degeneracies of the energy level scheme. For 3.5 Å radius nanotubes the confining potential was parabolic. For larger radii the potential had a Mexican hat shape; the H_2 molecule was off centred. Its translational vibrations were described as radial and tangential. For ortho-hydrogen ($J = 1$) one class of translation-rotation wavefunctions had the molecule, in a pure $M = 0$ state, aligned along the axis of the nanotube.

INS spectra of para- to ortho- and ortho- to ortho- transitions were calculated and compared with experiment. The manifold of calculated first ortho- to para- transition involved two features at 136 and 110 cm^{-1} with a 1:2 intensity ratio. This corresponds to a 26 cm^{-1} splitting between ($J=1$, $M=0$) and ($J=1$, $M= \pm 1$) a 1-D type model but complicated by rotation-translation coupling. The intensity weighted average of the $J = 1$ levels give an almost unperturbed value, 118 cm^{-1}, with an at most downward shift of 0.24 cm^{-1} from the free molecule value. This shift was smaller than that anticipated for H_2 in carbon nanotubes; H_2 molecules were therefore probably *not* inside the nanotubes. Radial phonon transitions occurred at 121 and 242 cm^{-1}. There were many lines due to transitions between different tangential phonon states. Indeed, the calculated spectrum was remarkably rich. The calculated ortho- to ortho-spectrum was similarly rich but dominated by transitions between tangential phonon states. Below 81 cm^{-1} (where experimental resolution of the peaks was possible) the tangential phonon transitions showed a maximum near to 31 cm^{-1}.

In the identification of dihydrogen binding sites it is advantageous to determine how the parameters of the INS spectra vary with the dihydrogen loading and temperature and we now summarise recent interpretations of such observations.

Table 6.1 INS of para-hydrogen adsorbed by single-walled carbon nanotubes.

ω/cm^{-1}	Assignment	H$_2$ binding site	Ref.
117 broad		Outer surface of nanotubes. Binding energy 6 kJ mol^{-1}.	17
109	$J, M(1, \pm1 \leftarrow 0,0)$ (a)	Valley between two adjacent	18 (b)(c)
123	$J, M(1,0 \leftarrow 0,0)$ (a)	tubes	
115	H$_2$ translation	Weaker bound H$_2$ on external	
118	$J(1 \leftarrow 0)$	surface	
110	$J, M(1, \pm1 \leftarrow 0,0)$(d)		16
136	$J, M(1, \pm1 \leftarrow 0,0)$(d)		
119	$J(1 \leftarrow 0)$(e)		
32	Tangential phonon transition		
123	Radial phonon transitions		
110	$J, M_J(1,0 \leftarrow 0,0)$ (f)		15 (c)
119	$J, M_J(1, \pm1 \leftarrow 0,0)$ (f)		
116	$J(1 \leftarrow 0)$	Weaker binding site	
118	Free H$_2$		

(a) Intensity 109 twice intensity 123. (b) para-hydrogen. (c) Note different assignments. (d) Computed for H$_2$ in a cylindrical potential.(e) Computed and experimental centre of gravity (f) Intensity 119 twice intensity 110.

The shift of the dihydrogen rotational peak from the free hydrogen value. It has been claimed that on single-walled carbon nanotubes (SWNT), nanofibres and activated carbons there was no shift or splitting of the dihydrogen rotational peak and that, therefore, the morphology of the carbon surfaces had little effect on the rotational potentials of hydrogen [15]. On the other hand a slight shift of the rotational transition −1.5 cm^{-1} was attributed to the large zero-point energy of the H$_2$ molecule[17].

Broadening of the dihydrogen rotational peak has been attributed variously to symmetry reduction at the hydrogen adsorption sites, which would lift the degeneracy of the rotational levels, or to a distribution of adsorption sites and hence of rotational energies [17].

Scattering intensities and hydrogen loading. The various binding sites of dihydrogen can be explored by observing changes in the INS peak intensities as a function of the dihydrogen loading [18][19]. Typical results are shown in Fig. 6.8 [18]. The scattering intensities varied with the hydrogen loading in a manner indicative of at least two different adsorption sites on the surfaces of the nanotube bundles, the system was a 2-D type. The intensity of the 109 and 123 cm^{-1} peaks both increased with increasing surface coverage and saturated at half-monolayer coverage. The adsorption site was considered to be the valley between two adjacent tubes and offered some barrier to azimuthal rotation of the adsorbed H_2 molecules. The 115 and 118 cm^{-1} peaks were assigned to less strongly bound H_2 molecules possibly on the external convex surface of the nanotubes and free to rotate and recoil. Similar results are reported by others [20] except that the intensities, and hence the assignments of the 110 and 119 cm^{-1} were reversed (see Table 6.1). The peak locations in these spectra were critically dependent on the chosen fitting procedures used to analyse the spectra.

Changes of the INS spectra with increasing temperature. The INS scattering intensity decreases with increasing temperature principally because of thermally activated desorption of hydrogen from the nanotube surface [17] or the para- to ortho-hydrogen conversion at higher temperature [21]. The width of the 117 cm^{-1} peak of H_2 molecules adsorbed by single-walled carbon nanotubes increased with increase of temperature from 4 to 35 K. The H_2 molecules remained in the sample up to 65 K and then started to desorb with increasing temperature. The residence time of dihydrogen on the nanotube (> 15 ps) was longer than on graphite (6 ps). The H_2 molecules were, therefore, considered to be more strongly bound probably in the interstitial tunnels of the nanotube bundles [21].

Metallic impurities [17] Some preparations of carbon nanotubes might contain magnetic impurities (e.g. cobalt or nickel) which will catalyse the conversion of ortho- to para-hydrogen with the consequence that low temperature population of the first excited level will be negligible [18].

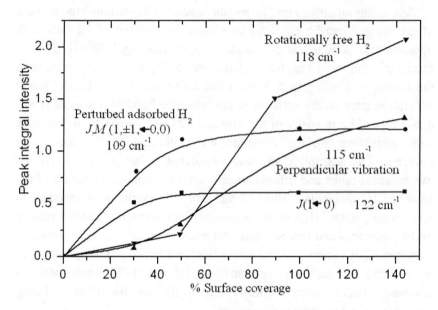

Fig. 6.8 Adsorption of para-hydrogen on single-walled carbon nanotubes. Peak intensities as a function of dihydrogen surface coverage. Redrawn from [18] with permission from the Institute of Physics.

6.3.2 Alkali metal intercalated graphite

In the graphite intercalation compounds alkali metal atoms are located between the graphite layers. The effect of intercalation is greatly to increase dihydrogen uptake, e.g. by tenfold [19]. The intercalated metal atoms increase the interplanar separation of the graphite so facilitating access of dihydrogen molecules to the internal surface of graphite. There have been a number of neutron scattering studies of the dynamics of the absorbed H_2 molecules.

Graphite intercalation compounds of the heavier alkali metals, $C_{24}M$ (M = K, Rb, Cs) are able to absorb dihydrogen into the metal layers to form compounds $C_{24}M(H_2)_2$. The molecules are centred in the plane of the metal atoms and oriented perpendicularly to the graphite plane [22]. INS spectra (at 10 to 85 K) included rotational tunnelling and were attributed [10] to a 1-D type model. The dihydrogen adsorbs at two sites: site-A, peaks at 11 cm^{-1} in $C_{24}Rb(H_2)_x$ ($x = 0$ to 2) and 9.5 cm^{-1} in $C_{24}Cs(H_2)_x$ ($x=0$ to 2); site-B, peaks at 5.3 cm^{-1} for the Rb and Cs compounds. In the Rb intercalate the 258 cm^{-1} feature was presented as a viable candidate for the transition to the next state ($J=1$, $M= \pm1$) and features at 89 cm^{-1} and 178 cm^{-1}, were discussed in terms of translational modes. The spectra show no signs of recoil, which is partly due to the low Q available on the INS instrument used, but the assignment of dihydrogen translational vibrations at high energies would also help to explain this effect. Higher energy features were observed at 295 cm^{-1} (A) and 355 cm^{-1} (B). The INS spectra of $C_{24}Rb$ with adsorbed H_2, HD and D_2 (15 K) have been modelled with the aid of a density functional calculation; see Table 6.2 [10].

Table 6.2 Assignment of the INS spectrum of $C_{24}Rb(H_2)_{1.0}$ at 15 K [10].

ω /cm^{-1}	Assignment
11	1, 0 (a)
21, 32	host lattice modes
89	in-plane translation
178	in-plane translation
258	1, $\pm$1(a)
355	out-of-plane translation

(a) rotational transition from the ground state (0, 0).

6.3.3 C_{60}

C_{60} molecules form an fcc lattice with two tetrahedral and one octahedral interstitial site. According to a neutron diffraction study adsorbed hydrogen molecules occupy octahedral sites and are randomly oriented [23]. INS spectra positions and assignments are reported in Table 6.3. The sorbed hydrogen was modelled as a quantum object trapped in a classical matrix, weakly 1-D in nature. Deviation from the free rotor model was small: 6 cm^{-1} splitting in the excited state for dihydrogen and dideuterium. The shift in the overall level was three times greater for dihydrogen than dideuterium and depended critically on the zero point motion. The coupling between the rotational and the vibrational motions of dihydrogen or dideuterium (zero-point motion) gave rise to a downward shift and additional splittings of the rotational levels.

Table 6.3 INS of H_2 and D_2 in C_{60} [23].

H_2 (40% (a))/C_{60}			D_2(40% (a))/C_{60}		
ω/ cm^{-1}	Assignment	Av. (shift) (b)	ω/ cm^{-1}	Assignment	Av. (shift) (b)
112, 118 (c)	$J(1\leftarrow0)$	116 (2.8)	60 (c)	$J(1\leftarrow0)$	58 (1.2)
108, 115, 120	Rotation + translation		54	$J(1\leftarrow0)$	
			74	Translation (d)	

(a) Of the maximum achieved loading. (b) Weighted average transition and shift relative to the gas phase value. (c) Neutron energy gain. (d) D_2 molecule vibrates against the C_{60} lattice.

The apparently simple spectrum observed under low resolution was found to be much richer at high resolution. One feature, at 234 cm^{-1}, was difficult to assign to rotational transitions and another, at 355 cm^{-1}, impossible. This was explained by considering the complete set of roton-phonon energy levels of the hydrogen molecule in a confining potential [24]. The observed neutron energy *gain* spectrum [23] destroys one translational phonon and converts one ortho-hydrogen to para-hydrogen thus accounting for the observed energy of *ca* 234 cm^{-1}, see Fig. 6.9. The temperature-dependence of the intensity of this transition between 30 K and 150 K corresponded with a thermal activation energy of 119 cm^{-1}. The translational vibrations of the dihydrogen are thus roughly accidentally degenerate with $2B_{rot}$. This unusual assignment is supported by the lack of recoil effects seen in the spectra.

6.4 Dihydrogen in microporous oxides including zeolites

There have been a number of INS studies of dihydrogen sorbed by microporous oxidic materials, including zeolites. As with carbon nanotubes, there is some interest in the use of these materials for dihydrogen storage. A number are also used as catalysts where the interaction with dihydrogen is the first step in their reduction. Our starting point, as a case study, is dihydrogen in a cobalt aluminophosphate.

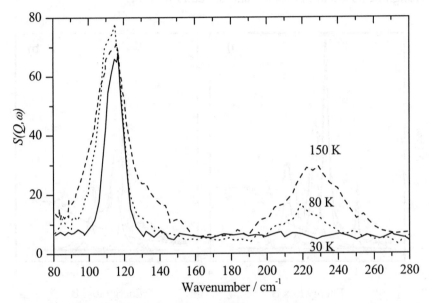

Fig. 6.9 Temperature-dependent neutron energy gain spectrum of dihydrogen in C_{60} (BT4, NIST). Reproduced from [23] with permission from the American Institute of Physics.

6.4.1 Dihydrogen in a cobalt aluminophosphate

Cobalt containing microporous materials are important selective oxidation catalysts in, for example, liquid phase oxidation of alkenes with dioxygen [25,26] and selective oxidation of cyclohexane to cyclohexanone [27]. The synthesis of, for example, cobalt aluminophosphate, CoAlPO⁻18, Co(III)$_{4.5}$(Al$_{13.5}$P$_{18}$O$_{72}$) involves reduction of cobalt(III) to cobalt(II) by dihydrogen. [28,29]. The CoAlPO⁻18 provided an opportunity to study the binding and activation of dihydrogen at a well defined site in a microporous oxide-type material [30, 31].

The INS spectrum of CoAlPO-18 dosed with dihydrogen at 20 K, *ca* 1 H_2 per cavity, is in Fig. 6.10(a) and after desorption between 25 K and 50 K of weakly adsorbed dihydrogen in Fig. 6.10(b) [30,31]. Assignments are in Tables 6.4 and 6.5 and Fig. 6.11.

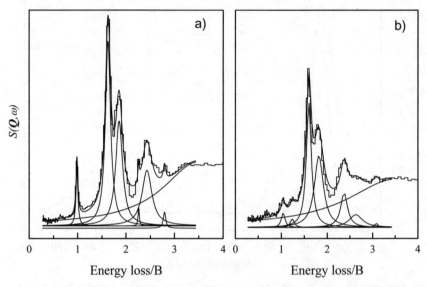

Fig. 6.10 INS spectrum of H_2 in CoAlPO-18 showing the rotational transitions [30], note the energy scale, units of the rotational constant *B*. The peaks are fitted to Lorentzians: (a) before desorption and (b) after desorption of weakly bound dihydrogen.

Table 6.4 Assignment of the rotational transitions in the INS spectrum of H_2 in CoAlPO-18 [30] *before* desorption of weakly bound H_2.

ω / cm^{-1}	ω / B (a)	Relative area	Transition (b)	Assignment
58	0.98	0.125	1, 0	Co Site (1-D type)
96	1.61	1.000	1, 1	Al site (2-D type)
107	1.79	0.896	1, −1	
133	2.24	0.055	1, 0	
142	2.39	0.584	1, 0	
165	2.85	0.116	1, 1	Co Site (1-D type)

(a) given to aid comparison with Figs 6.10 and 6.11. (b) from (0,0)

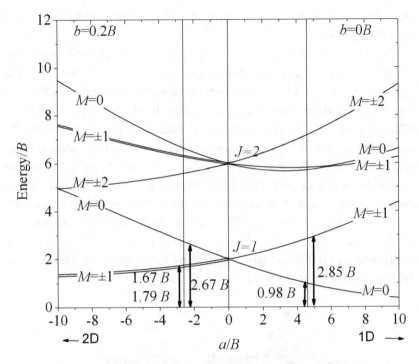

Fig. 6.11 Rotational states of constrained H_2 showing energy, in units of the rotational constant B, *vs* the perturbation parameter, with $b = 0.2\ B$, for $a < 0$ (2-D type) and $a > 0$ (1-D type). Vertical arrows locate peaks in the INS spectra of Fig. 6.10. Assignments are in Table 6.4 [30].

Table 6.5 Assignment of the rotational modes remaining in the INS spectrum of H_2 in CoAlPO-18 *after* desorption of weakly bound dihydrogen at 30 K.

$\omega\,/\,cm^{-1}$	Relative area	Transition (a)	Assignment
61	0.085	1, 1	Co Site (2-D type)
73	0.044	1, −1	
94	0.970	1, 1	Al Site (2-D type)
107	1.000	1, −1	
139	0.370	1, 0	Co Site (2-D type)
181	0.038	1, 0	

(a) from (0,0)

In Fig. 6.10(a) the lowest energy band corresponded to *ca* 1 B_{rot} (58.4 cm^{-1}). Since the lowest energy transition of unperturbed dihydrogen is

$2B_{rot}$, the sorbed hydrogen could not have been behaving as an unconstrained rotor.

The INS spectra were interpreted with the aid of the orientational potential models discussed above, Eq. 6.20 [10]; Fig. 6.4 has been amended, see Fig. 6.11, to show how the spectral transitions observed in Fig. 6.10 could be interpreted as arising from molecules trapped in different potentials on the surface and both 2-D and 1-D type models were invoked. The interaction energies were modelled by calculating the total energy of the system for a variety of geometrical configurations and trajectories using density functional theory. It was considered that the more strongly bound dihydrogen were those molecules interacting with CoAlPO-18 through the Co(III) centre. Both hydrogen atoms were involved and the molecule acted as a 2-D type. The H–H bond was lengthened by ca 2% and so weakened by the interaction. The proposed structure is shown in Fig. 6.12.

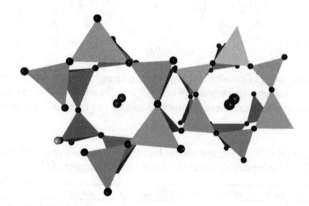

Fig. 6.12 The H_2 molecule in micoporoous CoAlPO-18. The tetrahedra represent alternate [MO_4] (M = Al or P) and [CoO_4] units. The H_2 molecules are shown as black spheres associated with cobalt.

6.4.2 Dihydrogen in zeolites

Hindered rotations of the hydrogen molecule adsorbed in zeolites represent a sensitive probe of their adsorption sites. Systematic studies of the INS spectra of hydrogen in different zeolites can lead to an approximate identification of these sites [32]. Some INS peaks and assignments are listed in Table 6.6.

The INS spectra of dihydrogen, dideuterium and hydrogen deuteride (HD) adsorbed in dehydrated calcium-exchanged NaZY to a loading of about one molecule per supercage were measured for both neutron energy loss and energy gain [32]. For HD there are no spin restrictions and HD readily decays into its rotational ground state; thus HD in contrast to H_2 and D_2 shows no neutron energy gain spectrum. INS peaks and probable assignments are given in Table 6.6.

Table 6.6 INS peaks (ω/ cm^{-1}) and assignments for dihydrogen in zeolites.

Ca/NaZY [32] (a)				Co/NaZA[33](b)		
H_2 (D_2, HD)	Assignment	Rotational barrier H_2 V_2/kJ mol^{-1}	Comment	H_2	Assignment	Comment
27	Hindered translation					
27	Hindered rotation	6.36		31	$J(1\leftarrow0)$	Rotational tunnel splitting
45 (-, 22)	$J(1\leftarrow0)$	4.27	H_2 adsorbed on Ca^{2+}, A-site			
62 (12, 41)	Rotation	3.06	H_2 adsorbed near Na^+	56		H_2 adsorbed near Na^+
89 (32, 62)		1.51				
141 (97,-)	Coupled perpendicular displacement and rotation			123		Vibration of bound H_2
				222		

(a) Part Ca exchanged NaZY. (b) Part Co-exchanged NaZA.

The hindered rotation was finally assigned using a 1-D type model (although several alternatives were discussed and the complexity of the spectra may hint at a failure of simple models like those developed here, §6.1).

Further insight into the interaction with the zeolite was gained by considering the isotope shift of HD transitions compared with dihydrogen. For HD the rotational constant depends on whether the HD molecule rotates about its centre of mass or its bond centre. For rotation about the centre of mass $B_{rot}^{HD} = 3/4\ B_{rot}^{HH}$; about the bond centre $B_{rot}^{HD} = 2/3\ B_{rot}^{HH}$. The observed ratio corresponded to the centre of mass axis, as would be expected for a physisorbed molecule.

Dihydrogen adsorbed by a partially cobalt-exchanged NaZA at 50 K to a concentration of 0.5 H_2 per supercage was considered to be bound end-on to the cobalt cations, pointing along a body diagonal in the direction of the electrostatic field lines of the cavity [33], a 1-D type model, as in Table 6.4. The bound hydrogen molecule performed 180° re-orientations with a barrier of 5.3 to 6.6 kJ mol^{-1}. The peak at 31 cm^{-1} (Table 6.6) is then the transition to ($J = 1$, $M = 0$).

6.4.3 Dihydrogen in Vycor, nickel(II) phosphate and a zinc complex

The INS spectra of dihydrogen in these materials provide further examples of the utility of INS for studying dihydrogen binding in micro- or nanoporous materials.

The peak at 118 cm^{-1}, was assigned to the essentially unperturbed free rotor, for dihydrogen at the pore centre of Vycor glass [34] (a prototypical mesoporous material which has been used extensively in studies of surface interactions due to its large interfacial pore surface area (ca 200 m^2 g^{-1})). Again shifted rotor peaks are interpreted to indicate rotational hindering, for example the peak at ca 81 cm^{-1} of dihydrogen in Vycor was assigned to rotationally hindered H_2 molecules strongly bound to the pore surface, for a 1-D type model [35]. These data were taken in neutron energy gain and were thus free of recoil effects.

For dihydrogen dosed nanoporous nickel(II) phosphate the rotational tunnelling peak appeared at 12 cm^{-1}, suggesting a 1-D type model [36]. This energy was lower than typical energies observed for dihydrogen adsorbed at cations in zeolites (Table 6.6) and suggested a stronger overall interaction. For higher loadings (from 1 to 3 and 6 H_2/unit cell), the 12 cm^{-1} peak broadened probably because of the presence of several similar binding sites around the Ni^{2+} cation.

INS spectroscopy of adsorbed dihydrogen molecules in the zinc complex [Zn$_4$(BDC)$_3$O] (BDC = 1,4-benzenedicarboxylate) indicated the presence of two well-defined binding sites, at zinc and at the BDC Linker 37. (Zinc complexes with a cubic three-dimensional extended porous structure will adsorb dihydrogen up to 4.5 wt% and have been proposed as hydrogen storage materials in, for example, fuel cell applications.) The authors provided only a sketch of the potential sensed by the adsorbed hydrogen but the 2-D type could be rejected from intensity arguments. The interpretive model is thus the 1-D type but the several adsorption sites serve to complicate this picture. Peak positions and the original assignments are given in Table 6.7.

Table 6.7 INS spectra and assignments for hydrogen absorbed by [Zn$_4$(BDC)$_3$O].

ω / cm^{-1}	Assignment(a)	Binding site
36	(2←1)	BDC
61	(2←1)	Zn
83	(1←0)	Zn
98	(1←0)	BDC
118	(1←0)	solid H$_2$
125	(2←0)	BDC
141	(2←0)	Zn

(a) These numbers, from [37], are neither J nor M

The rotational barrier associated with zinc was 1.67 kJ mol^{-1} and with BDC 1 kJ mol^{-1}, higher than single-walled carbon nanotubes (0.105 kJ mol^{-1}). The width of the rotational band (4 cm^{-1}) was much less than for carbon (20 cm^{-1}) showing a lower dihydrogen mobility and hence stronger binding in [Zn$_4$(BDC)$_3$O] than carbon.

6.5 Dihydrogen complexes

In the passage from physisorbed dihydrogen to chemisorbed hydride at some point there must be a stage where dihydrogen is bound to the active centre but the H–H bond is still present. For many years it was believed that this was a transition state, so could not be trapped and studied. This changed in 1984 with the announcement of a series of complexes that contained a sideways-on, η^2-bound, dihydrogen ligand [38]. The history of their discovery and a comprehensive review can be found in [39].

The bonding in the complexes is similar to the Dewar-Chatt-Duncanson description [40] of bonding in organometallic complexes where there is donation from a filled bonding orbital of the ligand to empty d-orbitals of the metal and back donation from filled d-orbitals to empty non-bonding or antibonding orbitals of the ligand. Dihydrogen is unique in that donation is from the filled σ-bonding orbital and back donation occurs to the empty σ^*-antibonding orbital. Both of these result in a weakening of the H–H bond, so the bond length increases from 0.746 Å in gaseous dihydrogen to typically 0.75–0.90 Å although 'super-stretched' bond lengths of 1.36 Å are known, almost double that of free dihydrogen. The increase in bond length would also require that the vibrational frequency of 4160 cm^{-1} should fall and this is found: values in the range 2000—3200 cm^{-1} are known, although 2600—3000 cm^{-1} are more typical [41].

The complexes provide a bridge between physisorbed and dissociated states of dihydrogen. As such they can be viewed from two directions: as an example of dihydrogen rotating in a strong potential or as a conventional complex. Since both viewpoints are describing the same phenomenon they should be equivalent. However, we shall see that both viewpoints are incomplete.

We will first consider the complexes as dihydrogen in a potential, the rotational model. The appropriate model is that of a perturbed planar rotor in a deep attractive field, the 2-DP type (§6.2.1.1). The energy level diagram is given in Fig.6.13. This a development of Fig. 6.4 that had both the parameters a and b small. In this case both a and b are large so the potential is now a waisted, figure-of-eight-like shape. This is chemically reasonable because the dihydrogen occupies a well-defined position in the complex and the 'cups' of the potential give the hydrogen positions.

Transitions between all the levels are allowed but because the intensities of para- to para- and ortho- to ortho- transitions depend on the coherent cross section of hydrogen they will be unobservable and only para- to ortho- or ortho- to para- transitions will be observed. Thus the expected transitions are: $\{J, M (1,-1\leftarrow0,0)\}$, $\{J, M (1,+1\leftarrow0,0)\}$, $\{J, M (2, -2\leftarrow1,-1)\}$, $\{J, M (2, +2\leftarrow1,-1)\}$ and $\{J, M (3, \pm3\leftarrow0,0)\}$. From Fig. 6.13 the first transition will be at very low energies (< 20 cm^{-1}) and even at very low temperature there will be a significant population of the $(1,-1)$ level and this gives rise to the two transitions to the $J = 2$ levels.

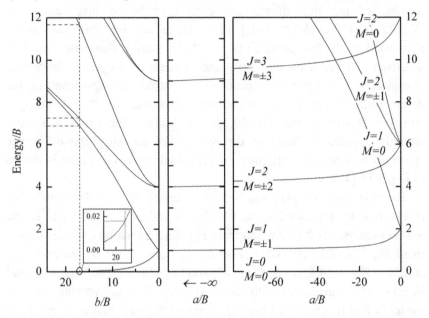

Fig. 6.13 The energies of the rotational states of dihydrogen as a perturbed planar rotor in a deep attractive field, the 2-DP type. The dashed vertical line shows the transitions expected for $[W(CO)_3(H_2)\{P(cyclohexyl)_3\}_2]$ with a tunnel splitting of 0.89 cm^{-1} (inset).

The lowest energy rotational transition is so low that it occurs in the energy regime expected for tunnelling transitions. The transition energies depend approximately exponentially on the barrier height hence they provide a sensitive probe of the bonding. The barrier height is a function of B thus extraction of the values requires a knowledge of the H–H bond distance. This varies from one complex to the next so structural information, generally from neutron diffraction, is needed. The observed frequencies mostly lie in the range 1–10 cm^{-1} with barriers to rotation of 1–20 kJ mol^{-1} [39,41].

Thus the spectrum in the vibrational spectroscopy region (> 100 cm^{-1}) would be expected to consist of four peaks (ignoring any contribution from the other ligands present). Transitions to higher J, M values will also occur but will be difficult to observe because of the effects of recoil and the Debye-Waller factor.

One of the first dihydrogen complexes to be discovered was [W(CO)$_3$(H$_2$){P(cyclohexyl)$_3$}$_2$] [38]. There are 66 hydrogen atoms associated with the phosphines so these would dominate the INS spectrum. To avoid the difficulty and expense of preparing deuterated phosphines, the difference spectrum was used: the spectrum of the dideuterium complex was subtracted from that of the dihydrogen complex. Assuming that the dihydrogen (dideuterium) modes are not strongly coupled to those of the rest of the complex, the difference spectrum will contain only those of the dihydrogen. The result is shown in Fig. 6.14 [42].

For this complex [42] r_{HH} = 0.82 Å hence B_{rot}^{HH} = 49.5 cm^{-1}, Eq. (6.11), (c.f. for gaseous H$_2$, r_{HH} = 0.746 Å and B_{rot}^{HH} = 59.3 cm^{-1}). The lowest transition is 0.89 cm^{-1} so from Fig. 6.13 transition energies of 340 (6.87 B), 359 (7.25 B), 576 (11.65 B), 689 cm^{-1} (13.9 B) are predicted and from Fig. 6.14 four bands are apparent at 325, 385, 462 and 640 cm^{-1} in reasonable agreement with the prediction. The disadvantage of this model is that it ignores all the vibrations of the coordinated dihydrogen except the torsion, it assigns the entire spectrum in terms of rotational transitions of dihydrogen.

The alternative view is to consider the complex as a compound and to treat the vibrations of the dihydrogen in the conventional manner, the vibrational model. Gaseous dihydrogen has six degrees of freedom: one internal (the H–H stretch), two rotational and three translational. When bound the rotational and translational degrees of freedom become frustrated and appear as internal modes of the complex as shown in Fig.6.15.

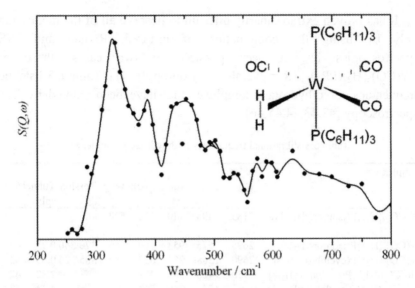

Fig. 6.14 The difference INS spectrum (FDS, LANSCE) of complexed dihydrogen. The points are $[W(CO)_3(H_2)\{P(cyclohexyl)_3\}_2] - [W(CO)_3(D_2)\{P(cyclohexyl)_3\}_2]$, Redrawn from [42] with permission from the American Chemical Society.

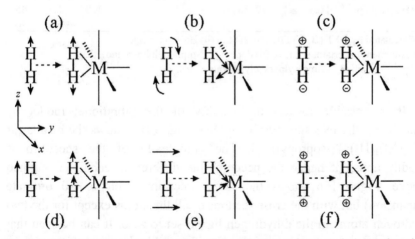

Fig. 6.15 How the six degrees of freedom of free dihydrogen become six internal modes of vibration of an η^2-coordinated dihydrogen molecule. (a) $\nu_{HH} \rightarrow \nu_{HH}$, (b) x axis rotation becomes the M–H$_2$ antisymmetric stretch (ν_{HMH}asym), (c) y axis rotation becomes the M–H$_2$ torsion, (d) z axis translation becomes the MH$_2$ rock, (e) y axis translation becomes the M–H$_2$ symmetric stretch (ν_{HMH}sym) and (f) x axis translation becomes the MH$_2$ wag.

It has proven extraordinarily difficult to observe all of the modes and only by using the combination of infrared, Raman and INS spectroscopies has it been possible in two cases, those of $[W(CO)_3(H_2)(PR_3)_2]$ (R = cyclohexyl, isopropyl) [41]. Table 6.8 lists the frequencies for dihydrogen complexes that have been studied by INS spectroscopy [42,43,44,45,46]

Table 6.8 Vibrational frequencies of dihydrogen complexes.

Complex	ω/cm^{-1}							
	ν_{HH}	ν_{HMH} asym	ν_{HMH} sym	Rock	Wag	Torsion	Tunnel split	Ref.
$[W(CO)_3(H_2)\{P(isopropyl)_3\}_2]$ (a) (b)	3183	1495	940	650	522	406		
$[W(CO)_3(H_2)\{P(isopropyl)_3\}_2]$	2695	1575	953	660	460	370/340	0.73	42
$[W(CO)_3(H_2)\{P(cyclohexyl)_3\}_2]$	2690	1568	951	650	462	385/325	0.89	42
$[Mo(CO)_3(H_2)\{P(cyclohexyl)_3\}_2]$	2950	1420	885	471			2.82	42
$[Cr(CO)_3(H_2)\{P(cyclohexyl)_3\}_2]$		1540	950	563			4.33	42
$[Fe(H)_2(\eta^2\text{-}H_2)\{P(ethyl)(phenyl)_2\}_3]$ (c)	2380		850	500	405	252/170	6.4	43
$[Fe(dppe_2)(H)(H_2)][BF_4]$ (d)			870	610	400	255/225	2.1	44
$[Rh(H)_2(H_2)(Tp^{3,5\text{-}Me})]$ (e)	2238					200	6.7	45 46

(a) Calculated by DFT (B3LYP/LANL2DZL) (b) Harmonic value.
(c)Assignment uncertain, see text, (d) dppe = (1,2-diphenylphosphino)ethane.
(e) Tp$^{3,5\text{-}Me}$ = tris-(3,5-dimethylpyrazolyl)borate.

It is possible to test the validity of the vibrational model by calculating the INS spectrum with DFT. Fig. 6.16a shows the results for $[W(CO)_3(H_2)\{P(isopropyl)_3\}_2]$. The complexity of the spectrum is readily apparent hence the need for the difference technique outlined above. Fig. 6.16b shows the same spectrum (with a ×10 ordinate expansion) but with the cross sections of all the atoms except for the two hydrogen atoms in the dihydrogen ligand set to zero. It can be seen that there are many modes that involve significant dihydrogen motion.

The calculated ν_{HH} of 3183 cm^{-1} is too high but this is the harmonic value, whereas the experimental value is an anharmonic value. This is also compounded by the tendency of DFT to overestimate frequencies. Scaling the frequency by the usual value of 0.97 and assuming that the anharmonicity correction is 5.2% as for gaseous dihydrogen gives a more reasonable value of 2926 cm^{-1}.

Inspection of the eigenvectors shows that the stronger modes generally agree with the literature assignments, see Table 6.8. However, it is also apparent that the torsion, wag and rock are strongly coupled to the W–C≡O linear bend deformation modes. This arises because as the heavy atoms move, to maintain the invariance of the centre-of-mass (a requirement for a normal mode) the hydrogens move in counterpoise.

Since they are much lighter, a correspondingly larger amplitude of motion is required and this results in significant INS intensity.

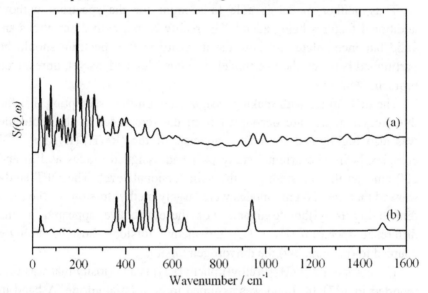

Fig. 6.16 The INS spectrum of complexed dihydrogen. in [W(CO)$_3$(H$_2$){P(isopropyl)$_3$}$_2$] calculated by DFT (B3LYP/LANL2DZL). (a) All atoms included and (b) only the dihydrogen ligand. Note that (b) is ×10 ordinate expanded relative to (a) and only the fundamentals (without phonon wings) are plotted, TOSCA parameters assumed.

Thus modes at 360 and 585 involve the wag, those at 386, 425 and 485 involve the rock and that at 457 cm^{-1} the torsion. A Wilson GF matrix method study of the cyclohexyl complex [42] came to similar conclusions, but considered that the modes were coupled to to the W–C stretches, which occur in the same region as the linear bends (§11.2.6). This was done by introducing an interaction force constant between the relevant modes, whereas the DFT results suggest the hydrogen's response is a simple mechanical reaction.

The disadvantage of the vibrational model is that it is unable to account for the splitting in the modes, particularly the torsion. This is treated in an *ad hoc* fashion by recognising that the ground state is split by the tunnelling transition that results in the interchange of the two hydrogen atoms in the dihydrogen ligand. The tunnel splitting is present in all the vibrational levels and increases with increasing transition energy. This 'failure' of DFT arises because it does not treat the proton as a quantum entity.

Thus neither model is able to account for the spectrum without additional features being added. The reality is that both descriptions are valid but incomplete and how the intensity in the spectrum should be partitioned between the two models is a complex and, as yet, unresolved problem.

The difficulties with making assignments on the assumption that the dihydrogen modes are decoupled from the remainder of the molecule was also highlighted by a DFT analysis of the $[Rh(H)_2(H_2)(Tp^{3,5-Me})]$ complex [46]. The original study [45] had assigned modes at 120 and 200 cm^{-1} as the transitions to the split torsional level. The DFT study showed that the 120 cm^{-1} modes were largely methyl torsions of the (3,5-dimethylpyrazolyl)borate ligand. For these to have appeared in the difference spectrum ($H_4 - D_4$), the methyl torsions must be strongly coupled to (presumably) the dihydrogen torsion.

The compound $[FeH_4\{P(ethyl)(phenyl)_2\}_3]$ is a curiosity that was first reported in 1971 [47] and was assumed to be a tetrahydride. A band in the infrared at 2380 cm^{-1} was noted but not assigned. A subsequent neutron diffraction structure [43] has shown that the complex is actually $[Fe(H)_2(\eta^2-H_2)\{P(ethyl)(phenyl)_2\}_3]$; the 2380 cm^{-1} band is the H–H stretch. Thus the complex contains both hydride and dihydrogen ligands and these readily interchange (fluxionality) to low temperature. The tunnel splitting is large, 6.4 cm^{-1}, which implies a rotational barrier of only 4.6 kJ mol^{-1}. This low barrier may be reesponsible for the fluxionality of the complex.

The INS spectrum of the complex is shown in Fig. 6.17. The bending modes of the hydrides are the strong bands at 680 and 720 cm^{-1}. The dihydrogen modes occur at somewhat lower energies than in the tungsten complexes, Table 6.8. This suggests a generally weaker metal-dihydrogen interaction, which would imply that the H–H stretch should be stronger than in the tungsten complex (2690 cm^{-1}) rather than much weaker (2380 cm^{-1}). The low frequency is also at odds with the H–H bond distance of 0.821 Å, the same as found in the tungsten complexes. In liquid xenon, the H–H stretch of $[Fe(CO)(\eta^2\text{-}H_2)(NO)_2]$ [48] is at 2690 cm^{-1} and the symmetric Fe–H$_2$ stretch is at 870 cm^{-1}.

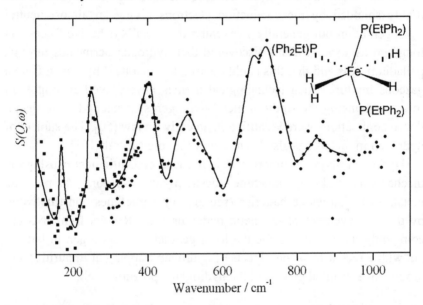

Fig. 6.17 The INS spectrum (FDS, LANSCE) of the $[Fe(H)_2(\eta^2\text{-}H_2)]$ fragment in $[Fe(H)_2(\eta^2\text{-}H_2)(PEtPh_2)_3]$. The points are the difference spectrum: $[Fe(H)_2(\eta^2\text{-}H_2)(PEtPh_2)_3]$ – $[Fe(D)_2(\eta^2\text{-}D_2)(PEtPh_2)_3]$. Redrawn from [43] with permission from the American Chemical Society.

To date (2004) there are no examples of dihydrogen ligated to a supported metal catalyst. However, there are tantalising hints that such species exist. Dihydrogen in Fe-ZSM [49] exhibits tunnel splittings of 4.2 and 8.3 cm^{-1} which are typical of the dihydrogen complexes. On the stepped surface of Ni(510), after it is reacted with hydrogen to generate a dense atomic hydrogen layer, there is good evidence that dihydrogen adsorbs intact at the steps. The prospects seem good that chemisorbed dihydrogen will be observed on a supported metal surface and clearly INS would be an excellent tool for its investigation.

6.6 Hydrogen in metals

Hydrogen dissolves in metals more or less easily to form a series of non-stoichiometric compounds. The hydrogen looses its electron to the metal in the case of the alkali metals (§6.7.2) or to the conduction band in the transition metals. In the transition metals the resulting proton is highly shielded by the conduction electrons giving an almost neutral hydrogen atom but generally weakening the metal's cohesive forces. As long ago as 1868 Graham discovered that hydrogen permeates through palladium foils and that this could be used to separate dihydrogen from a gaseous mixture. Such technological themes have played an important role in advancing the study of metal-hydrogen systems and more than a dozen applications were identified in an early review [50]. The subject of hydrogen in metals has also been reviewed more recently [51].

Hydrogen plays an important role in metal embrittlement causing the mechanical break-up, sometimes destructively, of, for example, the metals in rechargeable batteries (see below), sometimes constructively, by the improvement of magnetic materials [52]. Results from INS are commonly used to determine the local geometry of hydrogen in lattices but with a recent focus on aspects of hydrogen trapping at impurities and producing a detailed picture of the hydrogen's potential well.

6.6.1 The spectral characteristics

When hydrogen permeates a metal at room temperature, or above, it is evenly distributed throughout the bulk of the crystal (as a solid solution). Individual hydrogen atoms reside on interstitial sites and the crystal structure expands, whilst retaining its symmetry.

In close packed metal systems there are two sites available, see Fig 6.18: the constricted tetrahedral sites or the more spacious octahedral sites. However, even after only partly filling the lowest energy sites, the nature of which depends on the particular metal, other sites can begin to fill. Moreover, depending on its hydrogen content and temperature the hydrogen can 'precipitate' into a single region of the crystal, forming a specific phase. This often results in structural changes to the crystal and they then fracture [53].

Most metals have several phases (those of palladium are given, amongst other results, in Table 7.3). This can be of serious experimental concern since many phases are only distinguishable by their modest spectral differences.

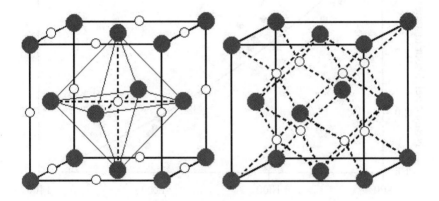

Fig. 6.18 The octahedral and tetrahedral sites available in a face centred cubic solid.

The volumes available to the hydrogen atoms are reflected in their characteristic frequencies, see Fig. 6.19. A truly tetrahedral site should give a single band at about 1200 cm^{-1} and an octahedral site a band at about 650 cm^{-1}. This easy distinction provides a simple, quick and often effective determination of hydrogen sites with high symmetry. It was the basis of much of the early INS work on hydrogen in metals.

6.6.1.1 Sites of low symmetry

Although dealing with high symmetry hydrogen sites is straightforward more sophisticated treatments become necessary as the symmetry is lowered. In niobium the tetrahedral sites in the metal are distorted to D_{2d} symmetry and the single transition at high frequencies, typical of tetrahedral sites is not observed. Rather in the solid solution regime, at low hydrogen concentration above 210 K, the INS spectrum consists of an A mode at 865 cm^{-1} and an E mode at 1315 cm^{-1}. (The weighted average, 1165 cm^{-1}, however, corresponds to the expected frequency for an F mode from a tetrahedral site.)

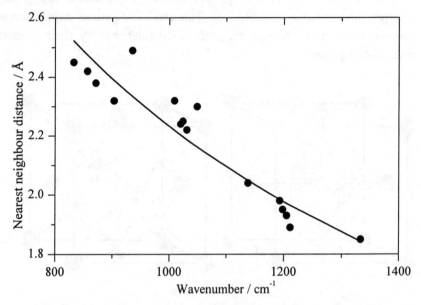

Fig. 6.19 The variation of the energy of the main INS spectral feature with nearest neighbour distance, r. The data are shown with a least squares fit, $\omega = 3334r^{-3/2}$. Reproduced from [53] with permission from Elsevier.

Further spectral structure is introduced when neighbouring oscillators interact. This leads to line spectra if the number of coupled oscillators is small and their local geometry defined, as with the metal hydrides (§6.7). Alternatively it leads to dispersion, when the number of coupled oscillators is large and the crystal geometry defined, see below [54—56]. When the geometry of oscillators is not well defined, regardless of coupling, heterogeneous band broadening effects are observed. Coupling effects are exposed by isotopic dilution experiments (§7.3.1.9). Occasionally quite exceptional dynamics are uncovered as with α-MnH$_{0.073}$, Fig. 6.20, where a very intense tunnelling mode was observed

[56]. The hydrogen tunnels between the two equivalent sites in a distorted octahedron.

In the ε-phase of niobium hydride, $NbH_{0.61}$, the additional structure observed in both the A and E modes remained in the spectrum of $NbH_{0.03}$ $D_{0.57}$ and the coupling model was dismissed [54]. An explanation in terms of minor structural differences between the neighbouring sites in this phase was favoured. The degree of detail that can be extracted from more complex materials than the simple metals is limited and the spectra of TbNiAl $H_{1.4}$ and UNiAl $H_{2.0}$ are cases in point [55].

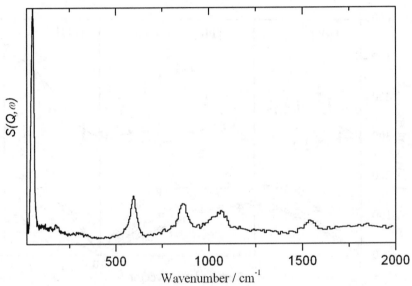

Fig. 6.20 The INS spectrum of α-$MnH_{0.0073}$, showing the very intense tunnelling transition at 50 cm^{-1}. Reproduced from [56] with permission from Elsevier.

The weaker the interaction between the metal and the hydrogen the more important are the inter-hydrogen forces in determining their dynamics. This leads to dispersion, a good example of which is that found in PdH, measured as its deuteride $PdD_{0.63}$ [58]. This dispersion is shown in Fig. 6.21 The low frequency acoustic modes, involving the Pd vibrations, have little hydrogen displacement and show only weakly in the INS spectrum of powdered PdH; however, the optic modes appear strongly, see Fig. 6.22 The relatively undispersed transverse optic modes, about 470 cm^{-1}, give the strongest feature whilst the dispersed longitudinal branch gives rise to a shoulder of intensity stretching out to 800 cm^{-1} [57].

Since the low frequency modes are so weak, phonon wing like combinations (§5.2.1.2) are irrelevant and the total mean square vibration of the hydrogen atom is dominated by the optic modes. Indeed, many hydrogen-in-metal systems are studied at room temperature with only modest spectral degradation. This single fact sets the INS spectra of hydrogen in metal systems clearly apart from those of molecular crystals.

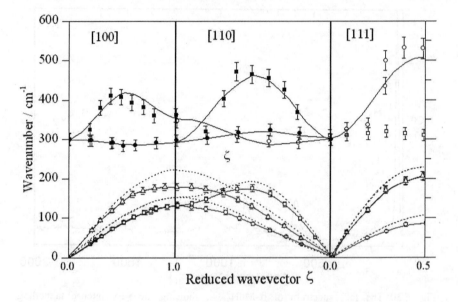

Fig. 6.21 The dispersion curves of PdD$_{0.63}$ the dashed lines show the dispersion curves of pure palladium. Notice how hydriding the metal has reduced the metal-metal force constants. Reproduced from [58] with permission from the American Physical Society.

6.6.2 Hydrogen trapping

Hydrogen in metals is trapped by all manner of lattice impurities, including other interstitials and lattice defects. Trapping reduces the mobility of hydrogen but its solubility is usually improved. This prevents precipitation, and the crystal degradation that it causes is avoided.

In tantalum hydride, $TaH_{0.086}$, at room temperature the solid solution, α-phase, hydrogen occupies D_{2d} 'tetrahedral' sites and shows two broad transitions in the INS, at 911 and 1315 cm^{-1} [59]. As the temperature falls the β-phase precipitates such that at *ca* 100 K this is complete, its spectrum is sharper and the lower transition is at 968 cm^{-1}.

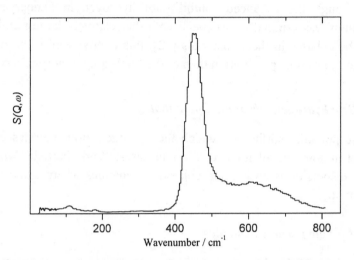

Fig. 6.22 The low energy INS spectrum of β-PdH. Reproduced from [57] with permission from the American Physical Society.

When nitrogen impurities are introduced into the tantalum (they reside on octahedral sites) the hydride, $TaN_{0.006}H_{0.003}$, behaves differently. Firstly, they retain their 'tetrahedral' frequencies, discounting suggestions that the hydrogen atoms have moved to octahedral sites. Moreover, this spectrum remains unchanged down to the lowest temperatures, 1.5 K, showing that the solid solution phase has not precipitated [59].

Substitutional impurities produce similar effects. The INS spectrum of $Nb_{0.95}Mo_{0.05}H_{0.03}$, at room temperature, shows evidence of hydrogen on 'tetrahedral' sites, with transitions at 863 and 1226 cm^{-1}. This compares well with INS results from the pure niobium, 863 and 1315 cm^{-1}. Similarly when cooled both $Nb_{0.95}Mo_{0.05}H_{0.03}$ and niobium produce precipitates, here the low frequency band moves to 944 cm^{-1}. At higher impurity levels, $Nb_{0.8}Mo_{0.2}H_{0.05}$, the precipitation is again frustrated by the hydrogen's interaction with the impurity [60].

The lanthanum in $LaNi_5$ is essentially a high level impurity in these technologically important battery materials, again serving to maintain hydrogen solubility. INS work on these systems showed that at low concentrations hydrogen occupies both 'octahedral' and 'tetrahedral' sites [61].

Although the enhanced solubility of hydrogen in nanocrystalline palladium was originally associated with the relatively large number of grain boundaries in these materials [62], this is now seen to be more a question of sample preparation than of the final sample morphology [63].

6.6.3 The hydrogen vibrational potential

The potential surface on which the hydrogen atom vibrates can be probed in some detail for two different cases; single particle dynamics where dispersion is absent and collective dynamics where dispersion is important.

6.6.3.1 Single particle dynamics

Single particle dynamics occurs where there are no interactions between hydrogen atoms in the metal so the oscillator mass, μ, is close to unity and the INS spectra extend over several harmonics (§5.2.1.2).

Table 6.9 The family of transitions for an isotropic oscillator and their relationship to the molecular spectroscopic and bound phonon conventions.

1 quantum	2 quanta		3 quanta		
				$(2, 1, 0)$	
$(1, 0, 0)$	$(2, 0, 0)$	$(1, 1, 0)$	$(3, 0, 0)$	$(2, 0, 1)$	
$(0, 1, 0)$	$(0, 2, 0)$	$(1, 0, 1)$	$(0, 3, 0)$	$(1, 2, 0)$	$(1, 1, 1)$
$(0, 0, 1)$	$(0, 0, 2)$	$(0, 1, 1)$	$(0, 0, 3)$	$(1, 0, 2)$	
				$(0, 2, 1)$	
				$(0, 1, 2)$	
F_2	F_2	$E + A_1$	F_2	$F_1 + F_2$	A_1
Phonon (a)	bi-ph	ph+ph	tri-ph	bi-ph+ph	ph+ph+ph

(a) Abbreviated ph.

There are only three vibrational modes, the hydrogen vibrations along the Cartesian coordinates (x, y, z) (§2.5.2.2). Transitions are assigned according to the final quantum number of each mode, thus the ground state is (0, 0, 0), the excitation of two single quantum transitions is (1, 1, 0) and the double excitation of a single vibration is (2, 0, 0). The family of excitations for an isotropic oscillator is shown in Table 6.9, along with their relationship to other conventions (§5.2.1.2). The results from the anisotropic, D_{2d}, system ε-NbH$_{0.7}$ are given in Table 6.10.

Table 6.10 The harmonically estimated and observed transitions in ε-NbH$_{0.7}$ [64].

ω / cm^{-1}			Assignment
Harmonic estimate	INS observed	Difference	
928	928	0	0,0,1
1291	1291	0	1,0,0 and 0,1,0
1856	1775	81	0,0,2
2219	2186	33	1,0,1 and 0,1,1
2784	2339	445	0,0,3
2582	2605	- 23	2,0,0 and 0,2,0
2582	2783	- 201	1,1,0
3147	3226	- 79	1,0,2 and 0,1,2
3147	3630	- 483	1,1,1
3873	4033	- 160	3,0,0; 0,3,0 and 2,1,0; 1,2,0

The observed transitions can be used to parameterise the potential, either according to simple models or, in terms of the symmetry adapted spherical harmonics for the most general potential [51]. The advantage of this approach is that the wavefunctions of the hydrogen in its excited states can be extracted. Further the full form of the Scattering Law (see Fig. 2.4) for any transition can be calculated. These can be compared with those observed on direct geometry spectrometers (§5.3.2.1) [64].

The results of such measurements would be unsurprising to the molecular spectroscopist where the system is, like ZrH$_2$, reasonably harmonic but some systems show significant deviation from harmonicity, like NbH$_{0.3}$, see Table 6.10 [64]. A spectral characteristic of these systems is the unusually weak transitions at the higher energies. It appears that the high energy wavefunctions are more extended than the equivalent harmonic wavefunctions. The simplest explanation is that, above some energy, the potential well opens out significantly.

6.3.3.2 Collective dynamics

In systems where inter-hydrogen forces are important, usually because of the close proximity of hydrogen neighbours, dispersion is clearly seen in the one quantum spectrum, as in Fig. 6.22. The higher order transitions from such systems would not then be expected to appear as sharp transitions, since the two phonon spectrum should resemble the self convolution of the one phonon spectrum (§2.6.3). Occasionally, however, sharp transitions are observed, at $E_{(2\leftarrow0)}$, just below the energy of $2E_{(1\leftarrow0)}$ If Δ is the full width at half height of the $(1\leftarrow0)$ transition. Then, provided that

$$2E_{(1\leftarrow0)} - E_{(2\leftarrow0)} > 2d \qquad (6.23)$$

is satisfied, the sharp leading transition may be assigned as a bound two phonon event, a bi-phonon. A bi-phonon is localised in the lattice, as is the excitation of an overtone in a molecule. Similarly, tri-phonons can be defined [65].

Structure in the high energy region of known dispersive systems can also result from an anisotropy appearing in the higher energy region of the vibrational potential. This is the account offered for the presence of an unexpected band that appears at 890 cm^{-1} in the two quantum region of PdH, since Eq. (6.23) is not satisfied. This band is effectively hidden in INS spectra of PdH powders but is exposed when Q lies along the [100] direction in single crystal experiments. The hydrogen site in PdH has O_h symmetry and its ground state symmetry is isotropic; anisotropy in the higher states was not expected. It is explained by the [100] axis lying in the direction of the hydrogen's closest Pd neighbour [57].

6.7 Metal hydrides

Hydrogen forms binary compounds of the type MH$_y$ or M$_x$H$_y$ of varying stability with all the main group elements except helium and neon. All of the lanthanoids and actinoids that have been studied also form hydrides as do most of the transition metals [66]. There is a rich variety of bonding types ranging from ionic hydrides through hydrogen-in-metals to covalent compounds. In this section we discuss salt-like binary (§6.7.1) and ternary metal hydrides (§6.7.2).

Binary and ternary metal hydrides are of interest for hydrogen storage applications [67]. One of the challenges associated with the use of hydrogen as a fuel is the need for a safe and efficient means of transportation and storage. Solid compounds where hydrogen is stored by a chemical reaction and released by heating are an attractive option. The requirements for such a material are stringent: hydrogen storage and release must occur at temperatures and pressures close to ambient and the amount of hydrogen stored must be significant. The United States Department of Energy has set a minimum quantity of 6.5 wt% hydrogen for any material to be considered commercially viable. The salt-like binary hydrides (§6.7.1) formed by the alkali metals and alkaline earths meet the hydrogen storage criterion but they have high hydrogen release temperatures and their chemical reactivity is a disincentive to their use. This has prompted a search for new hydrides with more favourable properties which has resulted in the synthesis of ternary (and higher) metal hydrides (§6.7.2).

6.7.1 Alkali metal hydrides

The binary alkali metal hydrides, MH (M = Li, Na, K, Rb, Cs) crystallise with the sodium chloride structure and are ionic [68]. Their INS spectra, Fig. 6.23, show peaks related to the density of transverse and longitudinal optic states due to the antiphase motions of the hydride ion and the M^+ cations in the lattice unit cell [69] (§4.3). The first overtones were also seen. The acoustic bands, at lower energy transfer, were weaker than the optical bands since the acoustic bands arise from in-phase motions of the hydride ion and the cations; the hydrogen mean square displacement is smaller in the acoustic than in the optical modes.

For lithium hydride an early (1981) lattice dynamics simulation [70] was in better agreement with the spectrum recorded on the TOSCA spectrometer than an older (1965) INS spectrum [71]. The INS spectra of the sodium, potassium, rubidium and caesium hydrides were accurately simulated by a totally *ab initio* method using density functional theory, pseudopotentials and a plane-wave basis. The agreement with the data was much better than with a lattice dynamics calculation using semi-empirical interatomic potentials.

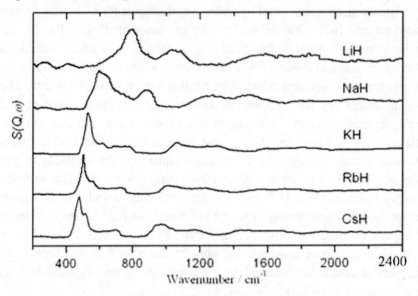

Fig. 6.23 INS spectra of alkali metal hydrides at 20 K [69].

These INS studies of the alkali metal hydrides provide an excellent example of the careful analysis of INS experimental data. It includes the application of corrections for multiple scattering, neutron absorption and heavy-ion scattering; the extraction of quantities related to the hydrogen dynamics (the hydrogen mean square displacement, mean kinetic energy and the hydrogen Einstein frequency) and provided the density of vibrational states for each type of atom, shown individually and *ab initio* modelling of the full INS spectra.

6.7.2 Ternary metal hydrides

Ternary metal hydrides have the general formula A_xMH_y (A = alkali or alkaline earth metal, M = metal, $1 \leq x \leq 4$, $2 \leq y \leq 9$), although considerable variation can occur. Thus compounds with lanthanides (particularly europium) as the counter ion are known, as are quaternary metal hydrides where there are two different counter ions. M is often a transition metal, although the compounds Na[BH$_4$] and Li[AlH$_4$] have long been used in organic chemistry as efficient reducing agents. There are also compounds with magnesium as the central metal.

The transition metal compounds are usually made by either reaction of the elements or reaction of an alkali metal hydride and the metal at elevated temperature and pressure. Very high temperatures and pressures (770 K and 1500–1800 bar) result in the formation of high oxidation states of second and third row transition metals. The materials have been extensively characterised by X-ray and neutron diffraction and several reviews are available [72,73,74,75]. There is a large diversity of structural types but the central [MH$_y$]$^{n-}$ ion generally has the classical structure of transition metal complexes; thus $y = 2$ is linear, $y = 3$ is triangular, $y = 4$ is tetrahedral or square planar, $y = 5$ is square-based pyramidal, $y = 6$ is octahedral and $y = 9$ is tricapped trigonal prismatic.

While several hundred ternary metal hydrides are now known, comparatively few have been investigated by vibrational spectroscopy. The hydrides are a textbook example of the complementarity of infrared, Raman and INS spectroscopies. The high symmetry in many of the complexes means that with infrared and Raman spectroscopies we are able to observe only some of the modes, but in most cases this restriction allows an unambiguous assignment of these bands. INS allows all the hydrogen-related modes to be observed and in combination with modelling completes the assignments. As an example we will consider Mg$_2$[FeH$_6$] [76].

At room temperature Mg$_2$[FeH$_6$] crystallizes in the cubic space group $F m\bar{3}m \equiv O_h^5$ (number 225) with four molecules in the unit cell. There is one formula unit in the Bravais cell and the ions all lie on special sites: the Mg^{2+} ions on tetrahedral T_d sites and the [FeH$_6$]$^{4-}$ ions on octahedral O_h sites. Using the correlation method [77] we can classify the vibrations as in Table 6.11. In Fig. 6.24 we show the infrared, Raman and INS spectra of Mg$_2$[FeH$_6$].

The assignment of the stretching modes is more complicated than might be expected. The band at 1873 cm^{-1} in the Raman and INS spectra is the A_{1g} mode, the band at 1740 cm^{-1} in the infrared and INS spectra is the T_{1u} mode.

Table 6.11 Classification of vibrations of $Mg_2[FeH_6]$ and $Mg_2[FeD_6]$.

Vibration	Representations (a)
Fe—H/D stretches	A_{1g} (R) + E_g (R) + T_{1u} (IR)
Fe—H/D bends	T_{2g} (R) + T_{1u} (IR) + T_{2u} (ia)
Libration	T_{1g} (ia)
Optic	T_{2g} (R) + T_{1u} (IR)
Acoustic	T_{1u} (ia)

(a) Activity shown in brackets: R = Raman active, IR = infrared active, ia = inactive in both Raman and infrared. Note *all* vibrations are allowed in the INS spectrum.

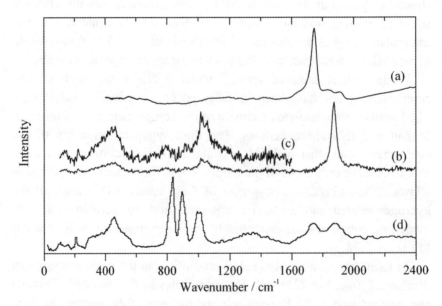

Fig. 6.24 (a) Infrared, (b) Raman, (c) Raman ×3.5 to show the librational mode at 460 cm^{-1} and the split T_{2g} mode at 1019/1057 cm^{-1} and (d) INS spectra of $Mg_2[FeH_6]$ [76].

In the Raman spectrum, the E_g mode is usually much weaker than the A_{1g} mode, so the failure to observe it, is not surprising; however, it should have significant INS intensity but only two bands are observed. We note that the ratio of the intensities of the A_{1g}, E_g and T_{1u} modes should be approximately 1:2:3. Thus for the INS feature at 1878 cm^{-1} to be of similar intensity to that at 1740 cm^{-1}, the A_{1g} and E_g modes must be accidentally degenerate.

The assignment of the bending modes is straightforward. The split band at 1019/1057 cm^{-1} in the Raman and INS spectra is the T_{2g} mode. The band at 899 cm^{-1} in the INS spectrum is the T_{1u} mode and should be present in the infrared spectrum, however, no band is observed even in concentrated samples (although a band is observed in the infrared spectrum of the deuterated compound [76]). The band at 836 cm^{-1} in the INS spectrum is the T_{2u} mode.

As all of the internal vibrations have been accounted for, it follows that the features below 600 cm^{-1} are the external modes and since they exhibit significant INS intensity it is likely that they relate to the metal hydride ion. The external modes are expected to occur (in order of energy): librational > (translational) optic > (translational) acoustic. Thus the natural assignment is that the strong, broad mode in the INS spectrum at 460 cm^{-1} that exhibits a deuterated counterpart at 324 cm^{-1} ($= 460/\sqrt{2}$) is the librational mode. A small rotation of the $[FeH_6]^{4-}$ ion would result in a large amplitude motion of all six hydrogen atoms, accounting for its high intensity and since it is the hydrogen atoms that move, deuteration would be expected to give the full isotope shift. In contrast to the libration, since a translation involves displacement of the ion from its lattice position, any isotope shift would be expected to be small: $(62/68)^{1/2}$ $= 0.95$. Hence the weak band near 220 cm^{-1} in the Raman and INS spectra that shows virtually no isotope shift is assigned to the T_{2g} optic translation, that involves mainly motion of the Mg^{2+} ions. In the absence of lower frequency infrared data, the assignment of the last two features in the INS spectrum is unclear. Some of the intensity below 200 cm^{-1} is due to the acoustic mode and dispersion in this mode could give rise to a feature with the observed shape. However, the remaining optic translational mode is also in this region and will contribute to the intensity.

This analysis has assigned all of the bands in the spectra on the basis of O_h symmetry for the parent molecules. For the perdeuterated complex the assignment is completely consistent with the spectra. For the hexahydrido complex there are some difficulties. These are: the splitting of the T_{2g} bending mode at ~ 1000 cm^{-1} in both the Raman and INS spectra, the presence of the formally forbidden T_{1g} librational mode in the Raman spectrum and the asymmetry of the T_{2u} mode in the INS spectrum suggestive of a small splitting. The presence of the splitting in both the room temperature Raman spectrum and the 20 K INS spectrum suggests that a phase change is *not* responsible for the effect. The lineshape of the INS bands could be accounted for by weak dispersion in the modes, but this would not account for the splitting in the Raman spectrum.

An alternative possibility is that the molecule undergoes a small axial distortion to D_{4h} symmetry. If the distorted axis was randomly oriented in the crystal then the average symmetry would be O_h, as observed crystallographically, but the local symmetry as probed by vibrational spectroscopy would be D_{4h}. Under D_{4h} symmetry, all the threefold degenerate T modes split to an E mode and either an A or B mode. Thus the T_{2g} bending mode splits to Raman allowed B_{2g} and E_g modes and the librational T_{1g} mode splits to A_{2g} and E_g modes, the latter of which is Raman allowed. These effects are only observed for the hexahydrido complex, thus deuteration must stabilise the symmetric form because the energy cost to displace the heavier deuterium atoms is too great. The major difficulty with this explanation is that there is no apparent driving force, a totally symmetric $^1A_{1g}$ electronic ground state should not split.

This analysis highlights the need for information from all three types of vibrational spectroscopy and the crucial role that the INS intensities play in the assignments. The assignments can be substantiated by generating the INS spectrum by the Wilson GF method and the result is compared to the experimental spectrum in Fig. 6.25. The agreement for the internal modes is good but the analysis has omitted the strongest feature in the spectrum, the librational mode at 460 cm^{-1}.

The external modes can be included by placing heavy masses (M) around the $[FeH_6]^{4-}$ ion and defining M–Fe stretches for the translational modes and M–H–Fe torsions for the librations. This device mimics the external modes as internal modes of the larger cluster. This works surprisingly well as shown by the analyses of a series of palladium complexes [78] and of $Mg_2[NiH_4]$ and $Rb_3[ZnH_5]$ [79].

However, there is no doubt that periodic DFT calculations are the way forward and Fig. 6.25c shows the results of such a calculation for $Mg_2[FeH_6]$. This gives the modes in the correct positions and includes the external modes. However, the shape of the librational mode shows that it is strongly dispersed and this is not taken into account in this type of calculation. Clearly the next step is to carry out a periodic DFT calculation across the complete Brillouin zone to include the dispersion, as was done for the alkali metal hydrides (§6.7.1).

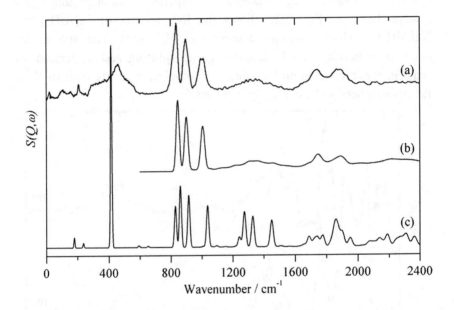

Fig. 6.25 INS spectra of $Mg_2[FeH_6]$: (a) experimental, (b) calculated from the Wilson GF method and (c) calculated by periodic DFT.

Periodic DFT provides a means of unambiguous assignment of the INS spectrum but the converse is also true: INS validates the theoretical studies. This idea was used to study the series of $A_2[PtH_6]$ (A = Li, Na, K, Rb, Cs) complexes [80]. The INS (see Figs. 5.9 and 5.10 for the spectra) and Raman spectra of the air and moisture sensitive rubidium salt were measured from the same sample in a custom-made quartz cell. A periodic DFT calculation using DMOL3 81 showed good agreement with the INS spectrum. Calculations were then carried out for the other salts. The results showed that the hydrogen in the complexes is essentially neutral and that the charge is shifted to the alkali metal ion. This occurs to a greater extent as the group is descended, accounting for the empirical observation that the heavier alkali metal salts are easier to synthesize. Interestingly, the, as yet, unknown lithium salt is predicted to be stable.

In Table 6.12 we list the compounds studied and some of their properties [76,78,79,82,83,84,85,86,87,88]. In most cases, the INS spectra show little or no evidence for dispersion, although this is not always the case. Fig. 6.26 shows the INS spectra of Li[AlH$_4$] and Na[AlH$_4$] , which contain tetrahedral [AlH$_4$]$^-$ ions. The spectra are complex indicating that both factor group splitting and dispersion are large in these systems and to obtain a complete analysis both of these factors will need to be explicitly included.

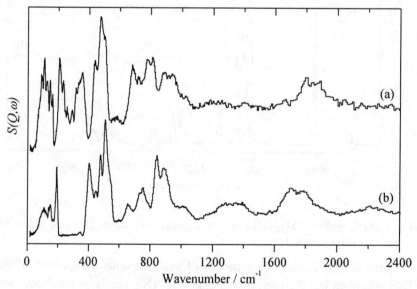

Fig. 6.26 INS spectra of (a) Li[AlH$_4$] and (b) Na[AlH$_4$] .

In Table 6.13 we list infrared active metal hydride stretching and bending frequencies [76,80,89,90,91,92,93,94,95]. Inspection of Table 6.12 and Table 6.13 highlight several trends. The hydrogen storage capacity is (unsurprisingly) largest for the lightest elements, thus $Li[BH_4]$ has a capacity of 18.5 wt%. Comparing $Mg_2[MH_6]$ (M = Fe, Ru, Os), the stretching frequencies increase on descending a group, as can also be seen for $Na[MH_4]$ (M = B, Al). Traversing a period results in a decrease in the frequency as shown by $Mg_2[[FeH_6]$, $Mg_2[CoH_5]$ and $Mg_2[NiH_4]$.

There is a marked dependence on the counter-ion as shown by the $A_2[PtH_6]$ (A = Na, K, Rb), $A_2[PdH_2]$ (A = Li, Na), borate and alanate series, higher frequencies are favoured by lighter elements. For $M_2[RuH_6]$ (M = Ba, Eu, Yb) which is the only case where traversing a period can be examined, the frequencies show a small increase.

This probably reflects a small decrease in ionic radius of the counter-ion. Higher oxidation states are also generally associated with higher frequencies this is particularly marked in the case of palladium, although much less so for platinum. For the octahedral complexes, the trends follow those of the lattice parameter and, by inference, the M–H length.

Table 6.12 Ternary metal hydrides studied by INS spectroscopy.

Compound	Hydrogen content / wt%	Oxidation state (a)	Complex ion structure	Symmetric stretch (b)	Ref.
$Na[BH_4]$	10.7	+3	Tetrahedral (D_{2d})	2334	82,83
$K[BH_4]$	7.5	+3	Tetrahedral (T_d)	2315	82,83
$Li[AlH_4]$	10.6	+3	Tetrahedral (T_d)	1838	84
$Na[AlH_4]$	7.5	+3	Tetrahedral (D_{2d})	1763	84
$Mg_2[FeH_6]$	5.5	+2	Octahedral (O_h)	1873	76
$Mg_2[CoH_5]$	4.5	+1	Square pyramidal (D_{4h})	1830/1632(c)	85
$Mg_2[NiH_4]$	3.6	0	Tetrahedral (C_1)	1691	79
$Rb_3[ZnH_5]$(d)	1.5	+2	Tetrahedral (D_{2d})	1302	79
$Li_2[PdH_2]$	1.6	0	Linear ($D_{\infty h}$)	1760	78
$Na_2[PdH_2]$	1.3	0	Linear ($D_{\infty h}$)	1660	78,86
$NaBa[PdH_3]$	1.0	0	Trigonal planar (D_{3h})	1370	78
$Ba_2[PdH_4]$	1.0	0	Tetrahedral (D_{2d})	1290	78
$K_2[PdH_4]$	2.1	+2	Square planar (D_{4h})	1840	78
$Rb_2[PtH_4]$	1.1	+2	Square planar (D_{4h})	1991	87
$Rb_2[PtH_6]$	1.6	+4	Octahedral (O_h)	2044	80
$Ba[ReH_9]$	2.7	+7	Tricapped trigonal prism (D_{3h})	1995	88

(a) oxidation state of the complex metal ion. (b) Totally symmetric stretch.(c) Axial and equatorial stretch respectively. (d) Complex is actually $(Rb^+)_3(H^-)[ZnH_4]^2$

For the $A_2[PtH_6]$ (A = Na, K, Rb, Cs) the lattice parameter increases from 7.34 to 8.95 Å while the M—H bond length only changes from 1.62 to 1.64 Å, consistent with computational studies [78] which show that the M—H bonding is largely covalent. It seems more likely that the effects are due to different degrees of charge transfer to the counter-ion.

Finally, it is worth noting that while the interest in these compounds is currently driven by potential hydrogen storage applications, there is undoubtedly a rich chemistry still to be explored. This is highlighted by the many organometallic (particularly phosphine) derivatives of the $[ReH_9]^{2-}$ ion. Many of the compounds are highly air and moisture sensitive showing intrinsic high reactivity. The availability of the complexes as soluble derivatives would open up applications in synthetic chemistry. The existence of the soluble $Mg_4(Cl,Br)_4(THF)_8[MH_6]$ (M = Fe, Ru) [18, 24] complex shows that this may well be feasible.

Table 6.13 Infrared active metal-hydride stretch and bend frequencies (ω / cm^{-1}) of octahedral ternary metal hydrides.

Compound	Hydrogen content / wt %	oxidation state (a)	M—H (D) stretch	M—H (D) bend	Ref.
$Mg_2[FeH_6]$	5.5	+2	1746 (1260)	899 (661)	76
$Mg_4[(Br,Cl)_4]($ $THF)_8[FeH_6]$	0.9	+2	1569/1514 (1138/1107)		89
$Mg_2[RuH_6]$	3.9	+2	1783 (1279)		90
$Mg_4[(Cl,Br)]_4$ $(THF)_8[RuH_6]$	0.8	+2	1562		94
$Ca_2[RuH_6]$	3.2	+2	1546 (1128)	896 (646)	90, 91, 92
$Sr_2[RuH_6]$	2.1	+2	1482	878	90, 91, 92
$Ba_2[RuH_6]$	1.6	+2	1438 (1023)		90, 91, 92
$Eu_2[RuH_6]$	1.5	+2	1480	878	90, 91, 92
$Yb_2[RuH_6]$	1.3	+2	1550	886	95
$Mg_2[OsH_6]$	2.5	+2	1849 (1325)	864	90
$Ca_2[OsH_6]$	2.2	+2	1637		90
$Sr_2[OsH_6]$	1.6	+2	1575		90
$Ba_2[OsH_6]$	1.3	+2	1505 (1078)		90
$Na_2[PtH_6]$	2.4	+4	(1278)	(645)	93
$K_2[PtH_6]$	2.2	+4	1748 (1251)	881 (632)	93
$Rb_2[PtH_6]$	1.6	+4	1743 (1235)	877 (631)	80, 93

(a) oxidation state of the complex metal ion

6.8 References

1   I.F. Silvera (1980). Rev. Mod. Phys., 52, 393–452. The solid molecular hydrogens in the condensed phase: Fundamentals and static properties.

2   W. Langel, D.L. Price, R.O. Simmons & P.E. Sokol (1988). Phys. Rev. B, 38, 11275–11283. Inelastic neutron scattering from liquid and solid hydrogen at high momentum transfer.

3   L. Pauling & E.B. Wilson (1963). *Introduction to Quantum Mechanics*, Dover Publications, New York, p. 264.

4   P.W. Atkins (1998). *Physical Chemistry*, 6[th]Edition, Oxford University Press, Oxford, p. 466.

5   P.F. Bernath (1995). *Spectra of Atoms and Molecules*, Oxford University Press, Oxford, p. 250.

6   J.A. Young & J.U. Koppel (1964). Phys. Rev., 135, A603–A611. Slow neutron scattering by molecular hydrogen and deuterium.

7   M. Celli, D. Colognesi & M. Zoppi (2002). Phys. Rev. E, 66, 021202. Direct experimental access to microscopic dynamics in liquid hydrogen.

8   G. Corradi, D. Colognesi, M. Celli & M. Zoppi (2003). Condensed Matter Physics, 6, 499–521. Mean kinetic energy and final state effects in liquid hydrogens from inelastic neutron scattering.

9   D. White & EN. Lassettre (1960). J. Chem. Phys., 32, 72–84. Theory of ortho–para hydrogen separation by adsorption at low temperature, isotope separation.

10   A.P. Smith, R. Benedeck, F.R. Trouw, M. Minkoff & L.H. Yang (1996). Phys. Rev. B, 53, 10187–10199. Quasi-two-dimensional quantum states of H_2 in stage-2 Rb-intercalated graphite.

11   J.A. MacKinnon, J. Eckert, D.F. Coker & A.L.R. Bug (2001). J. Chem. Phys., 114, 10137–10150. Computational study of molecular hydrogen in zeolite Na-A. II. Density of rotational states and INS spectra

12   A.C. Dillon, K.M. Jones, T.A. Bekkedahl, C.H. Kiang, D.S. Bethune & M.J. Heben (1997). Nature, 386, 377–379. Storage of hydrogen in single-walled carbon nanotubes.

13   A. Chambers, C. Park, R.T.K. Baker & N.M. Rodriguez (1998). J. Phys. Chem. B 102, 4253–4256. Hydrogen storage in graphite nanofibers.

14   P. Chen, X. Wu, J. Lin & K.L. Tan (1999). Science, 285, 91–93. High H_2 uptake by alkali-doped carbon nanotubes under ambient pressure and moderate temperatures.

15   H.G. Schimmel, G.J. Kearley, M.G. Nijkamp, C.T. Visserl, K.P. de Jong & F.M. Mulder (2003). Chem-Eur. J., 9, 4764–4770. Hydrogen adsorption in carbon nanostructures: Comparison of nanotubes, fibers, and coals

16   T. Yildirim & A.B. Harris (2003). Phys. Rev. B, 67, 245413. Quantum dynamics of a hydrogen molecule confined in a cylindrical potential.

17   C.M. Brown, T. Yildirim, D.A. Neumann, M.J. Heben, T. Gennett, A.C. Dillon, J.L. Alleman & J.E. Fischer (2000).Chem. Phys. Lett., 329, 311–316. Quantum rotation of hydrogen in single-wall carbon nanotubes.

18   P.A. Georgiev, D.K. Ross, A. De Monte, U. Montaretto-Marullo, R.A.H. Edwards, A.J. Ramirez-Cuesta & D. Colognesi (2004). J. Phys.: Condens. Matter, 16, L73–L78. Hydrogen site occupancies in single-walled carbon nanotubes studied by inelastic neutron scattering..

19   H.G. Schimmel, G. Nijkamp, G.J. Kearley, A. Rivera, K.P. de Jong & F.M. Muldera (2004). Materials Science and Engineering, B108, 124–129. Hydrogen adsorption in carbon nanostructures compared.

20   H.G. Schimmel, G.J. Kearley & F.M. Mulder (2004). Chem. Phys. Chem, 5, 1053–1055. Resolving rotational spectra of hydrogen adsorbed on a single-walled carbon nanotube substrate.

21   Y. Ren & D.L. Price (2001). Appl. Phys. Lett, 79, 3684–3686. Neutron scattering study of H_2 adsorption in single-walled carbon nanotubes.

22   W.J. Stead, I.P. Jackson, J.McCaffrey & J.W. White (1988). J. Chem. Soc, Faraday Trans. II, 1988, 84, 1669–1682 Tunnelling of hydrogen in alkali-metal-graphite intercalation compounds. A systematic study of $C_{24}Rb(H_2)_x$ and its structural consequences.

23   S.A FitzGerald, T.Yildirim, L.J Santodonato, D.A. Neumann, J.R.D. Copley, J.J Rush & F. Trouw (1999). Phys. Rev. B, 60, 6439–6451. Quantum dynamics of interstitial H_2 in solid C_{60}.

24   T. Yildirim & A.B. Harris (2002). Phys. Rev. B, 66, 214301. Rotational and vibrational dynamics of interstitial molecular hydrogen.

25   H.F.W.J. van Breukelen, M.E. Gerritsen, V.M. Ummels, J.S. Broens & J.H.C. van Hooff (1997). Stud. Surf. Sci. Catal., 105, Part A-C, 1029–1035. Application of CoAlPO-5 molecular sieves as heterogeneous catalysts in liquid phase oxidation of alkenes with dioxygen.

26   M. Hartmann & L. Kevan (1999). Chem. Rev., 99, 635–663. Transition-metal ions in aluminophosphate and silicoaluminophosphate molecular sieves: Location, interaction with adsorbates and catalytic properties.

27   T. Maschmeyer, R.D. Oldroyd, G. Sankar, J.M. Thomas, I.J. Shannon, J.A. Klepetko, A.F. Masters, J.K. Beattie & C.R.A. Catlow (1997). Angew. Chem. Int. Ed.., 36, 1639–1642. Designing a solid catalyst for the selective low-temperature oxidation of cyclohexane to cyclohexanone.

28   P.A. Barrett, G. Sankar, C.R.A. Catlow & J..M. Thomas (1996). J. Phys. Chem., 100, 8977–8985. X-ray absorption spectroscopic study of Bronsted, Lewis, and redox centers in cobalt-substituted aluminum phosphate catalysts.

29   J.M. Thomas, G.N. Greaves, G. Sankar, P.A. Wright, J.S. Chen, A.J. Dent & L. Marchese (1994). Angew. Chem. Int. Ed., 33, 1871–1873. On the nature of the active-site in a CoAlPO-18 solid acid catalyst.

30   A.J. Ramirez-Cuesta, P.C.H. Mitchell, S.F. Parker & P.A. Barrett (2000). Chem. Commun., 1257–1258. Probing the internal structure of a cobalt aluminophosphate catalyst. An inelastic neutron scattering study of sorbed dihydrogen molecules behaving as one- and two-dimensional rotors.

31 A.J. Ramirez-Cuesta, P.C.H. Mitchell, S.F. Parker (2001). J. Mol. Catal. A-Chem., 167, 217-224. An INS study of the interaction of dihydrogen with the cobalt site of cobalt alumino phosphate catalyst.

32 J. Eckert, J.M. Nicol, J. Howard & F.R. Trouw (1996). J. Phys. Chem., 100, 10646-10651. Adsorption of hydrogen in Ca-exchanged Na-A Zeolites probed by inelastic neutron scatttering spectroscopy.

33 J.M. Nicol,. J. Eckert & J. Howard (1988). J. Phys Chem, 92, 7117-7121. Dynamics of molecular hydrogen adsorbed in CoNa-A zeolite.

34 P.E. Sokol, D.W. Brown & S. Fitzgerald (1998). J. Low Temp. Phys., 113, 717-722. An inelastic neutron scattering study of H-2 in Vycor glass.

35 D.W. Brown, P.E. Sokol & S.A. FitzGerald (1999). Phys. Rev. B, 59, 13258-13256., Rotational dynamics of n-H_2 in porous Vycor glass.

36 P.M. Forster, J. Eckert, J.-S. Chang, S.-E. Park,.G. Férey & A.K Cheetham (2003).' J. Am. Chem. Soc., 125, 1309-1312. Hydrogen adsorption in nanoporous nickel(II) phosphates.

37 N.L. Rosi, J. Eckert, M. Eddaoudi, D.T. Vodak, J. Kim, M. O'Keeffe & O.M. Yaghi (2003). Science, 300, 1127-1129. Hydrogen storage in microporous metal-organic frameworks.

38 G.J. Kubas, R.R. Ryan, B.I. Swanson, P.J. Vergamini & H.J. Wasserman (1984). J. Am. Chem. Soc., 106, 451-452. Characterization of the first examples of isolable molecular hydrogen complexes, $Mo(CO)_3(PCY_3)_2(H_2)$, $W(CO)_3(PCY_3)_2(H_2)$, $Mo(CO)_3(PI-PR_3)_2(H_2)$, $W(CO)_3(PI-PR_3)_2(H_2)$— Evidence for a side-on bonded H_2 ligand.

39 G.J. Kubas (2001). *Metal dihydrogen and s-Bond Complexes*. Kluwer Academic, New York.

40 G.J. Kubas (2001). J. Organomet. Chem., 635, 37-68. Metal dihydrogen and σ-bond coordination: the consummate extension of the Dewar-Chatt-Duncanson model for metal-olefin π-bonding.

41 J. Eckert (1992). Spectrochim. Acta, 48A, 363-378. Dynamics of the molecular hydrogen ligand in metal complexes.

42 B.R. Bender, G.J. Kubas, L.H. Jones, B.I. Swanson, J. Eckert, K.B. Capps & C.D. Hoff (1997). J. Am. Chem. Soc., 119, 9179-9190. Why does D_2 bind better than H_2? A theoretical and experimental study of the equilibrium isotope effect on H_2 binding in a $M(\eta_2-H_2)$ complex. Normal coordinate analysis of $[W(CO)_3(PCy_3)_2 (\eta_2-H_2)]$.

43 L.S. Vandersluys, J. Eckert, O. Eisenstein, J.H. Hall, J.C. Huffman, S.A. Jackson, T.F. Koetzle, G.J. Kubas, P.J. Vergamini & K.G. Caulton (1990). J. Am. Chem. Soc., 112, 4831-4841. An attractive 'cis-effect' of hydride on neighbor ligands— experimental and theoretical-studies on the structure and intramolecular rearrangements of $[Fe(H)_2(\eta_2-H_2)(PEtPh_2)_3]$.

44 J. Eckert, H. Blank, M.T. Bautista & R.H. Morris (1990). Inorg. Chem., 29, 747-750. Dynamics of molecular hydrogen in the complex trans-$[Fe(\eta_2-H_2)(H)(PPh_2CH_2CH_2PPh_2)_2][BF_4]$ in the solid-state as revealed by neutron scattering experiments

45   J. Eckert, A. Albinati, U.E. Bucher & L.M. Venanzi (1996). Inorg. Chem., 35, 1292–1294. Nature of the Rh-H_2 bond in a dihydrogen complex stabilized only by nitrogen donors. Inelastic neutron scattering study of [TpMe$_2$RhH$_2$ (η_2-H$_2$)] (TpMe$_2$ = Hydrotris(3,5-dimethylpyrazolyl)borate).

46   J. Eckert, C.E. Webster, M.B. Hall, A. Albinati & L.M. Venanzi (2002). Inorg. Chim. Acta, 330, 240–249. The vibrational spectrum of Tp(3,5-Me)RhH$_2$(H$_2$)]: a computational and inelastic neutron scattering study.

47   M. Aresta, P. Giannoccaro, M. Rossi, A. Sacco (1971). Inorg. Chim. Acta, 5, 115–118. Hydrido complexes of iron(IV) and iron(II).

48   G.E. Gadd, R.K. Upmacis, M. Poliakoff & J.J. Turner (1986). J. Am. Chem. Soc., 108, 2547–2552. Complexes of iron and cobalt containing coordinated molecular dihydrogen: infrared evidence for [Fe(CO)(NO)$_2$ (H$_2$)] and [Co(CO) $_2$ (NO)(H$_2$)] in liquefied xenon solution.

49   B.L. Mojet, J. Eckert, R.A. van Santen, A. Albinati & R.E. Lechner (2001). J. Am. Chem. Soc., 123, 8147–8148. Evidence for chemisorbed molecular hydrogen in Fe-ZSM5 from inelastic neutron scattering.

50   G. Alefeld & J. Volkl (1978) Hydrogen in Metals II, Application-Oriented Properties, Springer-Verlag, Berlin, Chapter 1.

51   D.K. Ross (1997). Topics App. Phys. 73, 153–214, Neutron scattering studies of metal hydrogen systems.

52   V.A. Yartys, I.R. Harris & V.V. Panasyuk (2001). Materials Science, 37, 219–240. New metal hydrides: a survey.

53   A.I. Kolesnikov, V.E. Antonov, V.K Fedotov, G. Grosse, A.S. Ivanov & F.E. Wagner (2002). Physica B, 316–317, 158–161. Lattice dynamics of high pressure hydrides of the Group VI–VIII transition metals.

54   B. Hauer, R. Hempelmann, T.J. Udovic, J.J. Rush, E. Jansen, W. Kockelmann, W. Schafer & D. Richter (1998). Phys. Rev. B, 57, 11115–11124. Neutron scattering studies on the vibrational excitations and the structure of the ordered niobium hydrides: the ε-phase.

55   H.N. Bordallo, A.I. Kolesnikov, A.V. Kolomiets, W. Kalceff, H. Nakotte & J. Eckert (2003). J. Phys.: Condens. Matter, 15, 2551–2559. INS studies of TbNiAlH$_{1.4}$ and UNiAlH$_{2.0}$ hydrides.

56   V.E. Antonov, B. Dorner, V.K. Fedotov, G. Grosse, A.S. Ivanov, A.I. Kolesnikov, V.V. Sikolenko & F.E. Wagner (2002). J. Alloys Compounds, 330, 462–466. Giant tunneling effect of hydrogen and deuterium in α-manganese.

57   D.K. Ross, V.E. Antonov, E.L. Bokhenkov, A.I. Kolesnikov, E.G. Poynatovsky & J. Tomkinson (1998). Phys. Rev. B, 58, 2591–2595. Strong anisotropy in the inelastic neutron scattering from PdH at high energy transfer.

58   J.M. Rowe, J.J. Rush, H.G. Smith, M. Mostoller & H.E. Flotow (1974). Phys. Rev. Lett., 33, 1297–1300. Lattice dynamics of a single crystal of PdD$_{0.63}$.

59   M. Heene, H. Wipf, T.J. Udovic & J.J. Rush (2000). J. Phys.: Condens. Matter, 12, 6183–6190. Hydrogen vibrations and trapping in TaN$_{0.006}$H$_{0.003}$.

60   V.V. Sumin, H. Wipf, B. Coluzzi, A. Biscarini, R. Campanella, G. Mazzolai & F. M. Mazzolai (2001). J. Alloys Compounds, 316, 189–192. A neutron spectroscopy study of the local vibrations, the interstitial sites and the solubility limit of hydrogen in niobium-molybdenum alloys.

61   C. Schonfeld, R. Hempelmann, D. Richter, T. Springer, A.J. Dianoux, J.J. Rush, T.J. Udovic & S.M. Bennington (1994). Phys. Rev. B, 50, 853– 865. Dynamics of hydrogen in α-LaNi hydride investigated by neutron scattering.

62   U. Stuhr, H. Wipf, T.J. Udovic, J. Weissmueller & H. Gleiter (1995). J. Phys.: Condens. Matter, 7, 219–230. The vibrational excitations and the position of hydrogen in nanocrystalline palladium.

63   P. Albers, M. Poniatowski, S.F. Parker & D.K. Ross (2000). J. Phys.: Condens. Matter, 12, 4451–4463. Inelastic neutron scattering study on different grades of palladium of varying pretreatment.

64   Y. Fukai, *The Metal-hydrogen System* (1993), Springer Verlag, Berlin. Table 4.6.

65   A.I. Kolesnikov, I.O. Bashkin, A.V. Belushkin, E.G. Ponyatovsky, M. Prager & J. Tomkinson (1995). Physica B, 213–214, 445–447. Phonons and bound multiphonons in the γ-phases of TiH and ZrH– neutron spectroscopy studies.

66   N.N. Greenwood & A. Earnshaw (1997). *Chemistry of the Elements*, Butterworth-Heinemann, Oxford. Second Edition.

67   L. Schlapbach (ed.) (1992), *Hydrogen in Intermetallic Compounds II,* Topics in Applied Physics, Vol. 67, Springer, Berlin.

68   W.M. Mueller, J.P. Blackledge & G.G. Libowitz (1968). *Metal Hydrides*, Academic Press, New York.

69   D. Colognesi, A.J. Ramirez-Cuesta, M. Zoppi, R. Senesi & T. Abdul-Redah (2004). Physica B, 350, E983–E986. Extraction of the density of phonon states in LiH and NaH.

70   W. Dyck & H. Jex (1981). J. Phys. C.: Solid State (Phys.), 14, 4193–4215. Lattice-dynamics of alkali hydrides and deuterides with the NaCl type-structure.

71   M. G. Zemlianov, E. G. Brovman, N. A. Chernoplekov & Yu. L. Shitikov (1965). In *Inelastic Scattering of Neutrons*, IAEA, Vienna, Vol. II, p 431.

72   K. Yvon (1994). In *Encyclopedia of Inorganic Chemistry*, (ed.) R. B. King, Wiley, New York, p.1401. Hydrides; solid state transition metal complexes.

73   W. Bronger & G. Auffermann (1998). Chem. Mater., 10, 2723–2732. New ternary alkali metal-transition metal hydrides synthesized at high pressures: characterization and properties.

74   R.B. King (2000). Coord. Chem. Rev., 200, 813–829. Structure and bonding in homoleptic transition metal hydride anions.

75   K. Yvon (2003). Z. Kristallogr., 218, 108–116. Hydrogen in novel solid state hydrides.

76   S.F. Parker, K.P.J. Williams, M. Bortz & K. Yvon (1997). Inorg. Chem., 36, 5218–5221. Inelastic neutron scattering, infrared and Raman spectroscopic studies of $Mg_2[FeH_6]$ and $Mg_2[FeD_6]$.

77   W.G. Fateley, F.R. Dollish, N.T. McDevitt & F.F. Bentley (1972). *Infrared and Raman Selection Rules for Molecular and Lattice Vibrations: The Correlation Method*, Wiley-Interscience, New York.

78   M. Olofsson-Mårtensson, U. Häussermann, J. Tomkinson & D. Noréus (2000). J. Am. Chem. Soc., 122, 6960–6970. Stabilization of electron-dense palladium-hydrido complexes in solid-state hydrides.

79   S.F. Parker, K.P.J. Williams, T. Smith, M. Bortz, B. Bertheville & K. Yvon (2002). Phys. Chem. Chem. Phys., 4, 1732–1737. Vibrational spectroscopy of tetrahedral ternary metal hydrides: $Mg_2[NiH_4]$, $Rb_3[ZnH_5]$ and their deuterides.

80   S.F. Parker, S.M. Bennington, A.J. Ramirez-Cuesta, G. Auffermann, W. Bronger, H. Herman, K.P.J. Williams & T. Smith (2003). J. Am. Chem. Soc., 125, 11656–11661. Inelastic neutron scattering, Raman spectroscopy and periodic-DFT studies of $Rb_2[PtH6]$ and $Rb_2[PtD_6]$.

81   B. Delley (2000). J. Chem. Phys., 113, 7756–7764. From molecules to solid with the DMOL3 approach.

82   K.B. Harvey & N.R. McQuaker (1971). Can. J. Chem., 49, 3272–3278. Infrared and Raman spectra of potassium and sodium borohydride.

83   D.G Allis & B.S. Hudson, (2004). Chem. Phys. Lett., 385, 166–172. Inelastic neutron scattering spectra of $Na[BH_4]$ and $K[BH_4]$: reproduction of anion mode shifts *via* periodic DFT.

84   S.F. Parker, unpublished work.

85   S.F. Parker, U.A. Jayasooriya, J.C. Sprunt, M. Bortz & K. Yvon (1998). J. Chem. Soc. Faraday Trans., 94, 2595–2599. Inelastic neutron scattering, infrared and Raman spectroscopic studies of Mg_2CoH_5 and Mg_2CoD_5.

86   D. Noréus & J. Tomkinson (1989). Chem. Phys. Lett., 154, 439–442. Inelastic neutron scattering studies of a novel linear PdH_2 complex in Na_2PdH_2.

87   S.F. Parker & G. Auffermann, unpublished work.

88   S.F. Parker & K. Yvon, unpublished work.

89   D.E. Linn Jr, G.M. Skidd & E.M. Tippmann (1999). Inorg. Chim. Acta, 291, 142–147. Soluble complexes of $[FeH_6]^{4+}$.

90   M. Kritikos & D. Noréus (1991). J. Solid State Chem., 93, 256–262. Synthesis and characterization of ternary alkaline-earth transition-metal hydrides containing octahedral $[Ru(II)H_6]^{4-}$ and $[Os(II)H_6]^{4-}$ complexes.

91   R.O. Moyer Jr & J.R. Wilkins & P. Ryan (1999). J. Alloy Compd., 290, 103–109. An infrared study of some ruthenium containing mixed crystal quaternary metal hydrides.

92   H. Hagemann & R.O. Moyer Jr (2002). J. Alloy Compounds, 330, 296–300. Raman spectroscopy studies on $M_2[RuH6]$ where M=Ca, Sr and Eu.

93   D. Bublitz, G. Peters, W. Preetz, G. Auffermann & W. Bronger (1997). Z. Anorg. Allg. Chem., 623, 184–190. Darstellung, [195]Pt-NMR-, IR- und Raman-spektren sowie normalkoordinatenanalyse der komplexionen $[PtH_nD_{6-n}]^{4-}$, $n = 0$–6.

94   D.E. Linn, Jr., G.M. Skidd & S.N. McVay (2002). Inorg. Chem., 41, 5320–5322. Comparative preparations of homoleptic hydridic anions of iron and ruthenium using solution-based organometal hydrogenation techniques.

95   R.O. Moyer Jr, R. Lindsay & D.N. Marks (1978). In *Transition metal Hydrides*, ed. R. Bau, Advances in Chemistry Series 167, American Chemical Society, Washington D.C., pp 366–38. Results of reactions designed to produce ternary hydrides of some rarer platinum metals with europium or ytterbium.

Surface Chemistry and Catalysis

In this chapter we describe some applications of inelastic neutron scattering in surface chemistry, more particularly in studies of catalysts and adsorbed species [1]. Our emphasis will be on the spectroscopy. The subject matter is arranged broadly according to the type of catalyst: metals (§7.3), oxides (§7.4), zeolites and microporous materials (§7.5) and sulfides (§7.6) and, within each group, according to the reactant molecules. We start (§7.1) with a general discussion of surface vibrations.

7.1 Vibrations of atoms in surfaces and adsorbed species

Inelastic neutron scattering is not, in itself, a surface technique. Neutrons penetrate and so an INS spectrum is the sum of the spectra from neutrons scattered in the bulk of a substance and at its surface [2]. We see surface scattering when a substance has a large surface–to–volume ratio (surface area ca 20 m^2g^{-1}) and when the surface composition differs from the bulk composition, especially when the hydrogen concentration at the surface is greater than in the bulk, e.g. if the surface is protonated or a hydrogenous adsorbate is present. We now analyse the consequences for inelastic neutron scattering of the various ways in which atoms and molecules may interact with a surface. It is helpful to discuss internal and external modes separately. We recall that internal vibrations involve stretching and bending of chemical bonds in a molecule (vibrations that are analogous with gas-phase vibrations while bearing in mind that our reference state in INS spectroscopy is the solid phase), and external vibrations involve partial rotations and translations of whole molecules in a lattice. One way of investigating vibrations of atoms at a surface by INS is via the so-called riding modes associated with surface-bound hydrogen atoms or hydrogenous molecules. Riding modes arise from neutron scattering by a hydrogen atom attached directly or indirectly to the surface of a solid, e.g. a nickel particle, and moving synchronously with the surface atoms (as a horse rider moves with the horse). The hydrogen atom vibrates at a frequency very close to that of the surface atom; the scattering intensity of the vibration is amplified because the scattering is by hydrogen. How far the intensity of a surface mode is amplified by scattering from an adsorbed atom (we assume in the following that this atom is hydrogen) depends on the strength of binding of the adsorbed atom: when the binding is strong the riding mode is also strong. The effect can be understood from the following calculation on a simple model system.

The substrate (or adsorbent) was modelled in one dimension by a *linear* chain of five carbon atoms, (We could have chosen any other atom.) A terminal carbon atom represents a surface atom and the remaining atoms the bulk. There is only one force constant, the C–C stretching force constant (C—C of C_2 12.16 mdyn Å^{-1}). The dynamics were treated by the Wilson GF method. The substrate then had four vibrations; similar to longitudinal acoustic modes (LAM) (§10.1.2): LAM1,170 cm^{-1}; LAM2, 330 cm^{-1}; LAM3, 455 cm^{-1}; LAM4, 535 cm^{-1}. The calculated INS spectrum comprised four bands of equal, weak intensity.

Although the bands are of equal intensity, each carbon does not contribute equally to every LAM; the displacement of the terminal carbon (the surface atom) is greatest in LAM1 and least in LAM4. This could be demonstrated by arbitrarily ascribing to the surface atom a high scattering cross section, that of hydrogen was convenient (σ_H, 82.0 barn; cf. σ_C, 5.55 barn). The INS intensities of all the modes would then be stronger (because of the large cross section assigned to C_1) but their relative intensities are different.

We model hydrogen adsorption by attaching a hydrogen atom to the surface atom (terminal carbon): an extra band is produced, the stretching mode of hydrogen against the surface, C—H. This is usually a high frequency transition (*ca* 3000 cm^{-1}) but we may vary the frequency arbitrarily to model different strengths of hydrogen binding. The results of these calculations are shown in Fig. 7.1. When the stretching frequency of the hydrogen against the surface is relatively high, all the substrate modes increase in intensity but the modes involving the greatest surface atom motion are amplified most. In this example LAM1 involves the greatest surface atom motion and the intensity observed at that frequency (170 cm^{-1}) is the strongest of the substrate modes. The relative intensities of the substrate modes are approximately the same as when a terminal carbon was arbitrarily given the hydrogen scattering cross section. Here the increased intensity is a natural consequence of the dynamics.

For riding to be used to emphasise the vibrations of surface atoms from those of the bulk the stretching frequency of the hydrogen against the surface must be greater than about twice the maximum frequency of the substrate (($\omega_{sub})_{max}$) (§2.6.1). In this example the LAM4 frequency at 535 cm^{-1} represents (($\omega_{sub})_{max}$) and sets a limit of *ca* 1070 cm^{-1}, see Fig. 7.1. When the C–H stretching frequency is about the same as (($\omega_{sub})_{max}$) surface modes (motions of the terminal atom) are no more amplified than those of the bulk (motions of any other carbon). Note that there are still riding motions and that the spectrum of the bulk is still stronger than if hydrogen were not present but no part of the substrate spectrum is especially emphasised. An example is the wealth of surface modes made visible by the binding of spillover hydrogen at terminal carbons (§7.3.1.8). In contrast, the weakly bound spillover hydrogen had no riding modes. The riding modes in the INS spectra of hydrogenous molecular adsorbates, e.g. benzene, will in general fail to meet the high frequency stretching criterion discussed above. However, if the adsorbate is strongly bonded to the surface, e.g. ethylidyne, then the vibrations of the surface atoms may well become emphasised.

The reader should be aware that the above discussion is a simplified version of an extraordinarily complex dynamical problem. Our purpose has been to draw attention to certain aspects of surface vibrations which *may* be revealed by INS spectroscopy. While specific riding nodes can sometimes be associated with motions of the surface atoms of the substrate, to do so generally would be naïve.

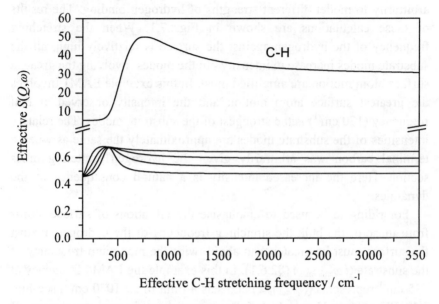

Fig. 7.1 Modelling the intensity amplification of surface atom vibrations by a one-dimensional linear chain of five carbon atoms. The chain has four vibrations, the LAMs, LAM1 has the highest intensity and LAM4 the lowest. When the very strong C—H stretching mode is at low frequencies this relative amplification is lost. See text for details.

7.2 Experimental methods

Experimental methods are described in Chapter 3 (§3.5.5). Here we refer to certain aspects of INS experimental methods which require special attention in measurements on catalysts: the mass of sample required if we are to obtain an acceptable signal–to–noise ratio, background subtraction, and the pre-treatment of the catalyst.

The mass of the sample must be sufficient to provide a spectrum distinct from any instrumental background and having an optimum signal–to–noise ratio. In experiments with catalysts, background scattering from the container and from the mass of the catalyst is significant. The mass of substance needed to obtain an INS spectrum with an acceptable signal–to–noise ratio is determined for a hydrogenous substance from its hydrogen content and the sensitivity of the spectrometer. For example, on the ISIS spectrometer TOSCA the requirement is to have in the beam *ca* 10 mmol H (as atoms) or 1 mmol H_2 (e.g. as adsorbed molecules). For a 40% wt Pt/C fuel cell catalyst which takes up hydrogen (as hydrogen atoms) to a ratio of *ca* 0.4 H/Pt we need *ca* 15 g of hydrogen-dosed catalyst in the neutron beam. Because of the quantity of material needed to obtain a spectrum INS measurements are made on catalyst powders rather than single crystals.

To obtain the INS spectrum of the surface of a catalyst or an adsorbed species (or both together) we have to subtract a background spectrum. The background may be the spectrum of the sample container, or the container with catalyst prior to any desired treatment or dosing. The spectrum after background subtraction is the *difference spectrum*. In this chapter the spectra discussed and presented will be difference spectra unless otherwise stated. The spectrum of a substance to be adsorbed is determined separately and represents the spectrum of a model compound used in the interpretation of the data (§5.1). Note that in INS experiments, because the spectrum is determined at 20 K, the reference state of a substance is the solid.

The catalyst must be pre-treated to remove extraneous sources of scattering, especially water adsorbed during exposure of the catalyst to air or retained after reduction of an oxide layer by dihydrogen. The pre-treatment is carried out with the catalyst already in the sample cell, heated in flowing helium, and evacuated to 10^{-7} mbar while any desorbed products are monitored to disappearance with a mass spectrometer—a procedure which may take some days.

7.3 INS studies of metal catalysts

In this section we discuss INS studies of metal catalysts where the main interest is the hydrogenation or hydrogenolysis of a hydrocarbon or other organic molecule. Such metal catalysts have at least two functions: to split the H_2 molecule providing reactive H atoms and to bind and activate a reactant molecule. In some cases hydrogen atoms may spillover from the catalyst to a support establishing a reservoir of hydrogen atoms (§7.3.1.8). Reactions of oxides including carbon monoxide and oxidations including the hydrogen—oxygen reaction are discussed in §7.4.

7.3.1 Metal–hydrogen vibrations and surface vibrational states

Bulk and surface spectra may be distinguished by comparing the spectra of particles of very different size (and hence surface/bulk ratio) and through the amplification of surface modes caused by hydrogen atoms riding on the surface atoms, the riding modes. We shall see that riding modes, which may not hitherto have been recognized as such, feature in a number of spectra. The early work on the study of surface vibrational states and riding modes by INS has perhaps not received the attention it merits.

From the vibrational spectrum of hydrogen on a metal catalyst one can establish the nature of the hydrogenous species (e.g. H or H_2), the binding site (e.g. the surface plane and the coordination number of the site), and the concentration of hydrogen at different sites. The spectra are modelled by fitting the experimental spectrum to a theoretical spectrum from a force field calculation or *ab initio* quantum mechanics. For the practical catalytic chemist the aim of INS spectroscopy is to identify the hydrogen binding sites and their participation in the catalytic chemistry. INS spectra may be used for fingerprinting molecules—assigning peaks to characteristic modes and hence to molecular species. This is similar to the use of infrared and Raman spectra except that in INS spectroscopy the role of the band intensity is more important.

A review of INS spectra of hydrogen on metallic (and other) catalysts covers the literature through the 1980s [3]. Peaks and assignments for nickel, palladium and platinum are given later in Tables 7.2 (§7.3.1.2), 7.3 (§7.3.1.5) and 7.4, 7.5 (§7.3.1.6) [4,5,6,7].

7.3.1.1 Nickel particles—surface vibrational states

Surface vibrational states of nickel particles have been investigated by comparing the INS spectra of nickel particles of different sizes [2]: macroparticles (polycrystalline granules 6—8 mm diameter) and microparticles (particle size 20—30 nm, surface area 22 m^2g^{-1}). The surface-to-volume ratio of the microparticles was a factor 4×10^5 greater than the macroparticles. The INS spectra are shown in Fig. 7.2. Assignments were made partly by comparison with other experimental data, e.g. from high resolution electron energy loss spectroscopy (HREELS), and partly with the aid of the phonon spectrum computed for a thin plate of nickel [2]. The spectrum was divided into the following contributions: A, the low energy region below 140 cm^{-1} containing the surface modes of nickel, i.e. vibrations within the surface layer; and, if an adsorbate layer is present, modes combining the motions of adsorbate atoms with those of the nickel atoms; B, the intermediate energy region *ca* 140–320 cm^{-1} containing the spectrum of bulk nickel for both macro- and micro-particles; C, the high energy region *ca* 320–2200 cm^{-1} containing the spectrum of an adsorbed layer.

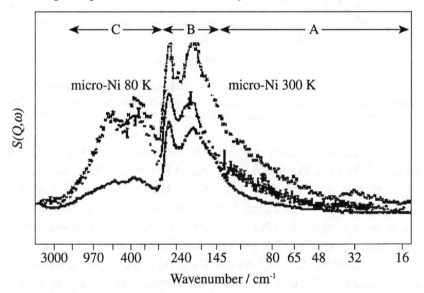

Fig. 7.2 INS spectra of powdered nickel (KDSOG-M, Dubna) [2]. Notice the non-linear and unconventional wavenumber scale. The characteristic regions of the spectra are labelled. Reproduced from [2] with permission from the Russian Academy of Science.

The spectrum of the adsorbed layer (water, arising from exposure of both samples of nickel particles to the atmosphere) was obtained by subtracting the spectrum of macro-nickel from that of micro-nickel [2], 8], see Fig. 7.3.

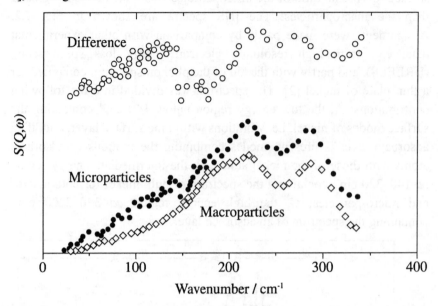

Fig. 7.3 INS of micro- and macronickel particles (KDSOG-M, Dubna), (bottom) and the difference spectrum (micro- minus macro, top) showing the spectrum of the riding. Reproduced from [8] with permission from Elsevier.

The vibrational density of states was calculated for nickel slabs up to 50 layers thick [8]. For the first and second layers, the vibrational density of states was different from that of the bulk reflecting their surface configuration. The calculated surface modes for the various surface planes were compared with experiment. The best agreement was for the (111) surface, see Fig. 7.4, evidence for predominant (111) faceting of the micro particles. Although by today's standards the spectra and the fits are poor, the work is of interest as an example of the use of INS in surface characterisation.

A more recent (2000) INS study of the vibrational density of states of nanocrystalline iron and nickel gave results consistent with the earlier work but with better quality spectra [9]. In comparison with reference coarse-grained specimens the nanocrystalline specimens exhibited a modest increase in the population of low-frequency modes, related to the presence of interface modes, and a distinct broadening of the transverse and longitudinal phonon:peaks, a consequence of reduced phonon lifetime due to the nanometric size of the crystallites. Peaks in the INS spectra were at 177 and 282 cm^{-1} for iron and 177 and 266 cm^{-1} for nickel [9].

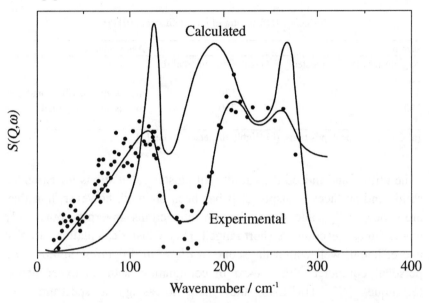

Fig. 7.4 Experimental (line with dots) and calculated (line) INS spectra for nickel micro particles (KDSOG-M, Dubna). Calculated data for Ni(111): layers 1 and 2 of a 50 layer nickel slab with force constant ratio surface/bulk 0.78. Reproduced from [8] with permission from Elsevier.

Calculations by semi-empirical quantum mechanics of the INS of water on nickel clusters also refer to the riding modes [10,11]. The cluster $Ni_{11}(H_2O)$, comprising a single layer of 11 nickel atoms with the water molecule bound to the central nickel atom, was modelled. The water molecule on top of the central nickel atom was at an optimised distance of 0.22 nm; it was not dissociated. The INS of nickel particles with adsorbed water molecules was assigned with the aid of the computed spectrum: peaks above 350 cm^{-1}, to two external and two internal vibrations of the adsorbed water molecules; peaks below 350 cm^{-1}, to adsorbent (i.e. nickel) vibrations enhanced by hydrogen atoms (the riding modes). Peaks are listed in Table 7.1 [10]. One water molecule caused a thousand-fold increase of the intensity of the vibrational spectrum of the nickel cluster. Similar calculations have been carried out for water on silica and silicon nitride [12,13]. The INS of coordinated water is discussed further in §7.4.1, see also §9.2.

Table 7.1 INS spectra of nickel clusters [10] (a).

ω/cm^{-1}			Assignments
Ni$_{11}$ cluster (a)	Ni$_{11}$ cluster + H$_2$O (b)	Experimental (b)	
25			
134	128	130	nickel surface vibrational
203	181	210	modes and riding modes
290	253	270	

(a) Calculated peak positions. (b) Highly dispersed nickel [2].

The utility, and indeed the validity, of cluster calculations for extended solids and surfaces is disputed. It has been claimed, however, that they are conceptually valid since 'electron interactions governing practically any chemical reaction are short ranged' [11]. Cluster calculations are fast and efficient, well within the scope of *ab initio* programs running on a personal computer. When used in conjunction with an experimental technique, INS, HREELS, they help to assign a spectrum. For computational studies of surface vibrational modes see also [14,15,16,17,18]. Force field lattice dynamics calculations for nickel slabs are reported in [14] and a theoretical treatment is given in [19,20,21]. The calculations use empirical nearest neighbour potentials, matching force constants to observed vibrational frequencies; e.g. for Ni a bulk force constant 0.379 mdyn Å^{-1} was matched to the maximum bulk frequency, 295 cm^{-1} [14].

7.3.1.2 Hydrogen on Raney nickel—Ni–H vibrations

There has been a number of studies of the INS of hydrogen on Raney nickel catalysts aimed at determining the binding sites of hydrogen atoms [4,5,6,7,22,23,24,25,26]. Raney nickel is prepared by etching away aluminium from a Ni-Al alloy with sodium hydroxide solution. The nickel remaining is a spongy mass with a high surface area (130 m^2 g^{-1}).

Table 7.2 INS spectra of hydrogen on Raney nickel (a).

ω/cm^{-1}	Assignment (b)	Surface plane	Site	Comment (f)	Ref.
320	lattice vibration			300 K	4
1120	Ni–H str				
940	as str (E)	Ni(111)	(c)	**300** and **80 K**	3,5
	H motion is	Ni(111)		*80* K	
	parallel to surface			(e)	
1140	s str(A_1)				
	Ni-H st				
590	s str(A_1)		(d)		6
780	as str (E)	Ni(110)	(c)	600 K	
(=(750 + 815)/2)					
1080 s	s str (A_1)				
600	s str	Ni(110)	(d)	H$_2$ adsorbed to	
780	as str (E)	Ni(111)	(c)	50% saturation	
940	as str (E)				
1100	s str (A_1)				
1300	s str		H bridge		
1900 or 2070	overtones combination (940+1130)				
(not observed)	as str (E)	Ni(110)	(c)	outgassed at 373 K, to	7
1100	s str(A_1)			residual H	
940	as str (E)	Ni(111)	(c)	373 K	
1130	s str (A_1)			300 K	
600	s str (A_1)		(d)		
1800	Ni-H s		a-top		
2070	combination 940+1130		(c)		

(a) Most recent measurements and assignments. See text for discussion. (b) as, antisymmetric; s, symmetric; str, stretching. (c) 3-fold hollow site, C_{3v} . (d) 4-fold hollow site. (e) INS intensity was proportional to the hydrogen loading. (f) temperatures shown in bold refer to the adsorption, those in italic refer to the INS experiment.

The nickel surface thus prepared retains oxygen and water: it is reduced in dihydrogen at *ca* 500 K and this treatment leaves hydrogen in the catalyst. INS peak positions and assignments are given in Table 7.2.

Spectra are shown in Fig. 7.5. Hydrogen is bound predominantly in three-fold (C_{3v}) sites characterised by INS peaks near 940 and 1140 cm^{-1}. [6].

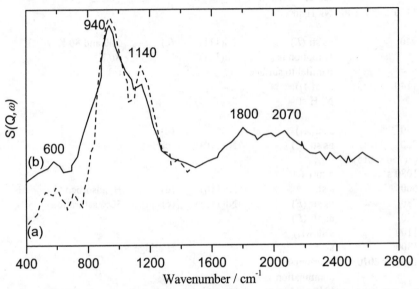

Fig. 7.5 INS difference spectra of hydrogen on Raney nickel. (a) Raney nickel (85 g, 7 nm, 14 m^2 g^{-1}) saturated H$_2$ at 300 K, INS 80 K (BT4, NIST) [3,5]. Reproduced from [5] with permission from the American Institute of Physics. (b) Raney nickel (15 g, 40 m^2 g^{-1}), H$_2$ 1.35 kPa, INS 300 K (IN1BeF, ILL). Reproduced from [7] with permission from Elsevier.

In a classic paper [4], the preparation of Raney nickel, adsorption of dihydrogen and dideuterium, and the theory and technique of time-of-flight INS measurements are described. The frequency distribution of the hydrogen or deuterium adatom motion contained one component (at 320 cm^{-1}) which is a riding motion involving the surface nickel atoms. A peak at higher wavenumber (1120 cm^{-1} for hydrogen and 800 cm^{-1} for deuterium) was assigned to a Ni–H or Ni–D vibration.

In another INS study of dihydrogen adsorbed by Raney nickel [23] INS bands were observed at 968 and 1129 cm^{-1}, and assigned by the Wilson GF method to hydrogen bound to three nickel atoms, Ni$_3$H. The dependence of the intensity on momentum transfer varied as Q^2 consistent with a one quantum process (§5.3.2.1). A peak at 1940 cm^{-1}

having an intensity varying as Q^4 was assigned to a two quantum process. A weak peak at 2175 cm^{-1} was assigned to a Ni–H stretching vibration of hydrogen bound to one nickel atom.

There has been a number of studies of the effect of varying dihydrogen pressure and temperature on the INS of hydrogen on Raney nickel. Above 150 K dihydrogen chemisorbed dissociatively predominantly in sites of threefold symmetry (see Table 7.2) [5]. The intensities of the 940 and 1140 cm^{-1} peaks were proportional to the dihydrogen adsorbed and their ratio was the same at all pressures and equal to two. Partial substitution of hydrogen by deuterium caused slight shifts in the peaks (to 985 and 1120 cm^{-1}) due to a repulsive hydrogen-deuterium interaction parallel to the surface.

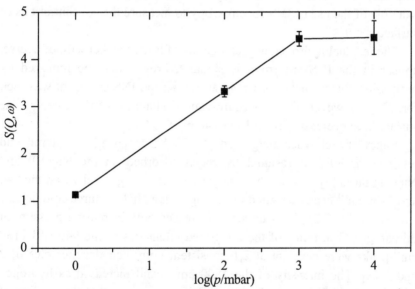

Fig. 7.6 INS of hydrogen on Raney nickel. Integrated intensity 600–1200 cm^{-1} *vs* dihydrogen pressure. Reproduced from [26] with permission from Elsevier.

Definitive assignments of Ni–H INS spectra rely on more recent work on the pressure dependence of the spectra. The effect of hydrogen pressure in the range 1 mbar to 10 bar on the formation of Ni–H species has been investigated [26]. At 1 mbar H$_2$ the INS spectrum showed an intense band at 940 cm^{-1}, a weaker band at 1130 cm^{-1} and a broad band centred at 2070 cm^{-1}. At 100 mbar an additional weak band was seen at 1850 cm^{-1}, which was assigned to the stretching mode of terminal hydrogen. Further increase of pressure to 10 bar caused band intensities to increase but otherwise did not change the spectrum. Thus the increase of pressure did not create new Ni–H species. From the frequencies and intensities the site concentrations of hydrogen were: C_{3v} on Ni(111), 60%; C_{3v} on Ni(110), 25%; C_{3v} on Ni(110), 10%; terminal H, 5%. There was no evidence for subsurface Ni–H species [27] at pressures up to 80 bar and no evidence for nickel hydride (peak at 720 cm^{-1}) [28,29]. The build-up of hydrogen on the surface could be followed by plotting the integrated intensity as a function of pressure, see Fig. 7.6. The saturation coverage (the horizontal portion of Fig. 7.6) was at 1.3–1.4 monolayer. The linear increase of INS intensity with increasing pressure is a specific example of the use of INS spectroscopy to measure the concentrations of surface species.

The residual hydrogen in preparations of Raney nickel will, of course, appear in the INS spectrum. Residual hydrogen may be removed by outgassing Raney nickel *in vacuo* at 520 K; the INS spectrum was then flat [23]. However, this procedure caused sintering of the catalyst, the surface area decreasing from 130 to 40 m^2 g^{-1}.

Raney nickel when outgassed at 373 K; remained unsintered and retained *ca* 30% of residual hydrogen adsorbed on the Ni(110) and Ni(111) faces [7]. Dosing the catalyst with dihydrogen enhanced the 940 and 1130 cm^{-1} peaks assigned to hydrogen in a Ni(111) three-fold hollow site (see Table 7.2). Their intensity increased with increasing pressure of dihydrogen. The ratios of the integrated intensities of the 940 and 1130 cm^{-1} peaks were constant at 2:1 consistent with their degeneracies of 2 and 1 [3]. The intensity of the 1800 cm^{-1} band increased as hydrogen adsorbed on nickel atoms (a-top sites). If we assume that the binding of hydrogen is stronger to the first occupied sites then hydrogen at the three -fold site is more strongly bound than a-top hydrogen. From the intensities it is estimated that a-top hydrogen comprises *ca* 15% of adsorbed hydrogen at saturation coverage.

Calculations of Ni–H vibration frequencies have helped in making assignments. An extended Hückel calculation of hydrogen on Ni(100),

Ni(110), Ni(111) gave peaks at 1350 cm^{-1} (a-top), 950 (bridge, two-fold), 600 (three- or four-fold). [30]. Experimental INS peaks at 940 and 1130 cm^{-1} were then assigned to hydrogen in three-fold sites and a peak at *ca* 600 cm^{-1} to hydrogen in four-fold sites. These assignments were not entirely in agreement with those from *ab initio* calculations on nickel clusters of up to 28 atoms: 1420, two-fold; 1212, three-fold; 592 cm^{-1}, four-fold [31].

7.3.1.3 Reaction of hydrogen and carbon monoxide on a nickel catalyst

The fate of adsorbed hydrogen atoms following a catalysed reaction may be determined by INS. Because of the time taken to accumulate an INS spectrum, the INS technique cannot be used directly to follow a chemical reaction. Generally the reaction will be allowed to proceed for a certain time and then frozen prior to the INS measurement.

The reaction between dihydrogen and carbon monoxide on Raney nickel giving methane has been studied by INS spectroscopy [24,25]. Dosing dihydrogen-saturated Raney nickel with carbon monoxide at 80 K caused little change in the spectrum, Fig. 7.7. Evidently, carbon monoxide and dihydrogen adsorbed independently and did not interact. Dihydrogen was evolved at 273 K with formation of a new hydrogenous species on the surface.

The formation of CH_x surface species was shown by conducting the CO/H_2 reaction in an INS cell, isolated from the gas flow and cooled to 20K. The INS spectra were similar after reaction at 300 and 450 K, reaction products were obtained only at 450 K. Peaks at 400 and 650 cm^{-1} were assigned to C–H fragments on the surface. On carbon-covered nickel the peaks shifted to 427, 589, 1000 cm^{-1} indicating the interaction of dihydrogen and carbon monoxide and the formation of CH species.

7.3.1.4 Palladium particles—surface vibrational states

The vibrational modes of hydrogen-doped nanocrystalline palladium have been studied by INS [23]. Two sizes of palladium particles were used: 17 nm with 2.9 at. % hydrogen and Pd black 4 nm with 4.0 at. % hydrogen. The hydrogen probed the vibrations of palladium atoms at surfaces and grain boundaries. The incoherent scattering from hydrogen was separated out from the coherent scattering from palladium by neutron spin analysis with polarised neutrons (the reader should consult the original paper and references therein for details of this technique). Surface palladium atoms contributed to additional modes at low frequencies (below *ca* 100 cm^{-1}). Palladium atoms within the grains were not thought to contribute to the low-frequency modes. The vibrational density of states, which for typical bulk materials is proportional to ω^3, was linear in ω to a cut-off near 100 cm^{-1}, implying that the vibrations are mostly restricted to the surface [32].

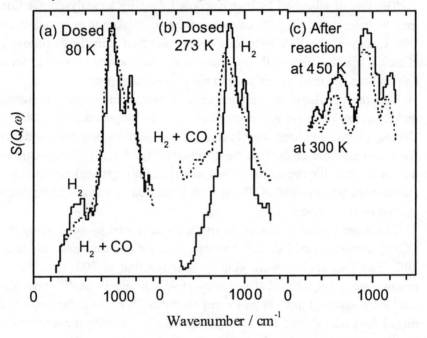

Fig. 7.7 INS spectra (BT4, NIST) (80 K) of H_2 and CO adsorbed on Raney nickel (a) Raney nickel (85 g, 14 m^2 g^{-1}) saturated with dihydrogen (53 mbar) at 300 K then dosed (1:1) with CO at 80 K. (b) Dihydrogen dosed to saturation coverage at 273 K, then dosed with CO at 80 K and brought to 273 K in equilibrium with gas phase. (c) After catalyst exposed to flowing gases (H_2:CO::7:1 by volume) at 300 K and 450 K. No conversion of CO to hydrocarbons observed at 300 K; 50% conversion at 450 K. Reproduced from [25] with permission from Elsevier.

7.3.1.5 Hydrogen and palladium—Pd hydrides and Pd–H vibrations

The ability of metallic palladium to dissociatively absorb hydrogen reversibly is well known. Hydrogen is first chemisorbed at the surface of the metal; at increased pressures hydrogen enters the palladium lattice and α- and β-phase palladium hydrides are formed. INS has been used to characterise both chemisorbed hydrogen and the hydrides (see Chapter 6).

The INS spectrum of hydrogen adsorbed by Raney palladium at 400 mbar corresponded to a mixture of β-PdH (480 cm^{-1}) and α-PdH (565 cm^{-1}) [27]. After dihydrogen adsorption at a lower pressure, 2 mbar, the INS showed peaks due both to the palladium hydrides and to adsorbed hydrogen. A broad peak at 975 cm^{-1} was assigned to the two unresolved vibrations of three-coordinated (C_{3v}) hydrogen [27].

According to an INS study of the β- and α-palladium hydride phases formed during absorption and desorption of hydrogen on palladium foil, coarse palladium powder and finely divided palladium [33], the retention of hydrogen by bulk palladium materials depended on their mechanical pre-treatment rather than their morphology; the powdered materials retained more hydrogen than palladium foil under comparable conditions. This finding is explained by the presence of trap sites, more for the powders than for the foil. This paper contains a useful summary of INS studies of the H/Pd system which is reproduced here, Table 7.3. [34,35,36,37,38,39,40,41,42,43,44,45,46,47,48,49].

Table 7.3 INS spectra of the palladium–hydrogen system.

Palladium sample type	ω/cm^{-1}	Assignment (a)	Additional features/cm^{-1}	Ref.
Single crystal	532	α-hydride	shift to 508 at 2.7 atom% H	34
Single crystal 0.2 atom% H	1089	α -hydride overtone		34
Foil	464	β-phase	high-energy shoulder at 645–725 (b)	35,36,37,38

Palladium sample type	ω/cm^{-1}	Assignment (a)	Additional features/cm^{-1}	Ref.
Raney Pd	480	β-phase	overtone at 1149	27,35
	565	α-hydride	second overtone at 1798, 604, 719, 771 and 974	
H on Pd(lll)	774			39
	1000			
Pd (100)	820	Pd–H s str		40,41
Pd(111)	915	Pd–H s str		
Pd (100),(111)	1700	Pd–H as str		
Pd black	557	α-hydride	758, 814, 968	42,43,44,45
	468	subsurface hydrogen	weak: 604, 895, 1452; 758, 814 and 968 (c)	
	968	Pd–H str perpendicular		46
	790	Pd–H str parallel		
	820	Pd–H s str		3,47
	915	Pd–H as str		
Pd Y zeolite	569	α-hydride		48
	480	β-phase		
	699	subsurface hydrogen		
	990	chemisorbed hydrogen		
Pd/activated carbon	450	β phase	829, 919	49
	610	α-hydride		
Pd/carbon black	452	β-phase		49
	619	α-hydride		

(a) str, stretching; as, antisymmetric; s, symmetric. (b) frequency falls as H concentration rises. (c) intensity increases with H-content

The INS of chemisorbed hydrogen and hydrogenous molecules has been reviewed [46]. The INS spectrum of hydrogen on palladium black was assigned as follows: 968 cm^{-1}, hydrogen on symmetric three-fold sites of Pd(111) with the Pd–H vibration perpendicular to the surface and non-degenerate; 790 cm^{-1}, Pd–H vibration parallel to the surface, doubly-degenerate with twice the intensity of the 968 cm^{-1} peak. From a Wilson GF analysis the best fit to the spectra was obtained for a Pd–H bond distance of 2.10 Å and a force constant 0.43 mdyn Å^{-1}. The 790 cm^{-1} peak was partially split into a peak at 758 cm^{-1} and a shoulder at 815 cm^{-1} owing to strong H–H lateral interactions. A peak at 468 cm^{-1} was assigned to hydrogen in a subsurface site. The surface chemistry, and hence details of the spectra, and the propensity of palladium to form bulk hydride phases can be affected by the presence of impurities, for example, aluminium left over from the preparation of Raney palladium.

7.3.1.6 Hydrogen on platinum—Pt–H vibrations

Assignments of Pt–H vibrations are given in Table 7.4.

Table 7.4 Pt–H vibrations—assignments.

ω/cm^{-1} INS	HREELS	Assignment (a)	References
530	548	Pt$_2$–H as str along $2\bar{1}\bar{1}$	3,48
925	863	Pt$_2$–H as str along $0\bar{1}\bar{1}$	
1340	1234	Pt$_2$–H s str along 111	
548		Pt$_3$–H s str (A_1)	48
1234		Pt$_3$–H as str (E)	

(a) Alternative assignments: see text for discussion. Pt$_2$–H, H in two-fold site; Pt$_3$–H, H in three-fold site; str, stretching; as, antisymmetric; s, symmetric.

For platinum powder before and after dosing with dihydrogen peak positions and assignments in the INS spectrum are given in Table 7.5 [50]. The platinum was in the form of a 55.7 g cylinder of compressed platinum powder which had been reduced in dihydrogen, evacuated and then dosed with dihydrogen.

Table 7.5 INS spectra of hydrogen on platinum, see text for details [50].

| Pt powder reduced by H_2 and evacuated | | Pt powder dosed with H_2 difference spectrum | |
ω/cm^{-1}	Assignments	ω/cm^{-1}	Assignments
320	residual H	400	residual H
1200		810	
2400		1770	
		55	H migrating
		30	
25	Pt lattice	70	
40		100	
90		145	
145		220	

Some hydrogen remained on the platinum even after evacuation (423 K, 4 d, 2.7×10^{-6} mbar). For platinum metal, peaks below *ca* 200 cm^{-1} were assigned to platinum lattice vibrations, and higher energy peaks to residual hydrogen. For dihydrogen-dosed platinum the spectrum was assigned with some hydrogen tightly bound to platinum and some weakly bound. Presumably, weakly bound H is created only after the strongly bound sites are full. The lower energy peaks were assigned to the motion of a hydrogen atom in a shallow potential well of *ca* 160 cm^{-1} [51].

In the INS of hydrogen on powdered platinum [52] peaks in the difference spectrum were observed at 70 and 150 cm^{-1}, correlating with peaks in the bulk vibrational spectrum of platinum, and at 400 cm^{-1}, assigned to a local vibrational mode of hydrogen coupled to platinum.

The vibrational frequencies were modelled by the Wilson GF method incorporating one internal coordinate for the Pt–H mode, a set of coordinates for the displacements of the surface atoms, and separate coordinates for the displacements of the atoms in the interior of the crystal. The model accounted qualitatively for the INS spectrum: a local mode coupled to the surface modes, lying above the high-frequency band-edge (§2.6.2) of the surface modes. The surface modes appeared as a side band to the local mode–a phonon wing (§2.6.3). According to the intensity near the high frequency edge of the platinum density of states, hydrogen was coupled to more than one platinum.

Three peaks have been reported in the low energy INS spectrum of hydrogen on platinum black [53]. A peak at 400 cm⁻¹ was assigned to a local mode of vibration of hydrogen on the surface. The peaks at 177 cm⁻¹ and 104 cm⁻¹ corresponded to the INS spectrum of platinum alone observed through the riding modes.

7.3.1.7 Hydrogen on carbon-supported platinum and platinum-alloys

There has been a number of studies of carbon-supported platinum and platinum-alloy catalysts. INS has been used in the characterisation of activated catalyst supports from different natural sources through patterns of edge termination reflected in the INS spectra in the out-of-plane C–H bending region near 880 cm⁻¹ [54,55].

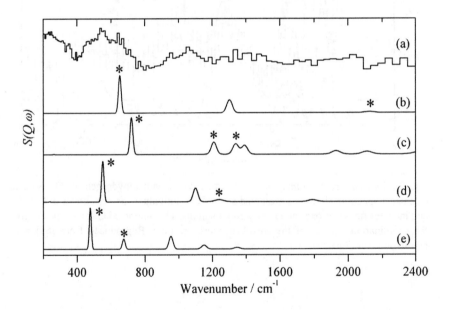

Fig. 7.8 (a) INS spectrum, 40 wt-% Pt/C, dosed with dihydrogen at 295 K and spectra computed for hydrogen in various surface sites: (b) a-top, (c) two-fold, (d) three-fold and (e) four-fold hydrogen coordination. Reproduced from [55] with permission from Elsevier.

High-metal-loading catalysts are used in fuel cells to catalyse the hydrogen-oxygen reaction. The Pt–H modes in the INS spectra of hydrogen on Pt 20 wt % and 40 wt % on carbon have been assigned with the Wilson GF matrix method, see Fig. 7.8. The spectra expected for hydrogen in various sites were modelled. The experimental spectra were fitted to Gaussian peaks, see Fig. 7.9, and the proportions of each type of site on the catalysts was calculated, see Table 7.6.

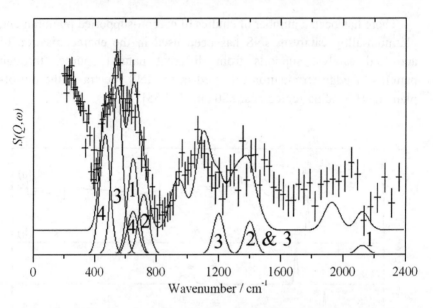

Fig. 7.9 INS spectrum (points), 40 wt-% Pt/C, dosed with dihydrogen at 295 K. The calculated spectrum (line) is a weighted sum of the individual species shown in Fig. 7.8 and includes first overtones and binary combinations. The numbered Gaussian peaks give the coordination number of the adsorbed hydrogen atom. Reproduced from [55] with permission from Elsevier.

Table 7.6 Percentages of each type of platinum site on the catalysts.

Site	Pt(20%)/C	Pt(40%)/C
a-top	16	19
4-fold bridge	21	26
2-fold bridge	29	22
3-fold bridge	34	33

7.3.1.8 Hydrogen and oxygen on platinum black—hydroxyl groups

Hydroxyl groups are formed when oxygen reacts with hydrogen on platinum black [56]. The reaction was followed with INS. Peaks assigned to residual hydrogen (839, 1226 cm^{-1}) were replaced, in the range observed, by a single peak at 1032 cm^{-1} assigned to the Pt–OH bending mode.

7.3.1.9 Hydrogen spillover on Pt/C and Ru/C catalysts

The presence of spillover hydrogen on the carbon support of Pt/C catalysts has been shown by INS. Practical metal catalysts consist of metal particles supported on high surface area oxide or carbon. Many catalytic reactions involve hydrogen and spillover of hydrogen is often implicated [57,58]. In spillover, hydrogen atoms, produced by dissociative chemisorption of dihydrogen on the metal part of a catalyst, diffuse from the metal to the catalyst support. On an oxide-supported metal catalyst, spillover hydrogen atoms react with the oxide forming OH species that can be identified by infrared spectroscopy. On a carbon-supported catalyst, spillover hydrogen may bind, for example at unsaturated edge sites, or constitute a layer of weakly bound, mobile hydrogen atoms.

Hydrogen spillover on carbon-supported platinum and ruthenium catalysts has been identified by INS spectroscopy, Fig. 7.10 [59]. The catalysts were commercial high loading (40 % wt metal) carbon-supported platinum and ruthenium fuel cell catalysts. The INS spectra of the catalysts dosed with hydrogen were determined in two sets of experiments: (a) using a standard circular cell with the catalyst (i.e. the metal component and the carbon support) in the neutron beam; (b) using an annular cell with catalyst pellets around the outside edge of the cell and not in the beam; but pure carbon support in the central region of the cell and in the beam. In this cell the only hydrogen which could contribute to the neutron scattering would be hydrogen which had dissociated and diffused on the carbon into the beam.

Two forms of spillover hydrogen were identified: hydrogen bound at edge sites of a graphite layer (formed after ambient dissociative chemisorption of dihydrogen), and a weakly bound layer of mobile hydrogen atoms (formed by surface diffusion of hydrogen atoms after dissociative chemisorption of dihydrogen at 500 K). The INS spectra exhibited characteristic riding modes of hydrogen on carbon and on platinum or ruthenium (§7.1). INS spectroscopy thus directly demonstrated hydrogen spillover to the carbon support of these metal catalysts.

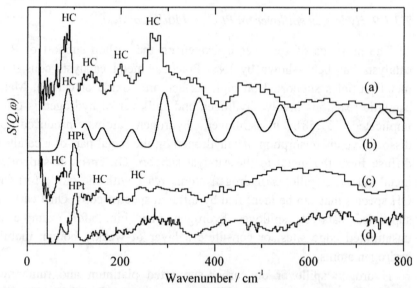

Fig. 7.10 INS spectra of (a) spillover hydrogen from a Pt/C catalyst with only carbon and hydrogen in the neutron beam (b) calculated carbon coronene spectrum; (c) hydrogen-dosed Pt/C catalysts with catalyst and hydrogen in the neutron beam; (d) Pt black reduced, evacuated and dosed with hydrogen. Peaks labelled HC are due to hydrogen on carbon and HPt to hydrogen on platinum. The HPt peak of the catalyst at *ca* 100 cm^{-1} corresponds to the strong peak in the platinum black spectrum. Reproduced from [59] with permission from the American Chemical Society .

The hydrogen-on-carbon INS spectra were assigned with reference to the observed and calculated INS spectra of the polycyclic aromatic hydrocarbon coronene, Fig. 7.10. Low wave number peaks in the spectra of the hydrogen-dosed catalysts were thereby assigned to out-of-plane displacements of edge carbon atoms. Lattice modes of the metal components of the catalysts were also observed amplified by bound hydrogen atoms. Thus through the amplification effect of hydrogen on the lattice modes of the catalyst, including the carbon support, INS spectroscopy was able uniquely to determine the location of hydrogen atoms and, in particular, to demonstrate hydrogen spillover.

7.3.1.10 Lateral interactions between hydrogen and deuterium atoms

INS spectra may be broadened and shifted through lateral interactions, dynamic coupling, between adsorbed hydrogen atoms on metal and other surfaces.

Substitution of hydrogen atoms by a majority of deuterium atoms changes the dynamics. The dispersion band of metal–deuterium vibrations now appears at much lower frequencies and the few remaining hydrogens are too well separated to interact (§2.2.1). We see band sharpening and a shift to higher frequency when deuterium is substituted for hydrogen, Fig. 7.11 [3,46]. The theoretical treatment of the isotope dilution effect uses mass defect theory [46,60]. Dilute hydrogen atoms are treated as isolated mass defects in a layer of deuterium atoms which comprise the surface host phase. The energy-dependent density of states of the deuterium layer, $g_D(\omega)$, is modified by the presence of hydrogen

$$\int \left(g_D(\omega) / \omega_H^2 \right) d\omega = \left(m_D / (m_D - m_H) \right) \omega_H^2 \tag{7.1}$$

Here ω_D is the hydrogen (defect) local mode energy and the m's are the masses of hydrogen and deuterium. The lighter hydrogen defect atom cannot vibrate within the dispersion-broadened phonon density of states of the surrounding heavier deuterium host atoms. Rather the hydrogen atom vibrates as a hydrogen-defect mode above the ω_{max} of the deuterium host (§7.1), see Fig. 7.11.

7.3.1.11 Dihydrogen on a ruthenium catalyst—the H_2 rotor

Rotational transitions of dihydrogen are readily observed in INS and are discussed in Chapter 6. The position of, and any splitting in, the $J(1 \leftarrow 0)$ rotational transition at *ca* 120 cm^{-1} can provide information on the strength of interaction of the dihydrogen molecule at, for example, a catalytic site. The interaction of dihydrogen with ruthenium at 300 K has been studied by *in situ* INS [61]. The concentration of hydrogen on the catalyst was proportional to the INS intensity. The $J(1\leftarrow0)$ transition was observed in the spectrum of dihydrogen adsorbed at 20–120 K on a Ru/C catalyst. Dihydrogen comprised two layers, one interacting with the catalyst, and a second, less strongly bound, on top of the first layer.

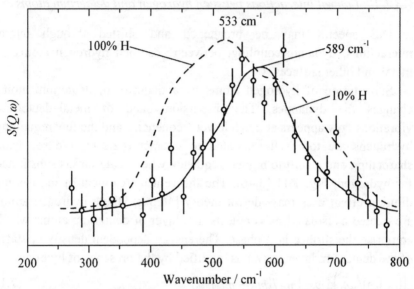

Fig. 7.11 INS spectra of hydrogen (BT4, NIST) showing dispersion (broken line) and hydrogen (10%) in deuterium, dispersion suppressed (circles and full line) at 80 K. The hydrogen peak shifts to higher energy and becomes narrower. Reproduced from [3,46] with permission from Elsevier.

The $J(1\leftarrow0)$ dihydrogen rotation appeared as a sharp strong peak at 120 cm^{-1} and the $J(2\leftarrow1)$ transition as a weak peak at 240 cm^{-1}. A broad peak at 535 cm^{-1} was assigned to the hydrogen recoil (§6.1.5). The spectrum was similar to that of solid dihydrogen [62,63]. When the dihydrogen-dosed catalyst was warmed to 120 K the dihydrogen rotor peak at 120 cm^{-1} disappeared as adsorbed dihydrogen dissociated.

7.3.2 Hydrocarbons on metal catalysts and reference hydrocarbons

The use of INS spectroscopy to study adsorbed hydrocarbons is complementary to the use of optical techniques. With INS we can exploit the frequencies and intensities of vibrational modes more readily. It is helpful to compare any spectrum of adsorbed species with the spectrum of the corresponding free molecule and we include accounts of the INS spectra of relevant reference compounds in this section. Fitting an INS spectrum to a force field which includes the metal binding site will tell us about the orientation of an adsorbed molecule.

7.3.2.1 Ethyne, C_2H_2, on platinum black

The catalytic hydrogenation of ethyne (acetylene, C_2H_2) to ethene proceeds through a surface reaction between adsorbed ethyne and hydrogen atoms. We use vibrational spectroscopy to learn about the interaction between adsorbed ethyne and the catalyst and to identify intermediates.

The INS spectrum of ethyne adsorbed by platinum black at 120 K was fitted to a force field with a deformed ethyne ($\angle$ H–C–C, 143.5°) [64]. The peak positions are close to those reported by HREELS [65] of ethyne on Pt(111) (the predominant exposed crystal face of platinum powder). The spectra may be assigned with reference to ethyne and the ethyne complexes $[Co_2(CO)_6(\mu_2\text{-}\eta^2\text{-}C_2H_2)]$ [66] and $[Os_3(\mu_2\text{-}CO)(CO)_9$ $(\mu_3\text{-}\eta^2\text{-}C_2H_2)]$ [67]. Spectra are shown in Fig. 7.12 and peaks are listed in Table 7.7. The structures and INS spectra of the ethyne complexes are discussed further in §7.3.2.2.

When the sample was heated to 300 K in 530 mbar of dihydrogen the spectrum changed (see Fig. 7.12). The additional peaks at 520 and 1370 cm^{-1} were associated with hydrogen but not assigned.

7.3.2.2 Ethyne complexes of cobalt and osmium

The complexes $[Co_2(CO)_6(\mu_2\text{-}\eta^2\text{-}C_2H_2)]$ [66] and $[Os_3(\mu_2\text{-}CO)(CO)_9(\mu_3\text{-}\eta^2\text{-}C_2H_2)]$ [67] are models for the bonding of ethyne to metals, e.g. Pd(111) and Pt(111) [68]. In the cobalt complex, the C–C group lies between the cobalt atoms above and at right angles to the Co–Co (Fig. 7.13(a)).

In the osmium complex, Fig. 7.13(b), the C–C group lies above and in a plane parallel to the Os$_3$ triangle with each C bound to two Os.

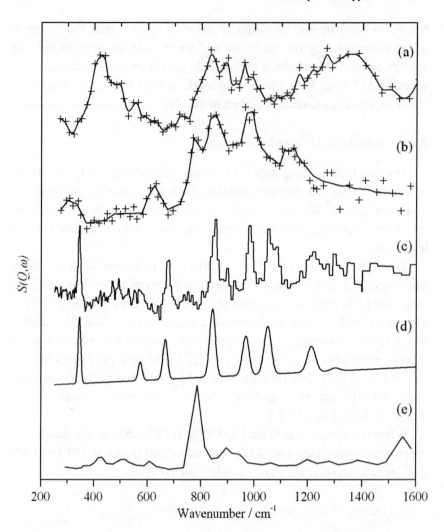

Fig. 7.12 INS spectra (BT4, NIST) of (a) ethyne on platinum black annealed at 300 K with 500 mbar of dihydrogen; (b) ethyne on platinum black adsorbed at 120 K, no dihydrogen. Reproduced from [64] with permission from the American Institute of Physics. $[Os_3(\mu_2\text{-}CO)(CO)_9(\mu_3\text{-}\eta^2\text{-}C_2H_2)]$: (c) experimental, (d) modelled with the Wilson GF method. Reproduced from [67] with permission from the PCCP Owner Societies. (e) $[Co_2(CO)_6(\mu_2\text{-}\eta^2\text{-}C_2H_2)]$, experimental (1NBeF, ILL). Reproduced from [66] with permission from the American Chemical Society. Note that the peak pattern of adsorbed ethyne (b) is similar to that of ethyne in the osmium complex (c). Note also the additional peaks near 500 and 1400 cm^{-1} when adsorbed ethyne was treated with hydrogen (a).

Table 7.7 Vibrational spectra (ω/ cm^{-1}) of ethyne gas, adsorbed and as complexes.

C$_2$H$_2$ gas [65]	C$_2$H$_2$/Pt black INS (HREELS) [64]	η^2-C$_2$H$_2$ (a) INS [66]	η^2-C$_2$H$_2$ (b) INS [67]	Assignment (c)
		165		OC–Co–C$_2$H$_2$ def
			190	CC ip def
	306 (339)		348	Os–C(π) str
		430		C$_2$H$_2$ torsion (d)
	484 (573)		570	Os–C(σ) s str
	629		678	Os–C(σ) as str
612				HC$_2$H trans def
729				HC$_2$H cis def
	774 (774)	776, 768		C–H s def and C$_2$H$_2$ torsion (e)
		894		C–H as def
	847		855	C–H op s def
	984 (984)		983	C–H op as def
	1145		1050	C–H ip s def
	(1307)		1213	C–H ip as def
1974		1403		C–C str
3295				C–H as str
3373				C–H s str

(a) [Co$_2$(CO)$_6$(μ_2-η^2-C$_2$H$_2$)] (b) [Os$_3$(μ_2-CO)(CO)$_9$(μ_3-η^2-C$_2$H$_2$)] (c) ip, in-plane; op, out-of-plane; str, stretching; def, deformation; s, symmetrical; as, antisymmetrical. (d) About axis through Co–Co and C–C bond centres. (e) Libration about C≡C.

The CO stretch and deformation modes, which appear strongly in the infrared and Raman spectra, give rise only to weak INS features since they are decoupled from the ethyne ligand. The INS spectrum isolates those modes due to ethyne.

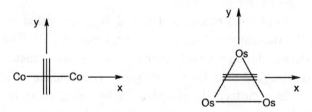

Fig. 7.13 Bonding of the ethyne ligand in [Co$_2$(CO)$_6$(μ_2-η^2-C$_2$H$_2$)] (left) and [Os$_3$(μ_2-CO)(CO)$_9$(μ_3-η^2-C$_2$H$_2$)] (right). The view is from above looking down the *z*-axis which is perpendicular to the page. The ethyne ligand lies above the metal atoms.

The experimental INS spectra of the Os and Co complexes and the spectrum of the Os complex modelled with the Wilson GF matrix method are shown in Fig. 7.12. Assignments for C_s symmetry are shown in Table 7.7. The vibrational modes of the osmium complex are illustrated in Fig. 7.14 (p. 311).

The osmium complex is a model for sideways ethyne adsorption on metal catalysts [69,70]. The pattern of peaks is similar to that of ethyne on platinum black (see Fig. 7.12). The frequencies were consistently higher than for adsorbed ethyne (see Table 7.7) implying stronger binding of ethyne to osmium in the complex than to platinum black.

For the cobalt complex INS peaks with significant intensity and assignments from the Wilson GF method are given in Table 7.7. The spectrum of the cobalt complex is remarkably different from that of the osmium complex (cf. Table 7.7 and Fig. 7.12); the wavenumbers of the C–H deformations are *ca* 300 cm^{-1} lower for the Co complex compared with the Os complex.

7.3.2.3 Ethene, $CH_2=CH_2$, and propene, $CH_3CH=CH_2$

Ethene (ethylene) is the prototypical alkene. Its mechanism of hydrogenation is a topic of continuing interest [69,70]. INS measurements have been especially useful in exploring the low frequency torsional motions of bound ethene and in the observation of spectra free from substrate interference.

The normal modes of the ethene molecule, adsorbed ethene and ethene complexes as derived from the infrared and Raman spectra of ethene, are shown in Fig. 7.15 [71].

The spectra of solid ethene at 4 K and 80 K consisted of weak and shifted peaks superimposed on a broad background [72]. The spectrum was broadened through recoil from the low–mass ethene molecule (§2.7.5) as observed also for dihydrogen. When ethene is bound in a complex or to a surface its effective Sachs-Teller mass (§2.6.5.1) is increased; the recoil spectrum is weaker and the peaks due to the vibrational modes are stronger and sharper. The vibrational peaks of gaseous and solid ethene and ethene adsorbed by carbon are listed in Table 7.8.

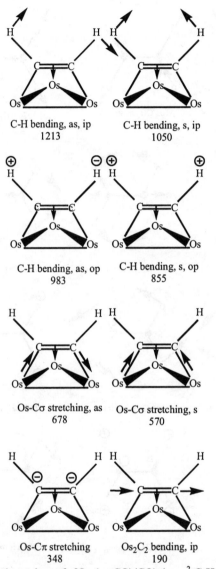

C-H bending, as, ip
1213

C-H bending, s, ip
1050

C-H bending, as, op
983

C-H bending, s, op
855

Os-Cσ stretching, as
678

Os-Cσ stretching, s
570

Os-Cπ stretching
348

Os₂C₂ bending, ip
190

Fig. 7.14 Vibrational modes of $[Os_3(\mu_2\text{-CO})(CO)_9(\mu_3\text{-}\eta^2\text{-}C_2H_2)]$ with appreciable hydrogen atom displacements and peaks (ω / cm^{-1}) in the INS spectrum (ip, in-plane bend, op out-of-plane bend) (cf Fig. 7.12 and Table 7.7). Reproduced from [67] with permission from the PCCP Owner Societies.

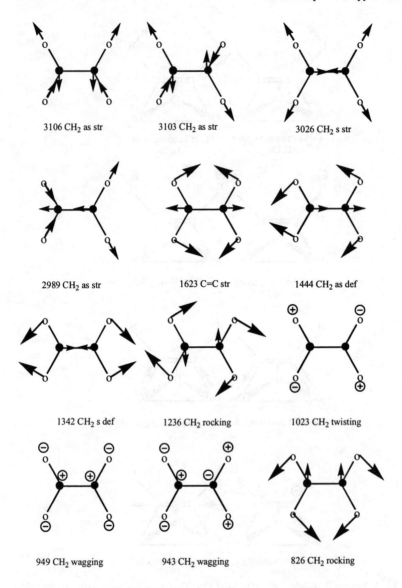

3106 CH$_2$ as str   3103 CH$_2$ as str   3026 CH$_2$ s str

2989 CH$_2$ as str   1623 C=C str   1444 CH$_2$ as def

1342 CH$_2$ s def   1236 CH$_2$ rocking   1023 CH$_2$ twisting

949 CH$_2$ wagging   943 CH$_2$ wagging   826 CH$_2$ rocking

Fig. 7.15 Approximate normal modes of vibration of ethene and observed transitions (ω/ cm^{-1}). The arrows show atom displacements in the plane of the figure; signs show displacements above ($\oplus$) and below ($\ominus$) the plane; str, stretching; def, deformation (bending); s, symmetric; as, antisymmetric.

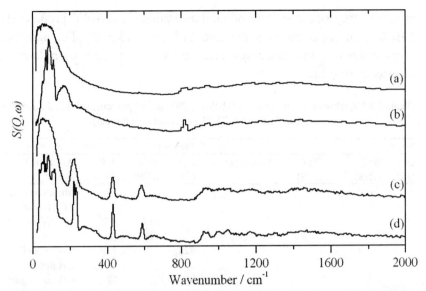

Fig. 7.16 INS spectra at 20 K of: (a) physisorbed and (b) solid ethene, (c) physisorbed and (d) solid propene. Reproduced from [73] with permission from the PCCP Owner Societies.

The INS spectra of ethene and propene adsorbed by carbon at 20 K show that both are physisorbed, forming disordered layers [73]. The interaction of molecules with carbon is weaker than the interaction between the molecules in the solid alkenes. Spectra are shown in Fig. 7.16 and peaks are listed and assigned in Table 7.8.

The spectra revealed a number of noteworthy features. A great deal of intensity was transferred into the phonon wings (a broad band centred at 1200 cm^{-1}). This is caused by high values of the mean-squared displacement of atoms in the external modes, $^{Tr}A_{ext}/3 = 0.045$ Å^2 (§2.6.3).

These are at the higher end of the range of values that yield unrecoiled INS spectra (cf solid benzene, 0.011 Å^2 (§5.2.2, §5.3.2)) reflecting the weak intermolecular forces in the solid alkenes and the yet weaker bonding to carbon. In the INS of adsorbed propene we see sharp features below 750 cm^{-1}, in contrast to ethene; the molecular recoil of propene is less than the recoil of ethene—a consequence of the greater mass of propene and because on low-bandpass spectrometers, transitions

at low energy occur at low momentum transfer. For solid propene the
CH_3 torsion appeared as a doublet, 217 and 228 cm^{-1}. For adsorbed
propene this torsion was a single broad peak. This mode is a single mode
in gaseous propene.

Table 7.8 Vibrational spectra (ω/cm^{-1}) below 2000 cm^{-1} of gaseous, solid and adsorbed
ethene and propene [72,73 and references therein].

Ethene				Propene			
Gas (IR)	Solid (INS)	On C (INS)	Assign (a)	Gas (IR)	Solid (INS)	on C (INS)	Assign (a)
	54,70,84, 108, 168						librations and Translations
				188	217 228	219	C-CH_3 torsion
				428		429	C=C-C def CH def
				575		586	CH op wag and C=C torsion
826	822	820	CH_2 rock				
				919 935		928	C—C str
943, 949	952	944	CH_2 wag	990		999	C=C torsion CH_3 rock
1023	1032	1021	CH_2 twist	1045		1057	CH_3 rock CH op wag
				1174		1164	CH_3 rock CH_2 rock
1236	1218	1132	CH_2 rock				
1342	1367		CH_2 scissors	1458		1452	CH_3 as def
1444	1444		CH_2 as def				
1623	1621		CC str				

(a) Assignments cf Fig. 7.15: def, deformation; str, stretching; as, antisymmetric; op,
out-of-plane.

The solid state splitting was attributed to factor group splitting
(§2.6.1.3). The absence of this splitting for adsorbed propene was
evidence for a disordered layer. Strictly, the collapse of the factor group
splitting arises because there is only one molecule per unit cell or
because intermolecular forces are weak. The frequency of the CH_3
torsion was greater by 31 cm^{-1} for adsorbed propene than for gaseous
propene—because the torsional motion is hindered due to adsorption.

7.3.2.4 Ethene on platinum black

INS spectra are shown in Fig. 7.17. The peak positions, Table 7.9, are close to those in the HREELS of ethene on Pt(111) (the predominant exposed crystal faces) [64]. The spectrum was modelled with a non-planar ethene (H–C–C–H dihedral 82.8°). Spectral changes as the temperature was raised were attributed to ethylidyne, CH_3C, by comparison with the spectra of a cobalt ethylidyne complex (§7.2.2.6).

Table 7.9 INS spectra (ω/cm⁻¹)of ethene and ethylidyne on platinum black [64].

Ethene/Pt, 120 K	Ethene/Pt, 300 K	Ethylidyne
468	467	403
557		548
653	645	
790	887	
952	984	1008
1049		
1194	1129	1161
1403	1371	1355
		1420

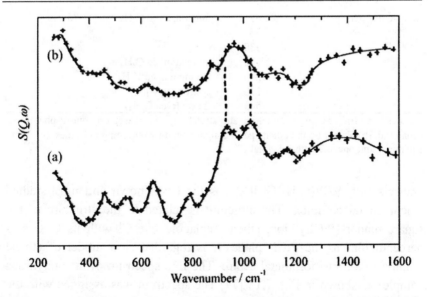

Fig. 7.17 INS spectrum of ethene on platinum black (BT4, NIST). (a) Adsorbed at 120 K. (b) Annealed at 300 K (without hydrogen). Reproduced from [64] with permission from the American Institute of Physics.

7.3.2.5 Ethene—$(\eta^2\text{-}C_2H_4)$ complexes

The INS spectra of ethene complexes of rhodium and iridium and of the platinum complex, Zeise's salt, $K[Pt(C_2H_4)Cl_3]$, help in the assignment of the the spectrum of adsorbed ethene.

The INS spectra of a series of ethene complexes were determined in the low frequency region (below 800 cm^{-1}) [74, 75, 76]. The interest was the torsional motions of ethene bound at a metal center. Peak positions and assignments are given in Table 7.10. The torsional force constants were calculated.

Table 7.10 Low frequency INS spectra of ethene complexes.

Compound	ω / cm^{-1}	Assignment (a)	Reference
$[\{Rh(\eta^2\text{-}C_2H_4)_2\}_2(\mu\text{-}Cl)_2]$ (b)	112	ip torsion	75
	226	op torsion	
$[\{Ir(\eta^2\text{-}C_2H_4)_2\}(\mu\text{-}Cl)_2]$ (b)	114	ip torsion	76
	163		
	243	op torsion	
$[Ir(\eta^2\text{-}C_2H_4)_4Cl]$ (c)	45	torsion, axial C_2H_4	76
	76		
	130		
	172	ip torsion, eq C_2H_4	
	240	op torsion, eq C_2H_4	
	290	s str(Ir–eq C_2H_4)	
	503	as str(Ir–eq C_2H_4)	

(a) s, symmetric; as, antisymmetric; str, stretch; ip, in-phase; op, out-of-phase; eq, equatorial. (b) Two π bound ethene per metal atom and two bridging chlorines. (c) Three equatorial and one axial ethene, axial Cl

Zeise's salt, $K[Pt(C_2H_4)Cl_3]H_2O$, is the best known and most-studied ethene–metal complex. The structure is related to the structure of the square planar $[PtCl_4]^{2-}$ ion, ethene replacing one Cl with its C=C axis perpendicular to the $PtCl_3$ plane. It is regarded as a model of ethene binding to a transition metal centre. The INS spectrum of the anhydrous complex is shown in Fig. 7.18 [77].The spectrum was assigned with the aid of a Wilson GF method calculation, see Table 7.11. However, a more recent spectrum, Fig. 1.1, shows that the factor group splitting of the main features is significant and that the isolated molecule approximation is inadequate. In this case a periodic DFT calculation is needed fully to model this system.

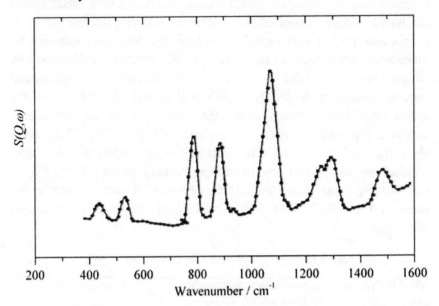

Fig. 7.18 INS spectrum (IN1BeF, ILL) of Zeise's salt, K[Pt(C$_2$H$_4$)Cl$_3$]. Reproduced from [77] with permission from Elsevier. Note that the spectrum is uncorrected, i.e. offset by *ca* 30 cm^{-1} by the transmission of the beryllium filter.

Table 7.11 INS of Zeise's salt.

ω /cm^{-1}	Assignment (a)
405	C$_2$H$_4$—Pt as str
493	C$_2$H$_4$—Pt s str
720	CH$_2$ twist
845	CH$_2$ rock
975	CH$_2$ wag
1010	CH$_2$ wag
1020	CH$_2$ twist
1180	CH$_2$ rock
1240	CH$_2$ scissor + CC str
1425	CH$_2$ s scissor
1515	CC str + CH$_2$ s scissor

(a) a, antisymmetric; s, symmetric; str, stretch.

7.3.2.6 Ethylidyne, CH₃C

7.3.2.6 Ethylidyne, CH$_3$C

The ethylidyne species, CH_3C, is formed by the dissociative chemisorption of ethene on metals. Although present on a metal during catalytic hydrogenation of ethene, ethylidyne is not an intermediate; it is a spectator [78]. Chemisorbed ethylidyne has been characterised by vibrational spectroscopy and low energy electron diffraction by comparison with the model compound tricobaltnonacarbonyl(ethylidyne), $[Co_3(CH_3C) (CO)_9]$. The INS of the cobalt ethylidyne complex aided the assignment of the vibrational spectrum especially in the low wavenumber region [79]. The torsion about the Co_3C–CH_3 bond was located (208 cm^{-1} rather than 383 cm^{-1} reported previously) and was found to be strongly mixed with the Co_3C–CH_3 bending mode. Peaks in the INS spectrum of the cobalt complex are listed in Table 7.12 along with the HREELS spectra of chemisorbed ethylidyne.

Table 7.12 Ethylidyne vibrational spectra (ω / cm^{-1}) [79].

[Co₃(CH₃C) (CO)₉]. INS	Surface-bound ethylidyne HREELS and RAIRS (a) on Pt(111)	on Ni(111)	Assignment(b)
180			(CoCo) as str (*E*)
208			(CH₃) torsion(*A₂*)
224	310		(CoC-CH₃) def(*E*)
242	160		(CoCo) s str (*A₁*)
401	430	362	(CoCCH₃) s str
555	600	457	(CoCCH₃) as str (*E*)
1004	980	1025	(CH₃) rocking(*E*)
1163	1124	1129	(CC) str (*A₁*)
1356	1339	1336	(CH₃) s def(*A₁*)
1420	1420	1410	(CH₃) as def(*E*)
2888	2884	2883	(CH₃) s str (*A₁*)
2930	2950	2940	(CH₃) as str(*E*)

(a) Reflection absorption infrared spectroscopy: refs. in [79]. (b) str, stretching; def, deformation; as, antisymmetric; s, symmetric

7.3.2.7 Methyl groups on a deactivated palladium catalyst

7.3.2.7 Methyl groups on a deactivated palladium catalyst

Spent technical palladium catalysts used in the hydrogenation of polyaromatic ketones were studied by INS to determine the surface species responsible for the deactivation [80]. The spent catalysts had only 10% of their regular hydrogenation activity. Catalyst deactivation can occur by many routes but common ones are poisoning and coke formation. The INS spectra of cokes are known and range from 'polymer-like' to almost pure graphite (§11.2.5). The INS spectrum of deactivated catalyst was very different from that of any form of coke and was much more reminiscent of a molecular compound. An intense band at 302 cm^{-1} was assigned to a methyl torsion. The INS spectrum was successfully modelled by the Wilson GF matrix method for a methyl group in an a-top position on palladium. The strong bands in the spectrum were assigned as follows: 179 cm^{-1}, Pd–Pd stretch; 302 cm^{-1}, CH$_3$ torsion; 420 cm^{-1}, Pd–C deformation; 580 cm^{-1}, CH$_3$ torsion; 973 cm^{-1}, CH$_3$ rocking; 1287 cm^{-1}, CH$_3$ deformation.

The intensities of the Pd–CH$_3$ bands correlated with the degree of catalyst deactivation, demonstrating that the surface methyl groups were the catalyst poison. This unique method of deactivation is presumably by site blocking.

7.3.2.8 Methane decomposition on supported nickel and ruthenium

In relation to the conversion of methane into higher hydrocarbons, INS was used to investigate the surface intermediate species— methylidyne, vinylidene and ethylidyne—formed during the decomposition of methane on Ru/Al$_2$O$_3$ and Ni/SiO$_2$ catalysts [81]. Species identified and assignments are given in Table 7.13. The ethylidyne assignments are in reasonable agreement with those in §7.3.2.6.

7.3.2.9 Cyclohexene

Cyclohexene was adsorbed by a nickel catalyst powder (surface area 15 m^2 g^{-1}) at 300 K at 1 mbar and 10^{-5} mbar pressure and on a nickel catalyst covered with hydrogen [82]. INS spectra are shown in Fig. 7.19. We note: (a) The INS spectrum of cyclohexene adsorbed by nickel at low pressure was different from the spectrum of solid cyclohexene. The spectrum was assigned to adsorbed benzene (peaks near 900, 1140 cm^{-1})

with a contribution from adsorbed hydrogen (970, 1130 cm^{-1}). Evidently cyclohexene had adsorbed dissociatively forming benzene and hydrogen. It was assumed that any cyclohexane formed by reaction of cyclohexene with surface hydrogen atoms had desorbed from the catalyst at low pressure. (b) After cyclohexene adsorption at the higher adsorption pressure cyclohexane was observed (550 cm^{-1}, C–C–C and C–C–H angle deformation; 1270 cm^{-1}, CH$_2$ deformation and wag) and benzene (enhanced intensity below 1000 cm^{-1}). (c) After cyclohexene adsorption on a nickel surface pre-covered with hydrogen, the spectrum corresponded to that of cyclohexane (see Fig. 7.19).

Table 7.13 INS spectra (ω/ cm^{-1}) of surface hydrocarbon species formed in the decomposition of methane on supported nickel and ruthenium catalysts.

Catalyst		Species and assignment (a)
Ni/SiO$_2$	Ru/Al$_2$O$_3$	
		Methylidyne, $\equiv$CH
890	830	δ(CH)
		Vinylidene, =CCH$_2$
184	200	δ(M–CC)
283	280	v_s(M–CCH$_2$)
375	360	v_{as}(M–CCH$_2$)
672	670	τ(CH$_2$)
1260	1220	ρ(CH$_2$)
1310	1300	δ(CH$_2$)
1600	1580	v(C=C)
		Ethylidyne, $\equiv$CCH$_3$
184	200	δ(M–CC)
125	120	τ(CH$_3$)
432	400	v_s(M–CCH$_3$)
605	570	v_{as}(M–CCH$_3$)
1040	1085	v_{as}(C–C)
1184	1165	ρ(CH$_3$)
1477	1485	δ_{as}(CH$_3$)

(a) δ, deformation; v, stretch (as, antisymmetric; s, symmetric); τ, torsion; ρ, rock.

7.3.2.10 Benzene

The hydrogenation of benzene to cyclohexane, a precursor for nylon, is one of the most important industrial hydrogenation processes [70]:

$$C_6H_6 + 3H_2 \rightarrow C_6H_{12}$$

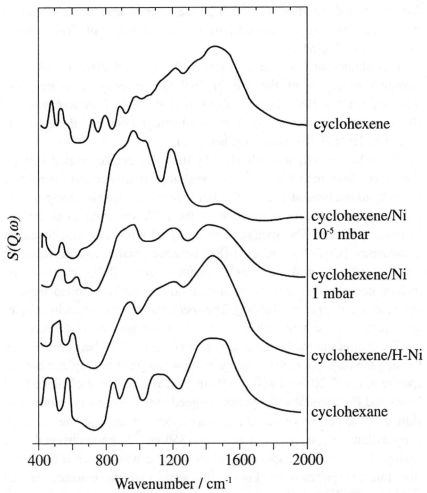

Fig. 7.19 INS spectra (IN1BeF, ILL) of cyclohexene at 77 K adsorbed on nickel powder at 300 K. Reproduced from [82] with permission from the American Chemical Society.

The hydrogenation is catalysed by nickel and other transition metals. Vibrational spectroscopy (infrared, HREELS, INS) has been applied to determining the orientation and binding of benzene on the catalyst surface. We summarise INS studies of benzene, adsorbed benzene and model complexes. The intensities and frequencies of the vibrational modes are computed by the Wilson GF method. The benzene molecule has 30 internal vibrations of which 20 are independent. These vibrations were described by a selection of the deformation of five internal coordinates. [83],[84].

INS studies of benzene on nickel [85, 86] and platinum [87] are excellent examples of the use of INS spectroscopy to characterise adsorbed species. Benzene was adsorbed at 300 K by bare Raney nickel (0.8 monolayer benzene), Fig. 7.20, and hydrogen-covered Raney nickel (0.5 monolayer H, 0.4 monolayer benzene).

When benzene-D_6 was adsorbed by the hydrogenated nickel catalyst the Ni–H vibration at 970 cm^{-1} became slightly narrower and the relative intensity of the band at 1130 cm^{-1} slightly increased but the changes were too small to invalidate subtraction of the H/Ni spectrum to obtain the benzene spectrum. The spectra were analysed with reference to model compounds: $[Cr(CO)_3(\eta^6\text{-}C_6H_6)]$ [88] benzene parallel to a surface, the ring interacting with a single surface atom [89]; chloro- and dichlorobenzene, benzene perpendicular to the surface, bound through one or two carbon atoms [85,86]. The spectrum of benzene on hydrogen-covered nickel was hardly changed from that of benzene on bare nickel.

The changes in the benzene spectrum consequent on benzene binding through the ring to a metal centre were observed (see the top and bottom spectra of Fig. 7.20) near 800–1000 cm^{-1}, the region of out-of-plane C–H bends and C–C torsions: the peaks assigned to the ν_{17} and ν_{18} modes had shifted to lower wave numbers, overlapping with ν_{10}; the in-plane deformations (ν_3, ν_9, ν_{15}, ν_{19}) at 1150–1350 cm^{-1} were unchanged. The changes in the benzene spectrum when benzene was adsorbed by nickel (the middle spectrum of Fig. 7.20) were towards benzene in the chromium model complex.

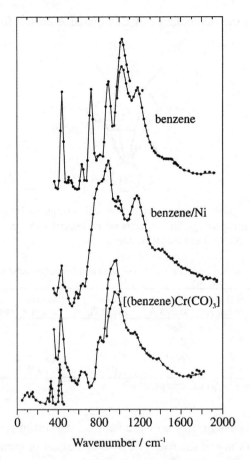

Fig. 7.20 INS spectra (IN1BeF, ILL) at 77 K of solid benzene, benzene adsorbed by nickel (0.8 monolayer) and [(benzene)Cr(CO)₃]. Reproduced from[82,89] with permission from Taylor and Francis Ltd and from [86,84] with permission from Elsevier.

The structure proposed for benzene on nickel is shown in Fig. 7.21 [86]. The force field which successfully simulated the INS spectrum of $[Cr(CO)_3(\eta^6\text{-}C_6H_6)]$ [89] was adapted for the calculation of the INS spectrum of benzene–Ni. The spectrum calculated for the structure of Fig. 7.21 was in excellent agreement with the observed spectrum of benzene on nickel. Force constants are listed in Table 7.14. We include also the results for benzene on platinum [87]. The distance of the centre of the benzene molecule was calculated in the force field fitting procedure from the dependence of this distance on the force constant of the C-Ni bond.

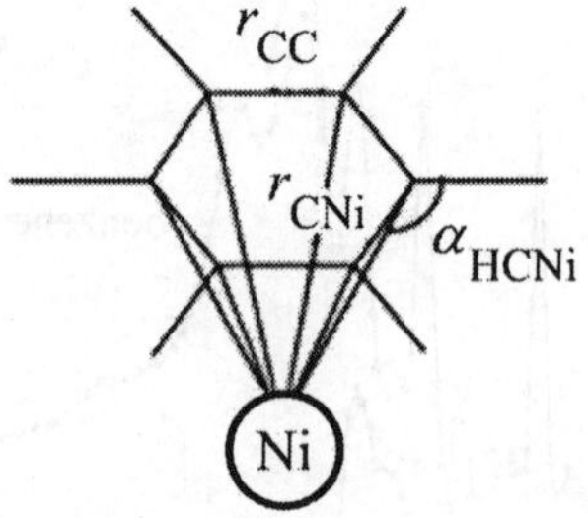

Fig. 7.21 Model structure of benzene interacting with a single nickel atom located below the centre of the benzene ring. The two extra internal coordinates, in addition to those of benzene, are the C–Ni bond and the HCNi angle.

Table 7.14 Calculated benzene force constants and benzene-metal separations (a).

System	f_{CC}/ mdyn $Å^{-1}$	$f_i f_{i+1}$/ mdyn $Å^{-1}$	$f_{C\text{-metal}}$/ mdyn $Å^{-1}$	$f_{HCmetal}$/ mdyn $Å^{-1}rad^{-2}$	Benzene metal / Å
Benzene, solid	6.0	0.84			
$[Cr(CO)_3(C_6H_6)]$	5.7	0.70	0.75	0.22	1.72
Benzene/Ni	5.0	0.60	1.5	0.13	2.5
Benzene/Pt	5.7	0.70	1.1	0.19	2.0

(a) Force constants, f ($f_i f_{i+1}$ CC cross term).

Finally, we discuss briefly how the neutron scattering results contribute to our understanding of the catalytic hydrogenation of benzene.

We see from the trend of decreasing force constants in Table 7.14 that the benzene molecule is perturbed more by interaction with nickel than with platinum. However the rate of hydrogenation of benzene on platinum is fivefold the rate on nickel [86,87]: we cannot, therefore, ascribe the catalysis to an activation of the adsorbed benzene. An alternative suggestion [86,87] is that the higher activity of platinum over nickel is due to the higher mobility of hydrogen atoms on platinum (by a factor of 3). The slow step is the initial reaction of C_6H_6 with hydrogen giving a reactive intermediate, C_6H_7, with a concentration too low to be detected by INS spectroscopy. The adsorbed benzene characterised by INS is a precursor.

7.3.3 Acetonitrile, CH₃CN—binding and hydrogenation

Amines are synthesized industrially by hydrogenation of liquid nitriles over a metal catalyst under high pressure. Over a Raney nickel catalyst at atmospheric pressure at 370–420 K acetonitrile, CH_3CN, was selectively hydrogenated to ethylamine, $CH_3CH_2NH_2$. The hydrogen species active in the gas phase hydrogenation of acetonitrile on a Raney nickel catalyst was the weakly bound hydrogen [7]. Raney nickel was outgassed at 373 K to avoid sintering and retained *ca* 30% of residual hydrogen which was located on the Ni(110) and Ni(111) faces. Dosing the catalyst with dihydrogen enhanced the 940 and 1130 cm^{-1} peaks assigned to hydrogen in a Ni(111) 3-fold hollow site (see Table 7.2). Their intensity increased with increasing pressure of dihydrogen showing increased occupancy of the 3-fold site (§7.3.1.2). In the hydrogenation of acetonitrile only weakly adsorbed hydrogen localised on top of the nickel atoms and on C_{3v} symmetry sites was active.

The orientation of acetonitrile adsorbed by Raney nickel at 323 K and 393 K has been determined by INS spectroscopy [90]. There are two extremes: acetonitrile lying parallel with the catalyst surface and bound through the CN group or perpendicular and bound through the N atom of the CN group. The strongest peak in the INS spectrum of solid acetonitrile, at 160 cm^{-1}, is a libration about an axis which is almost co-linear with (~ 9° off) the C–C≡N backbone. This corresponds to a rotation in the gas phase which has zero frequency. In the INS spectrum of acetonitrile adsorbed by Raney nickel at 393 K a broad band centred at 80 cm^{-1} had replaced the 160 cm^{-1} which was no longer observed. The 80 cm^{-1} band was assigned to a similar libration greatly modified by interactions with nickel atoms. It was, therefore concluded that the acetonitrile molecule was lying parallel to the surface. Had the acetonitrile molecule been perpendicular to the surface, i.e. bound to the surface only through the nitrogen atom, then one would have expected to observe a sharp transition as for the solid or, if the methyl group was free to rotate, a very broad recoil spectrum. It was noted that, according to the INS spectra, the surface concentration of hydrogen atoms did not increase; the adsorption was, therefore, associative and any C-H dissociation was negligible.

7.4 Oxides and oxide-supported catalysts

INS spectroscopy has been applied to characterising hydroxo and aquo components of oxide catalysts and adsorbed hydrogenous reactant molecules and intermediates. The INS spectra are complementary to infrared and Raman spectra, which have been used widely in the study of oxide catalysts. In an INS spectrum background scattering from an oxide is weak and can be subtracted from the spectrum; the lower energy region (below 600 cm^{-1}) is readily accessible. For an introduction to industrial applications of oxide catalysts see [91].

In this section we describe INS studies of molybdenum trioxide, a precursor of molybdenum disulfide catalysts (§7.5), and transition metal oxides which catalyse complete or partial oxidation of hydrocarbons, and copper zinc oxide catalysts, which catalyse methanol synthesis from carbon monoxide and dihydrogen (§7.3.3).

7.4.1 Molybdenum(VI) oxide on alumina—chemisorbed water

Molybdenum trioxide supported on alumina is the precursor of molybdenum disulfide based hydrodesulfurisation catalysts. Sulfiding of supported MoO$_3$ is facilitated by bound water. INS spectroscopy was used to determine the nature of the water in hydrated MoO$_3$/Al$_2$O$_3$ [92, 93]. The librational modes of co-ordinated water (Fig. 7.22) are observed in INS spectra when the water molecule is bound through the oxygen atom (§9.2).

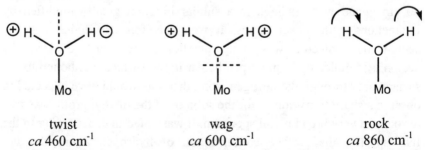

| twist | wag | rock |
| *ca* 460 cm^{-1} | *ca* 600 cm^{-1} | *ca* 860 cm^{-1} |

Fig. 7.22 Librational modes of coordinated water molecules. Dotted lines are axes, $\oplus$ and $\ominus$ refer to forward and reverse motions of hydrogen atoms normal to the plane of the paper; arrows are motions in the plane of the paper.

In a hydrated molybdenum oxide/alumina catalyst the water modes appeared as a broad peak with its maximum near 450 cm^{-1} and the first overtone near 900 cm^{-1}. The peak intensity was proportional to the molybdenum concentration in the catalyst; therefore the 450 cm^{-1} peak was assigned to water bound to molybdenum. The effective mass, 4 amu, m_{eff}, of water was calculated from the ratio of the intensities of the fundamental (S_{0-1})and first overtone (S_{0-2}), Eq. (5.10):

$$\frac{S_{0-2}}{S_{0-1}} = 2\left(\frac{1}{m_{eff}}\right)\exp\left(-\frac{1}{m_{eff}}\right)$$

This value indicates water coordinated to molybdenum.

When the calcined catalyst was exposed to air, so becoming hydrated, the Mo=O stretching vibration in the infrared shifted to lower wavenumber (1000 to 950 cm^{-1}) [94]. The shift to lower wavenumber was attributed to an increase of molybdenum coordination number from four to six as a result of binding of water molecules, Fig. 7. 23.

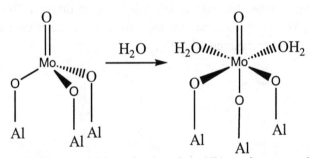

Fig. 7.23 Hydration of molybdate showing the addition of water molecules to a tetrahedral molybdate and formation of an octahedral hydrated molybdate.

The downward shift of the Mo=O vibration, observed in the Raman and infrared spectra of the catalyst, was concomitant with the development of the water twisting mode observed in the INS spectrum. The alternative possibility, that adsorption of water on calcined molybdate/ alumina was dissociative, was not consistent with the spectra. Dissociative chemisorption implies the formation of a Mo–OH group. Such groups have been identified by INS of hydrogen molybdenum bronzes by a band at 1257 cm^{-1} assigned to the Mo–OH deformation mode [95]. There was no band in this region in the Raman, infrared or INS spectra of the hydrated molybdate/alumina catalysts. The spectroscopic techniques are therefore complementary and both supported the idea that molybdenum in the calcined catalyst becomes hydrated on exposure to air.

7.4.2 Selective oxidation of propene—the allyl radical

The allyl radical, $C_3H_5 \bullet$, is an intermediate in the selective oxidation of propene to acrolein with oxygen, catalysed by bismuth molybdate and iron antimonate [69,91]

$$CH_3CH_2=CH_2 \rightarrow \bullet CH_2CH=CH_2 \rightarrow OCHCH_2=CH_2.$$

INS has been used to probe the binding of reactants and intermediates to an iron antimonate catalyst [96]. The spectra were assigned with reference to the spectra of compounds of known structure.

The allyl radical, $\bullet CH_2CH=CH_2$, may bind through one carbon (σ–allyl, **1**), through the double bond (η^2, **2**), or through three carbon atoms (η^3, **3**). The three structures have distinctive vibrational spectra as shown by model compounds [96,97].

The INS spectra of the σ-allyl compound allyl iodide, $CH_2=CHCH_2I$, and the η^3–allyl compound di-μ-chloro-bis(η^3–allylpalladium), [{Pd(η^3–C$_3$H$_5$)}$_2$(μ-Cl)$_2$] [94] and the INS below 800 cm^{-1} of nickel and palladium allyl complexes [97] have been reported. The structure of the palladium complex is shown in Fig. 7.24 [98]. The INS spectra are shown in Fig. 7.25. Peak positions and assignments are in Table 7.15.

A number of features of the spectra are noteworthy: (a) The allyl iodide peak near 140 cm^{-1} was especially sensitive to the physical state of allyl iodide. This peak was assigned to the fundamental torsional transition of the stable gauche conformer of allyl iodide (ν_{21}) observed in the infrared and Raman spectra of solid allyl iodide at 140 cm^{-1}, , the liquid at 106 cm^{-1}, and the gas at 106 cm^{-1}. The position of the torsional mode depended on the physical state of the allyl iodide and, hence, on the mutual interaction between the allyl iodide molecules. (b) Although iodine is a weak neutron scatterer, the C–I stretch was observed in the INS near 480 cm^{-1} with considerable intensity, the hydrocarbon moving against the heavy iodine atom.

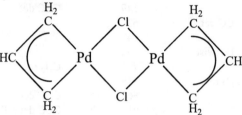

Fig. 7.24 Structure of di-μ-chloro-bis(η^3–allylpalladium) [98]. The PdCl$_2$Pd framework is planar. The planes of the allyl groups are at an angle of $112 \pm 9°$ to the PdCl$_2$Pd plane.

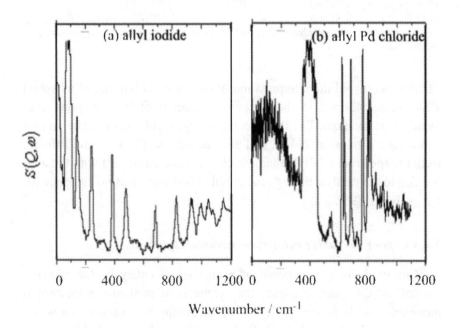

Fig. 7.25 INS spectra of (a) allyl iodide, (b) allyl palladium chloride [96].

Table 7.15 INS spectra (ω / cm^{-1}) of allyl iodide and allyl palladium chloride [96].

| Allyl iodide | | Allyl palladium chloride | |
ω / cm^{-1}	Assign (a)[99]	ω / cm^{-1}	Assign (a) [100]
76	lattice	100	lattice
143	as torsion		
242	CCl def		
384	CCC def	403	Pd–allyl as str
479	Cl str		
		549	CCC def
626	CCC def	631	
		647	
685	CH$_2$ twist	692	
		773	CH$_2$ rock
828	CH$_2$ rock	818	
		862	CCC def
		911	CCC def + Pd–allyl as str
927	CH$_2$ rock	940	CH$_2$ wag + CH$_2$ rock
		964	
998	CH def	1009	CH$_3$ twist
1050	CH$_2$ wag	1054	CCC s str
1148	CH$_2$ wag		

(a) Assignments, str, stretching; def, deformation (bending); as, antisymmetric; s, symmetric.

These spectra aid the interpretation of the mode of binding of adsorbed allyl species formed in the selective oxidation of hydrocarbons over metal oxide catalysts. The INS spectra of allyl iodide adsorbed by an iron antimonate catalyst at 293 K, and after heating to 353 K, were different from the spectrum of allylpalladium chloride and consistent with the allyl binding to the catalyst through the double bond there was no evidence for a η^3–allyl (3) [96].

7.4.3 Copper zinc oxide catalysts—methanol synthesis

Zinc oxide is a component of a number of catalysts, for example Cu/ZnO/Al$_2$O$_3$ which catalyses the synthesis of methanol from carbon monoxide and hydrogen [70]. INS spectroscopy was used to show the presence of Zn–H bonds in ZnO dosed with hydrogen. Hydrogen was chemisorbed on zinc oxide at room temperature; the INS spectrum showed Zn–H bending (829 cm^{-1}) and stretching (1665 cm^{-1}) modes and a O–H bend (1125 cm^{-1}). Hydrogen motions associated with lattice phonon modes were observed below 600 cm^{-1} [101].

In industrial methanol synthesis involving copper, formate species are surface intermediates. With INS, vibrational modes in the low energy region (<1000 cm^{-1}), which had not been identified in infrared and Raman studies, were observed for reference compounds (potassium and copper formate) [102] and for formate adsorbed by copper oxides [103] These characteristic bands included a doublet at *ca* 600 cm^{-1} assigned to librational modes of HCOO. The formate out-of-plane C–H bend near 1100 cm^{-1}, which is very weak in the infrared and Raman spectra, was observed with high intensity in INS. The INS spectrum of formate on copper (II) oxide resembled the spectrum of copper (II) formate implying a similar structure with formates acting as bridges between copper atoms.

7.4.4 Gold on titanium dioxide—the hydrogen–oxygen reaction

Nanosized gold clusters catalyse the conversion of propene to propene oxide using dioxygen and dihydrogen, and the synthesis of hydrogen peroxide directly from dioxygen and dihydrogen. The surface species formed in the reaction of dioxygen and dihydrogen on a highly dispersed Au/TiO$_2$ catalyst were identified by INS [104].This work was particularly notable because the experiment was carried out in-situ: the INS cell was used as the reactor vessel, when steady state conditions were achieved, the reaction was quenched in liquid nitrogen and the INS spectra were measured. In addition to water, these species were identified: hydrogen peroxide (OO stretching, 900 cm^{-1}, very weak, OOH symmetrical deformation, 1440 cm^{-1}), and a hydroperoxo, HO$_2$, radical (OO stretching, 1065 cm^{-1}, OOH deformation, 1505 cm^{-1}).

7.5 Zeolites

Zeolites are well known for their ability to absorb molecules selectively and as acid-base catalysts for, e.g. cracking, isomerisation and alkylation of hydrocarbons. They are aluminosilicates $(Si,Al)_xO_{2x}$ with an internal structure comprising linked channels and cages of molecular dimensions. Protons and other cations balance the negative charge of the framework. INS (in addition to infrared and nuclear magnetic resonance spectroscopy) has been used extensively to probe the Brönsted acidity and internal space of zeolites. We see deformation modes involving displacements of hydrogen atoms and framework modes coupled with hydrogen motions (i.e. riding modes). A recurring theme is the nature of the hydrogen–oxygen species in dehydrated and hydrated zeolites. We refer to the structures of individual zeolites as called for but it is not part of our purpose to describe zeolite structures and the reader should consult the literature for detailed descriptions of structures [105].

The INS spectra of zeolites have been calculated by force-field methods [106,107]. Note that the hydrogen atoms follow the displacements of the framework atoms due to the lattice modes (480–970 cm^{-1}). The lattice modes (Al–O and Si–O stretching and bending modes) are therefore observed in the INS spectra—another example of hydrogen riding modes.

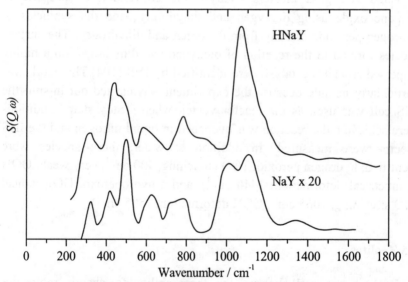

Fig. 7.26 INS spectra of zeolites H–NaY and NaY 20 K (IN1BeF, ILL). Note that the scattering intensity for NaY is *ca* 1/20 that of H–NaY. Reproduced from [106] with permission from Elsevier.

7.5.1 Sodium, ammonium and protonated zeolite Y

Type Y zeolite is a synthetic analogue of the mineral faujasite; the Si-to-Al ratio is between 1.5 and 3. The structure consists of a three-dimensional array of spherical supercages 13 Å in diameter. These spheres are connected by 7.4 Å windows. The sodalite cages are truncated octahedra of diameter 7.6 Å. The INS spectra of NaY and H–NaY, which are typical zeolite spectra, are shown in Fig. 7.26 and the peaks are listed and assigned in Table 7.16 [108,109].

Table 7.16 INS spectra (ω/ cm^{-1}) of Y zeolites [108,109].

NaY [108] (a)	H–NaY [108] (b)	HY [110] (c) (d)	HY de-alumin [110] (b) (d)	Assign (e)
303	298	319	298	pore opening
400	420	419	395	OH(br) op def
492	451	470	468	O–T–O def
613, 717	568, 758	565		framework H coupled
766		765	766	T–O s str
	863	860	833	Si–OH (t) ip def
1004	1060	1089	1052	OH(br) ip def
1113, 1350		1130	1169	T–O as str, .

(a) $Na_{56}Al_{56}Si_{136}O_{384}$. (b) H–NaY ($H_{44}Na_{12}Al_{56}Si_{136}O_{384}$) from NaY exchanged with NH_4Cl. (c) After repeated exchange of $Na_{51}Y$ with NH_4NO_3. (d) After de-alumination with $SiCl_4$ and repeated exchange with NH_4NO_3. (e) Assignments, op, out-of-plane; ip, in-plane; def, deformation (bending); (t), terminal OH; br, bridging OH; as, antisymmetric; s, symmetric; T, tetrahedal Al or Si.

The sodium ions of Na zeolite Y may be substituted by ammonium ions [110]. The degree of hydration of the NH_4Y depends on the thermal treatment: water is retained up to 373 K and lost at 473 K. The sodium ions may be replaced partly by ammonium ions and partly by caesium ions. Ammonium ions may also be introduced by treatment of the acidic zeolite with ammonia gas. The caesium ions occupy the supercages and the ammonium ions are restricted to the sodalite cages.

The INS spectra of these various materials are interpreted in terms of translational and librational motions of ammonium ions, ammonia and water molecules, Table 7.17.

Table 7.17 INS spectra (ω/ cm^{-1}) at 20 K of NH$_4$Y and CsNH$_4$Y, hydrated and dehydrated [109].

Water modes (a)		NH$_4^+$ ions (b)		
	Assignment	Sodalite cage(c) (d)	Supercage (d)	Assignment
105, 186	translation	97	153	translation
371, 417	libration	65		libration
			250, 331	weakly bound
			32, 48, 371, 468	strongly bound
		1476, 1718	1476, 1718	N–H bend

(a) From difference spectrum: NH$_4$Y evacuated at 373 K (hydrated) less NH$_4$Y evacuated at 473 K (dehydrated). (b) More strongly bound NH$_4^+$ retained at T > 473 K. (c) CsNH$_4$Z-Y. (d) NH$_4$Z-Y

7.5.2 Ammonium and protonated zeolite rho

Zeolite rho has attracted interest because of the structural changes which occur on dehydration and deammoniation of acidic rho-zeolites [111]. Four peaks were assigned in the INS spectrum of dehydrated rho zeolites [112,113]: 360 cm^{-1}, T–OH torsion of planar symmetric Al–O(H)–Si; 750 cm^{-1}, hydrogen motion associated with symmetrical T–O stretch; 1060 cm^{-1}, T–O–H out-of-plane bending; 1150 cm^{-1}, T–O–H in-plane bending (where T refers to Al or Si). Heating rho zeolites at 548 K generated a new peak at 260 cm^{-1} associated with a large amplitude hydrogen motion at a site populated at higher temperature (unassigned, but presumably a lattice vibration).

The thermal decomposition of ammonium rho-zeolites- has also been followed by inelastic and quasielastic neutron scattering [112]. As the deammoniation temperature was increased, the assigned vibrational features of the bound ammonium ion at 1452 cm^{-1} (bending modes), 306–363 cm^{-1} (ammonium-coupled pore-opening modes of the zeolite framework), and 80–113 cm^{-1} (hindered rotational modes) were found to diminish with the emergence of scattering features due to H–rho.

Table 7.18 INS spectra (ω / cm^{-1}) of H–mordenite and H–ZSM-5 before and after hydration.

Species	H–mordenite [114] (a)			H–ZSM-5 [115] (b)		
	Dry (c)	Wet (c)	Assign (g)	Dry (c)	Wet (c)	Assign (g)
H_2O		460	libration		450	twist+ rock
		910	libration			
H_2O (d)		60	translation		90	
		600				
		800			900	op def
					1650	def
H_3O^+		100	translation			
		1385	s def		1380	
		1670	as def		1650	
OH bridge	320	420	op def	320, 405		op def
	1060	1060	ip def	1080, 1210	1080	ip def
	565, 770		frame (f)	100, 447, 600, 690	155, 575, 785	frame (e)
					1170	op def (f)
					1380	ip def (f)

(a) 1 H_2O/OH (b) 3.5 H_2O/unit cell equivalent to 2 H_2O/H$^+$. (c) The sample is either Hydrated (Wet) or Dehydrated (Dry). (d) H-bonded water. (e) Framework modes coupled with H motions.(f) Perturbed. (g) Assignments, op, out of plane; ip, in plane; def, deformation (bend).

7.5.3 Hydrated H–mordenite and ZSM-5

Dosing an acidic zeolite with water forms aquo and hydroxo species which may be identified by INS, e.g. H–mordenite having one H_2O per OH group [114] and HZSM-5 dosed to various levels of water [115]. INS peaks of typical hydrated zeolites are listed in Table 7.18: For H–ZSM-5 the spectra were calculated [115] by an *ab initio* method for a model Si–O–Al–(OH)–Si cluster with bridging OH incorporating either water with hydrogen bonds to the OH or to the H_3O^+ ion. The hydrogen-bonded water model agreed better with the experimental spectrum than the hydroxonium model. The INS spectrum of hydrated mordenite was different from that of ice (§9.2.1); water clusters were, therefore, assumed to be absent from the mordenite. For high concentrations of water in HSM-5 (35 H_2O/unit cell) the INS spectrum was similar to that of ice (§9.2).

7.5.4 Molecules in zeolites

The interaction between the adsorption sites of a zeolite and a probe molecule perturbs the normal vibrations of the molecule and leads to frequency shifts compared to the gas phase which can be interpreted in terms of the strength and nature of the interactions with the adsorption sites of the zeolite. INS is advantageous for observing such interactions for hydrogenous molecules [116,117]. In the infrared spectra the out-of-plane bending modes and the ring vibrations coupled with CH in-plane bending are obscured by the zeolite framework modes below 1300 cm^{-1}.

7.5.4.1 Benzene in sodium zeolite Y

The INS spectra of solid benzene and benzene adsorbed in NaY at a concentration of one benzene molecule per supercage are shown in Fig. 7.27 [118]. The spectra were also calculated by a force field method. Peak positions and assignments are given for benzene in Table 7.19.

These spectra illustrate a number of features of INS spectra. In comparison with the infrared spectrum of benzene vapour, in solid benzene the out-of-plane C–H bending modes are shifted to higher wavenumbers (e.g. v_{11}, 673 to 694 cm^{-1}). The shift is a consequence of intermolecular forces in the solid. The in-plane modes are not shifted. Out-of-plane C–H bending modes are shifted further to higher wavenumbers for benzene in NaY (v_{11}, 700 cm^{-1}).

The Debye-Waller factor is smaller for adsorbed benzene than for solid benzene. This is due to a larger mean square amplitude of the external modes of adsorbed benzene (0.017 $Å^2$) than solid benzene (0.011 $Å^2$) because the benzene molecules are less strongly bound in the zeolite than in the solid. The external modes of adsorbed benzene are at lower frequency (see Table 7.20) and their contribution to the total displacement amplitude is increased. The intensity of the fundamentals is decreased and is transferred to multiphonon features with consequent broadening of the bands of adsorbed benzene (§5.3.2).

The adsorption geometry of benzene in the supercage is known from diffraction [119]. Its centre is on the cube diagonal with the ring oriented symmetrically over a sodium ion. From a simulation of the INS spectrum, which included Na–C interactions, a Na–C force constant was calculated. The value (0.06 mdyn Å^{-1}) was much smaller than for benzene bound to a transition metal centre (e.g. [Cr(CO)$_3\eta^6$-(C$_6$H$_5$)], 4.5 mdyn Å^{-1}) implying weaker binding of benzene to sodium consistent with the small perturbation of internal modes.

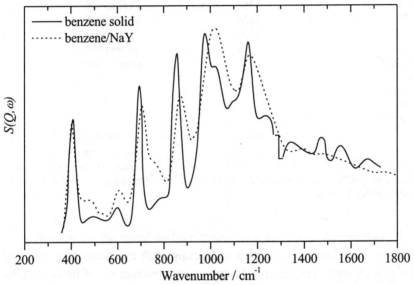

Fig. 7.27 INS of solid benzene at 5 K (IN1BeF, ILL) and benzene in NaY at 20 K. Reproduced from [118] with permission from Elsevier.

7.5.4.2 Furan in faujasite

For furan adsorbed by a series of alkali-metal-substituted zeolite Y and X the out-of-plane C–H modes are shifted much more than the in-plane modes [120].

Table 7.19 INS spectra ($\omega/$ cm^{-1}) of solid benzene and benzene adsorbed in NaY [118,121] (a).

Solid benzene(b)	Benzene/NaY (c)	Assignments (d)
	21	translation, rotation
	36	translation, rotation
80		lattice
407	405	star of David mode
494	465sh	phonon wing
605	608	C–C str + C–C–C def
694	700	C–H op def
780		phonon wing
855	875	C–H op def
977	1023	C–H op def
	1009	C–H op def
1168	1168	combination ip
1260		phonon wing
1349		
1476		
1560		
1677		
	1844	combination
	1985	combination

(a) The references do not provide complete listings of frequencies; we have expanded their lists. (b) 20 K. (c) 1 benzene per supercage. (d) str, stretching; def, deformation; op, out-of-plane; ip, in-plane.

The vibration frequencies were calculated *ab initio* for a model structure comprising a bare metal cation interacting with the furan ring. For both the experimental and calculated spectra the interaction of furan with the zeolite caused the out-of-plane vibrational modes to move to higher wave numbers, see Fig. 7.28. This result can be explained if one assumes an interaction, oriented perpendicular to the molecular plane of furan, between the zeolite cation and the π electron system of furan, which lead to an increase of the CH out-of-plane bending force constant. However, for the experimental spectra the wavenumbers of the out-of-plane modes then decrease in the series $Li^+ > Na^+ > K^+ > Cs^+$ whereas the computed values *increase*.

It was proposed that the trend in the experimental values is due to interaction of the furan's hydrogens with the oxygen atoms of the zeolite cage which become more basic from Li^+ to Cs^+ and so interact more strongly with the furan hydrogens (an interaction which is not included in the simple model calculation).

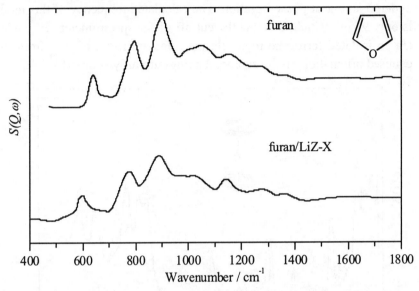

Fig. 7.28 INS spectra (IN1BeF, ILL) of furan at 5 K and furan adsorbed by lithium exchanged zeolite X, $Li_{93.1}Na_{2.4}Al_{95.5}Si_{97.5}O_{384}$ at 20 K. One molecule of furan per supercage. The out-of-plane bending modes near 750 and 870 cm^{-1} of solid furan are shifted to *ca* 780 and 890 cm^{-1} in the zeolite. Reproduced from [120] with permission from Elsevier.

7.5.4.3 Ferrocene in potassium zeolite Y

A catalytic center can be created in a zeolite cage by the incorporation of a transition metal compound, e.g. ferrocene, $[Fe(C_5H_5)_2]$, which may be the catalyst or the precursor. To assess the strength of the interaction between the encapsulated ferrocene and the zeolite the INS spectrum of ferrocene in the supercage of potassium zeolite Y has been compared with the INS of solid ferrocene [122]. The spectra are shown in Fig. 7.29 and the assignments, according to the INS and computed spectrum of solid ferrocene [123] in Table 7.20.

The differences between the spectra of encapsulated and solid ferrocene were small; the ferrocene-zeolite interactions are weak, comparable with the intermolecular interactions in solid ferrocene. The mode which is likely to be most sensitive to any distortion of the ferrocene molecule is the torsional mode in which the two C_5H_5 rings undergo an out of phase libration. The calculated frequency of this mode is only 9 cm^{-1} which is below the cut off of the spectrometer (20 cm^{-1}). (In an isolated ferrocene molecule the rings are rotated by 9° from the eclipsed orientation; in the zeolite, the rings are fully eclipsed.)

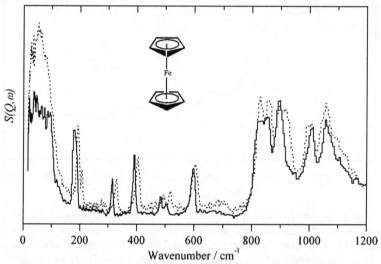

Fig. 7.29 INS spectrum of solid ferrocene (dashed line) and ferrocene in potassium zeolite Y at 20 K. Reproduced from [122] with permission from Springer.

Table 7.20 INS spectrum and assignments for solid ferrocene [123].

ω / cm^{-1}	Assignment (a)
(9)	calculated torsional C_5H_5 rings out of phase libration
180	ring–Fe–ring def + as ring tilt
315	ring–Fe str
391	s ring tilt
485	ring–Fe str + oscillating Fe
504	as ring tilt + ring–Fe–ring def
598	op ring def + CH def
820–860	C–H def
900	C–H def
1010	C–H def
1060	C–H def

(a) For a graphical representation and the calculated spectrum see [123]; as, antisymmetric; s, symmetric; str, stretching; op, out-of-plane; def, deformation.

7.6 Metal sulfide catalysts

Molybdenum disulfide catalysts, promoted by cobalt or nickel, are used to remove organosulfur compounds from crude petroleum by hydrogenolysis—hydrodesulfurisation or HDS [124,125,126]. These compounds are undesirable because they poison motor vehicle autoexhaust catalysts and burn to sulfur dioxide, an environmental pollutant. The most difficult compounds to desulfurise are sulfur heterocyclics: thiophene, benzothiophene, dibenzothiophene and their methyl substituted derivatives. A typical reaction is the removal of sulfur from thiophene:

$$\text{(thiophene)} \quad + \quad 4H_2 \quad \longrightarrow \quad C_4H_{10} \quad + \quad H_2S$$

Dihydrogen molecules adsorb and dissociate on the catalyst; the reactive surface species is the hydrogen atom [127]. In the technical catalysts used in the hydrotreatment of crude petroleum MoS_2 is supported on alumina and promoted by cobalt.

Inelastic neutron scattering has been applied since the early 1980s to characterise hydrogen and adsorbed species in the MoS_2 catalysts [128]. Related catalysts, tungsten disulfide [129] and ruthenium disulfide [130], [131] have been studied and also niobium and tantalum disulfides (where the interest is intercalated hydrogen in electrochemical applications [132]). We shall summarise the INS spectra of H–S and H–metal species in sulfides, undissociated dihydrogen, and adsorbed thiophene and benzothiophenes. Lattice vibrations of the sulfides are relevant because they appear in the hydrogen riding modes.

We should recall the structures of the disulfides [133]: NbS_2, MoS_2, and WS_2 have S–M–S layers with the metal atoms in six fold trigonal prismatic coordination by sulfur atoms; RuS_2 has the iron pyrites structure with Ru in octahedral coordination by S_2 groups in a NaCl-type structure. The MoS_2 structure is shown in Fig. 7.30. Note the presence of basal plane and edge sites; the edge sites are considered to be the reactive centres of MoS_2 [126].

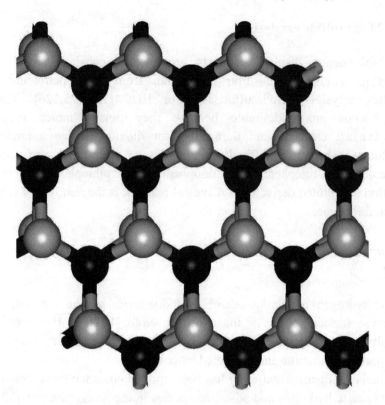

Fig. 7.30 Structure of MoS$_2$. Mo atoms (darker spheres) are linked to S atoms (lighter spheres) in six-membered rings. The view shows a layer through the basal plane, and crystallite edges with coordinatively unsaturated Mo and S atoms where hydrogen atoms may bind [133].

For hydrogen-treated MoS$_2$, possible hydrogen species are shown in Fig. 7.31 with their *calculated* vibration frequencies according to the atom superposition and electron delocalization molecular orbital method [134]. Typical spectra of hydrogen-treated RuS$_2$ and MoS$_2$ [130,131] are shown in Fig. 7.32.

A recent spectrum of MoS$_2$ [135], which includes the lattice modes, is shown in Fig.7.33. Early spectra of MoS$_2$ [126,128] and WS$_2$ [127,129] are similar to our MoS$_2$ spectrum but of poorer quality. Peaks are listed and assigned in Table 7.21 [128,129,130,135,136,137].

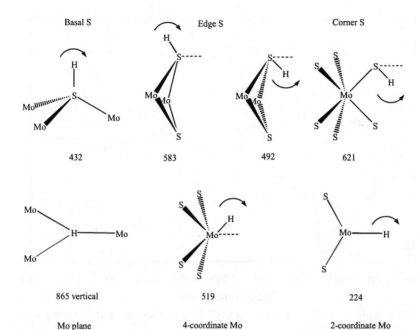

Fig. 7.31 Proposed binding sites of hydrogen on MoS_2 and calculated bending frequencies in wavenumbers.

Table 7.21 INS of hydrogen on sulphide catalysts: (ω / cm^{-1})
(I) Hydrogen adsorbed by MoS_2.

MoS_2 [135] (a)	MoS_2 [128,130]	[136]	$NiMoS_2$ [137]	Assignments (b)
125, 155, 195, 275,	170 or 220	105, 186,		H riding on
395, 455, 530	350 or 400	315		MoS_2 lattice
610, 645, 675	640–660 (650)		650	H–S def
	850			220+640
			795	H–Ni?
	1350 (1400)			650 × 2
			1895	H–Ni?
	1980			650 × 3
	2500			H–S stretch

(II) Hydrogen adsorbed by WS_2 and RuS_2 catalysts.

WS_2 [129]	RuS_2 [128]	Assignment (b)
	540	H–Ru def
695	650, 720	H–S def
	1400	695 × 2

(a)Surface was reduced in hydrogen. (b) def, deformation.

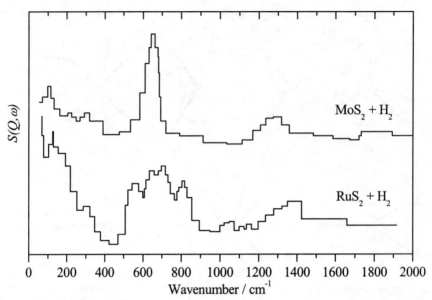

Fig. 7.32 INS spectra of hydrogen adsorbed (0.5 bar) by RuS$_2$ (reduced in hydrogen at 473 K) [131] and MoS$_2$ [130].

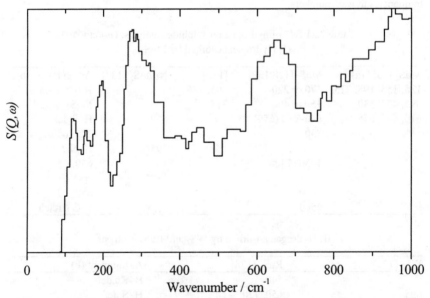

Fig. 7.33 INS spectrum of MoS$_2$ reduced in dihydrogen [135].

7.6.1 S–H vibrational modes

Peaks observed at 650–700 cm^{-1} on the hydrogen-reduced sulfides are assigned to the S–H bending modes (i.e. vibrations parallel to the surface). They are two-fold degenerate. Adsorption of dihydrogen on amorphous MoS_2 at high pressure (50 atm) and at temperatures above 423 K gives, in addition to the 660 cm^{-1} peak, a peak at *ca* 400 cm^{-1} attributed to a second adsorption site (which was not characterised) [138]. For details of the evolution of the spectra during hydrogen dosing of the sulfides the original papers should be consulted [130,138].

For tungsten disulfide dosed with hydrogen, in addition to the S–H bending at 694 cm^{-1}, there are broad weak overtone peaks at 1380 and 2074 cm^{-1}, and a broad, weak peak between 2500 and 2700 cm^{-1} assigned to the S–H stretch (i.e. perpendicular to the surface) [129].

The 650 cm^{-1} S–H bend peak is also observed on a nickel-promoted MoS_2 catalyst [137]. On alumina-supported catalysts this peak was also observed but the spectra were obscured by strong scattering from hydroxyl groups on alumina. The intensity of the 650 cm^{-1} peak was weaker than expected from hydrogen uptake measurements, possibly a result of spillover from MoS_2 to alumina [137].

7.6.2 Metal–hydrogen vibrational modes

Metal–hydrogen vibrations have been reported for hydrogenated RuS_2 (540 and 820 cm^{-1}) [130] but metal–hydrogen vibrations have not been observed for MoS_2 or WS_2.

Metal–hydrogen bonds have been characterized by neutron diffraction and INS in hydrogenated niobium and tantalum disulfides. The hydrogen intercalation compounds H_xMS_2 (M = Ti, Nb, Ta) have been prepared by cathodic reduction of the corresponding binary sulfides, MS_2, in aqueous mineral acids. Hydrogen atoms in a single crystal of hydrogenated niobium disulfide, H_xNbS_2 were located by single crystal neutron diffraction [132]. Each hydrogen atom is located in a three-centre metal bond inside the disulfide layers. The Nb–H bond length was 1.93 Å. The INS spectra are then interpreted as Nb–H fundamental vibrations parallel to the *c*-axis (Nb–H bending, 744 cm^{-1})

and perpendicular to the c-axis (Nb–H stretch, 1544 cm^{-1}). The conclusions for tantalum disulfide were similar. The three-centre metal-hydrogen bond is established only for $x > 0.1$ in the H$_x$MS$_2$ compounds. For a NiMoS$_2$ catalyst Ni–H vibrations have been assigned (see Table 7.21) [137].

7.6.3 Lattice vibrations and hydrogen riding modes

The early INS spectra of MoS$_2$ and hydrogen-dosed MoS$_2$ [128] have features assigned to MoS$_2$ lattice modes which are seen more clearly in later spectra [136]. The lattice vibrations of MoS$_2$ are seen at 105, 186 and 315 cm^{-1}. For hydrogen dosed MoS$_2$, when the spectrum of MoS$_2$ was subtracted the 186 and 315 cm^{-1} peaks disappeared but the 105 cm^{-1} peak remained—a MoS$_2$ lattice mode amplified by hydrogen riding. In hydrogenated MoS$_2$ two lattice modes are seen at 176 and 365 cm^{-1} (in addition to the S–H bend at 635 cm^{-1}). The lattice modes of hydrogenated MoS$_2$ are shifted compared with MoS$_2$ and the intensity of the higher frequency mode is much enhanced. Hydrogen riding modes (150- 350 cm^{-1}) are clearly observed in the spectrum of reduced MoS$_2$, see Fig. 7.33.

7.6.4 Computed INS spectra

The lattice modes of MoS$_2$, pure and hydrogenated were calculated in order to interpret experimental INS spectra [139]. The local vibrations of molybdenum and sulfur in displacements parallel and perpendicular to the layer are below 500 cm^{-1}. At low pressures (H$_{0.011}$MoS$_2$) hydrogen atoms bond directly above a single S atom. The local modes of the hydrogen atom lie at 2540 cm^{-1} (S–H stretch) and 666 cm^{-1} (S–H bend). The scattering intensity is dominated by neutrons scattered from hydrogen atoms vibrating parallel to the layer. The density of states for such vibrations resembles the calculated vibrational density of states of the sulfur atom parallel to the layer which gives rise to the higher frequency band (ca 350 cm^{-1}). The intensity of this band, after hydrogen adsorption, is accordingly much increased. The possibility that for H$_{0.067}$MoS$_2$ (obtained at high hydrogen pressures) the high intensity observed at 500 cm^{-1} might be due to the occupancy of an additional site at higher hydrogen concentrations was investigated. A threefold site with each hydrogen bound to three molybdenum atoms was proposed. For this site the hydrogen local modes were at 513 cm^{-1} (perpendicular) and 1207 cm^{-1} (parallel, two-fold degenerate); in this case the calculated scattering intensity is greater for perpendicular vibrations of the hydrogen atom (although this is contrary to expectations based on the degeneracy of the modes). The corresponding lattice mode (equivalent to hydrogen riding on Mo) is at *ca* 225 cm^{-1} and there is virtually no intensity below 200 cm^{-1} in agreement with the low and high pressure difference spectra [138]. However an alternative explanation in terms of a lattice contraction due to hydriding has not been ruled out by these calculations.

More recently, S–H and Mo–H vibrational frequencies have been calculated by the atom superposition and electron delocalization molecular orbital method [134]. Structures and vibration frequencies are shown in Fig. 7.31. The most stable chemisorption form is heterolytic at edges of the crystal layers. Under high dihydrogen pressure and at sufficiently high temperatures for hydrogen to diffuse over the anion surface (a 96,800 cm^{-1} barrier is calculated), hydrogen can diffuse from edge sites. The calculated basal plane S–H bending vibration was 431 cm^{-1} and values for edge and corner S–H and one Mo–H edge bond were in the 500-600 cm^{-1} range. Diffusion of hydrogen atoms over the sulfur anion basal planes was proposed; the uncoordinated molybdenum edge sites will become saturated and at sufficiently high hydrogen pressures the temperature is high enough to overcome the anion-to-anion diffusion barrier. This diffusion at T > 423 K leads to the formation of S–H bonds on the basal planes with lower bending frequencies and is responsible for the observed INS peak at about 400 cm^{-1}. The hydrogenated MoS$_2$ species is considered to be a conducting bronze, H$_x$MoS$_2$, x >1.

7.6.5 Adsorbed dihydrogen

The rotational spectrum of undissociated dihydrogen molecules may be observed by INS on sulfide catalysts as on carbon-supported metal catalysts. On MoS$_2$ a peak near 120 cm^{-1}, assigned to the $J(1{\leftarrow}0)$ transition of the adsorbed dihydrogen molecule, was observed at hydrogen pressures greater than 10 bar [136] and at 1 bar [135]. Line broadening was explained by a distribution of hindering potentials at different sites. There is a contribution to the spectrum from molecular recoil: the two features near 320 and 1600 cm^{-1} are the recoil shifted transitions $J(1\leftarrow0)$ and $J(3\leftarrow0)$ (§6.2).

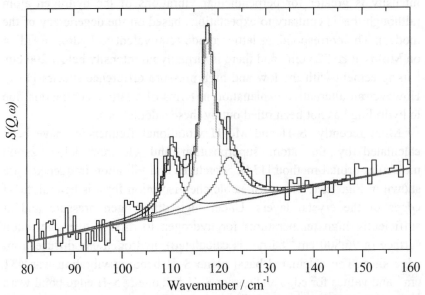

Fig. 7.34 INS spectrum of adsorbed para-hydrogen on MoS$_2$ showing the unperturbed dihydrogen rotor (118 cm^{-1}) and a second site where the degeneracy is lifted by the interaction with the surface (110, 122 cm^{-1}) [135].

On RuS$_2$ dihydrogen was observed at 0.5 bar [130]. The peak was split into two components as a consequence of the anisotropic interaction of H$_2$ molecules with RuS$_2$.

The splitting of the dihydrogen rotor peak is a measure of the strength of interaction of dihydrogen with the catalyst. By modelling the interaction of dihydrogen with the catalyst it is possible to use dihydrogen to explore the surface sites of sulfide catalysts (c.f. dihydrogen interacting with graphite (§6.3.1)and a CoAlPO catalyst (§6.4.1)) but these calculations have not yet been done for sulfides. In a preliminary study, the INS of para-hydrogen on MoS$_2$ before and after reduction in hydrogen has been determined, see Fig. 7.34. The splitting of the hydrogen rotor signal is less than for dihydrogen on RuS_2 indicating a weaker anisotropy of dihydrogen on MoS_2.

7.6.6 Thiophene and related compounds

Thiophene is the simplest sulfur heterocyclic that can be used as a model compound for desulfurisation. The earliest stage of the HDS process is, the adsorption of thiophene. The INS spectra of adsorbed thiophene [140] were obtained. The challenge is to interpret changes, if any, of the INS spectrum due to adsorption in terms of the interaction of thiophene with the catalyst. The INS spectra of thiophene complexes of known structure were used to aid in the interpretation of the INS spectrum of adsorbed thiophene, Fig. 7.35.

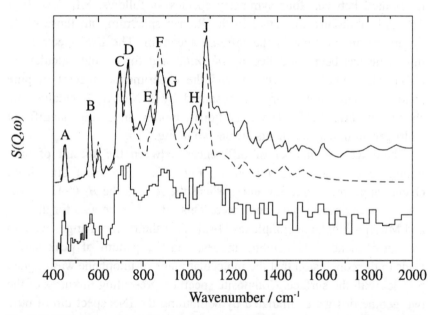

Fig. 7.35 INS spectra of pure thiophene (top, solid line), the spectrum calculated with the Wilson GF method (middle, broken line) and thiophene on a reduced and sulfided Mo(14%)/Al$_2$O$_3$ catalyst (bottom, histograms). The peaks are labelled according to the scheme of Fig. 7.36, A to J in increasing energy. Reproduced from [140] with permission from the PCCP Owner Societies.

Two extreme model compounds could be used to represent the binding of thiophene to the catalyst. In the first, thiophene was bound to molybdenum through the ring (with the molecule aligned parallel with the surface), analogous to η^5–thiophene in the compounds [Cr(CO)$_3$(η^5–thiophene)] and [Mn(CO)$_3$(η^5–thiophene)][CF$_3$SO$_3$]. In the second, thiophene was bound to molybdenum through the sulfur atom (in an upright like geometry), analogous to η^1–S-thiophene in the compound [Fe(CO)$_2$(η^5–C$_5$H$_5$)(η^1S-thiophene)][BF$_4$]. The possibility that the adsorbed molecules adopt a mixture of different binding geometries can be accommodated by combining the characteristic INS spectrum of each geometry in proportions which reproduce the observed INS spectrum of the adsorbed species.

Thiophene has C_{2v} symmetry and the 21 fundamental vibrations are distributed between four symmetry species as follows: $8A_1+3A_2+7B_1+3B_2$. All vibrations are active in the Raman spectrum, but those of A$_2$ symmetry are inactive in the infrared spectrum. The INS spectrum of thiophene has been modelled by *ab initio* and force field calculations [141]. The calculated spectrum and the experimental spectra of pure solid thiophene and thiophene adsorbed by a Mo/Al$_2$O$_3$ catalyst are shown in Fig. 7.35. The vibrational modes involving significant hydrogen atom displacements are shown in Fig.7.36.

There were a number of differences between the spectra of pure thiophene and adsorbed thiophene, Fig. 7.35. The most significant change appeared about 720 cm^{-1}, where the A_2 and the B_2 C–H out-of-plane bending modes have coalesced. This effect is seen also for the Cr– and Mn–η^5–thiophene complexes [140]. That the major changes were to the out-of-plane C–H bending modes, and the pattern of peak shifts, leads to the conclusion that the molecular plane of thiophene was aligned parallel with the surface. Composite spectra representing mixtures of the two geometries were simulated by combining the INS spectrum of pure thiophene (representing the molecule aligned with the molecular plane parallel with the surface) and the Fe–η^1–S thiophene complex. (representing upright molecules) in differing proportions. The observed spectrum of adsorbed thiophene was best simulated with a contribution of *ca* 10% thiophene in the upright form. The conclusion that thiophene is aligned predominantly with the molecular plane parallel with the catalyst surface has, however, been challenged in an infrared study of adsorbed thiophene[142,143]. However, the model with the molecular plane mostly parallel to the surface is consistent with a scanning tunnelling microscopy study of thiophene on MoS_2 [144,145,146]. The S-bound thiophene molecule is presented as tilted towards the surface. As such it is consistent with *ab initio* calculations, which showed that the preferred orientation of thiophene at a positive centre is indeed tilted [147].

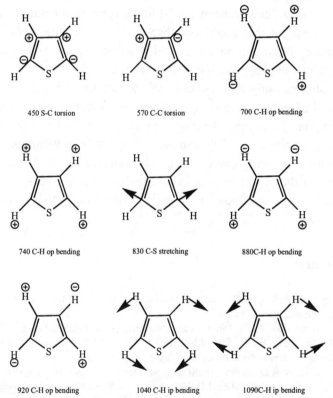

Fig. 7.36 The fundamental vibrational modes of thiophene involving significant hydrogen atom displacement. A plus ((⊕) sign represents a displacement above, and a minus (⊖) sign below, the plane of the ring; arrows represent displacements in the plane of the ring; ip, in-plane; op, out-of-plane.

7.7 Conclusion

In this chapter we have presented and reviewed the extensive body of work that neutron vibrational spectroscopy applied to catalysis represents. It is one of the largest, most important and influential bodies of neutron work to appear in the literature and much of the original work remains of considerable interest. This is partly because catalyst experiments are difficult to perform and require considerable commitment of neutron time and other resources.

The re-analysis of old data, so successfully applicable to pure compounds, is unlikely to be as useful in the context of catalytic studies, since the metadata that could establish the nature of the surface to be modelled are unlikely to have survived in sufficient detail. It could well be that a *tabula rasa* approach is best for future work.

Early studies using this spectroscopic technique took place against a background of technique development and many were limited to the low energy region. Technique development is now complete and modern instruments, irrespective of their nature, chopper or low-bandpass, have access to the full energy range with good resolution. In combination with the advent of the synthetic approach to understanding INS spectra, it is clear that INS spectroscopy of catalytic systems has much to offer the community.

7.8 References

1   P.C.H. Mitchell (1994). Acta Phys. Hung., 75, 131–140. Inelastic neutron scattering studies of catalysts.

2.   N.N. Besperstov, A.Yu. Muzychka, I. Natkaniec, P.B. Nechitailov, E.F. Sheka & Yu.L. Shitikov (1989). Sov. Phys. JETP, 69, 989–995. Inelastic scattering of neutrons by surface oscillations of highly disperse nickel.

3   T.J. Udovic & R.D. Kelley (1988). In *Hydrogen Effects in Catalysis*, (Ed.) Z. Paal & P.G. Menon, p.167. Marcel Dekker, Inc., New York. Neutron scattering studies of hydrogen in catalysts.

4   R. Stockmeyer, H.M. Conrad, A. Renouprez & P. Fouilloux (1975). Surf. Sci., 49, 549–566. Study of hydrogen chemisorbed on Raney-nickel by neutron inelastic spectroscopy.

5 R.R. Cavanagh, R.D. Kelley & J.J. Rush (1982). J. Chem. Phys., 77, 1540–1547. Neutron vibrational spectroscopy of hydrogen and deuterium on Raney nickel.

6 H. Jobic & A. Renouprez (1984). J. Chem. Soc. Faraday Trans., 80, 1991–1997. Inelastic neutron scattering spectroscopy of hydrogen adsorbed on Raney nickel.

7 F. Hochard, H. Jobic, J. Massardier & A.J. Renouprez (1995). J. Mol. Catal. A–Chem., 95, 165–172. Gas phase hydrogenation of acetonitrile on Raney nickel catalysts: reactive hydrogen.

8 E. Preuss, M. Wuttig, E. Sheka, I. Natkaniec & P. Nechitaylov (1990). J. Electron Spectrosc., 54(55), 425–443. Amplitude weighted density of bulk and surface vibrations: ultrafine nickel particles.

9 E. Bonetti, L. Pasquini, E. Sampaolesi, A. Deriu & G. Cicognani (2000). J. Appl. Phys., 88, 4571–4575. Vibrational density of states of nanocrystalline iron and nickel.

10 E. Sheka, E. Nikitina, V. Khavryutchenko, V. Zayetz, I. Natkaniec, P. Nechitailov & A. Muzychka (1991). Physica B, 174, 187–191. Neutron-scattering from water adsorbed on ultrafine nickel particles.

11 E.F. Sheka (1991). Physica B, 174, 227–232. Computational INS spectroscopy of dispersed catalysts.

12 V.D. Khavryuchenko & E.F.Sheka (1994). J. Struct. Chem., 35, 215–223. Computational modelling of amorphous silica .1. Modelling the starting structures––a general conception

13 E.F. Sheka, V.D. Khavryuchenko & I.V. Markichev (1995). Russ. Chem. Rev., 64, 389–414. Technological polymorphism of disperse amorphous silicas—inelastic neutron-scattering and computational modelling.

14 S. Lehwald, F. Wolf & H. Ibach (1987). J. Electron Spectrosc., 44, 393–396. Surface vibrations on Ni(110): the role of surface stress.

15 J.C. Ariyasu, D.L. Mills, K.G. Lloyd & J.C. Hemminger (1984). Phys. Rev. B, 30, 507–518. Lifetime of adsorbate vibrations: the role of anharmonicity.

16 J.M. Szeftel, S. Lehwald, H. Ibach, T.S. Rahman, J.E. Black & D.L. Mills (1983). Phys. Rev. Lett., 51, 268–271. Dispersion of adsorbate vibrational modes—the c (2x2) oxygen overlayer on Ni(100).

17 T.S. Rahman, D.L. Mills, J.E. Black, J.M. Szeftel, S. Lehwald & H. Ibach (1984). Phys. Rev. B, 30, 589–603. Surface phonons and the c (2x2) oxygen overlayer on Ni(100): theory and experiment.

18 M.-L. Xu, B.M. Hall, S.Y. Tong, M. Rocca, H. Ibach, S. Lehwald & J.E. Black (1985). Phys. Rev. Lett., 54, 1171–1174. Energy dependence of inelastic electron scattering cross section by surface vibrations: experimental measurement and theoretical interpretation.

19 R.E. Allen, G.P. Alldredge & F.W. de Wette (1971). Phys. Rev. B, 4, 1648–1659. Studies of surface vibrational modes. I. General formulation.

20 R.E. Allen, G.P. Alldredge & F.W. de Wette (1971). Phys. Rev. B, 4, 1661–1681. Studies of surface vibrational modes. II. Monatomic fcc crystals.

21 R.E. Allen, G.P. Alldredge & F.W. de Wette (1971). Phys. Rev. B, 4, 1682–1697. Studies of surface vibrational modes. III. Effect of an adsorbed layer.

22   C.J. Wright (1976). J. Chem. Soc. Farad. Trans. I, 73, 1497–1500. Alternative explanation of the inelastic neutron scattering from hydrogen adsorbed by Raney nickel.

23   A. Renouprez, P. Fouilloux, G. Coudurier, D. Tocchetti & R. Stockmeyer (1977). J. Chem. Soc. Farad. Trans. I, 74, 1–10. Different species of hydrogen chemisorbed on Raney nickel studied by neutron inelastic spectroscopy chemisorbed hydrogen and hydrogenous molecules.

24   R.D. Kelley, J.J. Rush & T.E. Madey (1979). Chem. Phys. Lett., 66, 159–164. Vibrational spectroscopy of adsorbed species on nickel by neutron inelastic scattering.

25   R.D. Kelley, R.R. Cavanagh & J.J. Rush (1983). J. Catal., 83, 464–468. Coadsorption and reaction of H_2 and CO on Raney nickel: neutron vibrational spectroscopy.

26   H. Jobic, G. Clugnet & A. Renouprez (1987). J. Electron Spectrosc., 45, 281–290. Neutron inelastic spectroscopy of hydrogen adsorbed at different pressures on a Raney nickel catalyst.

27   H. Jobic, J.P. Candy, V. Perrichon & A. Renouprez (1985). J. Chem. Soc. Farad. Trans. I, 81, 1955–1961. Neutron-scattering and volumetric study of hydrogen adsorbed and absorbed on Raney palladium.

28   J. Eckert, C.F. Majkzrak, L. Passell & W.B. Daniels (1984). Phys. Rev. B., 29, 3700–3702. Optic phonons in nickel hydride.

29   R. Wisniewski, R. Dimitrova, 1. Natkaniec & J. Wasicki (1985). Solid State Commun., 54, 1073–1075. Neutron-scattering study of NiH_x at 77 K.

30   S. Andersson (1978). Chem. Phys. Lett., 55, 185. Vibrational excitations and structure of molecular hydrogen, deuterium and hydrogen-deuteride adsorbed on Ni(100).

31   T.H. Upton, W.A. Goddard III & C.F. Melius (1979). J. Vac. Sci. Technol., 16, 531–536. Theoretical studies of nickel clusters and chemisorption of hydrogen.

32   U. Stuhr, H. Wipf, K.H. Andersen & H. Hahn (1998). Phys. Rev. Lett., 81, 1449–1452. Low-frequency modes in nanocrystalline Pd.

33   P. Albers, M. Poniatowski, S.F. Parker & D.K. Ross (2000). J. Phys.: Condens. Matter, 12, 4451–4463. Inelastic neutron scattering study on different grades of palladium of varying pretreatment.

34   W. Drexel, A Murani, D. Tocchetti, W. Kley, I. Sosnowska & D.K. Ross (1976). J. Phys. Cherm. Solids, 37, 1135–9. The motions of hydrogen impurities in a palladium-hydride.

35   D.K. Ross, P.F. Martin, W.A. Oates & R. Khoda Bakhsh (1979). Z. Phys. Chem., 114, 221–30. Inelastic neutron scattering measurements of optical vibration frequency distribution in hydrogen–metal systems.

36   A. Renouprez, R. Stockmeyer, & C.J. Wright (1979). J. Chem. Soc. Faraday Trans. I, 75, 2473–80. Diffusion of chemisorbed hydrogen in a platinum–zeolite.

37   A. Rahman, K. Skold, C. Pelizzari, S.K. Sinha & H.E. Flotow (1976). Phys. Rev. B, 14, 3630–3634. Phonon spectra of nonstoichiometric palladium hydrides.

38   I.J. Braid, J. Howard & J. Tomkinson (1983). J. Chem. Soc. Faraday Trans. II, 79, 253–262. Inelastic neutron-scattering study of hydrogen adsorbed on impure palladium black.

39   H. Conrad, M.E. Kordesch, R. Scala & W. Stenzel (1986). J. Electron Spectrosc., 38, 289–298. Surface resonances on Pd(111)/H observed with HREELS.

40   J. Howard, T.C. Waddington & C.J. Wright (1978). Chem. Phys. Lett., 56, 258–62. The vibrational spectrum of hydrogen adsorbed on palladium black measured using inelastic neutron scattering spectroscopy.

41   H. Jobic, J-P. Candy, V. Perrichon & A. Renouprez (1985). J. Chem. Soc. Faraday Trans. I, 81, 1955–1961. Neutron-scattering and volumetric study of hydrogen adsorbed and absorbed on Raney palladium.

42   C.J. Wright (1985). In *Catalysis. A Specialist Periodical Report*, (Ed.) G.C. Bond & G. Webb, pp. 46–74. Vol. 7. The Royal Society of Chemistry, London. Catalysts characterization with neutron techniques.

43   C.J. Wright & C.M. Sayers (1983). Rep. Prog. Phys., 46, 773–815. Inelastic neutron-scattering from adsorbates.

44   R. Gomer, R. Wortman, & R. Lundy (1957). J. Chem. Phys., 26, 1147–64. Mobility and adsorption of hydrogen on tungsten.

45   D.G. Hunt & D.K. Ross (1976). J. Less-Common Met., 49, 169–91. Optical vibrations of hydrogen in metals.

46   J.M. Nicol (1992). Spectrochim. Acta,A,48, 313–327. Chemisorbed hydrogen and hydrogenous molecules.

47   C.M. Sayers & C.J. Wright (1984). J. Chem. Soc. Faraday Trans. I, 80, 1217–1220. Hydrogen adsorbed on nickel, palladium and platinum powders.

48   A.M. Baro, H. Ibach, & H.D. Bruchmann (1979). Surf. Sci., 88, 384–398. Vibrational modes of hydrogen adsorbed on platinum (111): adsorption site and excitation mechanism.

49   C.M. Sayers (1984). Surf. Sci., 143, 411–422. Hydrogen adsorption on platinum.

50   H. Asada, T. Toya, H. Motohashi, M. Sakamoto & Y. Hamaguchi (1975). J. Chem. Phys., 63, 4078 – 4079. Study of hydrogen adsorbed on platinum by neutron inelastic scattering spectroscopy.

51   T.Ito &T. Kadowaki (1975). Phys. Lett., 54A, 61–2. NMR of hydrogen absorbed in palladium black.

52   J. Mahanty, D.D. Richardson & N.H. March (1976). J. Phys. C: Solid State Phys., 9, 3421–3436. Neutron inelastic scattering from covered surfaces: hydrogen on platinum.

53   R.K. Thomas (1978). In *Molecular Spectroscopy*, (Ed.) R.F. Barrow, D.A. Long & J. Sheridan, p.312. A Specialist Periodical Report, Vol. 6; The Chemical Society: London. Inelastic and quasielastic neutron scattering spectroscopy; and references therein.

54   P.W. Albers, J. Pietsch, J. Krauter& S.F. Parker (2003). Phys. Chem Chem Phys, 5, 1941–1949. Investigations of activated carbon catalyst supports from different natural sources.

55   P. Albers, E. Auer, K. Ruth & S.F. Parker (2000). J. Catal., 196, 174–179. Inelastic neutron scattering investigation of the nature of surface sites occupied by hydrogen on highly dispersed platinum on commercial carbon black supports.

56   T.J. Udovic, R.R. Cavanagh & J.J. Rush (1988). J. Am. Chem. Soc., 110, 5590–5591. Neutron spectroscopic evidence for adsorbed hydroxyl species on platinum black.

57   G.M. Pajonk (2000). Appl.Catal. A, 202, 157–169. Contribution of spillover effects to heterogeneous catalysis.

58   W.C. Conner (1993). Stud. Surf. Sci. Catal., 77, 61–68. Spectroscopic insight into spillover.

59   P.C.H. Mitchell, A.J. Ramirez-Cuesta, S.F. Parker, J. Tomkinson & D. Thompsett (2003). J. Phys. Chem. B, 107, 6838–6845. Hydrogen spillover on carbon-supported metal catalysts studied by inelastic neutron scattering. Surface vibrational states and hydrogen riding modes.

60   S.W. Lovesey (1980). In Frontiers in Physics, (Ed.) D. Pines,Vol. 49. Benjamin Cummings, Reading, Massachusetts. Condensed matter physics: dynamic correlations.

61   P.C.H. Mitchell, S.F. Parker, J. Tomkinson & D. Thompsett (1998). J. Chem. Soc. Faraday Trans., 94, 1489–1493. Adsorbed states of dihydrogen on a carbon supported ruthenium catalyst. Inelastic neutron scattering study.

62   P.N. Jones, E. Knozinger, W. Langel., R.B. Moyes & J. Tomkinson (1988). Surf. Sci., 207, 159–176. Adsorption of molecular-hydrogen at high-pressure and temperature on MoS_2 and WS_2 - observed by inelastic neutron-scattering.

63   P.J.R. Honeybone, R.J. Newport, W.S. Howells, J. Tomkinson, S.M. Benningon & P.J.. Revell (1991). Chem. Phys. Lett., 180, 145–148. Inelastic neutron-scattering of molecular-hydrogen in amorphous hydrogenated carbon.

64   R.R. Cavanagh, J.J. Rush, R.D. Kelley & T.J. Udovic (1984). J. Chem. Phys., 80, 3478–3484. Adsorption and decomposition of hydrocarbons on platinum black. Vibrational modes from NIS.

65   G. Herzberg (1945). Molecular Spectra and Molecular Structure: Infrared and Raman of Polyatomic Molecules, Vol. 2. Van Nostrand, New York, p.181.

66   H. Jobic, C.C. Santini & C. Coulombeau (1991). Inorg. Chem., 30, 3088–3090. Vibrational modes of acetylenedicobalt hexacarbonyl studied by neutron spectroscopy.

67   S.F. Parker, P.H. Dallin, B.T. Keiller, C.E. Anson & U.A. Jayasooriya (1999). Phys. Chem. Chem. Phys., 1, 2589–2592. An inelastic neutron scattering study and re-assignment of the vibrational spectrum of $[Os_3(CO)_9(\mu_2^-CO)(\mu_3^-\eta^{2-}C_2H_2)]$, a model compound for chemisorbed ethyne.

68   N. Sheppard & C. de la Cruz (1998). Adv. Catal., 42, 181–313. Vibrational spectra of hydrocarbons adsorbed on metals—Part II. Adsorbed acyclic alkynes and alkanes, cyclic hydrocarbons including aromatics, and surface hydrocarbon groups derived from the decomposition of alkyl halides, etc.

69   G.C. Bond (1987). Heterogeneous Catalysis: Principles and Applications, 2nd edition, Clarendon Press, Oxford.

70   R.J. Farrauto & C.H. Bartholomew (1997). *Fundamentals of Industrial Catalytic Processes*, Chapman and Hall, London.

71   K. Nakamoto (1997). *Infrared and Raman Spectra of Inorganic and Coordination Compounds*. Part B: Applications in Coordination, Organometallic and Bioinorganic Chemistry. 5th edition, John Wiley, New York.

72   H. Jobic (1984). Chem. Phys. Lett., 106, 321–324. Neutron inelastic-scattering from ethylene—an unusual spectrum.

73   D. Lennon, J. McNamara, J.R. Phillips, R.M. Ibberson & S.F. Parker (2000). Phys. Chem. Chem Phys., 2, 4447 – 4451. An inelastic neutron scattering spectroscopic investigation of the adsorption of ethene and propene on carbon.

74   R.E. Ghosh, T.C. Waddington & C.J. Wright (1973). J. Chem. Soc. Faraday Trans. II, 69, 275–281. Characterisation of the torsion potential for ethylene ligands using inelastic neutron scattering.

75   J. Howard, T.C. Waddington & C.J. Wright (1976). J. Chem. Soc. Faraday Trans. II, 72, 513–523. Interactions between cis-ethylene ligands studied by inelastic neutron scattering.

76   J. Howard & T.C. Waddington (1978). J. Chem. Soc. Faraday Trans. II, 74, 1275–1284. Ligand–ligand interactions and low frequency vibrations in [Ir(C_2H_4)$_4$Cl] and [Ir(C_2H_4)$_2$Cl]$_2$ studied by inelastic neutron scattering and optical spectroscopy.

77   H. Jobic (1985). J. Mol. Struct., 131, 167–175. A new inelastic neutron-scattering study of Zeise's salt, K[Pt(C_2H_4)Cl$_3$], and a more confident assignment of the vibrational frequencies.

78   B.C. Gates (1992). *Catalytic Chemistry*, John Wiley, New York.

79   S.F. Parker, N.A. Marsh, L.M. Camus, M.K. Whittlesey, U.A. Jayasooriya & G.J. Kearley (2002). J. Phys. Chem. A, 106, 5797–5802. Ethylidyne tricobalt nonacarbonyl: Infrared, FT-Raman, and inelastic neutron scattering spectra.

80   P. Albers, H. Angert, G. Prescher, K. Seibold & S.F. Parker (1999). Chem. Commun., 1619–1620. Catalyst poisoning by methyl groups.

81   S. Chinta, T.V. Choudhary, L.L. Daemen, J. Eckert & D.W. Goodman (2002). Angew Chem. Int. Ed.., 41, 144–146. Characterization of C_2 (C_xH_y) intermediates from adsorption and decomposition of methane on supported metal catalysts by in situ ins vibrational spectroscopy.

82   J.P. Candy, H. Jobic & A.J. Renouprez (1983). J. Phys. Chem., 87, 1227–1230. Chemisorption of cyclohexene on nickel—a volumetric and neutron inelastic spectroscopy study.

83   J.Anderson. J.J. Ullo & S. Yip (1986). Physica B & C, 136, 172–176. Inelastic neutron scattering by benzene: direct spectral analysis using molecular dynamics simulation.

84   J. Tomkinson (1988). In *Neutron Scattering at a Pulsed Source*, (Ed.) R.J. Newport, B.D. Rainford & R. Cywinski, pp. 324 – 343, Adam Hilger, Bristol. Inelastic scattering spectroscopy of hydrogen in metals and molecules.

85   H. Jobic, J. Tomkinson, J.P. Candy, P. Fouilloux & A. Renouprez (1980). Surf. Sci., 95, 496. The structure of benzene chemisorbed on Raney nickel: a neutron inelastic spectroscopy determination.

86   A.J. Renouprez, G. Clugnet & H. Jobic (1982). J. Catal., 74, 296–306. The interaction between benzene and nickel—a neutron inelastic spectroscopy study.

87   H. Jobic & A. Renouprez (1981). Surf. Sci., 111, 53–62. Neutron inelastic spectroscopy of benzene chemisorbed on Raney platinum.

88   M.F. Bailey & L.F. Dahl (1965). Inorg. Chem., 4, 1314–19. Three-dimensional crystal structure of benzenechromium tricarbonyl with further comments on the dibenzenechromium structure.

89   H. Jobic, J. Tomkinson & A. Renouprez (1980). Mol. Phys., 39, 989–995. Neutron inelastic scattering spectrum and valence force field for benzenetricarbonylchromium.

90   F. Hochard, H. Jobic, G. Clugnet, A. Renouprez & J. Tomkinson (1993). Catal. Lett., 21, 381–389. Inelastic neutron-scattering study of acetonitrile adsorbed on raney-nickel.

91   R.J. Farrauto & C.H. Bartholomew (1997). Fundamentals of Industrial Catalytic Processes, Blackie Academic and Professional, London.

92   P.C.H. Mitchell & J. Tomkinson (1991). Catal. Today, 9, 227–235. Inelastic neutron scattering as a tool for in situ studies of catalysts–surface OH groups of aluminium and iron oxide hydroxides.

93   P.C.H. Mitchell, J. Tomkinson, J.G. Grimblot & E. Payen (1993). J. Chem. Soc. Faraday Trans., 89, 1805–1807. Bound water in aged molybdate/alumina hydrodesulfurisation catalysts: an inelastic neutron scattering study.

94   E. Payen, S. Kasztelan & J. Grimblot (1988). J. Mol. Struct., 174, 71–76. In situ laser Raman spectroscopy of the sulfiding of $WO_3(MoO_3)gamma-Al_2O_3$ catalysts.

95   P.G. Dickens, J.J. Birtill & J.C. Wright (1979). J. Solid State Chem., 28, 185–193. Elastic and inelastic neutron studies of hydrogen molybdenum bronzes.

96   P.C.H. Mitchell, M. Bowker, N. Price, S. Poulston, D. James & S.F. Parker (2000). Top. Catal., 11, 223–227. Iron antimony oxide selective oxidation catalysts—an inelastic neutron scattering study.

97   J. Howard & T.C. Waddington (1978). J. Chem. Soc. Faraday Trans. II, 74, 879–888. Low frequency vibrations of some p-allyl complexes of Ni and Pd studied by incoherent inelastic neutron scattering.

98   W.E. Oberhansli & L.F. Dahl (1965). J. Organometal. Chem., 3, 43–54. Structure of and bonding in $[(C_3H_5)PdCl]_2$.

99   J.R. Durig, Q. Tang & T.S. Little (1992). J. Raman Spectrosc., 23, 653–666. Raman and infrared-spectra, conformational stability, barriers to internal-rotation, vibrational assignment, and ab initio calculations for 3-iodopropene.

100   D.M. Adams & A. Squire (1970). J. Chem. Soc. A, 1808 – 1813. Vibrational spectra of the chloro- and bromo-π-allylpalladium dimers and of the π-2-methylallyl analogues.

101   J. Howard, I.J. Braid & J. Tomkinson (1984). J. Chem. Soc. Faraday Trans. I, 80, 225–235. Spectroscopic studies of hydrogen adsorbed on zinc oxide (Kadox 25).

102 P.C.H. Mitchell, R.P. Holroyd, S. Poulston, M. Bowker & S.F. Parker (1997). J. Chem. Soc. Faraday Trans., 93, 2569–2575. Inelastic neutron scattering of model compounds for surface formates. Potassium formate, copper formate and formic acid.

103 S. Poulston, R.P. Holroyd, M. Bowker, S.F. Parker & P.C.H. Mitchell (1998). Surf. Sci., 402/404, 599–603. An inelastic neutron scattering study of formate on copper surfaces.

104 C. Sivadinarayana, T.V. Choudhary, L.L. Daemen, J. Eckert & D.W. Goodman (2004). J. Am. Chem. Soc., 126, 38–39. The nature of the surface species formed on Au/TiO$_2$ during the reaction of H$_2$ and O$_2$: an inelastic neutron scattering study.

105 Structure commission of the International Zeolite Association (2003). *Database of Zeolite Structures*. http://www.iza-structure.org/databases/.

106 W.P.J.H. Jacobs, J.H.M.C. van Wolput, R.A. van Santen & H. Jobic (1994). Zeolites, 14, 117–125. A vibrational study of the OH and OD bending modes of the Brönsted acid sites in zeolites.

107 H. Jobic, K.S. Smirnov & D. Bougeard (2001). Chem. Phys. Let., 344, 147–133. Inelastic neutron scattering spectra of zeolite frameworks–experiment and modelling.

108 H. Jobic (1991). J. Catal., 131, 289–293. Observation of the fundamental bending vibrations of hydroxyl groups in HNaY zeolite by neutron inelastic scattering.

109 W.P.J.H. Jacobs, H. Jobic, J.H.M.C. van Wolput & R.A. van Santen (1992). Zeolites, 12, 315–319. Fourier-transform infrared and inelastic neutron scattering study of HY zeolites.

110 W.P.J.H. Jacobs, R.A. van Santen & H. Jobic (1994). J. Chem. Soc. Faraday Trans., 90, 1191–1196. Inelastic neutron scattering study of NH$_4$Y zeolites.

111 J.B. Parise, L. Abrams, T.E. Gier, D.R. Corbin, J.D. Jorgensen & E. Prince (1984). J. Phys. Chem., 88, 2303–2307. Flexibility of the framework of zeolite rho–structural variation from 11 to 573 K–a study using neutron powder diffraction data.

112 T.J. Udovic, R.R. Cavanagh, J.J. Rush, M.J. Wax, G.D. Stucky, G.A. Jones & D.R. Corbin (1987). J. Phys. Chem., 91, 5968–5973. Neutron scattering study of NH$_4^+$ dynamics during the deammoniation of NH$_4$-rho zeolite.

113 M.J. Wax, R.R. Cavanagh, J.J. Rush, G. D Stucky, L. Abrahams & D.R. Corbin (1986). J. Phys. Chem., 90, 532–534. A neutron scattering study of zeolite-rho.

114 H. Jobic, M. Czjzek & R.A. van Santen (1992). J. Phys. Chem., 96, 1540–1542. Interaction of water with hydroxyl groups in H-mordenite–a neutron inelastic scattering study.

115 H. Jobic, A. Tuel, M. Krossner & J. Sauer (1996). J. Phys. Chem., 100, 19545–19550. Water in interaction with acid sites in H-ZSM-5 zeolite does not form hydroxonium ions. A comparison between neutron scattering results and *ab initio* calculations.

116 H. Jobic (2000). Physica B, 276/278, 222–225. Inelastic scattering of organic molecules in zeolites.

117   F. Jousse, S.M. Auerbach, H. Jobic & D.P. Vercauteren (2000). J. Phys. IV France,
      10, 147–150. Simulations of low frequency vibrations of adsorbed molecules in
      zeolites.
118   H. Jobic & A.N. Fitch (1997). Stud. Surf. Sci. Catal., Pts. A–C, 105, 559–566.
      Vibrational study of benzene adsorbed in NaY zeolite by neutron spectroscopy.
119 . A.N. Fitch, H. Jobic & A. Renouprez (1986). J. Phys. Chem., 90, 1311–1318.
      Localization of benzene in sodium-Y zeolite by powder neutron diffraction.
120   I.A. Beta, H. Jobic, E. Geidel, H. Bohlig & B. Hunger (2001). Spectrochim. Acta
      A, 57, 1393–1403. Inelastic neutron scattering and infrared spectroscopic study of
      furan adsorption on alkali-metal cation-exchanged faujasites.
121   H. Jobic, A. Renouprez, A.N. Fitch & H.J. Lauter (1987). J. Chem. Soc. Faraday
      Trans. I, 83, 3199–3205. Neutron spectroscopic study of polycrystalline benzene
      and of benzene adsorbed in Na-Y Zeolite.
122   E. Kemner, A.R. Overweg, U.A. Jayasooriya, S.F. Parker, I.M. de Schepper & G.J.
      Kearley (2002). Appl. Phys. A, 74, S1368–S1370. Ferrocene-zeolite interactions
      measured by inelastic neutron scattering.
123   E. Kemner, I.M. de Schepper, G.J. Kearley & U.A. Jayasooriya (2000). J. Chem.
      Phys., 112, 10926–10929. The vibrational spectrum of solid ferrocene by inelastic
      neutron scattering.
124   P.C.H. Mitchell (1977). In Catalysis. A Specialist Periodical Report, (Ed.) C.
      Kimball, pp. 204–234. Vol. 1. The Chemical Society, London. Reactions on
      sulfide catalysts.
125   P.C.H. Mitchell (1981). In Catalysis. A Specialist Periodical Report, (Ed.) C.
      Kimball & D.A. Dowden, pp. 175–184. Vol. 4. The Royal Society of Chemistry,
      London. Sulfide catalysts: characterization and reactions including
      hydrodesulfurisation.
126   H. Topsoe, B.S. Clausen & F.E. Massoth (1996). Hydrotreating Catalysis,
      Springer, Berlin.
127   M. Breysse, E. Furimsky, S. Kasztelan, M. Lacroix & G. Perot (2002). Catal. Rev.,
      44, 651–735. Hydrogen activation by transition metal sulfides.
128   C.J. Wright, C. Sampson, D. Fraser, R.B. Moyes & P.B. Wells (1980). J. Chem.
      Soc. Faraday I, 76, 1585–1598. Hydrogen sorption by molybdenun sulphide
      catalysts.
129   C.J. Wright, D. Fraser, R.B. Moyes & P.B. Wells (1981). Appl. Catal., 1, 49–58.
      The adsorption of hydrogen and hydrogen sulphide on tungsten sulphide—
      Isotherm and neutron scattering studies.
130   M. Lacroix, H. Jobic, C. Dumonteil, P. Afanasiev, M. Breysse & S. Kasztelan
      (1996). Stud. Surf. Sci. Catal. Part A&B., 101, 117–126, Role of adsorbed
      hydrogen species on ruthenium and molybdenum sulfides. Characterization by
      inelastic neutron scattering, thermoanalysis methods and model reactions.

131 H. Jobic, G. Clugnet, M. Lacroix, S.B. Yuan, C. Mirodatos & M. Breysse (1993). J. Am. Chem. Soc., 115, 3654–3657. Identification of new hydrogen species adsorbed on ruthenium sulfide by neutron spectroscopy. H. Jobic, M. Lacroix, T. Decamp & M. Breysse (1995). J. Catal., 157, 414–422. Characterization of ammonia adsorption on ruthenium sulfide. Identification of amino species by inelastic neutron scattering.

132 C. Riekel, H.G. Rezmk, H. Schollhom & C.J. Wright (1979). J. Chem. Phys., 70, 5203—55212. Neutron diffraction study on formation and structure of D_xTaS_2, and H_xNbS_2.

133 A.F. Wells (1984). *Structural Inorganic Chemistry,* Oxford University Press; Oxford. 5th edition. K.D. Bronsema, J.L.de Boer & F. Jellinek (1986). Zeit. Anorg. Allgem. Chem., 540, 15–17. On the structure of molybdenum diselenide and disulfide.

134 A.B. Anderson, Z.Y. Al-Saigh & W.K. Hall (1988). J. Phys. Chem., 92, 803–809. Hydrogen on MoS_2. Theory of its heterolytic and homolytic chemisorption.

135 P.C.H. Mitchell & A.J. Ramirez-Cuesta (2003). Unpublished work.

136 P.N. Jones, E. Knozinger, W. Langel, R.B. Moyes & J. Tomkinson (1988). Surf. Sci., 207, 159–176. Adsorption of molecular hydrogen at high pressure and temperature on MoS_2 and WS_2 observed by inelastic neutron scattering.

137 P. Sundberg, R.B. Moyes & J. Tomkinson (1991). Bull. Soc. Chim. Belg., 100, 967–976. Inelastic neutron scattering spectroscopy of hydrogen adsorbed on powdered-MoS_2, MoS_2 -alumina and nickel-promoted MoS_2.

138 C. Sampson, J.M. Thomas, S. Vasudevan & C.J. Wright (1981). Bull. Soc . Chim. Belg., 90, 1215–1224. A preliminary investigation of the sorption of hydrogen at high pressure by MoS_2.

139 C.M. Sayers (1981). J. Phys. C: Solid State Phys., 14, 4969–4983. The vibration spectrum of H bound by molybdenum sulphide catalysts.

140 P.C.H. Mitchell, D.A. Green, E. Payen, J. Tomkinson & S.F. Parker (1999). Phys. Chem. Chem. Phys., 1, 3357–3363. Interaction of thiophene with a molybdenum disulfide catalyst—an inelastic neutron scattering study.

141 G.D. Atter, D.M. Chapman, R.E. Hester, D.A. Green, P.C.H. Mitchell & J. Tomkinson (1997). J. Chem. Soc., Faraday Trans., 93, 2977–2980. Refined *ab initio* inelastic neutron scattering spectrum of thiophene.

142 T.L. Tarbuck, K.R. McCrea, J.W. Logan, J.L. Heiser & M.E. Bissell (1998). J. Phys. Chem. B, 102, 7845–7857. Identification of the adsorption mode of thiophene on sulfided Mo catalysts.

143 P. Mills, S. Korlann, M.E. Bussell, M.A. Reynolds, M.V. Ovchinnikov, R.J. Angelici, C. Stinner, T. Weber & R. Prins (2001). J. Phys. Chem. A, 105, 4418–4429. Vibrational study of organometallic complexes with thiophene ligands: Models for adsorbed thiophene on hydrodesulfurization catalysts.

144 J.V. Lauritsen, S. Helveg, E. Laegsgaard, I. Stensgaard, B.S. Clausen, H. Topsoe & E. Besenbacher (2001). J. Catal., 197, 1–5. Atomic-scale structure of Co–Mo–S nanoclusters in hydrotreating catalysts.

145   S. Helveg, J.V. Lauritsen, E. Laegsgaard, I. Stensgaard, J.K. Norskov, B.S. Clausen, H. Topsoe & F. Besenbacher (2000). Phys. Rev. Let., 84, 951–954. Atomic-scale structure of single-layer MoS$_2$ nanoclusters.

146   J.V. Lauritsen, M. Nyberg, R.T. Vang, M.V. Bollinger, B.S. Clausen, H. Topsoe, K.W. Jacobsen, E. Laegsgaard, J.K. Norskov & F. Besenbacher (2003). Nanotechnology, 14, 385– 389. Chemistry of one-dimensional metallic edge states in MoS$_2$ nanoclusters.

147   P.C.H. Mitchell, G.M. Raos, P.B. Karadakov, J. Gerratt & D.L. Cooper (1995). J. Chem. Soc. Faraday Trans., 91, 749–758. Catalytic chemistry of furan and thiophene—*ab initio* calculations, using the spin-coupled valence-bond method, of the interaction of furan and thiophene with a positively charged center.

8

Organic and Organometallic Compounds

Several chapters in this book are focussed on specific applications of INS spectroscopy. Integral to these studies has been the use of small molecules, usually organic, as reference materials and model compounds. This may give a somewhat misleading impression of INS spectroscopy in that it suggests that the study of organic compounds is almost incidental to the main thrust of work. This is contradicted by the fact that on TOSCA at ISIS more than 50% of the spectra measured are of organic compounds.

The chapter starts with an example of how the analysis of the spectra of organic compounds has evolved as illustrated by maleic anhydride (§8.1). The succeeding sections consider: alkanes, including cyclic and polycyclic alkanes (§8.2), alkenes and alkynes (§8.3), aromatics including heteroaromatics (§8.4), oxygenated compounds (§8.5), nitrogen containing compounds (§8.6) and concludes with organometallic compounds (§8.7).

8.1 Analysis of the INS spectra of organic compounds

One advantage of studying organic compounds is that it is usually possible to obtain infrared, Raman and INS spectra of the material and this should always be the aim. The complementarity of the methods has been stressed previously and is evident from Fig. 8.1: modes that are strong in one form of spectroscopy are weak or absent in the others.

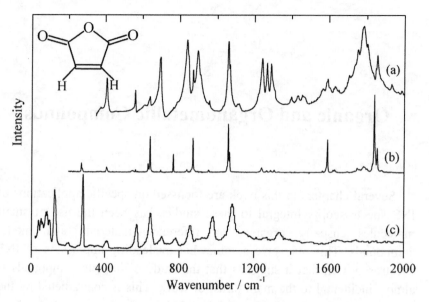

Fig. 8.1 Vibrational spectra of maleic anhydride: (a) infrared, (b) Raman and (c) INS spectrum.

8.1.1 Group frequencies

Historically, the first step in the spectroscopic analysis is an assignment in terms of group frequencies. Compilations of these are readily available for infrared and Raman spectra [1]; however, this is not so for INS spectra. This is largely because it is only in the last decade that sufficiently high resolution INS spectra have become available that would enable the creation of such tables. Fortunately, the *frequency* information in the infrared and Raman tables is still valid for INS spectra, it is the *intensity* information that is inapplicable.

Maleic anhydride will be used as example [2] (see Fig. 8.1 for the structure). Using the tables the spectra can be assigned as follows. The symmetric and antisymmetric C–H stretches are seen in the infrared and Raman spectra at 3132 and 3124 cm^{-1}. The symmetric and antisymmetric C=O stretches are at 1857 and 1785 cm^{-1} respectively. The assignment is based on the reversal of intensities between the infrared and Raman spectra and the absence in the INS spectrum. The C=C stretch is observed as a strong band at 1595 cm^{-1} in the Raman spectrum. The C–H in-plane bends are clearly seen in the INS spectrum at 1338 and 1084 cm^{-1} as are the C—H out-of-plane bends at 978 and 859 cm^{-1}. The bands in the 800–600 cm^{-1} region are assigned to ring deformations and those in the 600–200 cm^{-1} range to C=O in-plane and out-of-plane bends. The modes below 200 cm^{-1} are assigned as lattice modes.

The analysis may seem rudimentary but for many years it formed the basis of vibrational spectroscopy. With additional information from isotopic substitution and polarisation studies it provides a successful means of analysis. For complex systems such as biomolecules this type of analysis is still used. The method offers a simple, useful check on the spectra recorded. It is also needed when the unexpected happens and the observed spectrum does not match expectations: it provides clues for further investigation.

8.1.2 The Wilson GF method

The vibrational frequency of a normal mode is determined by the kinetic and potential energies of the system. The kinetic energy depends on the geometrical arrangement of the atoms in the molecule and their masses. The potential energy arises from the interactions between the atoms, thus containing information about the bonding present. This is expressed in the force constants of the system.

The method adopted to obtain the force constants from the vibrational frequencies treated the molecule as a system of point masses connected by springs that obeyed Hooke's law, so the system was purely harmonic. The approach was codified in a classic book [3] and is known as the Wilson GF matrix method. The basis of the method is described in §4.2.2 and in more detail in [4,5]. The key equation is:

$$|GF - E\lambda| = 0 \qquad (8.1)$$

where the G matrix is determined by the geometry of the system, the F matrix is the force constant matrix, E is the unit matrix and λ is related to the vibrational wavenumber ω_v (Eq. (4.17)):

$$4\pi^2\omega_\nu^2 = \lambda_\nu \tag{8.2}$$

(note that λ is conventionally written λ [3,5]). The usual procedure is to diagonalise GF to obtain frequencies (the eigenvalues) and atomic displacements (the eigenvectors). As described in Chapter 4, the latter are used to generate the INS intensities. A minimisation routine then attempts to estimate better force constants using the difference between the observed and calculated INS spectra.

The significant difference between analysis of INS spectra and of infrared and Raman spectra is that the INS intensities can be used in the refinement. This is important because for a molecule with N_{atom} atoms the F matrix has of the order $(3N_{atom} - 6)^2/2$ independent components, whereas there are only $(3N_{atom} - 6)$ frequencies, thus the problem is severely under-determined. The use of INS intensities potentially doubles the number of observables.

The problem is usually still under-determined so isotopic substitution is used to increase the number of available frequencies (the F matrix is assumed to be unchanged by isotopic substitution). In addition, some of the force constants are arbitrarily set equal to zero. It is the off-diagonal force constants that are the most difficult to determine: these result in coupling of different types of motion. It is generally assumed that modes that are separated by hundreds of wavenumbers can be treated independently. Thus the C–H stretching modes are assumed to be decoupled from the rest of the molecule. Symmetry is also used extensively: in maleic anhydride there is no symmetry operation that mixes in-plane and out-of-plane modes so the corresponding off-diagonal force constant must be zero. There is also an expectation that interaction constants will fall in the range ± 0.5 mdyn Å^{-1}; those that fall outside this range are usually regarded with suspicion.

The process is laborious; it is not guaranteed to succeed and the force constants so obtained are not necessarily unique. It can produce excellent results so providing confidence in the assignments. The force constants obtained may be transferable, allowing comparison with other systems and giving insight into the bonding. Fig. 8.2a and 8.2b compare the experimental and calculated spectra using such a program [4].

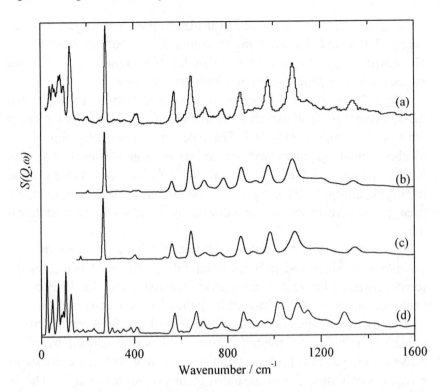

Fig. 8.2 INS spectra of maleic anhydride [2]: (a) experimental, (b) calculated using the Wilson GF matrix method, (c) from a DFT calculation (using the B3LYP functional with the 6-31G(d,p) basis set) of the isolated molecule and (d) from a periodic DFT calculation of the complete unit cell.

8.1.3 Ab initio methods

As described in Chapter 4, it is straightforward to obtain vibrational frequencies and displacements in the modes from computational packages. For maleic anhydride, three increasing levels of theory were explored: molecular mechanics using the AM1 force field and density functional theory using the BP86 functional with the 6-21G and the B3LYP functional with the 6-31G(d,p) basis sets. The agreement between calculated and experimental results for both the structure and the frequencies improved as the quality of the calculation improved.

With the B3LYP / 6-31G(d,p) calculation the bond lengths were all within 0.01Å and the bond angles within 1° of the experimental gas phase structure. In the solid state the molecule occupies a C_1 site in the lattice but is essentially unchanged from the gas phase.

Fig. 8.2c shows the INS spectrum calculated from these results and the agreement is again excellent. This degree of agreement is typical of what can be achieved with DFT. The experience of several groups is that to obtain good agreement without the use of large scaling factors, this level of theory is needed. In general, for molecules containing carbon, hydrogen, nitrogen and oxygen, the use of larger basis sets does not improve the agreement much and drastically increases the computational time.

It is possible, albeit complex, to extract the force constants from an *ab initio* calculation and so is not usually done. Table 8.1 compares the force constants for maleic anhydride obtained from the Wilson GF matrix method with those obtained from the *ab initio* calculation.

Comparison of the empirical and *ab initio* force fields reveals some interesting points. The empirical force field was generated based on conventional practice (use of symmetry, off-diagonal force constants assumed small, modes well-separated in energy, do not couple). This is to be contrasted with the *ab initio* force field. The *only* terms that are zero are those that are enforced by symmetry, (in-plane and out-of-plane coordinates do not mix because there are no symmetry elements that couple them). As shown in Table 8.1, the diagonal terms from the DFT are generally smaller than with the empirical force field, while the off-diagonal terms are often much larger and values greater than 1.0 occur. In essence, it appears that the *ab initio* calculation is more 'democratic' with less distinction between the diagonal and off-diagonal terms than is customary with empirical force fields.

This raises the question 'which is correct'? As judged by the INS spectra that each force field generates, both are equally good, see Fig. 8.2. The potential energy distribution from both force fields is similar. Thus from a practical viewpoint there is little to choose between them. From the purist view, the *ab initio* force field is better because it produces a unique solution in as far as the *ab initio* calculation is correct.

Table 8.1 Comparison of selected force constants of maleic anhydride obtained from the Wilson GF matrix method wth those from a DFT calculation using the B3LYP functional with the 6-31G(d,p) basis set.

Coordinate(a)	Force constant / mdyn Å^{-1}	
	Wilson GF	DFT
O–C torsion	0.179	0.100
C–C torsion	0.247	0.112
C=C torsion	0.291	0.230
C–O–C ip bend	1.585	1.125
O–C–C ip bend	1.618	0.696
C–C=C ip bend	1.157	0.697
C=C–H ip bend	0.347	0.394
C–H op bend	0.225	0.231
C=O op bend	0.750	0.505
C–H stretch	5.324	5.830
C=O stretch	9.300	14.338
C=C stretch	8.898	6.706
C–O stretch	4.371	3.193
C–C stretch	4.990	3.036
C–O stretch/C–C stretch int	1.231	-0.036
C–O stretch/C=C stretch int		1.755
C=O oop/C=O op int	-0.130	0.033
C–H oop/C–H op int	-0.035	-0.015

(a) ip= in-plane, oop = out-of-plane, int = interaction

Note that implicit in an *ab initio* calculation is the use of the isolated molecule approximation (i.e. gas-phase-like with no intermolecular interactions) and the harmonic approximation, whereas the empirical force field is conveniently compact but is not unique. For this reason, analysing small to medium size molecules with empirical force fields is rapidly becoming extinct. Current practice is to use the *ab initio* calculated spectrum and omit extracting the force constants.

The analysis so far has all been based on the isolated molecule approximation. For maleic anhydride this is valid as the agreement between the gas phase calculation and the solid state spectrum shows. However, it does not account for some of the most intense bands in the spectrum, those below 200 cm^{-1} in the lattice mode region. It is possible to assign the bands, if it is assumed that the librational modes are more intense and occur at higher frequency than the translational modes. It also requires the assumption that the factor group splitting is small (there are four molecules in the unit cell) and that the dispersion is negligible.

In principle, periodic DFT calculations make no assumptions (apart from the harmonic approximation) so should provide unambiguous assignments. Practically, complete analyses including the dispersion (such as that carried out for the alkali metal hydrides §6.7.1) are limited to systems with a few atoms in the unit cell. Thus calculations based on frequencies at the Γ point (Brillouin zone centre) are generally used, since these are much less demanding computationally. The results for such a calculation for maleic anhydride are shown in Fig. 8.2d. It can be seen that reasonable agreement is obtained, even in the lattice mode region, although the absence of dispersion in the calculation makes the region look 'spikier' than the experimental data. The calculation confirms the assignments made empirically [2]: librations at higher energy than translations and the librations are in the reverse order of their moments of inertia.

The conclusion of this section is that the analysis of the vibrational spectra of organic molecules up to 30 atoms or so is almost routine and is best carried out using DFT at an appropriate level (B3LYP with the 6-31G basis set generally works well). This assumes that the isolated molecule approximation is valid. For systems where there are significant intermolecular interactions, particularly hydrogen bonding, periodic DFT is the best approach, although the number of atoms in the unit cell is a consideration. Currently (2004) systems with more than 20 atoms require a week or more to compute on a desktop computer. If Moore's law continues to hold true then these limitations will rapidly disappear.

8.2 Alkanes and cycloalkanes

The INS spectra of the normal (unbranched) n-alkanes ($n = 5$—25)) are discussed in §10.1.2 and [6,7]. The light alkanes, ethane, propane and butane have also been investigated [8,9]. Methane has been extensively studied by tunnelling spectroscopy both in the solid [10] and as an adsorbate e.g. [11–13].

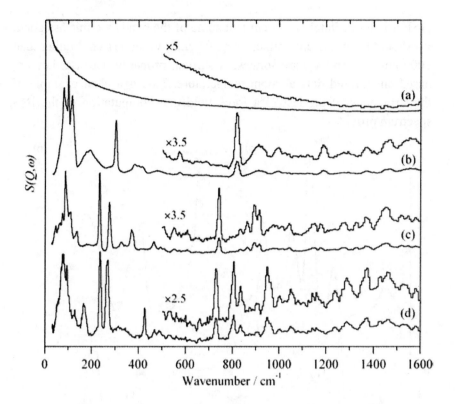

Fig. 8.3 INS spectra of *n*-alkanes: (a) methane, (b) ethane, (c) propane and (d) butane.

The INS spectra of methane, ethane, propane and butane are shown in Fig. 8.3 and the reason for the lack of INS spectra of methane is evident. The spectrum is dominated by its molecular recoil (§2.6.5) and all spectroscopic information is washed out. This is a consequence of its light mass and the very weak intermolecular interactions: methane boils at 112K. As the molecular weight of the alkane increases, it becomes possible to distinguish features at increasing wavenumber. Also modes measured at low wavenumber on a low-bandpass spectrometer, are measured at low Q and since recoil is proportional to Q^2, spectral quality improves (§5.2.2).

A branched alkane which has been studied by INS is neopentane (2,2-dimethylpropane, $C(CH_3)_4$) [14]. Fig. 8.4 shows a comparison of the INS spectrum and that calculated using a classic Wilson GF force field

[15]. It is immediately apparent that some of the bands are mis-assigned: ν_{12} should be at 281 cm^{-1} rather than 203 cm^{-1}, ν_{11} at 951 cm^{-1} rather than 1000 cm^{-1}; ν_8 and ν_{19} are somewhat shifted, probably because they are based on infrared data at room temperature. This is a clear example of the stringency of the test of a force field that computation of the INS spectrum provides.

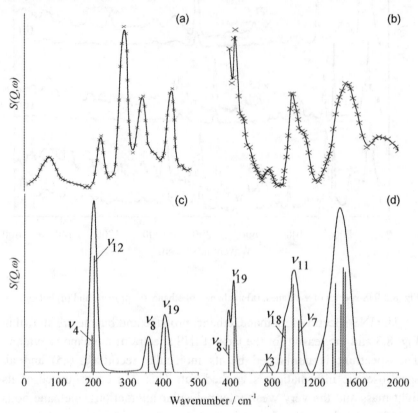

Fig. 8.4 INS spectrum of neopentane recorded at (a) 30K (IN4, ILL) and (b) at 80K (IN1BeF, ILL). (c) and (d) are spectra of the same regions calculated with the classic Wilson GF force field [15]. Reproduced from [14] with permission from Academic Press.

As Figs. 8.3 and 8.4 show, the methyl torsion is a rare example of a reliable group frequency mode in INS spectroscopy. Most organic compounds that contain a methyl group bonded to an sp^3 hybridized carbon atom have a strong band at 250 ± 10 cm^{-1}. This is often the most intense band in the spectrum.

Of the monocyclic alkanes, cyclopropane has been studied by INS [16]. The work was able to resolve an uncertainty in the assignment by showing that two modes were accidentally degenerate.

More polycyclic alkanes have been treated by INS spectroscopy. The spectrum of of adamantane (tricyclo[3.3.1.1^{3,7}]decane, $C_{10}H_{16}$) is shown in Fig. 4.6. Norbornane and some of its mono- and dimethyl derivatives (see Fig. 8.5 for the structures) were extensively studied and force fields developed using the Wilson GF matrix method [17—19]. Given the low symmetry and relatively large number of atoms present, this was a *tour de force* of spectroscopy.

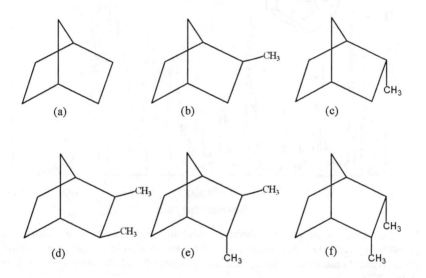

Fig. 8.5 Structure of norbornane and its derivatives that have been studied by INS spectroscopy. (a) Norbornane, (b) *exo*-2-methylnorbornane, (c) *endo*-2-methylnorbornane, (d) *exo*-2-*exo*-3-dimethylnorbornane, (e) *exo*-2-*endo*-3-dimethylnorbornane and (f) *endo*-2-*endo*-3-dimethylnorbornane.

The Platonic solids cubane C_8H_8 [20,21] and dodecahedrane $C_{20}H_{20}$ [22] have been investigated by INS spectroscopy. The latter, see Fig. 8.6, provides an interesting example of the strengths and limitations of INS spectroscopy. The isolated molecule has icosahedral, I_h, symmetry and the 114 normal modes of vibration are classified into $2A_g + T_{1g} + 2T_{2g} + 4G_g + 6H_g + 3T_{1u} + 4T_{2u} + 4G_u + 4H_u$ symmetry types. Thus there are only 30 discrete vibrational frequencies due to the high average degeneracy. Of these only the $3T_{1u}$ modes are infrared active and only the $2A_g$ and $6H_g$ modes are Raman active. Thus 19 of the 30 modes of vibration are unobservable optically as long as the molecule retains its high symmetry

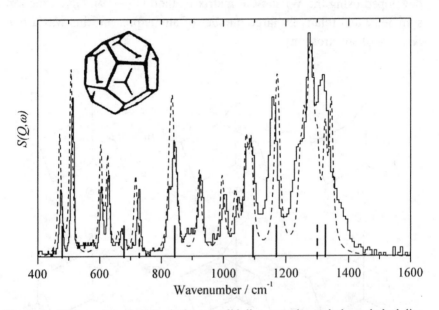

Fig. 8.6   INS spectra of dodecahedrane: solid line experimental data, dashed line calculated by DFT. The short vertical lines show the position of the gas phase infrared (dashed) and Raman (solid) allowed modes. Reproduced from [22] with permission from Wiley-VCH.

As would be expected for a molecule where the isolated molecule approximation is valid, there is reasonable agreement between the observed and calculated spectrum. Further, all the modes are observed (although not necessarily resolved) irrespective of whether they are infrared or Raman allowed. However, in the solid state I_h symmetry is not an allowed site symmetry so the molecule deforms slightly to the very rarely seen T_h symmetry. This has the effect of retaining the centre of symmetry so there are still no coincidences between the infrared and Raman spectra but causing the G and H modes to split to $(A + T)$ and $(E + T)$ modes respectively. This results in splitting of some of the modes and additional modes being present in the infrared and Raman spectra. These effects are present in Fig. 8.6 but are beyond INS resolution at present.

8.3 Alkenes and alkynes

The INS spectra of ethene [23,24] and propene [24] are discussed in §7.3.2.3 and shown in Fig. 7.16. The spectra are dominated by the effects of molecular recoil. This is less of a problem for propene because it has internal vibrations at lower energy (and hence on low-bandpass spectrometers, lower Q) than ethene. With the much heavier tetrabromoethene [25] this does not occur but the small cross section means that a large (8 g) sample was needed. Tetracyanoethene has been studied by coherent INS [26]. The bicyclic alkene norbornene [27] has been studied by INS because it is the parent compound for a class of advanced composites.

An alkyne which has been studied is 2-butyne (dimethylacetylene, $CH_3-C\equiv C-CH_3$) [28]. The crystal contains two molecules in the unit cell and the focus of the work was on the interactions between the methyl groups and how their influence on the methyl rotational tunnelling spectrum and the librational and torsional modes.

8.4 Aromatic and heteroaromatic compounds

The archetypal aromatic compound benzene [29] is comprehensively discussed in §5.3 and the spectrum shown in Fig. 5.5, that of C_6D_6 is shown in Fig. 8.1. Benzene has also been widely used as an adsorbate in catalyst studies (§7.3.2.10).

The INS spectrum of toluene is shown in Fig. 3.31. 1,4-dimethylbenzene (*p*-xylene) [30,31] has been studied in order to investigate the spectral effects of interacting methyl groups. INS spectra of the other isomeric xylenes have also been investigated [32].

1,3,5-trimethylbenzene (mesitylene) [33] and the 2,4,6-trihalo-mesitylenes (bromo, chloro and iodo) [34,35] have been studied. The trihalomesitylenes provide a textbook example of how the combination of crystallography, spectroscopy and modelling can provide a coherent picture of the structure and dynamics and allow insights into the intra- and intermolecular forces present [36].

1,2,4,5-tetramethylbenzene (durene) [37,38] was investigated as a quantitative probe of its environment. Analysis of the spectrum using three different periodic DFT codes gave good agreement for the internal modes in all cases. The external modes were reasonably well reproduced although less well than the internal modes. The differences in the external modes were shown to be the result of significant anharmonicity in the external modes.

3- and 4-fluorostyrene [39,40] have been studied. In solution, the former exists as a mixture of two conformers and to account for the solid state spectrum it was necessary to include both conformers.

Monocyclic aromatic compounds exhibit a strong characteristic band at a group frequency near 400 cm^{-1}, this is the shown in Fig. 8.7.

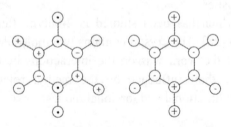

Fig. 8.7 The form of both components of the, E_{2u}, group frequency mode at ca 400 cm^{-1} in aromatic compounds (•, atoms at nodes; the larger the ± symbol the greater the atomic displacement).

A number of polycyclic aromatic compounds have been investigated by coherent INS of the perdeuterated compounds including naphthalene [41] and anthracene [42]. Perylene was used as a model compound for the aromatic regions of coal [43].

A range of nitrogen containing heteroaromatic compounds has been extensively characterised [44—49]. The structures are shown in Fig. 8.8. These molecules are of interest because derivatives are commonly found in biological systems.

Thiophene [50] and furan [51] are important model systems for catalytic studies and are considered in §7.6.6 and §7.5.4.2 respectively.

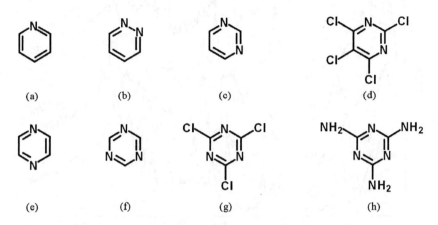

Fig. 8.8 Structure of nitrogen containing heteroaromatic compounds that have been studied by INS spectroscopy. (a) pyridine [44], (b) pyridazine [45], (c) pyrimidine [46], (d) 2,4,5,6-tetrachloropyrimidine [47], (e) pyrazine [48], (f) *s*-triazine [48], (g) 1,3,5-trichloro-*s*-triazine [48] and (h) melamine [49].

8.5 Oxygen containing compounds

Oxygen containing compounds are important industrially where they are used as solvents and feedstocks for catalytic processes (§7.4.3). The interaction of methanol, ethanol, 1-propanol and 1-butanol with zeolites was studied [52]. The spectra of the solid alcohols are shown in Fig. 8.9. Alcohols readily form glasses and the lack of spectral detail in the external mode region of ethanol, propanol and butanol suggests that this has occurred in their spectra. DFT calculations of the spectra of dimers and trimers of the molecules gave reasonable agreement with the spectra.

The simplest aromatic alcohol, phenol, has not been investigated by INS although the chloro derivative pentachlorophenol [53] has been studied. The molecule forms a number of unusually strongly hydrogen bonded complexes with substituted pyridines and this is undoubtedly the first step in the investigation of these complexes.

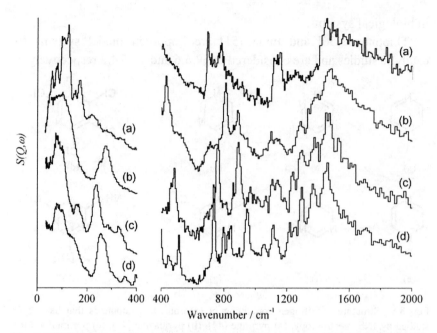

Fig. 8.9 INS spectra of (a) methanol, (b) ethanol, (c) 1-propanol, (d) 1-butanol [52]. The 400–2000 cm^{-1} region is ×3 ordinate expanded relative to the 0–400 cm^{-1} region.

Aldehydes, ketones and ethers have yet to be studied by INS spectroscopy. Formic acid, has been extensively investigated [54–56]. The low temperature phase consists of chains of hydrogen bonded molecules. Fig. 8.10 shows the power of selective deuteration to provide unequivocal assignments for the modes, which are given in Table 8.2.

Table 8.2 Assignments of the INS spectra (ω/ cm^{-1}) of formic acid isotopomers. Reproduced from [54] with permission from the American Chemical Society.

Assignment(a)	HCOOH	HCOOD	DCOOH
C–H ip bend	1420m	1400m	
C–O stretch	1230m		1241w
C–H oop bend	1080s	1075s	1080w
O–H torsion	991s		1001s
O=C–O ip bend	718m	740w	709m
(0–2) chain axis libration	510m	500m	
(0–1) chain axis libration	251vs	246vs	
C.....H stretch	151w	156vw	151w

(a) ip = in-plane, oop = out-of-plane, s = strong, m = medium, w = weak, v = very,

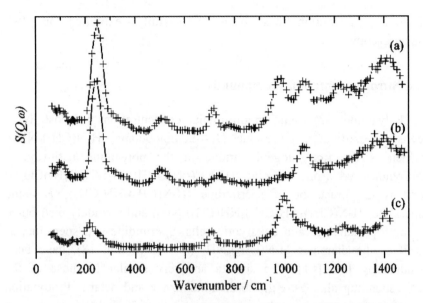

Fig. 8.10 INS spectra of isotopomers of formic acid recorded with a low-bandpass spectrometer (now defunct). (a) HCOOH, (b) HCOOD, and (c) DCOOH. Reproduced from [54] with permission from the American Chemical Society.

Sodium and potassium formate, together with copper formate (anhydrous and as the tetrahydrate) have been used as model compounds for surface formates [56] and §7.4.3. Copper formate tetrahydrate undergoes an antiferroelectric transition at 235.5 K. The INS spectra clearly show that above this temperature the water molecules are disordered and below it they are ordered [57].

Potassium oxalate monohydrate [58] forms chains of alternating oxalate and water moieties. INS of single crystals has been used to assign the water translational and librational modes, Fig. 9.5.

Maleic anhydride [2] is discussed in §8.1 and nadic anhydride in [27]. Potassium hydrogenmaleate forms a cyclic structure containing a strongly hydrogen bonded hydrogen atom midway between the two oxygen atoms. The assignment of the vibrational spectra has proven controversial [57] (§9.3.1.5).

Benzoic acid [60] forms centrosymmetric dimers in the solid state and is of interest because of the debate on the mechanism of the proton transfer in the dimer (§9.3.1.3). It is also the only example where ^{18}O

isotopic substitution has been successfully exploited in INS spectroscopy.

8.6 Nitrogen containing compounds

Polyamines are found in millimolar concentrations in most living cells. In particular, putrescine (1,4-diaminobutane, $H_2N(CH_2)_4NH_2$), which is the first biogenic amine in the polyamine pathway, is biosynthesised from arginine ($H_2NC(=NH)NH(CH_2)_3CH(NH_2)CO_2H$) and is the precursor of spermidine ($H_2N(CH_2)_3NH(CH_2)_4NH_2$) and spermine ($H_2N(CH_2)_3NH(CH_2)_4NH(CH_2)_3NH_2$) and is widely distributed in both prokaryotic and eukaryotic cells. Spermidine and spermine, as well as the diamines $H_2N(CH_2)_nNH_2$ (n = 2–10 and 12), have been studied by INS [61]. The spectra are very similar to those of the corresponding alkanes, e.g. 1,6-diaminohexane and octane. Protonation of the amine groups results in a dramatic shift in the NH_2 torsion on protonation as shown in Fig. 8.11.

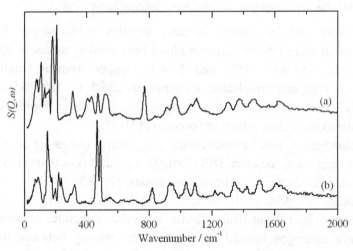

Fig. 8.11 INS spectra of. (a) 1,6-diaminoethane and (b) 1,6-diaminoethane dihydrochloride. Note the shift in the NH_2 torsion on protonation from 178/206 cm⁻¹ to 468/489 cm⁻¹.

Amino acids are discussed in §10.3.3. N-methylformamide [62] and N-methylacetamide [63] (§10.3.1) have also been studied.

8.7 Organometallic compounds

Many organometallic compounds have been used as model compounds for surface species; these are discussed in Chapter 7 and listed in Table 8.3. Vibrational spectra of Zeise's salt are shown in Fig. 1.2 (see also §7.3.2.5).

A few compounds have been studied for other reasons. These include trimethylaluminium [64] and trimethylgallium [65] which are of interest because of their use in metal organic chemical vapour deposition (MOCVD) processes. The compounds were also studied because the interactions between methyl groups in the same molecule are a continuing source of fascination. Tetramethyltin was studied for the same reason [66].

Despite having been studied for more than 40 years there were still uncertainties in the assignment of the vibrational spectra of ferrocene which were finally resolved by a combined INS and DFT study [67].

Table 8.3 Organometallic compounds that have been used as model compounds for surface species as studied by INS spectroscopy.

Stoichiometric name	Formula	Ref.
Hexacarbonyl(ethyne)dicobalt	$[Co_2(CO)_6)(\mu_2-\eta^2-C_2H_2)]$	68
Decacarbonyl(ethyne)triosmium	$[Os_3(CO)_9(\mu_2-CO)(\mu_3-\eta^2-C_2H_2)]$	69
Potassium trichloro(ethene)platinate(1−) (Zeise's salt)	$K[Pt(C_2H_4)Cl_3]$	70
Bis(allyl)dichlorodipalladium	$[\{Pd(\eta^3-C_3H_5)\}_2(\mu-Cl)_2]$	71
(Benzene)tricarbonylchromium	$[Cr(CO)_3(C_6H_6)]$	72
Dicarbonyl(cyclopentadienyl)thiophene iron tetrafluoroborate	$[Fe(CO)_2(C_4H_4S)(\eta^5-C_5H_5)][BF_4]$	73
Tricarbonyl(thiophene)manganese trifluoromethylsulphonate	$[Mn(CO)_3(C_4H_4S)][CF_3SO_3]$	73
Tricarbonyl(thiophene)chromium	$[Cr(CO)_3(\eta^5-C_4H_4S)]$	73
Nonacarbonyl(ethylidyne)tricobalt	$[Co_3(CH_3C)(CO)_9]$	74
Copper(II) formate	$Cu(HCO_2)_2$	56

8.8 References

1   D. Lin-Vien, N.B. Colthup, W.G. Fateley & J.G. Grasselli (1991). *Infrared and Raman Characteristic Frequencies of Organic Molecules*, Academic Press, Boston.

2   S.F. Parker, C.C. Wilson, J. Tomkinson, D.A. Keen, K. Shankland, A.J. Ramirez-Cuesta, P.C.H. Mitchell, A.J. Florence & N. Shankland (2001). J. Phys. Chem. A, 105, 3064–3070. Structure and dynamics of maleic anhydride.

3   E.B. Wilson Jr, J.C. Decius & P.C. Cross (1955). *Molecular Vibrations*, Dover: New York.

4   G.J. Kearley (1995). Nucl. Instrum. and Meth. A, 354, 53–58. A review of the analysis of molecular vibrations using INS.

5   K. Nakamoto (1997). *Infrared and Raman Spectra of Inorganic and Coordination Compounds*. Part A: Theory and Applications in Inorganic Chemistry, 5th Edition John Wiley, New York.

6   D.A. Braden, S.F. Parker, J. Tomkinson & B.S. Hudson (1999). J. Chem. Phys., 111, 429–437. Inelastic neutron scattering spectra of the longitudinal acoustic modes of the normal alkanes from pentane to pentacosane.

7   J. Tomkinson, S.F. Parker, D.A. Braden & B.S. Hudson (2002). Phys. Chem. Chem. Phys., 4, 716–721. Inelastic neutron scattering spectra of the transverse acoustic modes of the normal alkanes.

8   W.B. Nelligan, D.J. LePoire, C.-K. Loong, T.O. Brun & S.H. Chen (1987). Nucl. Inst. and Meth., A254, 563–569. Molecular spectroscopy of n-butane by incoherent inelastic neutron scattering.

9   W.B. Nelligan, D.J. LePoire, T.O. Brun & R. Kleb (1987). J. Chem. Phys., 87, 2447–2456. Inelastic neutron scattering study of the torsional and CCC bend frequencies in the solid n-alkanes, ethane, hexane.

10 M. Prager, W. Press, B. Asmussen & J. Combet (2002). J. Chem. Phys., 117, 5821–5826. Phase III of methane: crystal structure and rotational tunneling.

11 J.Z. Larese (1998). Physica B, 248, 297–303. Neutron scattering studies of the structure and dynamics of methane absorbed on MgO(100) surfaces.

12 J.Z. Larese, D. Martin y Marero, D.S. Sivia & C.J. Carlile (2001). Phys. Rev. Lett., 87, art. no. 206102. Tracking the evolution of interatomic potentials with high resolution inelastic neutron spectroscopy.

13 I.A. Krasnov, B. Asmussen, C. Gutt, W. Press, W. Langel & M. Ferrand (2000). J. Phys.: Cond. Matter, 12, 1613–1626. Dynamics of methane molecules in porous TiO_2.

14 H. Jobic, S. Sportouch & A. Renouprez (1983). J. Mol. Spectrosc., 99, 47–55. Neutron inelastic-scattering spectrum and valence force-field for neopentane.

15 R.G. Snyder & J.H. Schactschneider (1965). Spectrochim. Acta 21, 169–195. A valence force field for saturated hydrocarbons.

16 C. Coulombeau & H. Jobic (1990). J. Mol. Struct., 216, 161–170. Neutron inelastic scattering study of the vibrational modes of aziridine and cyclopropane.

17 Y. Brunel, C. Coulombeau, C. Coulombeau, M. Moutin & H. Jobic (1983). J. Am. Chem. Soc., 105, 6411–6416. Optical and neutron inelastic-scattering study of norbornane— a new assignment of vibrational frequencies.

18 Y. Brunel, C. Coulombeau, C. Coulombeau & H. Jobic (1985). J. Phys. Chem., 89, 937–943. Optical and neutron inelastic-scattering study of 2-methylnorbornanes.

19 Y. Brunel, C. Coulombeau, C. Coulombeau & H. Jobic (1986). J. Phys. Chem., 90, 2008–2015. Optical and neutron inelastic-scattering study of 2,3-dimethylnorbornanes.

20 P.M. Gehring, D.A. Neumann, W.A. Kamitakahara, J.J. Rush, P.E. Eaton & D.P. VanMeurs (1995). J. Phys. Chem., 99, 4429–4434. Neutron scattering study of the lattice modes of solid cubane.

21 T. Yildrim, Ç. Kiliç, S. Ciraci, P.M. Gehring, D.A. Neumann, P.E. Eaton & T. Emrick (1999). Chem. Phys. Lett., 309, 234–240. Vibrations of the cubane molecule: inelastic neutron scattering study and theory.

22 B.S. Hudson, D.A. Braden, S.F. Parker & H. Prinzbach (2000). Angew. Chem. Int. Ed., 39, 514–516. The vibrationally inelastic neutron scattering spectrum of dodecahedrane: Experiment and DFT simulation.

23 H. Jobic (1984). Chem. Phys. Lett., 106, 321–324. Neutron inelastic-scattering from ethylene— an unusual spectrum.

24 D. Lennon, J. McNamara, J.R. Phillips, R.M. Ibberson & S.F. Parker (2000). Phys. Chem. Chem Phys., 2, 4447–4451. An inelastic neutron scattering spectroscopic investigation of the adsorption of ethene and propene on carbon.

25 R. Mukhopadhyay & S.L. Chaplot (2002). Chem. Phys. Lett., 358, 219–223. Phonon density of states in tetrabromoethylene: lattice dynamic and inelastic neutron scattering study.

26 S.L. Chaplot, A.Mierzejewski, G.S. Pawley, J. Lefebvre & T. Luty (1983). J. Phys. C Solid State ., 16, 625–644. Phonon dispersion of the external and low-frequency internal vibrations in monoclinic tetracyanoethylene at 5K.

27 S.F. Parker, K.P.J. Williams, D. Steele & H. Herman (2003). Phys. Chem. Chem. Phys., 5, 1508–1514. The vibrational spectra of norbornene and nadic anhydride.

28 O. Kirstein, M. Prager, M.R. Johnson & S.F. Parker (2002). J. Chem. Phys., 117, 1313–1319. Lattice dynamics and methyl rotational excitations of 2-butyne.

29 H. Jobic & A.N. Fitch (1997). Stud. Surf. Sci. and Catal., 105, 559–566. Vibrational study of benzene adsorbed in NaY zeolite by neutron spectroscopy.

30 J. Kalus, M. Monkenbusch, I. Natkaniec, M. Prager & J. Wolfrum (1995). Mol. Cryst. and Liq. Cryst. A, 268, 1–20. Neutron and Raman scattering studies of the lattice and methyl group dynamics in solid p-xylene.

31 I. Natkaniec, J Kalus, W. Griessl & K. Holderna-Natkaniec (1997). Physica B, 234, 104–105. Lattice and methyl-group dynamics in solid p-xylene with different deuterated molecules.

32 I. Natkaniec, K. Holderna-Natkaniec, J. Kalus & V.D. Khavryutchenko (1999). In Neutrons and Numerical Methods-N(2)M, (Ed.) M.R. Johnson, G.J. Kearley, and H.G. Büttner, pp.191–194. A. I. P. Conf. Proc. Vol.479, Neutron spectrometry and numerical simulations of low-frequency internal vibrations in solid xylenes.

33 L. Cser, K. Holderna-Natkaniec, I. Natkaniec & A. Pawlukojc A (2000). Physica B, 276, 296–297. Neutron spectroscopy and QC modeling of the low-frequency internal vibrations of mesitylene.

34 J. Meinnel, W. Häusler, M. Mani, M. Tazi, M. Nusimovici, M. Sanquer, B. Wyncke, A. Heidemann, C.J. Carlile, J. Tomkinson & B. Hennion (1992). Physica B, 180, , 711–713.Part B. Methyl tunnelling in trihalogeno-trimethyl-benzenes.

35 Meinnel F. Boudjada, J. Meinnel, A. Cousson, W. Paulus, M. Mani & M. Sanquer
    (1999). In *Neutrons and Numerical Methods–N(2)M*, eds. M.R. Johnson, G.J.
    Kearley, and H.G. Büttner, pp.217–222. A. I. P. Conf. Proc. Vol.479,
    Tribromomesitylene structure at 14 K methyl conformation and tunnelling.
36 M. Plazanet, M.R. Johnson, A. Cousson, J. Meinnel & H.P. Trommsdorff (2002).
    Chem. Phys., 285, 299–308. Molecular deformations of halogeno-mesitylenes in the
    crystal: structure, methyl group rotational tunneling, and numerical modeling.
37 M.A. Neumann, M.R. Johnson, P.G. Radelli, H.P. Trommsdorff & S.F. Parker
    (1999). J. Chem. Phys., 110, 516–527. Rotational dynamics of methyl groups in
    durene: A crystallographic, spectroscopic and molecular mechanics investigation.
38 M. Plazanet, M.R. Johnson, J.D. Gale, T. Yildirim, G.J. Kearley, M.T. Fernández-
    Diaz, E. Artacho, J.M. Soler, P. Ordejón, A. Garcia & H.P. Trommsdorff (2000).
    Chem. Phys., 261, 189–203. The structure and dynamics of crystalline durene by
    neutron scattering and numerical modelling using density functional methods.
39 J.M. Granadino-Roldán, M. Fernández-Gomez, A. Navarro, L.M. Camus & U.A.
    Jayasooriya (2002). Phys. Chem. Chem. Phys., 4, 4890–4901. Refined, scaled and
    canonical force fields for the cis- and trans-3-fluorostyrene conformers. An interplay
    between theoretical calculations, IR/Raman and INS data.
40 J. M. Granadino-Roldán , M. Fernández-Gómez , A. Navarro & U.A. Jayasooriya
    (2003). Phys. Chem. Chem. Phys., 5, 1760–1768. The molecular force field of 4-
    fluorostyrene: an insight into its vibrational analysis using inelastic neutron
    scattering, optical spectroscopies (IR/Raman) and theoretical calculations.
41 I. Natkaniec, E.L. Bokhenkov, B. Dorner, J. Kalus, G.A. Mackenzie, G.S. Pawley,
    U. Schmelzer & E.F. Sheka (1980). J. Phys. C: Solid State Phys., 13, 4265–83.
    Phonon dispersion in d_8-naphthalene crystal at 6K.
42 B. Dorner, E.L. Bokhenkov, S.L. Chaplot, J. Kalus, I. Natkaniec, G.S. Pawley, U.
    Schmelzer & E.F. Sheka (1982) J. Phys. C: Solid State., 15, 2353–2365. The 12
    external and the 4 lowest internal phonon dispersion branches in d_{10}-anthracene at
    12K.
43 F. Fillaux, R. Papoular, A. Lautie & J. Tomkinson (1995). Fuel, 74, 865–873.
    Inelastic neutron scattering study of the proton dynamics in coals.
44 F. Partal, M. Fernández-Gómez, J. J. López-González, A. Navarro & G.J. Kearley
    (2000). Chem. Phys., 261, 239–247. Vibrational analysis of the inelastic neutron
    scattering spectrum of pyridine.
45 A. Navarro, J. Vaquez, M. Montejo, J.J.L. Gonzalez & G.J. Kearley (2002). Chem.
    Phys. Lett., 361, 483–491. A reinvestigation of the v_7 and v_{10} modes of pyridazine
    on the basis of the inelastic neutron scattering spectrum analysis.
46 A. Navarro, M. Fernández-Gómez, J.J. López-González, M. Paz Fernández-
    Liencres, E. Martínez-Torres, J. Tomkinson & G.J. Kearley (1999). J. Phys. Chem.
    A, 103, 5833–5840. Inelastic neutron scattering spectrum and quantum mechanical
    calculations of the internal vibrations of pyrimidine.
47 A. Navarro, M. Fernández-Gómez, M.P. Fernández-Liencres, C.A. Morrison,
    D.W.H. Rankin & H.E. Robertson (1999). Phys. Chem. Chem. Phys., 1, 3453–3460.
    Tetrachloropyrimidine: molecular structure by electron diffraction, vibrational

analysis by infrared, Raman and inelastic neutron scattering spectroscopies and quantum mechanical calculations.

48 G.J. Kearley, J. Tomkinson, A. Navarro, J.J. López-González, & M. Fernández-Gómez (1997). Chem. Phys., 216, 323–335. Symmetrised quantum-mechanical force-fields and INS spectra: s-triazine, trichloro-s-triazine and pyrazine.

49 M.P. Fernández-Liencres, A. Navarro, J.J. López-González, M. Fernández-Gómez, J. Tomkinson & G.J. Kearley (2001). Chem. Phys., 266, 1–17. Measurement and *ab initio* modeling of the inelastic neutron scattering of solid melamine - Evidence of the anisotropy in the external modes spectrum..

50 G.D. Atter, D.M. Chapman, R.E. Hester, D.A. Green, P.C.H. Mitchell & J. Tomkinson (1997). J. Chem. Soc. Faraday Trans., 93 2977–2980. Refined *ab initio* inelastic neutron scattering spectrum of thiophene.

51 I.A. Beta, J. Herve, E. Geidel, H. Bohlig & B. Hunger (2001). Spectrochim. Acta A, 57, 1393–1403. Inelastic neutron scattering and infrared spectroscopic study of furan adsorption on alkali-metal cation-exchanged faujasites.

52 R. Schenkel, A. Jentys, S.F. Parker & J.A. Lercher (2004). J. Phys. Chem. B, 108, 7902–7910. Investigation of the adsorption of methanol on alkali metal cation exchanged zeolite X by inelastic neutron scattering.
    R. Schenkel, A. Jentys, J.A. Lercher & S.F. Parker (2004). J. Phys. Chem. B, 108, 15013–15026. INS and IR and NMR spectroscopic study of C_1-C_4 alcohols adsorbed on alkali metal-exchanged zeolite X.

53 A. Pawlukojć, I. Natkaniec, I. Majerz & L. Sobczyk (2001). Spectrochim. Acta, 57, 2775–2779. Inelastic neutron scattering studies on low frequency vibrations of pentachlorophenol.

54 C.V. Berney & J.W. White (1977). J. Am. Chem. Soc., 99, 6878–6880. Selective deuteration in neutron-scattering spectroscopy: formic acid and deuterated derivatives.

55 D.H. Johnson, C.V. Berney, S. Yip & S.-H. Chen (1979). J. Chem. Phys., 71, 292–297. Analysis of neutron incoherent inelastic scattering of normal and deuterated formic acid. 1. Planar chain model.

56 P.C.H. Mitchell, R.P. Holroyd, S. Poulston, M. Bowker & S.F. Parker (1997). J. Chem. Soc. Faraday Trans., 93, 2569–2577. Inelastic neutron scattering of model compounds for surface formates. Potassium formate, copper formate and formic acid.

57 T. Omura, C. Moriyoshi, K. Itoh, S. Ikeda, H. Fukazawa (2002). Ferroelectrics, 270, 1561–1566. Structural change of $Cu(HCOO)_2 4H_2O$ associated with the antiferroelectric phase transition.

58 R.G. Delaplane, H. Küppers, J. Noreland & S.F. Parker (2002). Appl. Phys. A, 74 [Suppl.] S1366–S1367. Dynamics of the water molecule in potassium oxalate monohydrate as studied by single crystal inelastic neutron scattering.

59 F. Fillaux, N. Leygue, J. Tomkinson, A, Cousson and W. Paulus (1999). Chem., Phys., 244, 387–403. Structure and dynamics of the symmetric hydrogen bond in potassium hydrogen maleate: a neutron scattering study.

60   M. Plazanet, N. Fukushima, M.R. Johnson, A.J. Horsewill and H.P. Trommsdorff
     (2001). J. Chem. Phys., 115, 3241–3248. The vibrational spectrum of crystalline
     benzoic acid: Inelastic neutron scattering and density functional theory calculations.

61   M.P.M. Marques, L.A.E.B. de Carvalho and J. Tomkinson (2002). J. Phys. Chem.
     A, 106, 2473–2482. Study of biogenic and alpha, omega-polyamines by combined
     inelastic neutron scattering and Raman spectroscopies and by *ab initio* molecular
     orbital calculations.

62   P. Bour, C.N. Tam, J. Sopková & F.R. Trouw (1998). J. Chem. Phys., 108, 351–358.
     Measurement and *ab initio* modeling of the inelastic neutron scattering of solid N-
     methylformamide.

63   G.J. Kearley, M.R. Johnson, M. Plazanet & E. Suard (2001). J. Chem. Phys., 115,
     2614–2620. Structure and vibrational dynamics of the strongly hydrogen-bonded
     model peptide: N-methylacetamide.

64   M. Prager, H. Grimm, S.F. Parker, R. Lechner, A. Desmedt, S. McGrady & E.
     Koglin (2002). J. Phys.: Condens. Matter, 14, 1833–1845. Methyl group rotation in
     trimethylaluminium.

65   M. Prager, J. Combet, S.F. Parker, A. Desmedt & R.E. Lechner (2002). J. Phys.:
     Condens. Matter, 14, 10145–10157. The methyl rotational potentials of $Ga(CH_3)_3$
     derived by neutron spectroscopy.

66   G.J. Kearley & J. Tomkinson (1992). Physica B, 180/181, 700–702. The inelastic
     neutron spectrum of $Sn(CH_3)_4$.

67   E. Kemner, I.M. de Schepper, G.J. Kearley & U.A. Jayasooriya (2000). J. Chem.
     Phys., 112, 10926–10929. The vibrational spectrum of solid ferrocene by inelastic
     neutron scattering.

68   H. Jobic, C.C. Santini & C.Coulombeau (1991). Inorg. Chem., 30, 3088–3090.
     Vibrational   modes   of   acetylenedicobalt   hexacarbonyl   studied   by   neutron
     spectroscopy.

69   S.F. Parker, P.H. Dallin, B.T. Keiller, C.E. Anson & U.A. Jayasooriya (1999). Phys.
     Chem. Chem. Phys., 1, 2589–2592. An inelastic neutron scattering study and
     re-assignment of the vibrational spectrum of $[Os_3(CO)_9(\mu_2\text{-}CO)(\mu_3\text{-}\eta^2\text{-}C_2H_2)]$, a
     model compound for chemisorbed ethyne.

70   H. Jobic (1985). J. Mol. Struc., 131, 167–175. A new inelastic neutron scattering
     study of Zeise's salt, $K[Pt(C_2H_4)Cl_3]$, and a more confident assignment of the
     vibrational frequencies.

71   P.C.H. Mitchell, M Bowker, N. Price, S. Poulston, D. James & S.F. Parker (2000). .
     Top. Catal., 11, 223–227. Iron antimony oxide selective oxidation catalysts—an
     inelastic neutron scattering study.

72   H. Jobic, J. Tomkinson & A. Renouprez (1980). Mol. Phys., 39, 4, 989–999.
     Neutron   inelastic   scattering   spectrum   and   valence   force   field   for
     benzenetricarbonylchromium.

73   P.C.H. Mitchell, D.A. Green, E. Payen, J. Tomkinson & S.F. Parker (1999). Phys.
     Chem. Chem. Phys., 1, 3357–3363. Interaction of thiophene with a molybdenum
     disulfide catalyst: an inelastic neutron scattering study.

74 S.F. Parker, N.A. Marsh, L.M. Camus, M.K. Whittlesey U.A. Jayasooriya & G.J. Kearley (2002). J. Phys. Chem. A, 106, 5797–5802. Ethylidyne tricobalt nonacarbonyl: Infrared, FT-Raman and inelastic neutron scattering spectra.

9

Hydrogen Bonding

In a hydrogen bond, here written H-bond, a hydrogen atom occupies a more or less central position between, usually, two electronegative atoms. This is conventionally represented as a linear $A–H---B$ combination of the hydrogen donor system, $A–H$, and the acceptor, B. Relevant properties of the different strengths of H-bonds are given in Table 9.1 [1].

Table 9.1 Properties of strong, moderate and weak H-bonds, after [1].

Properties	Strong H-bonds	Moderate H-bonds	Weak H-bonds
Bond energy / kJ mol^{-1}	4 to 10	1 to 3	< 1
Bond nature	mostly covalent	mostly electrostatic	electrostatic
Bond linearity, $A–H----B$	always linear	mostly linear	sometimes linear
Bond lengths / Å			
A–H	1.2 to 1.5	ca 1.0	ca 1.0
H---B	1.2 to 1.5	1.5 to 2.2	2.2 to 3.2
A·····B	2.2 to 2.5	2.5 to 3.2	3.2 to 4.0

The dynamics of the hydrogen atom in an H-bond is controlled by the potential surface between the heavy AB atoms and spectroscopy is undertaken with the objective of elucidating the fine details of that surface. Because of the high cross section of the hydrogen atoms these dynamics are uniquely accessible using INS spectroscopy. (The strong infrared response of the H-bond stretch is, rightly, regarded as an excellent indicator of the presence of H-bonds. However, its very strength introduces opto-electric effects that make the spectra difficult to interpret, neutron spectroscopy avoids this problem.)

This chapter focuses on results from INS for the stronger and moderate H-bonds where, as explained below, the dynamics of hydrogen transfer should be accessible. Since most known examples of moderate H-bonds occur in oxygen-oxygen systems they will assume a particular prominence in this chapter. H-bonds also determine, to a great extent, the tertiary and quaternary structure essential to achieve biological activity (§10.3).

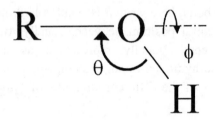

Fig. 9.1 The internal deformations of an alcohol group, the in-plane deformation θ and the libration ϕ, which become the in-plane and out-of-plane deformations, respectively, upon H-bond formation.

After briefly reviewing the spectral consequences that result for the formation of H-bonds (§9.1) the extensive body of work on water is presented (§9.2), including water in minerals (§9.2.1), its protonated species (§9.2.2), the ices (§9.2.3) and water at bio-interfaces (§9.2.4). The next section (§9.3) covers proton transfer, which is important for systems, across all the scientific disciplines from Materials Science to Biology. This section includes the dicarboxylates as models (§9.3.1), proton conductors (§9.3.2) and unusual species (§9.3.3).

9.1 Spectroscopic consequences of hydrogen bonding

Apart from the geometric and energetic distinctions, seen in Table 9.1, the different strengths of H-bond also exhibit a range of spectral characteristics. We can put these into a broader context by discussing the changes introduced by the impact of H-bonding on the spectra of the non-associated donor and acceptor.

9.1.1 General considerations

It is usual to describe the vibrations of the hydrogen atom in H-bonds by relating them to vibrations that existed before the bond formed. Take, for instance, the reaction between an alcohol and an aldehyde.

R—O—H + O=CH—R' → R—O—H---O=CH—R'

In this reaction the diffusional motion of the two moieties is internalised within the new system and is now described as the ν(H---O) stretch with typical values of 200 cm^{-1}. The two R–OH deformations are also changed. A typical structure is shown in Fig. 9.1, here the in-plane deformation, θ, would be expected about 500 cm^{-1} but this stiffens to become the in-plane deformation, δ(O–H---O). It is the OHO angle that deforms in this new vibration and, typically, it occurs in the region from 900 (weak H-bond) to 1500 cm^{-1} (strong H-bond). The twist, ϕ, which may be at very low frequencies in the unassociated alcohol becomes the out-of-plane deformation, γ(O–H---O), occurring from 600 (weak H-bond) to 1200 cm^{-1} (strong). The most dramatic change is reserved for the ν(O–H) stretch.

In an unassociated alcohol the ν(O–H) stretch is usually observed about 3200 cm^{-1}, but it falls to, say, 2900 cm^{-1} upon weak H-bond formation. This can be rationalised as a lengthening and weakening of the original O–H bond as the hydrogen is attracted away from the donor toward the acceptor. However, this picture is too simplistic; of greater impact is the dramatic increase in the mechanical anharmonicity of the H-bond. This lowers the stretching frequency without much affecting the O–H distance, r(OH) [2,3]. There is a well known relationship between the oxygen-oxygen distance, R(O—O), and the ν(O–H) stretch [4].

The strong γ(O–H---O) band is usually easily identified in the INS. This vibration involves almost exclusively the motion of the hydrogen atom against the undeforming molecule. It has an oscillator mass close to unity, which gives it its strong INS intensity.

The relationship between γ(O–H---O), in cm^{-1}, and R(OO), in Å, is conveniently summarised in Eq. (9.1) and shown in Fig. 9.2 [5].

$$R(OO) = 3.01 - 0.00044(\gamma) \tag{9.1}$$

Mineral systems have their γ(O–H---O) about 100 cm^{-1} higher than this relationship would imply [6].

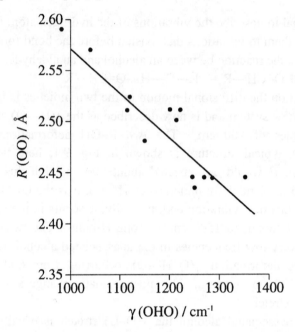

Fig. 9.2 The correlation between $R(OO)$, Å, and γ(O–H---O), cm^{-1}, determined, in part, by INS. Redrawn from [5] with permission from the American Institute of Physics.

9.1.2 Symmetric hydrogen bonds

In symmetric homonuclear systems either of the heavy atoms could have been the donor (or acceptor) and the hydrogen potential is thus, at least in the gas phase, a symmetric double well. In weak symmetric H-bonds the hydrogen atom will always stay with the donor because the energy barrier to accessing the other well is, practically, the dissociation energy of the covalent O–H bond. The much weaker O---H bond would break before the barrier could be crossed. However, as the bond strengthens the O–O distance falls and the two potential minima approach one another forcing the central barrier to fall. Until, eventually, the two energy minima overlap and a single, central well emerges as the potential surface experienced by the strongest H-bonds.

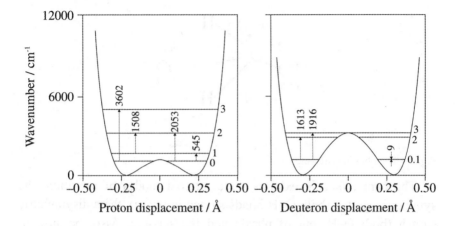

Fig. 9.3 Potential forms for the H-bond in so-called chromous acid, CrO(OH), left and deuterated, right. Reproduced from [7] with permission of Taylor & Francis Ltd.

A description of the double well O–H–O bond in so-called chromous acid (in reality a Cr(III) compound, CrO(OH)) has been derived in agreement with some early INS work [7], see Fig 9.3. In symmetric H-bonds the vibrations are best described as, the symmetric stretch, ν_s(O–H–O), and the antisymmetric stretch, ν_{as}(O–H–O): where $\nu_s (\bar{O} - H - \bar{O})$ the two heavy atoms move in anti-phase; $\nu_{as} (O - \ddot{H} - O)$, the proton moves between the heavy atoms.

In strong symmetric H-bonds ν_s(O–H–O) can rise to about 500 cm⁻¹ and ν_{as}(O–H–O) can fall to around 800 cm⁻¹. Thus, the main spectral features, at least of moderate and strong H-bonds, fall in the range from 200 to 1500 cm⁻¹, which is optimal for neutron spectroscopy.

9.2 Water

All of the important chemical and biological properties of water, and most of the physical properties, are related to its unique ability to H-bond, both to other molecules and to itself. It is not only a hydrogen acceptor but also a hydrogen donor, but weakly so in the gas phase [8].

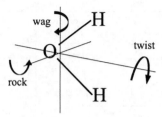

Fig. 9.4 The librations of water.

The bare molecule exhibits three internal modes of vibration, the symmetric stretch (both O–H bonds stretch in phase), the antisymmetric stretch (both O–H, out of phase) and the scissors. Also, because of water's low Sachs-Teller mass (§2.6.5.1) the external vibrations, shown diagrammatically in Fig. 9.4 appear at frequencies up to ca 600 cm^{-1}. In a few specific cases involving single crystals the assignment of the external vibrations of an isolated water molecule can be determined by aligning the Q and the u vectors (§2.5.2). This was nicely achieved in $K_2C_2O_4.H_2O$ where the H_2O translational vibrations along x, y and z were assigned at 97, 210 and 215 cm^{-1} respectively. The three torsions were the rock, at 656 cm^{-1}, and the wag and twist, undifferentiated at 746 cm^{-1}, Fig. 9.5 [9].

9.2.1 Isolated water molecules in mineral lattices

The role of water in lattices is of great interest to the mineralogist and over 500 minerals contain water in more or less important bonding scenarios. The view of water as a static element within a mineral was challenged by several spectroscopic techniques including INS [10]. Especially in weakly bonding environments the water molecules are disordered and diffraction experiments are unable to determine the hydrogen positions, INS techniques can provide useful information in these circumstances.

Zeolites form an important subset of minerals that, in their protonated form, are used as acid catalysts. One open question has been the exact nature of the water on these catalysts because the acidity of the Brönsted site may have been sufficient to generate hydroxonium ions, H_3O^+ (§9.2.2). This problem was tackled using INS to study ZSM-5 in combination with *ab initio* calculations based upon a simple structural unit [11, 12] (§7.4.1, §7.5.3).

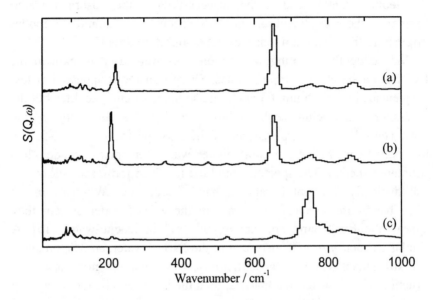

Fig. 9.5 The INS of single crystals of $K_2C_2O_4.H_2O$ with the Q vector, (a) across the plane of the molecule, (b) along the C_2 axis of the molecule and (c) perpendicular to the plane of the water molecule. Redrawn from [9] with permission from Springer-Verlag.

In the absence of water, typical vibrational frequencies were found for the protonated zeolite but an unanticipated degree of heterogeneity was also discovered. With the addition of one water molecule per Brönsted site, single-load, the ZSM-5 spectrum was considerably modified and strongly resembled earlier work on mordenite. The interpretation of this spectrum was, however, quite different from the previous work and lead to its reassessment. A difference spectrum showing only the INS of water was compared with the results of two calculated spectra based on either a H-bonded water molecule or a H_3O^+. The comparison clearly favours the H-bonded water molecule model and the presence of hydroxonium ions was discounted.

As the concentration of water was increased, multi-load, the signal strength grew monotonically, as did the general scattering background. Interestingly, even for the single-load spectrum not all the OH sites were bound to a water molecule but some 40% were found to point away from the zeolite cavities and to be inaccessible to the adsorbed water. Similarly, the assignment of the INS spectrum of alunite to modes originating from oxonium ions was not straightforward [13].

Independently vibrating water molecules were used to explain the INS from a series of A-type zeolites. Sharp transitions appeared at low frequencies in the lithium, 63 cm^{-1}, and sodium, 29 cm^{-1}, zeolites but the potassium and calcium zeolites showed no bands at all in this region [14]. The relatively sharp features of the INS spectrum of ZSM-5 with low water content gave way at higher water content to broader, less structured spectra. The spectra resembled ice Ih in form, see below, but had distinctly different librational band frequencies. Whereas, in ice these bands are at about 600 cm^{-1}, in the ZSM-5-water system they appear about 500 cm^{-1} and about 400 cm^{-1} in leucite-water [15]. A similar frequency drop is also seen in the INS of water, *ca* 3%, on silica gel [16]. This is related to the earlier observation that the more 'open' the structure of bulk water in a material then the lower will be the librational band frequencies [17]. The structural aspects of water in confined geometries has been reviewed recently [18].

Small-pored zeolites offer the possibility of eliminating any complications introduced by water-water interactions present in large-pored systems. In small pores only the specific cation-water interaction is important, although the framework-water interaction also remains this is generally regarded as diffuse and non-specific. The INS spectra of natrolite, wairakite, scolecite and bikitaite zeolites, with the silicate apophylite, were analysed to determine the relative importance of H-bonds in these systems [19]. It has been shown that even very simple model representations of the complex zeolite structures, first applied to natrolite [20], were very effective in reproducing the INS spectra of several systems. Cation-water stretching modes were assigned about 80 to 120 cm^{-1} with the H-bond (H---O) stretch occurring about 170 to 280 cm^{-1}. The spectra were all rather well delineated and relatively sharp features were observed. These zeolite-water systems were thus regarded as ordered. This is in contrast to earlier work on the systems that showed broad bands, probably from water in disordered arrangements [21].

9.2.2 The protonated species of water

Of all the possible autoionisation products of water the best known cations are the hydroxonium, H_3O^+, and the $H_5O_2^+$, species. These ions are commonly reported in the structure of solid acids but spectroscopically they have remained somewhat elusive. This stems not from an inability to recognise the presence of the ions from the overall spectral response but rather to produce specific and reliable assignments for individual features. Occasionally rather unusual species like the tetrahedral H_3O^+ (written $(H_{0.75})_4O^+$) [22] or H_4O^{2+} [23] have been suggested.

The assignment problem is real for all vibrational techniques but it is particularly unfortunate for INS, which should be the *sine qua non* of relevant spectroscopic techniques. Early work on the tungstophosphoric and -silicic acids brought together infrared, Raman and INS data around two alternative assignment schemes [24]. A complete normal coordinate approach was used to test the two models and their predicted INS intensities and yet failed to select a convincingly definitive scheme. Meanwhile, there have been one general review [25] and two studies specifically on the dehydration of 12-tungstophosphoric acid [26, 27]. Although these latter gave a pleasing degree of INS reproducibility between difficult samples of quite different origin measured on different spectrometers there remains as yet little else. The currently accepted characteristic bands for H_3O^+ and $H_5O_2^+$, as found in tungstophosphoric acid, are given in Table 9.2.

The structure and dynamics of these ions depend critically upon the local environmental details [6] and they are implicated in the mechanism of proton conduction in several materials.

The overall assessment given in a recent review is probably still the simplest description of these ions, they are best regarded as clusters of one or more molecules of water associated with a more or less independent 'slave' proton [28].

Table 9.2 The characteristic bands of H_3O^+ and $H_5O_2^+$[26,27] (a).

373 K (6 H_2O) ω / cm^{-1}	473 K ($\approx$ 1H_2O) ω / cm^{-1}	573 K (>0.5 H_2O) ω / cm^{-1}	Assignment
83			H_3O^+ rotation
180			H_2O translation
300-650			Lattice modes
485			$H_5O_2^+$ OH...O sym str
	460		H_3O^+
		518	
618		618	$H_5O_2^+$ rotation
	685		H_3O^+
830			$H_5O_2^+$ rotation
1070			$H_5O_2^+$ OH...O asym str
	1000—1080		H_3O^+
		1149	free proton
1750			deformation

(a) Tungstophosphoric acid at temperatures and compositions (water molecules per unit cell) shown.

9.2.3 Water–water solids—the ices

Neutron diffraction studies have played the major role in our present understanding of the solid phases of water ice [29]. However, the role of INS has been limited because of its inability to study small samples. As is seen from Fig. 9.6 the phase diagram of the water ices is complex and its full exploration can only be achieved by the application of very high pressures. This is technologically impossible using the large sample volumes currently required for INS work.

The vibrational results obtained at the very highest pressures have come from the Raman technique, coupled with advanced *ab initio* calculations [30 and references].

INS results from some high-pressure phases have been limited to 'recuperated' phases, which are samples prepared in the laboratory and not *in situ*. However, the breadth of the INS work relating to the accessible phases of ice remains impressive and important. The body of this work has recently been reviewed by the workers principally responsible for its creation [31] and the role played by classical lattice and molecular dynamics in its interpretation has also been reviewed [32]. There is a considerable body of work on ice phenomena in amorphous ices where the thrust of the work is related to the physics of glasses [33 and references]; and the clathrate hydrates where the thrust is the physics of host-guest coupling [34 and refs within]. These areas of interest are omitted here.

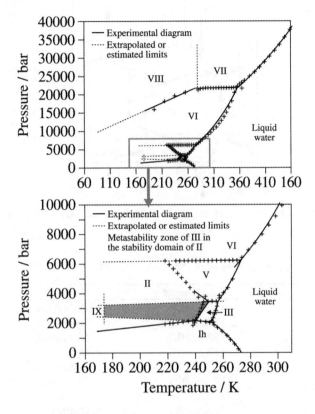

Fig. 9.6 The phase diagram of ice. Reproduced from [35] with permission from Elsevier.

The phase diagram of the water ices is very rich and over thirteen phases are currently known; most are indicated in Fig. 9.6 [35]. The structure of the common form of ice, Ih, is shown in Fig. 9.7 [32]. A list, not exhaustive, of the published INS spectra is given in Table 9.3 (see also Appendix 4) and for the spectra of several phases see Fig. 9.8 [31].

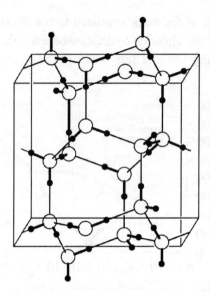

Fig. 9.7 The local structure of ice Ih. Reproduced from [32] with permission from Elsevier.

Table 9.3 Résumé of INS work on the phases of ice [31].

Phase	Spectrometer	Comments
Ih	TFXA, HET	Disordered
Ic	TFXA	Disordered
II	TFXA, HET	Disordered
V	TFXA	Ordered
VI	TFXA, HET	Ordered
VIII	TFXA, HET	Disordered
IX	TFXA	Ordered
XI	TFXA, HET	Ordered

It is important to note that some phases of ice are particularly difficult to prepare, as in ice XI and ice XII, and sample purity is subject to their particular preparation conditions and sample history [35]. This offers one plausible explanation for the slight differences reported between INS spectra of ancient ices recovered from the depths of the Arctic and Antarctic. Altogether, however, there are but minor differences between either ancient ice and, or, ice Ih [31, 36].

The INS spectroscopy of water ice is particularly demanding of the current world-wide suite of neutron spectrometers. Excellent energy transfer resolution is required, to provide data capable of discriminating between competing dynamical models. However, low Q values are also needed, to reduce the impact of the Debye-Waller factor from this very light molecular species (H_2O has a Sachs-Teller mass about 2 amu).

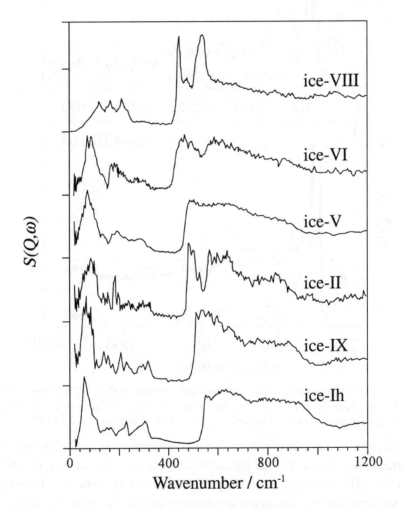

Fig. 9.8 The INS spectra of several phases of ice. Reproduced from [31] with permission of Elsevier.

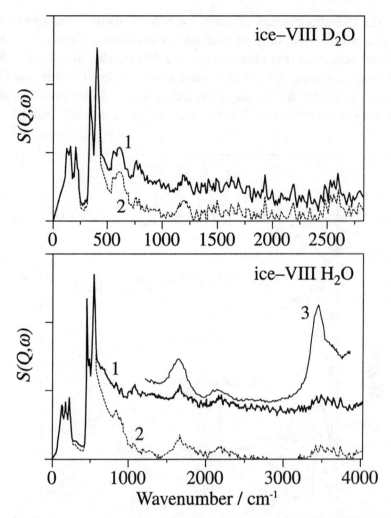

Fig. 9.9 The INS spectrum of ice VIII: top, D_2O; bottom H_2O. 1: TFXA spectrum; 2:TFXA spectrum, as 1, but excluding multiple excitation contributions; 3: the HET spectrum. Reproduced from [37] with permission from the American Institute of Physics.

This can be seen in Fig. 9.9 where the phonon-wings and the higher overtones are shown and the fundamental transition spectra of H_2O and D_2O ice-VIII are shown separately. The effect of an improved Debye-Waller factor can be seen in the spectrum of D_2O. It has a heavier Sachs-Teller mass and the bands appear at lower frequency and so lower Q.

The impact of reducing the experimental Q values (by working on the direct geometry spectrometer HET) is also seen in a comparison of Figs. 9.8 and 9.9 [37]. The increased signal from the scissoring mode about 1650 cm^{-1} is clear but is accompanied by a consequent loss of resolution (§5.4.4).The important experimental results obtained from the INS of ice are: the splitting of the LO-TO modes in certain phases of ice, the increase of band-widths in disordered ices and the degree of proton ordering in ice-Ih.

The low frequency spectra of a single crystal of ice-Ih, obtained with Q perpendicular and parallel to the c-crystal axis, is shown in Fig. 9.10, after [38]. Since the crystal is not cubic, the finding that the perpendicular and parallel spectra are identical, and equivalent to that of a powder, is unusual. It implies that the degree of order amongst the protons is greater along the c-crystal axis than in the other directions. Similar results were also obtained from single crystal studies of ice-XI [31]. (The degree to which these results suffered from 'leakage' is unknown (§2.5.2.1).)

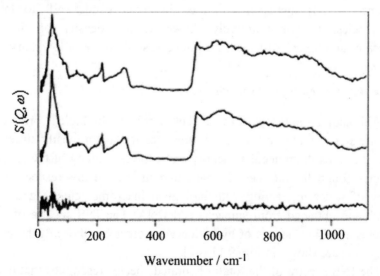

Fig. 9.10 The INS spectra of single crystals of ice- Ih taken with Q parallel, above, and perpendicular, below, to the *c*-axis. The bottom spectrum is the difference spectrum. Reproduced from [38] with permission from the American Institute of Physics.

The optic mode region of the spectrum shows two distinct triangular features at 226 and 298 cm^{-1}. This region involves the (O.....H) stretch and for a tetrahedral bonding motif in the lattice, see Fig. 9.8, only one feature would be expected. Early attempts to calculate this aspect of the INS spectrum introduced the model of a 'two-strengths' H-bond [32]. However, this phenomenological approach is unsatisfactory and considerable work now focuses on refining water–water potentials. These were originally parameterised for the liquid state but can be applied to lattice dynamics or molecular dynamics [39]. Alternatively *ab initio* methods have also been used [40]. Unfortunately, the majority of the calculational effort on ice concerns the more exotic phases, presently inaccessible to neutron spectrometers, and the data on the low pressure phases has been passed over in silence [30,41]. Currently no calculational approach is capable of reproducing all of the INS spectral features of any ice, although all models provide some general agreement.

This body of work provides insight into the nature of the 'open' structures of ice (§9.2.1). As the strength of the O...O bond decreases the distance $R(OO)$ increases and the librational bands shift to lower frequencies. Counter-intuitively, however, the density of the ice increases and 'open' structures are not necessarily those of low density.

9.2.4 Water at biological interfaces

Although H-bonds perform a central role in biological processes, when water is found at biological interfaces it plays a mostly structural role. It retains its molecular identity and expresses many of the spectral features found for the ices. Of particular interest in this respect is the study of water interacting with biological macromolecules, 'grana' (a membrane obtained from spinach) and DNA. The INS spectra of both systems at various stages of hydration were obtained, those of the grana membrane are shown in Fig. 9.11 [42].

The INS spectra of the totally hydrated species resembles that of ice Ih, see Fig.9.10, when the contribution from the dry sample is subtracted. However, at modest degrees of hydration the INS spectrum of the samples show significant differences from ice Ih. Using a simple model, where the water present was treated as either interfacial or bulk in nature, estimates were made of their amounts in each system. It was suggested that something between 1 and 4 layers of water comprise the interfacial component.

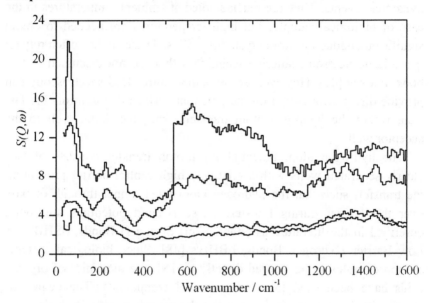

Fig. 9.11 The INS spectra of hydrated grana membrane. Whilst the 100% hydration (upper spectrum) resembles ice Ih, Fig. 9.8, at lower hydration the spectra are modified. Reproduced from [42] with permission from the American Chemical Society.

In DNA the transition from interfacial to bulk water occurs after *ca* 1 layer but grana membranes retain interfacial water up to *ca* 2 layers on each surface. Similarly the water in close contact with gelatine and chitosan did not have the ice Ih structure [43].

The INS from water on other biologically relevant samples has been compared to the INS of high-density ice [44]. Beyond the assessment of the extent of interfacial water and its possible nature the interpretation of the INS data is very difficult and several papers focus on the physics of the glass transition found in these systems [45]. Only modest progress has been made with more quantitative biological approaches involving modelling [46].

9.3 Proton transfer

Proton transfer mechanisms rely on several, generally slow, dynamical events. They are much studied at ambient temperatures in the case of biological samples but high temperatures are needed to obtain significant conduction from crystalline solids. These studies often rely on quasielastic neutron scattering techniques that are not discussed in this book, but see [47]. However, at low temperatures INS spectroscopy can provide direct information on the potential surface that controls the fast step, where the hydrogen atom transfers from the donor well to the acceptor well.

The hydrogen atom potential for proton transfer consists of two clearly separated wells with a relatively high central barrier preventing the transfer, such that the hydrogen atom must tunnel through or pass over this barrier during transfer. These H-bond potentials are found described in the literature as either, Short Strong H-Bonds, SSHB's, or Low Barrier Hydrogen Bonds, LBHB's [48]. Since biological systems are too complex to be studied directly by INS the dynamics of organic acids have been used to serve as model compounds. The degree of complexity is thus much reduced but the essentials remain. Below we consider some recent INS results from such model systems.

9.3.1 The dicarboxylate model systems

Several neutral carboxylic acids as well as their acid salts (hydrogencarboxylates) form molecular crystals using H-bonded structural motifs. Principal amongst these are simple linear dimers and cyclic dimers, see Appendix 4. Their importance lies not in the INS results from any individual system but the support that can be obtained from the body of results for the different models available for their interpretation. There are, at least, two approaches to interpreting the INS spectra of dicarboxylates [49].

The gross outlines of the proton's vibrational potential are similar in both approaches. This is an asymmetric double well, which is deep enough to explain the OH stretch at about 3000 cm^{-1}, and that are separated by a more or less substantial barrier, Fig. 9.12.

What differentiates the models is how the double well potential is linked to other vibrations of the molecule or lattice. The two approaches are, first; the *phonon assisted tunnelling* model and, second; the *independent proton* model.

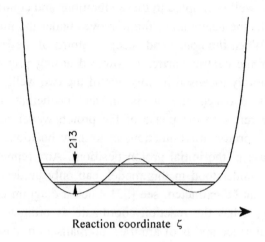

Reaction coordinate ζ

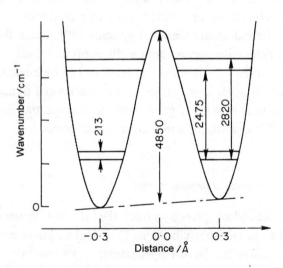

Fig. 9.12 Diagrammatic representations of the potential wells for the hydrogen atom in a H-bond as understood within the *phonon assisted tunnelling* model, shown above, and the *independent proton* model, below.

9.3.1.1 The phonon assisted tunnelling model

In the phonon assisted tunnelling model the hydrogen of the H-bond is assisted in its attempt to tunnel through the central barrier, which prevents proton transfer, by the vibrations of the system [50]. The H-bond double well is coupled to these vibrations and continuously distorts and adapts as the heavy atom frame moves under the influence of these vibrations. When the donor and acceptor atoms of the H-bond approach one another the central barrier is lowered easing proton transfer. The proton is entirely localised in only one of the two wells at any one time and tunnels between the wells in an incoherent process. (Here, incoherence refers to the phase of the proton wavefunction. The final phase of the proton wavefunction, in its new position, is not coherent with its phase in the initial proton position). Any representation of the potential, as understood in this model, can only be descriptive, even if parameters can be extracted, see [52]. Such a diagram is shown in Fig. 9.12, above, it has the reaction coordinate as ordinate but this is not linear in real space and is of arbitrary normalisation. Also, the height of the central barrier has no simply intuitive meaning. This model lends itself to the interpretation of a wide variety of proton transfer data drawn from many different techniques and systems [50]. Since the hydrogen atom is never delocalised across both wells but is entrained as a localised particle by the system's vibrations, its motions can be treated classically. The results of *ab initio* calculations of the vibrational dynamics of the system should thus provide a good guide to the spectral interpretation. (The *ad hoc* introduction of some anharmonic corrections may also be needed.)

9.3.1.2 The independent proton model

In the independent proton model the H-bond potential well is disconnected from all other vibrations [51]. Unlike the case of electrons in molecules, where the Born-Oppenheimer approximation applies, this disconnection cannot be on the grounds of significantly differing energy scales. The energies of H-bond vibrations are very similar to those of the heavy atoms. Rather the separation of the two sets of dynamics is achieved by allowing the spins of two protons in the dicarboxylate to be weakly coupled in the ground state. Thus the Pauli principle can be invoked to factorise the total wavefunction into two non-interacting parts, bosons and fermions [51]. This isolates the proton system and reduces the transfer process to one of low-dimensionality (possibly even one-dimensional) involving an 'effective potential' that is static. The tunnelling is coherent with the proton existing simultaneously in both wells, even if its expectation value shows it predominantly residing in only one. Any representation of the potential, as understood in this model, can hope to be significantly more than simply diagrammatic. Such a potential is shown in Fig. 9.12; notice that the coordinates have clear intuitive meanings. Considering H-bonds in isolation has a long history in the vibrational spectroscopy of solid state H-bonds but its application to the analysis of data from other techniques on more complex systems is generally discounted [50]. Any comparison with the output from *ab initio* calculations of the vibrations of the whole system is irrelevant since the hydrogen atom is not entrained in those modes and only the H-bond potential needs to be considered. Indeed, since each proton exhibits the full range of quantum mechanical properties, the use of classical mechanics to describe its motion is inappropriate. (As with the previous model the *ad hoc* introduction of anharmonic corrections may also be needed.)

9.3.1.3 Benzoic acid

The two models have recently come into sharp contrast since they have both been applied to the analysis of the same INS spectrum of benzoic acid [51,53]. The molecular structure of this system is shown in Fig. 9.13. These are the only INS data on O–H–O bonds to be fully analysed using the *phonon assisted tunnelling* model. However, it is probable that the INS spectra of most systems could be subjected to a similar analysis.

In the *phonon assisted tunnelling* model the H-bond stretch of benzoic acid is anharmonically linked to the entire thermal bath of internal and external vibrations. No individual vibration has any exceptional role in this almost stochastic process and in particular the vibration giving rise to a strong infrared band at 56 cm^{-1}, is not specially involved. The *ab initio* calculations conducted on a benzoic acid lattice were used to produce the first entirely consistent assignment scheme of the low frequency region of its spectrum. Calculations also revealed a modest degree of dispersion in the O....O stretching mode, at about 150 to 200 cm^{-1} [54]. This was used to explain the weak shoulder appearing in the INS spectrum about 180 cm^{-1}.

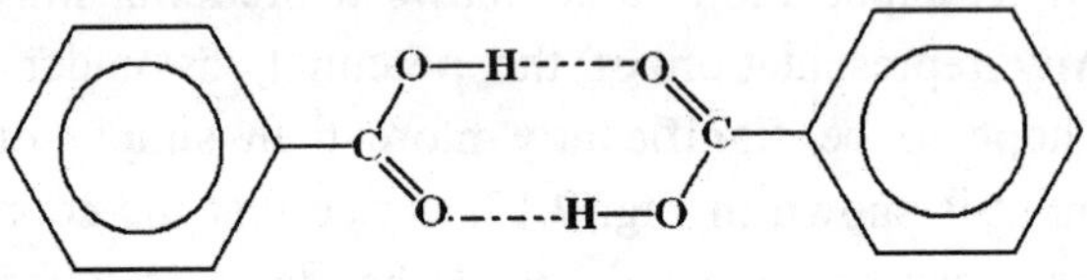

Fig. 9.13 The molecular structure of benzoic acid.

In the *independent proton* model, the INS spectrum of benzoic acid consists of a few strong features, associated with tunnel split states and weak ancillary bands. The strong INS bands observed are typical of the unit mass oscillators found in this model. The model also gives frequencies close to the observed values using reasonable values for the force constants. The role of the heat bath lies in the population of extended coherent states, the phonons, that it generates. The phonon states are quantum mechanically mixed with states in the double well. The strongly temperature dependant vibration that produces an infrared band about 56 cm^{-1} plays a special role in this respect. The weak shoulder on the O....O stretching mode, about 180 cm^{-1}, was assigned to the coherent tunnelling transition [51].

9.3.1.4 Potassium hydrogencarbonate

Apart from some early work on powders, potassium hydrogen carbonate has been mostly examined using the direct geometry spectrometer MARI to study oriented single crystals [55]. Because of the unique crystal structure of the carbonate three different crystals were cut. The total momentum transfer, Q_{total}, could then be defined in terms of the directions of displacement of the three main vibrations of the H-bond. These are the deformations δ and γ, and the antisymmetric stretch, here written ν, so $Q_{total} = Q_{\nu} + Q_{\gamma} + Q_{\delta}$. The magnitude of both Q_{total} and the energy transferred was varied in the study of each crystal. The scattering law for this single crystal becomes:

$$S(Q,\omega)_{total} = \frac{\sigma_H}{4\pi}\left\{S(Q,\omega)_{v'}^n \times S(Q,\omega)_{\gamma}^n \times S(Q,\omega)_{\delta}^n\right\}$$

$$S(Q,\omega)_v^n = \frac{\left[\left(Q_v \cdot {}^v u\right)^2\right]^n}{n!} \exp\left(-Q_v \cdot {}^v u\right)^2 \qquad v = v', \gamma, \delta \qquad (9.2)$$

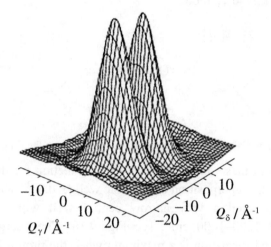

Fig. 9.14 The plot of $S(Q_\delta, Q_\gamma, 960\ cm^{-1})$, from single crystals of KHCO$_3$ (MARI, ISIS).Reproduced from [55] with permission from the American Institute of Physics.

In this manner the momentum variation of the main vibrations of the H-bond were separated. As an example, the out-of-plane deformation, γ, at 960 cm^{-1} could be seen to produce scattering only if some component of the momentum was transferred parallel to its vibrational displacement, see Fig. 9.14. The experimental profiles were fitted to the form of Eq. (9.2) and values for the mean square hydrogen displacement were determined. These values were close to those of pure hydrogenic vibrations involving no mixing with other modes. Even the δ-mode, often found to be heavily mixed with the O—C=O in-plane deformations in other carboxylates, was shown to be pure [55].

9.3.1.5 Potassium hydrogenmaleate

The structure of potassium hydrogenmaleate, Fig. 7.14, is closely related to the other dicarboxylates but here the H-bond is intramolecular. Growing single crystals of molecules that were deuterated at the carbon positions uniquely isolated the INS of this symmetric O–H–O bond [56]. The resulting INS spectrum provided conventional values for the γ- and δ-modes but was unexpectedly rich in features below 1000 cm^{-1}, associated with the ν_{as} mode.

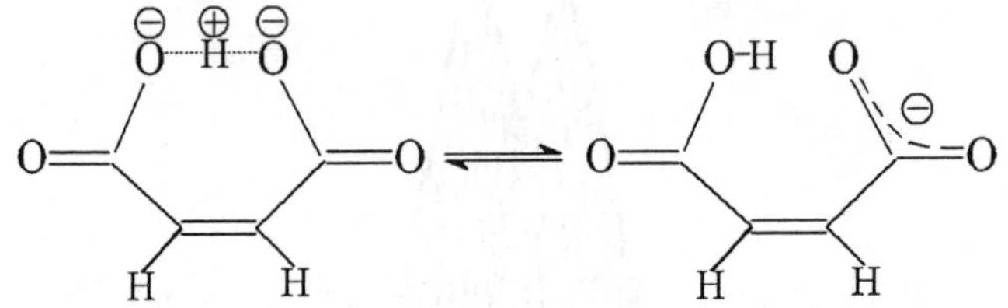

Fig. 9.15 Resonant Valence Bond descriptions of the H-bond in the maleate ion.

It was from these data that the potential function for the proton was modelled as the superposition of a double well ($\pm$ 0.8 Å) with a barrier of 150–200 cm^{-1} and a very narrow central well with a dissociation threshold of 1200–1400 cm^{-1}, see Fig. 9.16. The energy gain upon hydrogen bond formation is a maximum when the proton is localised in the narrow central well in the totally symmetric structure.

This structure is stabilised by electrostatic interactions but the energy gain vanishes rapidly for even small displacements of the proton away from its central position. This is a consequence of the rearrangement of the electrical charges and results in a covalent O–H bond of about 1 Å. If the abscissa of Fig. 9.16 naively represents the hydrogen potential in an unchanging heavy atom geometry the upper minima would correspond to unacceptably short O–H distances, *ca* 0.4 Å.

It was hypothesised that as the proton moves off-centre the planarity of the ring is lost and the donor carboxylate group rotates out of the plane, relaxing the strain energy [56]. The effective potential along the stretching coordinate presented in Fig. 7.16 is a snapshot of the proton dynamics before the rotation of the carboxylic group creates asymmetry in the potential surface.

The strained planar geometry is stabilised by H-bonding only in the ground state of the stretching-mode. In any other state the ring twists, the O–H bond assumes a typical covalent length, the H----O distance increases to about 2 Å and the hydrogen bond is effectively 'broken'. This requires a dissociation energy of only 0.4 kJ mol^{-1}. The maleate ion thus demonstrates all the characteristics of a 'very strong H-bond' that is nonetheless easily dissociated by modest thermal activation, or solvent effects. It was further suggested that whilst the proton ground state is 'H-bonding' the excited states could be regarded as 'H-antibonding' and that this may represent the situation in all strong H-bonds.

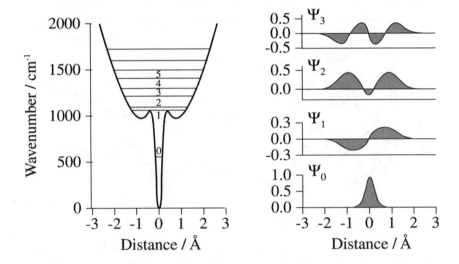

Fig. 9.16 Potential function, left, and wave functions, right, for the of the hydrogen in the v_{as}(OHO) mode of the maleate ion. Reproduced from [56] with permission from Elsevier.

This model stresses the view of the hydrogen atom as a quantum particle acting independently of the heavy atoms and playing a role somewhat like the electrons in a molecular orbital. Systems containing formally symmetric hydrogen bonds could, in this model, have similar dissociation thresholds, and overall potential functions, almost independently of their chemical nature.

9.3.2 Proton conducting materials

The local, non-diffusive, dynamics of the hydrogen in proton conducting solids can give useful information. Especially in doped and disordered systems were diffraction techniques have failed to locate the hydrogen atoms. An early study of the high pressure phases of $CsHSO_4$ showed that strong correlations existed between the vibrations of the H-bonded hydrogen atoms and, dependent on the phase, these dynamics were decoupled from the heavy atom vibrations [57]. The use of INS to investigate subtle phase changes in proton conducting materials is also known [58]. There have also been several studies of related systems, sulphates [58], selenates [59], arsenates [60] and phosphates [61].

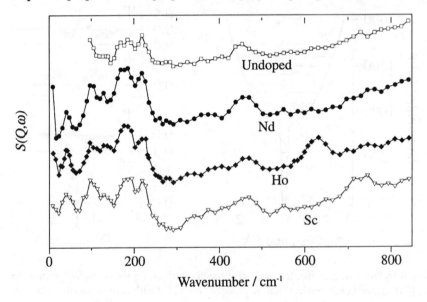

Fig. 9.17 The INS spectra (FANS, NIST) of the differently doped samples of strontium cerate, as discussed in the text. Redrawn from [62] with permission from Elsevier.

It is generally held that metal oxide perovskites with extrinsic oxygen vacancies react with atmospheric water. This entrains hydrogen into the lattice and leads to their significant proton conducting properties. The location of these hydrogen atoms was studied as a function of a series of dopants in a cerium based ceramic. The dopants were niobium, holmium and scandium [62]. The Sc and Ho doped material exhibited a strong $\gamma(OH...O)$ feature about 850 cm^{-1} and this transition moved to lower frequency as the ionic radius of the dopant grew, see Fig. 9.17.

However, for ionic radii greater than the metal of the host lattice (and here Nd>Ce) the hydrogen content of the ceramic falls and, since other INS features remain strong, this stems from the loss of sites associated with the 850 cm^{-1} band. A special hydrogen site is, therefore, available for small dopants.

The interpretation of the spectral details was improved by comparing results from two model calculations, one for an undoped cerium oxide lattice and the other for a Sc doped lattice. The main hydrogen site was found to be at the Ce, giving features about 650 cm^{-1}. The minority population was next to the doped site, giving features about 850 cm^{-1} [63]. The temperature dependence of the $\gamma(OH...O)$, 805 cm^{-1}, in RbHSO$_4$ was also measured.

Studies on a rechargeable battery material, bismuth doped manganese oxide, were discussed in terms of two unresolved contributions, about 540 cm^{-1} and 815 cm^{-1} [64]. (They show no evidence for the isotropic-protons of §9.4.1.) The intensity of the former band decreased whilst the latter increased as more protons were pushed into the lattice. These observations were related to specific water species in the material [64]. INS studies of the electro-active material, NiO(OH), are known [65,66].

9.4 Unusual protonic species

Although not specifically related to H-bonding these unusual species do involve materials similar to others discussed in this section. There are two of particular interest; the isotropic-proton [22] and the free-proton [67]. Both species have been detected in several electrically important materials and so may prove to be of some general interest.

9.4.1 The isotropic proton

Isotropic, or nearly so hydrogen potentials in lattices are normally found only for hydrogen-in-metal systems (§6.6), and materials like electro-deposited MnO_2 would normally be considered far too complex to have such high symmetry sites. However, a related material, manganite (MnOOH) has an INS spectrum that is quite devoid of the usual spectral signatures for H-bonds and in their stead a single harmonic sequence is found. The observed INS intensities of the fundamental, 1124 cm^{-1}, and its first overtone, ca 2200 cm^{-1}, reflect that of a unit mass oscillator (§5.2.1.2). In the electroactive material analogous bands were found, somewhat broadened, at the same frequencies and increasing in intensity with the proton content [67]. An isotropic potential was also used to explain the INS of dehydrated tungstophosphoric acid [26] (§9.2.2).

9.4.2 The free proton

The second unusual proton species is the free proton. This behaves like a recoiling molecule but the hydrogen mass is so small it gives a unique signal on low-bandpass spectrometers (see Fig. 5.4). In these electroactive materials any features observed, on low-bandpass spectrometers, appear above a fixed level of scattering. In γ-MnO_2 this level, or continuum, remains unchanged as the system is drained of protons but the intensities of the superimposed bands fall. Moreover, the level disappears completely when the samples have been thoroughly dehydrogenated and so the scattering is from some hydrogenous species and is not a background arising from the matrix. The hydrogen atoms thus access more or less large volumes of space within the inorganic matrix, they are not bound in the normal chemical sense.

The full spectral form is seen in the data obtained on a chopper spectrometer, see Figs. 9.18 and 9.19 [68]. The continuum is a very broad response that tracks the unit-mass recoil line and is by far the strongest spectral component. Especially since the unbound scattering cross section of hydrogen, 20 barn (§2.1), should be used in calculations of this effect. Analysis of this spectrum has proved very difficult because the width of the response, Fig. 9.19, far exceeds conventional predictions. Since neutrons cannot determine the electrical nature of the scatterer directly H^+, H^0, or H^- are all possibly present.

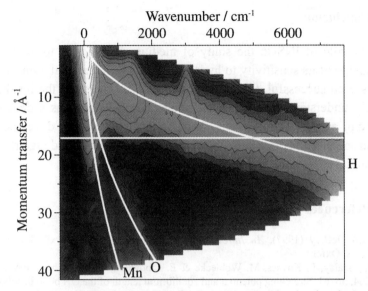

Fig. 9.18 The INS spectrum (MARI, ISIS) of manganese dioxide, showing the response of a *free* proton. The horizontal line at 17 Å^{-1} is the line of the cut shown in Fig. 9.19. Reproduced from [68] with permission from Elsevier.

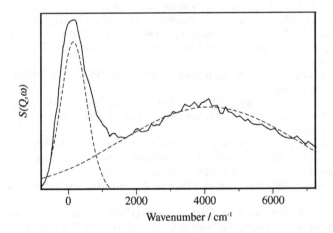

Fig. 9.19 The INS spectrum of manganese dioxide, taken along the $Q = 17$ Å^{-1} cut, see Fig. 9.18, showing the very broad response from the *free* protons, centred around 4000 cm^{-1}. Reproduced from [68] with permission of Elsevier.

9.5 Conclusion

The role of INS in the study of medium H-bonds is to exploit the advantage of its sensitivity to hydrogen acting as a classical atom. In this it has been successful, the interpretive model is clear and the results are readily understandable. However, where the hydrogen has a developed quantum aspect the interpretive models are inadequately developed, as for strong H-bonds, the situation remains open: the H-bond eigenvectors remain insufficiently exploited

9.6 References

1   G.A. Jeffrey (1997). *An Introduction to Hydrogen Bonding*, Oxford University Press, Oxford.

2   V. Szalay, L. Kovacs, M. Wohlecke & E. Libowitzky (2002). Chem. Phys. Lett., 354, 56–61. Stretching potential and equilibrium length of the OH bond in solids.

3   J.M. Besson, P. Pruzan, S. Klotz, G. Hamel, B. Silvi, R.J. Nelmes, J.S. Loveday, R.M. Wilson & S. Hull (1994). Phys. Rev. B, 49, 12540–12550. Variation of interatomic distances in ice-VIII to 10 Gpa.

4   G. Gilli & P. Gilli (2000). J. Mol. Struct., 552, 1–15. Towards a unified hydrogen-bond theory.

5   J. Howard, J. Tomkinson, J. Eckert, J.A. Goldstone & A.D. Taylor (1983). J. Chem. Phys., 78, 3150–3155. Inelastic neutron scattering studies of some intramolecular hydrogen bonded complexes: a new correlation of gamma–(OHO) vs R(OO).

6   A Beran, G. Giester & E. Libowizky (1997). Miner. Petrol., 61, 223–235. The hydrogen bond system in natrochalcite-type compounds— an FTIR spectroscopic study of the H_3O unit

7   M.C. Lawrence & G.N. Roberston (1981). Mol. Phys., 43, 193–213. Proton tunnelling in chromous acid.

8   K.R. Leopold, G.T. Fraser, S.E. Novick, & W. Klemperer (1994). Chem. Rev., 94, 1807–1827. Current themes in microwave and infrared spectroscopy of weakly bound complexes.

9   R.G. Delaplane, H. Kuppers, J. Noreland & S.F. Parker (2002). Appl. Phys. A, 74, S1366–S1367. Dynamics of the water molecule in potassium oxalate monohydrate as studied by single crystal inelastic neutron scattering.

10   B. Winkler (1996). Phys. Chem. Minerals, 23, 310–318. The dynamics of H_2O in minerals.

11   H. Jobic, A. Tuel, M. Krossner & J. Sauer (1996). J. Phys. Chem., 100, 19545–19550. Water in interaction with acid sites in H-ZSM-5 zeolite does not form hydroxonium ions. A comparison between neutron scattering studies and *ab initio* calculations.

12   H. Jobic, M. Czjzek & R.A. van Santen (1992). J. Phys. Chem., 96, 1540–1542. Interaction of water with hydroxyl groups in H-mordenite: a neutron inelastic scattering study.

13 G.A. Lager, G.A. Swayze, C-K Loong, F.J. Rotella, J.W. Richardson & R.E. Stoffregen (2001). Can. Mineral., 39, 1131–1138. Neutron spectroscopic study of synthetic alunite and oxonium-substituted alunite.

14 F.R. Trouw & D.L. Price (1999). Annu. Rev. Phys. Chem., 50, 571–601. Chemical applications of neutron scattering.

15 M.T. Dove (2002). Eur. J. Miner.., 14, 203–224. An introduction to the use of neutron scattering methods in mineral sciences.

16 A.I. Kolesnikov, J.C. Li & S.F. Parker (2002). J. Mol. Liq., 96/97, 317–325. Liquid-like dynamical behaviour of water in silica gel at 5K. V. Crupi, D. Majolino, P. Magliardo, V. Venuti & A.J. Dianoux (2002). Appl. Phys. A, 74, S555–S556. Low-frequency dynamical response of confined water in normal and supercooled regions obtained by IINS.

17 J.D.F. Ramsay, H.J. Lauter & J. Tomkinson (1984). J. de Phys.-Paris, 45 (NC7), 73–79. Inelastic neutron scattering of water and ice in porous solids.

18 J. Dore (2000). Chem. Phys., 258, 327–347. Structural studies of water in confined geometry by neutron diffraction.

19 C.M.B. Line & G.J. Kearley (2000). J. Chem. Phys., 112, 9058–9067. An inelastic incoherent neutron scattering study of water in small-pored zeolites and other water-bearing minerals.

20 C.M.B. Line & G.J. Kearley (1998). Chem. Phys., 234, 207–222. The librational and vibrational spectra of water in natrolite, $Na_2Al_2Si_3O_{10}$. $2H_2O$, compared with *ab initio* calculations.

21 H. Fuess, E. Stukenschmidt & B.P. Schweiss (1986). Ber. Bunsen. Phys. Chem, 90, 417–421. Inelastic neutron scattering of water in natural zeolites.

22 Ph. Colomban & J. Tomkinson (1997). Solid State Ionics, 97, 123–134. Novel forms of hydrogen in solids: The 'ionic' proton state and the 'quasi-free' proton.

23 F. Fillaux (1999). J. Mol. Struct., 512, 35–47. New proton dynamics in solids revealed by vibrational spectroscopy with neutrons.

24 G.J. Kearley, R.P. White, C. Forano & R.T.C. Slade (1989). Spectrochim. Acta A, 46, 419–424. An analysis of the vibrational frequencies and amplitudes of the H_5O^+ ion in $H_4SiW_{12}O_{40}.6H_2O$ (TSA.6H$_2$O) and $H_3PW_{12}O_{40}.6H_2O$ (TPA.6H$_2$O)..

25 D.J. Jones & J. Roziere (1993). Solid State Ionics, 61, 13–22. Complementarity of optical and incoherent inelastic neutron scattering spectroscopies in the study of proton conducting materials.

26 U.B. Mioc, Ph. Colomban, M. Davidovic & J. Tomkinson (1994). J. Mol. Struct., 326, 99–107. Inelastic neutron scattering study of protonic species during the thermal dehydration of 12-tungstophosphoric hexahydrate.

27 N. Essayem, Y.Y. Tong, H. Jobic & J.C. Vedrine (2000). Appl. Catal. A, 194/195, 109–122. Characterisation of the protonic sites in $H_3PW_{12}O_{40}$ and $Cs_{1.9}H_{1.1}PW_{12}O_{40}$: a solid-state H-1, H-2, P-31 MAS-NMR and inelastic neutron scattering study on samples prepared under reaction conditions..

28 Ph. Colomban (1999). Ann. Chim. Sci. Mat., 24, 1–18. Latest developments in proton conductors.

29 C.C. Wilson (2000). *Single Crystal Neutron Diffraction from Molecular Materials,* World Scientific, Singapore.

30 A. Putrino & M. Parrinello (2002). Phys. Rev. Lett., 88, art no. 176401. Anharmonic Raman spectra in high-pressure ice from *ab initio* simulations.

31   J.C. Li & A.I. Kolesnikov (2002). J. Mol. Liq., 100, 1–39. Neutron spectroscopic investigation of the dynamics of water ice.

32   J.C. Li & J. Tomkinson (1999). In *Molecular Dynamics -from Classical to Quantum Methods*, (Ed.) P.B. Balbuena, J.M. Seminario (Elsevier Scientific) Theo. Comp. Chem., 7, 471–532. Interpretation of inelastic neutron scattering spectra for water ice, by lattice molecular dynamics.

33   Y. Madokoro, O. Yamamuro, H. Yamasaki, T. Matsuo, I. Tsukushi, T. Kamiyama & S. Ikeda (2002). J. Chem. Phys., 116, 5673–5679. Calorimetric and neutron scattering studies on the boson peak of lithium chloride aqueous solution glasses.

34   C. Gutt, J. Baumert, W. Press, J.S. Tse & S. Janssen (2002). J. Chem. Phys., 116, 3795–3799. The vibrational properties of xenon hydrate: An inelastic incoherent neutron scattering study.

35   L. Mercury, P. Vieillard & Y. Tardy (2001). Appl. Geochem., 16, 161–181. Thermodynamics of ice polymorphs and 'ice-like' water in hydrates and hydroxides.

36   H. Fukazawa, S Mae, S. Ikeda & O. Watanabe (1998). Chem Phys Lett., 294, 554–558. Proton ordering in Antarctic ice observed by Raman and neutron scattering.

37   J.C. Li, C. Burnham, A.I. Kolesnikov & R.S. Eccleston (1999). Phys Rev. B, 59, 9088–9094. Neutron spectroscopy of ice VIII in the region 20–500 meV.

38   J.C. Li (1996). J. Chem. Phys., 105, 6733–6755. Inelastic neutron scattering studies of hydrogen bonding in ices.

39   S.L. Dong, Y. Wang & J.C. Li (2000). Chem. Phys., 270, 309–317. Potential lattice dynamical simulations of ice.

40   S. Jenkins & I. Morrison (2001). J. Phys. Condens. Matter, 13, 9207–9229. The dependence on the structure of the projected vibrational density of states of various phases of ice as calculated by *ab initio* methods.

41   L. Ojamae, K. Hermansson, R. Dovesi, C. Roetti & V.R. Saunders (1994). J. Chem. Phys., 100, 2128–2138. Mechanical and molecular properties of ice VIII from crystal-orbital *ab initio* calculations.

42   S.V. Ruffle, I. Michalarias, J.C. Li & R.C. Ford (2001). J. Am. Chem. Soc., 124, 565–569. Inelastic incoherent neutron scattering studies of water interacting with biological   macromolecules.
     R.C. Ford, S.V. Ruffle, A.J. Ramirez-Cuesta, I. Michalarias, I. Beta, A. Miller & J.C. Li (2004). J. Am. Chem. Soc., 126, 4682–4688. Inelastic incoherent neutron scattering measurements of intact cells and tissues and detection of interfacial water.

43   G.I. Wu, I.A. Beta, J.B. Ma & J.C. Li (2002). Appl. Phys. A, 74, S1267–S1269. Inelastic neutron scattering studies of water on bio-polymers.

44   F. Cavatorta, N. Angelini, A. Deriu & G. Albanese (2002). Appl. Phys A, 74, S504–S506. Vibrational dynamics of hydration water in amylose.

45   C. Branca, S. Magazu, G. Maisano, F. Migliardo, G. Romeo, S.M. Bennington, B. Fak, E. Bellocco & G. Lagana (2002). Appl. Phys., 74, S459-S460. Analysis of the changes of vibrational properties of water in the presence of disaccharides.
     C. Branca, S. Magazu, F. Miligliardo, G. Romeo, V. Villari, U. Wanderlingh & D. Colognesi (2002). Appl. Phys. A, 74, S452–S453. Neutron scattering study of the vibrational behaviour of trehalose in aqueous solutions.

46   M. Tarek & D.J. Tobias (2002). Phys. Rev. Lett., 89, art no. 275501. Single particle and collective dynamics of protein hydration water: A molecular dynamics study.
     A.R. Bizzarri & S. Cannistrato (2002). J. Phys. Chem. B, 106, 6617–6633. Molecular dynamics of water at the protein solvent interface.

47 R.E. Lechner (2001). Solid State Ionics, 145, 167–177. Proton conduction mechanism in $M_3 H(XO_4)_2$ crystals: the trigonal asymmetric hydrogen bond model.

48 S. Humbel (2002). J. Phys. Chem. A, 106, 5517–5520. Short strong hydrogen bonds: A valence bond analysis.

49 V.V. Krasnoholovets, P.M. Tomchuk & S.P. Lukyanets (2003). Adv. Chem. Phys., 125, 351–548. Proton transfer and coherent phenomena in molecular structures with hydrogen bonds.

50 V.A. Benderskii, D.E. Marakov & C.A. Wight (1994). *Chemical dynamics at low temperatures*, Wiley, New York. (Adv. Chem. Phys., LXXXVIII)

51 F. Fillaux, M.H. Limage & F. Romain (2002). Chem. Phys., 276, 181–210. Quantum proton transfer and interconversion in the benzoic acid crystal: vibrational spectra, mechanism and theory.
F. Fillaux (1998). Physica D, 113, 172–183. The Pauli principle and the vibrational dynamics of protons in solids, a new spin related symmetry.

52 V.A. Benderskii, E.V. Vetoshkin & H.P. Trommsdorff (2001). Chem. Phys., 271, 165–182. Tunnelling splittings in vibrational spectra of non-rigid molecules, X. Reaction path Hamiltonian as zero-order approximation.

53 M. Plazanet, N. Fukushima, M.R. Johnson, A.J. Horsewill & H.P. Trommdorff (2001). J. Chem. Phys., 115, 3241–3248. The vibrational spectrum of crystalline benzoic acid; inelastic neutron scattering and density functional theory calcualtions.
M. Neumann, D.F. Broughham, C.J. McGloin, M.R. Johnson, A.J. Horsewill & H.P. Trommsdorff (1998). J. Chem. Phys., 109, 7300–7311. Proton tunnelling in benzoic acid at intermediate temperatures: Nuclear magnetic resonance and neutron scattering studies.

54 M.R. Johnson & H.P Trommsdorff (2002). Chem. Phys. Lett., 364, 34–38. Dispersion of vibrational modes in benzoic acid crystals.

55 S. Ikeda, S. Kashida, H Sugimoto,Y. Yamada, S.M. Bennington & F. Fillaux (2002). Phys. Rev. B, 66, art no 184302. Inelastic neutron scattering study of the localised dynamics of protons in $KHCO_3$ single crystals.
S. Ikeda & F. Fillaux (1999). Phys. Rev. B, 59, 4134–4145. Incoherent elastic-neutron-scattering study of the vibrational dynamics and spin-related symmetry of the protons in the $KHCO_3$ crystal.

56 F. Fillaux, N. Leygue, J. Tomkinson, A. Cousson & W. Paulus (1999). Chem. Phys., 244, 387–403. Structure and dynamics of the symmetric hydrogen bond in potassium hydrogen maleate: a neutron scattering study.

57 A.V. Belushkin, M.A. Adams, A.I. Kolesnikov & L.A. Shuvalov (1994). J. Phys. C: Condens. Matter, 6, 5823–5832. Lattice dynamics and effects of anharmonicity in different phases of cesium hydrogen sulphate.

58 A.V. Belushkin, T. Mhiri & S.F. Parker (1997). Physica B, 234/236, 92–94. Inelastic neutron scattering of $Cs_{1-x}(NH_4)_x HSO_4$ mixed crystals.

59 A.V. Belushkin, J. Tomkinson & S.A. Shuvalov (1993). J. de Phys. II, 3, 217–225. Inelastic neutron scattering study of proton dynamics in $CsH(SeO_4)_2$ and $Rb_2H(SeO_4)_2$.

60 N. Le Calve, B. Pasquier & Z. Ouafik (1997). Chem. Phys., 222, 299–313. Vibrational study by inelastic neutron scattering, infrared adsorption and Raman scattering of potassium, rubidium and cesium dihydrogen arsenate crystals:, Comparison with thallium dihydrogen arsenate.

61   A.V. Belushkin & M.A. Adams   (1997).   Physica B, 234/236, 37–39. Lattice dynamics of KH_2PO_4 at high pressure.

62   C. Karmonik, T.J. Udovic, R.L. Paul, J.J. Rush, K. Lind & R. Hemplemann (1998). Solid State. Ionics, 109, 207–211. Observation of the dopant effects on hydrogen modes in $SrCe_{0.95}M_{0.05}:H_xO_{3-\delta}$ by neutron vibrational spectroscopy.

63   T. Yildirim, B. Reisner, T.J. Udovic & D.A. Neumann (2001). Solid State Ionics, 145, 429–435. The combined neutron scattering and first-principles study of solid state protonic conductors.

64   F. Fillaux, C. Cachet, S.F. Parker, J. Tomkinson, A. Quivy & L.T. Yu (2000). J. Electrochem. Soc., 147, 4184–4188. Inelastic neutron scattering studies of the proton dynamics in bi-doped manganese oxides.

65   R. Baddour-Hadjean, F. Fillaux & J. Tomkinson (1995). Physica B, 213/214, 637–639. Proton dynamics in $\beta Ni(OH)_2$ and $\beta NiOOH$.

66   J. Eckert, R. Varma, L. Diebolt & M. Reid (1997). J. Electrochem. Soc., 144, 1895–1899. Effects of cycling conditions of active material from discharged Ni positive plates studied by inelastic neutron scattering spectroscopy.

67   F. Fillaux, C.H. Cachet, H. Ouboumour, J. Tomkinson, C. Levy-Clement & L.T. Yu (1993). J. Electrochem. Soc., 140, 585–591. Inelastic neutron scattering study of the proton dynamics of manganese oxides I. γ-MnO_2 and manganite. F. Fillaux, ,H. Ouboumour, C.H. Cachet, J. Tomkinson, C. Levy-Clement & L.T. Yu (1993). J. Electrochem. Soc., 140, 592–598. Inelastic neutron scattering study of the proton dynamics of manganese oxides II. Proton insertion in electrodeposited MnO_2.

68   F. Fillaux, S.M. Bennington & L.T. Yu (1996). Chem. Phys., 209, 111–125. Inelastic neutron-scattering study of free proton dynamics in γ-MnO_2.

10

Soft Condensed Matter—
Polymers and Biomaterials

Soft condensed matter is a rapidly growing field that includes areas such as polymers, biopolymers and biological materials. In this chapter we describe some of the applications of INS spectroscopy in these areas.

There is an enormous body of work on quasielastic neutron scattering from polymers [1,2]. There is a smaller literature on neutron vibrational spectroscopy of polymers but this has had a significant impact on the characterisation of these materials. Crystalline or semi-crystalline polymers are the most important class of polymers commercially. The most-studied and technologically most important of these is polyethylene and this will be considered in some depth and we will highlight the use of the n-alkanes as model compounds (§10.1.2). We will then see how these concepts can be transferred to polypropylene (§10.1.3), nylon (§10.1.4), and conducting polymers (§10.1.5). Non-crystalline polymers (§10.2) have been much-less studied by INS. As examples, we will consider polydimethylsiloxane (§10.2.1) and advanced composites (§10.2.2).

Biological materials (§10.3) are difficult systems for INS to study. The amount of sample is often very limited, 0.1 g is a 'large' sample for biologists, and the large number of atoms means that there are many vibrational modes which leads to dense, congested spectra. INS has been effective in understanding the dynamics of constituents of biopolymers such as nucleic acids (§10.3.1) and there are hints that proteins may yield useful information (§10.3.2). Water in biological materials is considered elsewhere (§7.2.4). Bone is an important area of research and INS is well suited to the study of the inorganic components of bone (§10.3.3).

10.1 Crystalline polymers

10.1.1 Polyethylene and the n-alkanes

Polyethylene $-(CH_2-CH_2)_n-$ is the largest tonnage plastic; in 1995 world-wide production was 3.8×10^7 tonnes. It has been produced commercially since 1939 for a wide range of applications from electrical insulation to packaging and pipes. Its usefulness derives from its low cost, easy processability, chemical resistance and the ability to tailor its properties by the method of manufacture and the inclusion of co-monomers (usually *n*-alk-1-enes, where $n = 3-12$).

Chemically, polyethylene is the simplest possible polymer, however, the apparent simplicity of the molecular formula $(CH_2)_n$ belies the real complexity of the material. To understand the physical properties of the polymer it is necessary to consider at least a two-phase model consisting of crystalline blocks in an amorphous matrix [3]. The crystalline component ranges from *ca* 50% to >95%; generally it increases with increasing molecular weight and decreases with increasing chain branching. Vibrational spectroscopy has played a key part in the characterisation and understanding of the polymer since it is sensitive to both regions and also to the effects of finite chain length and side branches [4].

Polyethylene can adopt a number of crystal forms depending on the method of preparation. The most common form is a centrosymmetric orthorhombic form, space group $Pnma \equiv D_{2h}^{16}$ (number 62), with two formula units (CH_2-CH_2) in the unit cell rotated at 90° with respect to each other as shown in Fig. 10.1. Each chain adopts an *all-trans* planar zigzag conformation. This can also be described as a 2_1 helix with one complete turn for each two CH_2 units.

The normal modes at the centre of the Brillouin zone, (§4.3.1), ($k = 0$) can be deduced by a conventional factor group analysis [4,5]. Since there are two chains in the unit cell, equivalent modes on each chain couple to give an in-phase and out-of-phase pair of modes (Davydov splitting). Above 600 cm^{-1} the modes can be described using standard group frequency correlations; CH_2 rock 720–1170 cm^{-1}, C–C stretch

$1050-1150$ cm^{-1}, CH$_2$ twist $1050-1300$ cm^{-1}, CH$_2$ wag $1175-1400$ cm^{-1}, CH$_2$ scissors $1400-1500$ cm^{-1} and C-H stretch $2800-3000$ cm^{-1}.

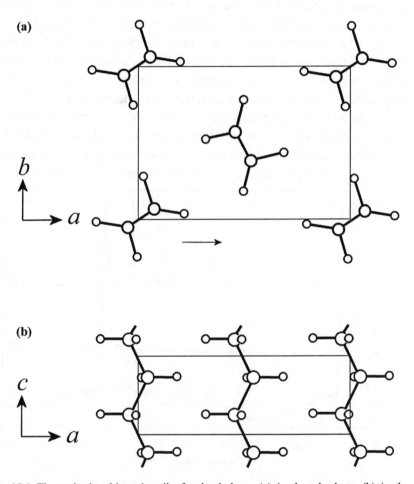

Fig. 10.1 The orthorhombic unit cell of polyethylene: (a) in the *ab* plane, (b) in the *ac* plane showing the *all-trans* planar zigzag conformation of crystalline polyethylene. The 2_1 helix (one complete turn for each two CH$_2$ units) is also apparent.

Below 600 cm^{-1} are the ν_5 in-plane C—C—C bending modes (longitudinal acoustic modes, LAMs), the ν_9 out-of-plane C—C—C bending modes (transverse acoustic modes, TAMs) and the 12 lattice modes. These comprise in-phase and out-of-phase pairs of librations about the three axes, giving six modes, three optic translation and three acoustic translation modes. This region is more difficult to study by infrared and Raman spectroscopies. In contrast, as Fig. 10.2c shows, it gives rise to intense INS features and here INS has made significant contributions. Early INS work on polyethylene is reviewed in [6,7].

Fig. 10.2 shows the infrared, Raman and INS spectra of polyethylene in the 0–1600 cm^{-1} region. The infrared and Raman spectra show no coincidences since the crystal is centrosymmetric. The INS spectrum is very different; the absence of selection rules means that the features present in both optical spectra are apparent, but the major difference is that the regions between the infrared and Raman bands are 'filled-in' in the INS spectrum. This is because INS spectroscopy gives information at all values of k, not just those at zero as for the optical spectroscopies.

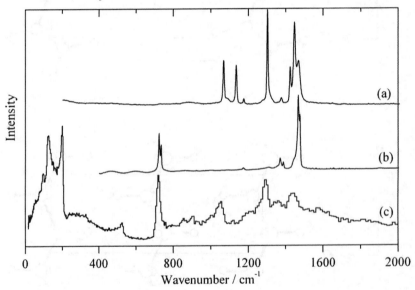

Fig. 10.2 Vibrational spectra of polyethylene in the 0–2000 cm^{-1} region. (a) INS, (b) infrared and (c) Raman.

10.1.1.1 Dispersion curves of polyethylene

The variation of the energy of a vibration with wavevector k is the dispersion of the mode and arises from coupling between oscillators in different unit cells. Since for polyethylene adjacent unit cells contain the same polymer chain, the presence of dispersion is to be expected. Dispersion curves of materials are conventionally plotted over the first half of the Brillouin zone. In the case of a polymer the dispersion curve can be plotted in two ways: either as a function of the phase angle, φ (§2.6.1.1), between adjacent atoms (or, in the case of polyethylene, CH_2 groups) or as a function of k the phase difference between adjacent translational units. Since the unit cell of polyethylene contains two chain segments each consisting of two CH_2 groups, the cell is twice as long for the adjacent translational units as that for the adjacent oscillator and consequently the Brillouin zone of the former is half the size of the latter. Thus dispersion curves as a function of k can be obtained from those as a function of φ by folding back the right-hand half of the plot. For infrared and Raman spectroscopy, the allowed modes are those at $k = 0$, the Brillouin zone centre. In terms of φ the allowed modes are those at 0 and π since the dispersion curves in this case span a complete Brillouin zone from centre-to-centre. Equivalently, the selection rules require that vibrations in each translational unit are totally in phase. This is clearly satisfied at $\varphi = 0$, but since the *all-trans* conformation of polyethylene is a 2_1 helix (one complete turn for each two CH_2 units) it follows that a phase difference of π between adjacent oscillators will result in a phase difference of 2π between adjacent translational units and hence the $k = 0$ requirement is satisfied.

The dispersion curves of polyethylene have been the subject of intense study by several groups. The best available are those of Barnes and Franconi [8]. The model is a purely empirical 'ball-and-springs' force-field with Williams-type potentials to include the intermolecular forces. The potentials have the form:

$$V(r) = \frac{A}{r^6} + B \exp(-C r) \tag{10.1}$$

where A, B and C are fitted parameters.

The vibrational density of states (§2.6.2) may be considered to be the one-dimensional projection of the dispersion curves onto the energy axis (formally, it is proportional to $(dk/d\omega)$). Thus sharp maxima (van Hove singularities (§2.6.2.2)) in the vibrational density of states will occur at energies corresponding to flat portions (critical points) of the dispersion curves. As shown in §2.5.1, the INS intensity, S, is a weighted measure of the vibrational density of states.

The dispersion curves can be used to qualitatively predict the overall form of the INS spectrum. Fig. 10.3 shows a comparison of the dispersion curves for the optic branches and the INS spectrum of polyethylene at 30K for the 600—2000 cm^{-1} region. The agreement is excellent; maxima occur at the energies corresponding to phase differences of 0 and π (which are the infrared and Raman active wavevectors). The strong band at ~ 1500 cm^{-1} is assigned to the overlap of the v_2 CH$_2$ scissors mode and the first overtone of the 720 cm^{-1} methylene rocking band. The broad feature at 900 cm^{-1} is mainly due to the phonon wing (§2.6.3) of the 720 cm^{-1} feature; there may also be a contribution from the minimum in the v_4 dispersion curve. The predicted splitting in the 720 cm^{-1} methylene rocking band due to in-phase and out-of-phase coupling of motions on adjacent chains in the unit cell is not observed because of the instrumental resolution.

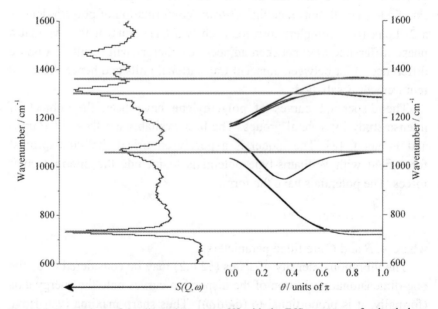

Fig. 10.3 Comparison of the dispersion curves [8] with the INS spectrum of polyethylene for the region 600–1600 cm^{-1}. The horizontal lines show the van Hove singularities indicating where peaks in the INS spectrum are expected.

In the region below 600 cm⁻¹, the two acoustic branches v_5 and v_9 occur. The factor group splitting of polyethylene results in each branch giving two sub-branches. Each of the four sub-branches terminates at $k = 0$, and π, thus there are 8 modes to be considered. Three of these are the pure translations, hence have zero energy (at $k = 0$ and π) and are unobservable, one is inactive in both the infrared and Raman spectrum, two are Raman active and two are infrared active. However, the entire branches are directly observable with INS spectroscopy. Fig. 10.4 shows a comparison of the dispersion curves and the INS spectrum in the 0 - 600 cm⁻¹ region. The maximum frequency of the v_5 dispersion curve, $(\omega_{v5})_{max}$ occurs at 550 cm⁻¹ as compared to the experimental value of 525 cm⁻¹. (The curves are calculated from data obtained at 90K, however, the experimental band positions are invariant with temperature in the range 5—100K). Additional features are observed at 200 cm⁻¹ $(\omega_{v9})_{max}$, 130 cm⁻¹ (v_{5a} at $k = 0$), 97 cm⁻¹ (v_{9b} at $k = 0$) and 53 cm⁻¹ (v_{5b} at $k = 0$). The last of these is significant since it is both infrared and Raman inactive.

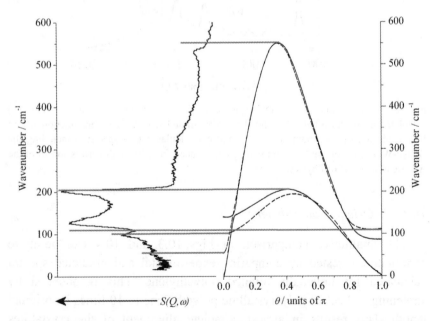

Fig. 10.4 Comparison of the dispersion curves [8] with the INS spectrum of polyethylene for the region 0–600 cm⁻¹. The grey horizontal lines show the van Hove singularities thus indicate where peaks in the INS spectrum are expected.

It is worth noting that extended systems can also be modelled by density functional theory (DFT) and a calculation [9] for the internal modes of an isolated chain gave good agreement with the experimental data as shown in Fig. 10.5. A calculation of the complete unit cell is tractable and will represent the next advance in the understanding of the vibrational spectra of polyethylene.

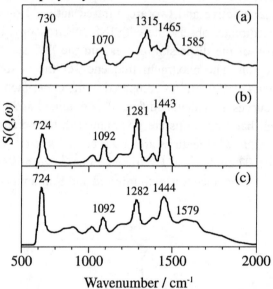

Fig. 10.5 Comparison of the INS spectrum of polyethylene (a) experimental, (b) calculated at the SVWN/6-31G* level and convoluted with a Gaussian function whose full width at half-maximum is 30 cm^{-1} and (c) calculated INS spectrum including the Debye-Waller factors and phonon wings. Reproduced from [9] with permission from the American Institute of Physics.

10.1.1.2 Oriented polyethylene

The qualitative comparison of Figs. 10.3 and 10.4 can be more quantitatively tested by comparing experimental and predicted spectra calculated for uniaxially aligned polyethylene. This is obtained by stretching a sheet of polycrystalline polyethylene to 10 times its original length. This results in almost complete alignment of the crystallites c-axis with the draw direction, with the a and b axes remaining randomly oriented perpendicular to the c-axis.

In Fig. 10.6 we show the experimental spectra with Q aligned perpendicular to the c-axis (see Fig. 10.6 dashed line, transverse spectrum) and parallel to the c-axis (see Fig. 10.6, solid line, longitudinal spectrum). The inset shows the calculated low energy region [10]. It can be seen that the trends are well reproduced, but the detail is incorrect. In particular, the intensity of the feature at 130 cm^{-1} is seriously underestimated for the longitudinal orientation in the calculations. Although Q is not strictly perpendicular to the c-axis in the lower energy regions (it is at *ca* 75° in the range 50 - 100 cm^{-1}) it is still very orientation sensitive. This is demonstrated by the clear polarisation effect of the feature at 97 cm^{-1}. The largest difference between the calculated and experimental spectra is the relative intensities of the transverse and longitudinal spectra, the calculations indicate that the longitudinal spectrum should have very low intensity whereas the experimental spectrum shows a significant intensity in this region. The sample was highly oriented (>90%) so incomplete orientation in the sample may be excluded.

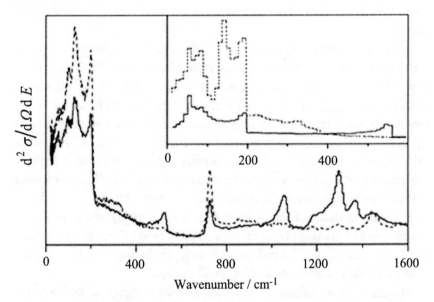

Fig. 10.6 Experimental INS spectra of polyethylene with Q aligned perpendicular (dashed line, transverse spectrum) and parallel (solid line, longitudinal spectrum) to the c-axis respectively. The inset shows a calculation for the low energy region [10].

To understand the reasons for the differences between the experimental spectra and that predicted by the models, it is necessary to consider the data from which the models are built. The only way to directly measure the dispersion curves of a material is by coherent INS spectroscopy. For hydrogenous polyethylene, this method fails because the background caused by the incoherent scattering from hydrogen completely swamps the coherent signal (§2.1.1). For perdeuterated polyethylene, the larger coherent and smaller incoherent cross sections of deuterium (see Appendix 1) has allowed the ν_5 acoustic branch to be mapped by coherent INS [11].

10.1.2 The n-alkanes

Information on the dispersion curves of polyethylene has been obtained indirectly by studying the normal, unbranched, alkanes. The n-alkanes form a homologous series, are oligomers of polyethylene and have the same masses (-CH_2- repeated n times). As more CH_2 units are added they are constrained by the same forces in chains that simply grow longer.

Let us consider the dynamics of a single, isolated, idealised chain, n (CH_2) units long. Simplistically this would be linear with every (CH_2) unit a distance a apart, see Fig. 4.7, but in the case of the alkanes is actually in a zigzag *all-trans* configuration. Since the external forces are absent only the internal modes need be considered. Because of symmetry different species of vibration produce band sequences (e.g. C–H stretch, CH_2 wag, CH_2 twist etc…) that can occupy the same frequency regions. The number of oscillators gives the maximum number, N_{mode}, of modes of each type of oscillation that a chain can support, and $\nu = 1, 2,... N_{mode}$. Thus for an n-alkane, C_nH_{2n+2} there are: $2(n-2)$ methylene C-H stretch modes, $(n-2)$ CH_2 wag, twist or rock modes, $(n-1)$ C–C stretch modes, $(n-2)$ in-plane C–C–C bend modes and, excluding the two methyl torsions, $(n-3)$ out-of-plane C–C–C bend modes.

Standing wave solutions of different wavelength, λ, embody the dynamics of each type of motion and their displacements away from equilibrium positions (§4.3.2). Each of these solutions is identified by the number of half-waves, ν, that can be accommodated in the length of the chain, the wavevectors, k_v. (We may suppress vector aspects of k since it can only propagate in one direction.) Each wavevector in this ideal chain has a particular frequency, $(\omega_v)_{ideal}$. The wavevectors fix the phase relationship, for a given type of vibration, between neighbouring $-CH_2-$ units in the chain (§2.6.1.1, §4.3.2) where:

$$k = \frac{2\pi}{\lambda} = \frac{2\pi}{2(\lambda/2)} = \frac{\pi}{(\lambda/2)} \tag{10.2}$$

But the sum of the lengths of the half-waves is the same as the length of the alkane and, for the C–C stretch,

$$v(\lambda/2) = (n-1)a \quad \text{where} \quad v = 1, 2, 3, \cdots, N_{mode} \tag{10.3}$$

We substitute Eq. (10.3) into Eq. (10.2):

$$k_v = \left(\frac{v}{n-1}\right)\frac{\pi}{a} \quad v = 1, 2, 3, \cdots, N_{mode} \tag{10.4}$$

Since $N_{mode} \leq n-1$; so $0 \leq k_v \leq \pi/a$ and Eq. (10.4) is seen to be equivalent to Eq. (4.46) but expressed in discrete, as opposed to continuous, terms (§2.6.1.2). In the chemical literature the wavelengths are expressed in terms of a (thus, this factor disappears from our formulation and the k_v are given in radians) and $(n-1)$ is replaced by the more general term $(N_{mode}+1)$. The k_v values are occasionally given specific names and the LAM-numbers refer to the k_v values of the C–C–C in-plane deformations (collectively termed the 'v_5' vibrations) with LAM-1 being the excitation with the longest wavelength. The highest v-number, N_{mode} always has the shortest wavelength. Thus, if pentane were an ideal alkane, then with $n = 5$, the three CH_2 wagging modes will occur at $k_1 = \pi/4$, $k_2 = 2\pi/4$ and $k_3 = 3\pi/4$. These would constitute observations at three points on the dispersion curve of the CH_2 wag. Measurement of the spectrum of an ideal dodecane, $n = 12$, would measure at ten different values of k. With sufficient alkanes, it is possible to map the entire dispersion curve. The process can then be repeated for the other modes. This method was used [12] in a *tour de force* of optical spectroscopy to map the dispersion curves of all the modes of polyethylene in the region 600—1600 cm^{-1}. The C–H stretching modes show little dispersion so the method is not useful in this region.

The same method had also been applied to the low energy C–C–C in-plane bending, v_5 and the out-of-plane bending modes, v_9, but with less success. Experimentally, it is more difficult to access this region by infrared and, to a lesser extent, Raman spectroscopy. A much larger problem is that not all of the modes are allowed and even those that are allowed do not necessarily have any significant intensity. This is illustrated in Fig. 10.7, which compares the Raman and INS spectrum of tetratetracontane, $C_{44}H_{90}$. The rapid decrease in intensity with decreasing LAM wavelength and the systematic absence of modes is obvious. In contrast, the INS spectrum shows *all* the internal and lattice modes, the problem becomes one of assignment.

10.1.2.1 The longitudinal acoustic modes

The LAM modes involve deformation of the in-plane C–C–C angles of the alkane. There are only $n-2$ such angles and so, in Eq. (10.2) $N_{mode} = n-2$. To assign the LAMs a twofold approach was adopted. The n-alkanes have either C_{2v} (n = odd) or C_{2h} (n = even) symmetry. The LAMs belong to the A_1 or B_1 representations (n = odd) or A_g or B_u representations (n = even). For the first mode, LAM-1 ($v = 1$) with $k_1 = \pi/(n-1)$, the eigenvectors exhibit a node at the chain centre with '+' and '–' anti-mode, one at each end of the chain, LAM-1 must be totally symmetric. LAM-2 has two nodes and so must be antisymmetric, LAM-3 must be totally symmetric again since it has three nodes and so on. Thus the modes must follow the sequence:
$A_1\, B_1\, A_1\, B_1\, ...A_1$ (n = odd)
or
$A_g\, B_u\, A_g\, B_u\, ...B_u$ (n = even)
Density functional theory was used to calculate the frequencies and symmetry assignments. As a check, the available Raman data were used, since for the n = even alkanes only the odd LAM modes are allowed.

The harmonic dynamical treatment of a model of the n-alkanes as a zigzag chain of point mass beads with unconstrained ends and an infinitely strong stretching force constant (so that only CCC bending occurs) has been presented [13]. The analytic result for the infinite chain for which the frequencies of the LAMs are:

$$\omega_v = \Gamma_{zig} \left(1 + \cos k_v \pi\right) \left(\frac{1 - \cos k_v \pi}{\left(1 + \cos\theta \cos k_v \pi\right)}\right)^{\frac{1}{2}} \left(\omega_{LAM}\right)_{max}$$

where

$$\left(\omega_{LAM}\right)_{max} = \frac{1}{\pi}\sqrt{\frac{f_a}{m_{CH_2}}} \qquad (10.5)$$

Here, θ is the zigzag (CCC) angle, f_a is the angle bending force constant and m_{CH_2} is the mass of the CH_2 unit. The normalisation constant, Γ_{zig}, is 0.857 for a tetrahedral θ-angle at the carbon, so that the maximum frequency of this type of vibration is $(\omega_{LAM})_{max}$. Notice that Eq. (10.5) was not derived under the cyclic boundary conditions used to obtain Eq. (4.46), which restricts solutions to the dynamical equations to be whole wavelengths, as with benzene (§2.6.1.1), and only even numbers of nodes are available, the A modes would thus be excluded. This is inadequate since modes of both A and B character are present in the alkanes.

For the infinite chain $v \rightarrow \infty$ and k_v becomes continuous, $0 \leq k \leq 1$, radians. This model has a single parameter, the band-head frequency, $(\omega_{LAM})_{max}$ and it may be applied, as an approximation, to the finite chain case where k_v takes on the values of Eq. (10.4). It is convenient to index the LAM modes for each alkane using the mode number v and to specify the corresponding, ideal, value of k_v. The frequencies of the LAMs for finite length alkanes are expected to be close to the value calculated from Eq. (10.5). Application of the infinite chain limit to the finite chain case assumes that the alkanes are a uniform linear material with no specific effects associated with the chain ends.

By combining the information drawn from *ab initio* (mode symmetry and frequency) and the infinite chain model (k_v and frequency), the LAM modes above 200 cm^{-1} can be unambiguously assigned [14]. Table 10.1 lists the assignments from pentane ($n = 5$) to pentacosane ($n = 25$). Fig. 10.8 shows the observed and calculated LAM modes, the dashed line is a polynomial fit and is the best available approximation to the v_5 dispersion curve of polyethylene. The INS data for $\kappa > 0.7$ is absent because it is not possible to assign the modes owing to the spectral congestion with the TAMs and lattice modes. This does not occur for the DFT data, since the LAMs and TAMs have different symmetries.

Before proceeding, it is worthwhile to check that the assumption that the n-alkanes are good models for polyethylene is valid. Fig. 10.9 shows a comparison of the INS spectrum of n-alkanes of increasing chain length with that of polyethylene. The rapid convergence of the n-alkane spectra to a very close resemblance of that of polyethylene is apparent.

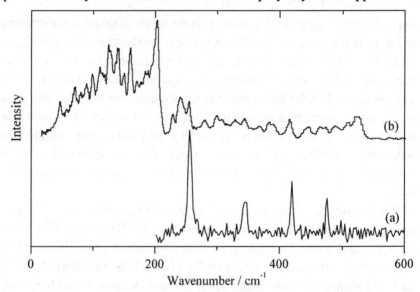

Fig. 10.7 Raman (a) and INS (b) spectra of tetratetracontane, $C_{44}H_{90}$. In the Raman spectrum the rapid decrease in intensity with increasing LAM number and the systematic absence of modes due to symmetry restrictions is apparent.

More quantitatively, the same method used for the assignment of the hydrogenous n-alkanes can be used for the assignment of the experimental frequencies and the frequencies calculated by DFT of $C_{16}D_{34}$ and compared [16] to the experimental data for the ν_5 mode of perdeuterated polyethylene [11]. The solid line is a quintic least squares fit to the DFT results for perdeuterated-n-hexadecane as a guide to the eye. It can be seen that there is excellent agreement between all three sets of data. This provides experimental validation of the assumption that the forcefields derived from the n-alkanes are transferable to polyethylene. Together, Figs. 10.9 and 10.10 provide compelling evidence for the use of the *n*-alkanes as models for polyethylene.

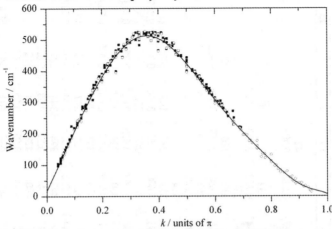

Fig. 10.8 The observed LAM modes (■) and those calculated by DFT (○). The dashed line is a polynomial fit and is the best estimate of the ν_5 dispersion curve of polyethylene.

Table 10.1 Observed and calculated LAM modes (cm^{-1}) of the *n*-alkanes from pentane ($n = 5$) to docosane ($n = 22$) [14]. Frequencies in parentheses are Raman data from [15].

$n\downarrow$		$\nu\rightarrow 1$	2	3	4	5	6	7	8	9
5	Exp	400	400	190						
	DFT	395	397	177						
	k	0.25	0.50	0.75						
6	Exp	370	466	305						
	DFT	366	464	295	132					
	k	0.20	0.40	0.60	0.80					
7	Exp	308	483	420	250					
	DFT	301	481	416	244	98				
	k	0.17	0.33	0.50	0.67	0.83				
8	Exp	281	471	471	342					
	DFT	274	470	473	338	196	79			
	k	0.14	0.29	0.43	0.57	0.71	0.86			
9	Exp	250	457	497	401	294				
	k	0.12	0.25	0.38	0.50	0.62				
10	Exp	229	401	499	475	364	229			
	DFT	222	399	502	474	359	240	134	55	
	k	0.11	0.22	0.33	0.44	0.56	0.67	0.78	0.89	
11	Exp	219	382	495	414	310				
	k	0.10	0.20	0.30	0.40	0.50				

Table continued on next page.

Table 10.1 cont.

n↓		ν→1	2	3	4	5	6	7	8	9	10	11	12	13
12	Exp.	(195)	345	445	510	488	384	273	178	98	42			
	DFT	187	342	444	511	490	382	273	178	98	42			
	k	0.091	0.182	0.273	0.364	0.455	0.545	0.636	0.727	0.818	0.909			
13	Exp.	(183)	339	449	519	506	428	336	271	152	82	30		
	DFT	175	329	445	514	500	420	329	238					
	k	0.083	0.167	0.250	0.333	0.417	0.500	0.583	0.664	0.750	0.833	0.917		
14	Exp.	(169)	320	434	509	509	454	380	294	220	136	75	32	
	DFT	161	316	432	509	512	453	375	288	212				
	k	0.077	0.154	0.231	0.308	0.385	0.462	0.539	0.615	0.692	0.769	0.846	0.923	
15	Exp.	(158)	293	399	473	518	504	434	347	268				
	k	0.071	0.143	0.214	0.286	0.357	0.429	0.500	0.571	0.643				
16	Exp.	(150)	280	390	470	513	513	455	390	314	237	169	108	60
	DFT	141	276	387	470	512	519	455	385	311				
	k	0.067	0.133	0.200	0.267	0.333	0.400	0.467	0.533	0.600	0.667	0.733	0.8	0.867
17	Exp.	(141)	260*	378	462	515	515	482	417	349	288	221		
	k	0.063	0.125	0.188	0.250	0.313	0.375	0.438	0.500	0.563	0.625	0.688		0.813
18	Exp.	(133)	245*	356	436	490	521	521	465	400	329	255*		
	k	0.059	0.118	0.176	0.235	0.294	0.353	0.412	0.471	0.529	0.588	0.647		0.765
19	Exp.	(126)	237*	341	425	481	517	517	480	425	369	299	229	177
	DFT	120	242	339	424	486	524	520	478	423	361	297		
	k	0.056	0.111	0.167	0.222	0.278	0.333	0.389	0.444	0.500	0.556	0.611	0.667	0.722
20	Exp.	(122)	222	324	420	484	523	523	494	449	394	344	281	
	k	0.05	0.11	0.16	0.21	0.26	0.32	0.37	0.42	0.47	0.53	0.58	0.63	
21	Exp.	(115)	219	312	392	454	497	520	520	480	428	374	312	
	k	0.05	0.10	0.15	0.20	0.25	0.30	0.35	0.40	0.45	0.50	0.55	0.60	
22	Exp.	(113)		289	383	448	492	522	522	492	448	401	344	304
	k	0.05	0.09	0.14	0.19	0.24	0.29	0.33	0.38	0.43	0.48	0.52	0.57	0.62

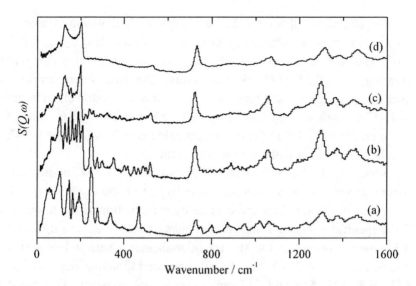

Fig. 10.9 Comparison of the INS spectra of *n*-alkanes, C_nH_{2n+2}, (a) octane ($n = 8$), (b) tetracosane ($n = 24$), (c) hexacontane ($n = 60$) with that of (d) polyethylene.

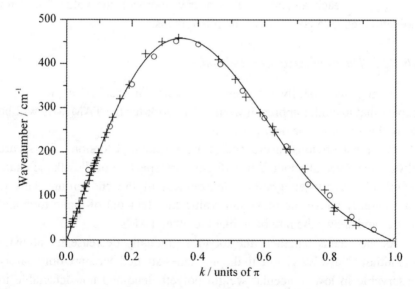

Fig. 10.10 Comparison of experimental data for the ν_5 mode of deuterated polyethylene (+) and the experimental LAM frequencies of $C_{16}D_{34}$ (○). The solid line is a quintic least squares fit to the DFT data for perdeutero-*n*-hexadecane [16].

Fig. 10.8 shows that $(\omega_{LAM})_{max}$ is always reached, within experimental error, at the same k-value irrespective of the alkane, $k(\omega_{LAM})_{max} = 0.37 \pm 0.01$. This can be compared with the value obtained for polyethylene, $k(\omega_{LAM})_{max} = 0.34$ [11]. Within errors, the two determinations of $k(\omega_{LAM})_{max}$ are the same. From Eq. (10.5), for a tetrahedral chain with $\theta = 101.46°$ ($\cos(\theta) = -1/3$) $k(\omega_{LAM})_{max} = 0.355$. For $\theta = 113.6°$ (the CCC angle predicted by DFT) the maximum value is at $k(\omega_{LAM})_{max} = 0.347$, in excellent agreement with the observed value.

According to Eq. (10.5) the frequency of all LAM modes with a particular value of k is fixed and independent of the chain length. That this is not the case is best appreciated by examination of the frequencies for a specific value of k that occurs many times. One example is $k = 0.333$ which occurs in the spectra of n-alkanes where $n-1$ is a simple multiple of 3 ($n = 7, 10, 13, 16, 19$). The corresponding frequencies are 483, 499, 501, 513 and 517 cm^{-1}, clearly not constant. This trend is reproduced by the *ab initio* calculations, where the corresponding DFT values are 481, 502, 500, 519 and 524 cm^{-1}. The trend is for the $(\omega_{LAM})_{max}$ of each alkane to more nearly approach the value found for polyethylene, 525 cm^{-1}, as the chain lengthens.

10.1.2.2 The transverse acoustic modes

Having successfully assigned the spectra of the LAMs, we might expect that a similar approach would also work for the TAMs. As will be seen, this proves to be only partly true.

The most obvious discrepancies are the methyl torsions. These are always observed at about 250 ± 10 cm^{-1}, irrespective of the alkane chain length or the crystallographic environment of the end group. This is considerably above the maximum value for other out-of-plane torsional modes and shows them to be unlike the other TAMs.

As the chains lengthen, the torsions gradually decrease in intensity and although always visible in the n-alkanes studied here, are only barely observable in low molecular weight polyethylene and not detectable in the most common grades. In the shorter alkanes their overtone intensities ($\sim$490 cm^{-1}) are weak but can be extracted. The overtone to fundamental intensity ratio ($S_{(2\leftarrow 0)}/S_{(1\leftarrow 0)}$) for hexane was measured, 0.14 ± 0.04, which implies (§5.2.1.2) that the oscillator mass could be as high as 14 amu. The torsion of a simple methyl group about its C–C bond would have a much smaller oscillator mass, about 3 amu. Therefore, the two methyl torsions engage CH_2-units towards the chain's ends and are certainly not restricted to librations of the terminal methyl group alone. However, their effective masses are only about half that of the other TAMs, which implies that they involve the displacement of a significant fraction of the chain but not the whole chain. Each vibration is localised towards one end of the molecule and cannot be described in terms of a simple phase relationship between all the CH_2-units. They are, thus, independent of k and play a role similar to light mass substitutional defects in the density-of-states of simple crystals.

The two methyl torsions of the shorter alkane chains couple to produce an in-phase and out-of-phase pair of bands. The longer the chain the weaker this coupling and so the less the frequency splitting of the pair. In the limit, the splitting should completely collapse. The methyl torsion of one end of a chain would then be unaware of the nature, or even the presence, of the other end of the chain and both methyl torsions oscillate independently. Only one accidentally degenerate band is observed. In the alkanes this collapse has already occurred in undecane. Naively then, all methyl torsions, $n > 11$, will occur at the same frequency. Unfortunately this is not observed.

In the DFT calculations the degenerate methyl torsions occur at either 251 cm^{-1} or 247 cm^{-1}, dependent upon whether the n-alkane has n-even ($n \geq 14$), or n-odd ($n \geq 15$). This difference also manifests itself in the observed INS results ($12 > n > 20$): for n-even these have methyl-torsion frequencies at 255 cm^{-1} and, for n-odd, at 250 cm^{-1}. Further, the average frequency values for the shorter alkanes, with coupled methyl torsions, fall close to the appropriate n-even, or n-odd, frequency. There is an obvious difficulty with the suggestion that many ($n > 11$) alkanes are so long that their end of chain vibrations are independent but, nonetheless, that the nature (n-odd, or -even) of a chain's ends determines these dynamics. This apparent inconsistency is resolved if we recognise that, for the purpose of their physical properties, the n-odd and n-even alkanes are quite different and should not be naively compared. Thus the methyl torsions and the TAM dispersion curves for the two types of alkane are different, subtly different but yet different. This difference, as revealed by the *n*-odd, *n*-even, frequency alternation of the methyl torsions, is so clear-cut because these vibrations, alone of all the LAMs and TAMs, have no dependence on *k*. The alternation is present in the other dynamics but is hidden by our inability to assign real, as opposed to ideal, values of *k* to these vibrations. Since the *ab initio* calculations refer to isolated molecules these are our best estimates of the gas-phase methyl torsion frequencies, where only internal forces are acting. Condensation into the solid state has a modest impact on these frequencies, as can be seen from the INS results for the methyl torsion. This amount to some $\pm$ 10 cm^{-1} and is reflected in the fact that *n*-odd alkanes (n > 9) have orthorhombic crystal structures and the *n*-even alkanes (n > 6) are triclinic, see below. Obviously the major contribution to the vibrational potential remains the intramolecular force field.

The presence of several molecules bound together in a molecular crystal generates an external force field and new sequences of vibrations are created, the external modes. These run down from about 80 cm^{-1}, see below, and again the lowest attainable frequency is a function of the number of oscillators. The number of molecules in a crystal is generally very large and the dispersion curve of the external modes becomes truly continuous, the very lowest frequencies are available and the density-of-states is not composed of sharp individual transitions. However, the spectral features associated with the TAMs remain sharp since the original wavevectors retain their significance.

There are three translational dispersion curves and three rotational dispersion curves. In the absence of rotation-translation coupling these vibrations are also distinctly separate. However, external modes and internal modes of the same character will become mixed, the closer they are in frequency the greater will be the mixing. Atomic displacements are no longer controlled by exclusively internal or external forces but by both. Modes of a mostly internal (or mostly external) character now involve displacements of the centre of mass (or deformations of the molecular frame). This is a general feature of all molecular crystals but it is usual to ignore such complications when the frequency of the lowest internal modes is significantly above the highest frequency of the external modes, ω_{max}. Mode mixing is thereby reduced but not eliminated. In the *n*-alkanes the two frequency regions overlay each other and this decoupling mechanism is not available, the lower TAM modes are totally mixed with the external modes. A pair of mixed-mode vibrations is produced from the two originally unmixed modes. The lowest frequency vibration of the pair retains most external character and remains almost unchanged in frequency. The other mode, of mostly TAM character, has a frequency too high to be transmitted by the lattice. Such 'internal' vibrations are localised over the molecule and are excluded from the external density-of-states, as a consequence of the non-crossing rule. The TAM dispersion relationship changes and the TAM frequencies take values, $\omega_{TAM} > \omega_{max}$.

Mixing mostly modifies those TAMs with the longest and the shortest λ since, as isolated molecules, these had the lowest vibrational frequencies. The high frequency TAM vibrations (those at intermediate *k*) are less altered by the mixing. Although the impact on the TAMs is most obvious the same effects are found for the LAMs. Here, however, the external vibrations are at the very lowest frequencies and only LAM-1 of the longer alkanes will show any significant changes.

In the region around $k = 0.5$, especially in the longer chains, several vibrations with frequencies close to $(\omega_{TAM})_{max}$ occur and the spectrum becomes congested into a band-head reminiscent of polyethylene.

In Table 10.2 we list the observed and calculated frequencies for the TAMs of the *n*-alkanes ($n = 5$ to 12). They are gathered onto the common plot shown in Fig. 10.11. It is seen that the shorter alkanes have their upper frequency band-heads $(\omega_{TAM})_{max}$ below or about 180 cm^{-1} but the longer alkanes have them at frequencies about 205 cm^{-1}. A steady increase is seen in the value of the $(\omega_{TAM})_{max}$ as the *n*-alkane chain length increases, from pentane to hexane and heptane. This frequency variation is very similar to that observed for the $(\omega_{LAM})_{max}$ in the INS spectra of the same samples. It is a consequence of the finite length of the alkanes and the failure of the idealised wavevectors as descriptions of the atomic displacements in short alkanes. The longer the alkane the more nearly it is approximated by the idealised wavevector and the less its dynamics differs from close members of the homologous series.

The value $k(\omega_{TAM})_{max}$, where an alkane achieves $(\omega_{TAM})_{max}$, is seen from Fig. 10.11 to remain constant across the series. Again this is very reminiscent of the behaviour of the LAMs. This idealised wavevector value, from Fig. 10.11 is estimated to be $k(\omega_{TAM})_{max} = 0.46 \pm 0.02$. The difference between the observed value of 0.46 and the model value of 0.5 is a result of the dynamical coupling of the torsional modes with the much higher energy CH_2 rocking modes.

Table 10.2 Band maxima (cm^{-1}), wavevectors (k / rad) of the TAMs of n-alkanes.

$n\downarrow$		$v\rightarrow1$	2	3	4	5	6	7	8
5	INS	152	136						
	DFT	113	117						
	k	0.33	0.66						
6	INS	124	178	94					
		132	185	105					
	Average (a)	128	182	100					
	DFT	103	149	74					
	k	0.25	0.50	0.75					
7	INS	125	178	167	101				
		143	186	167	112				
	Average (a)	134	182	167	107				
	DFT	82	151	147	68				
	k	0.2	0.4	0.6	0.8				
8	INS	141	186	196	167				
		150	196	205	167				
	Average (a)	146	191	201	167				
	DFT	67	137	165	113	49			
	k	0.17	0.33	0.5	0.67	0.83			
9	INS	141	175	203	195	151	134		
	DFT	69	121	166	161	103	45		
	k	0.14	0.29	0.43	0.57	0.71	0.86		
10	INS	139	164	(194)	199	185	148	122	
	DFT	56	109	157	165	130	79	31	
	k	0.13	0.25	0.38	0.5	0.63	0.75	0.88	
11	INS	106	141	167	201	184	152	128	93
	DFT	61	102	142	173	168	126	75	33
	k	0.11	0.22	0.33	0.44	0.56	0.67	0.78	0.89
12	INS	136	143	176	196	204	189	160	136
	DFT	52	92	132	164	172	144	103	58
	k	0.1	0.2	0.3	0.4	0.5	0.6	0.7	0.8

(a)Average of Davydov splitting

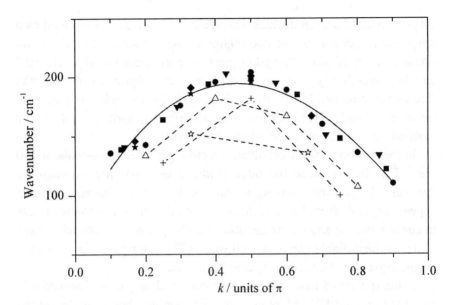

Fig. 10.11 The band positions (ω / cm^{-1}) of all of the TAMs of the *n*-alkanes from pentane to dodecane, shown plotted against their ideal wavevectors, k, see Table 10.2. Except for pentane, hexane and heptane all the results fall on or close to a common curve, the lines are drawn to guide the eye. The effects of Davydov splitting have been averaged out. (Key: pentane open star, hexane cross, heptane open triangle, octane filled diamond, nonane filled triangle, decane filled square, undecane filled star, dodecane filled circle).

The mixed nature of TAMs provides a convincing explanation for the large number of bands observed in the region from 100 to 210 cm^{-1}. However, it fails to explain why there are far too many modes in this spectral region of some samples. This is very obvious in the INS spectrum of hexane. Here only three individual TAM modes are anticipated (ignoring the two methyl torsions) but a total of seven bands is observed. One of these must be the LAM-5 band at about 132 cm^{-1}, leaving twice as many observed bands as that predicted for the known crystal structure of hexane. It is probable that the experimental procedure, of rapid cooling to the low temperatures used in INS spectroscopy, has occasionally produced phases with two molecules to the asymmetric unit. This is supported by a recent investigation of the low temperature (< 50K) structure of octane that found a transition to a unit cell with double the number of molecules [17] and suggested that this also occurs in other alkanes. Dispersion is no longer represented by a single curve but consists of two related curves, corresponding to the in-phase and out-of-phase Davydov pairs. The assignments of Table 10.2 can be immediately seen to be a significant improvement over the frequencies obtained for the isolated molecule by *ab initio* methods, also given in the table. This poor result for *ab initio* methods stems from the isolated molecule approach used in these calculations.

In the spectra of the short alkanes a clear gap opens below about 100 cm⁻¹. This lies between the internal-like TAMs and the external-like spectrum. In pentane, hexane, heptane, octane and dodecane this gap appears beyond 80 cm⁻¹. Generally, however, the gap is not so distinct, as in nonane, decane and the longer alkanes, where it is quite absent. (There is also no identifiable gap in polyethylene.) The external vibrations of the alkane crystalline lattice are expected below the gap.

In the spectra of those alkanes for which a clear gap is observed it is possible to regard 80 cm⁻¹ as ω_{max} and the spectrum below as that of the lattice vibrations. A complicating feature of the gap region is the presence of lattice vibration overtones, or multiphonon contribution. However, since the lattice fundamentals are not a sequence of individual transitions their overtone intensity appears as a broad contribution. It builds up rapidly from the very lowest frequency to its strongest underneath the TAMs and provides a decreasing background for the LAM region, it has died away completely by about 600 cm⁻¹. Beyond dodecane the lack of a clear gap makes the upper frequency bound of the external vibrations unidentifiable and we simply assume that the earlier cut-off ($\omega_{max} = 80$ cm⁻¹) remains representative.

The shape of the external spectra of chains longer than dodecane clearly fall into two classes dependent upon the alkane number being even or odd. Even numbered alkanes retain a simple shape (at least to within the limits that the absence of a gap imposes upon our interpretation). This rises smoothly from the lowest frequencies, flattens at about 60 cm⁻¹, to fall abruptly at 80 cm⁻¹, see Fig. 10.12. The odd numbered alkanes have a quite different shape to their lattice mode spectra. It appears to end at a lower ω_{max} frequency and is now composed of two equally strong features. The first, a band at about 48 cm⁻¹, and the other just before the cut-off, at about 70 cm^{-1}, see Fig. 10.12. This separation into two types of lattice spectra clearly correlates with the crystal structures that have long been known to be different for the *n*-even and *n*-odd alkanes (reviewed in [18,19]). The *n*-odd > 9 have orthorhombic structures (Z = 4) and the *n*-even > 6 have triclinic structures (Z = 1). This would suggest that the lattice spectrum of pentane (orthorhombic, Z = 4) should resemble other *n*-odd lattice spectra, which it does not, and that the spectra of heptane and nonane (triclinic, Z = 2) should actually resemble spectra of the *n*-even alkanes, which they do. (The exceptional dynamics of pentane possibly reflects the quite unusual nature of its orthorhombic structure.)

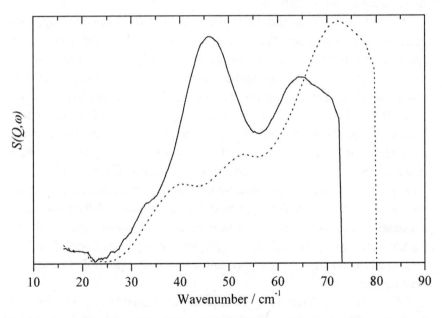

Fig. 10.12 A comparison of the lattice mode region of the *n*-even (continuous line) and the *n*-odd (dashed line) alkanes.

The odd to even alternation of physical properties of the *n*-alkanes has been noted previously and was attributed to the *n*-even members having an optimal packing arrangement of molecules. The *n*-odd structures are less dense, their crystals melt at relatively low temperatures and they have lower lattice energies, which all agrees with the INS observations.

10.1.2.3 What have we learned?

The extensive body of work on polyethylene and the *n*-alkanes illustrates many of the advantages of INS spectroscopy. The absence of selection rules enables *all* of the modes to be observed, this allows detection of the optically forbidden v_{5b} mode 53 cm^{-1} (v_{5b} at $k = 0$). The LAM and TAM modes in the *n*-alkanes are also observed, those most easily seen by INS are unobservable (because of low intensity) by optical methods. Thus the complementarity of infrared, Raman and INS spectroscopies is highlighted.

The information gained from the three forms of spectroscopy has been used to construct the dispersion curves of polyethylene. INS spectroscopy is unique in that it provides a stringent test of the models because it gives information across the entire Brillouin zone rather than just at the zone centre as do infrared and Raman spectroscopy. As Figs. 10.3, 10.4 and 10.6 show, the agreement is generally good but in the low energy region where the effects of the intermolecular interactions are most apparent, there is still a need for improvement. Clearly, the next step forward is to model polyethylene with periodic-DFT methods.

Good models are needed because information on important properties such as heat capacity and elastic modulus can be derived from the force constants. The elastic modulus data is particularly useful since it allows the ultimate tensile strength of polyethylene to be determined. Based on present estimates of this, it is apparent that it is still possible to improve existing materials.

The *n*-alkanes have long been used as valuable models for the dynamics of polyethylene. The INS work confirms their utility but also highlights their limitations. Most work has neglected the effect of the finite length and the chain ends but both are present in the real compounds and modify the dynamics.

10.1.3 Polypropylene

The vibrational spectroscopy of polypropylene, and all the other polyolefins, has been much less-studied than polyethylene—they are chemically and structurally more complex; a homologous series of model compounds, $H–(CH_2–CHCH_3)_n–H$, analogous to the n-alkanes is not available. Thus information on the dispersion curves is very sketchy. Polypropylene exists in three isomeric forms, depending on the arrangement of the methyl groups along the backbone as shown in Fig. 10.13. Atactic polypropylene is an amorphous polymer, isotactic and syndiotactic polypropylene are semi-crystalline with different structures. The vibrational spectra of the three isomers are different as can be seen from Fig. 10.14. Isotactic polypropylene is the most important commercially and so, unsurprisingly, is the most studied [21,22].

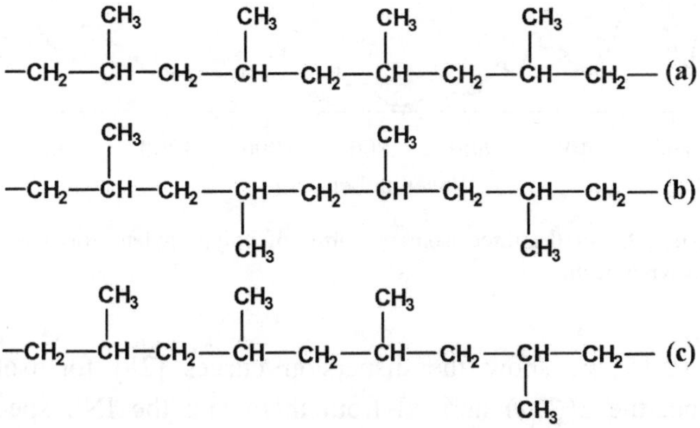

Fig. 10.13 The three isomeric forms of polypropylene: (a) isotactic (all the methyl groups are on the same side), (b) syndiotactic (the methyl groups alternate) and (c) atactic (random arrangement of methyl groups).

There have been several Wilson GF calculations of the isolated chain for both syndiotactic [20] and isotactic polypropylene [21,22]. There is also a lattice dynamics calculation for the full unit cell of isotactic polypropylene [23]. All of these are at $k = 0$ where the infrared and Raman active modes occur and as a result, these modes are well assigned [4]. There are two calculations of the dispersion curves [24,25] for a single helical chain of isotactic polypropylene and these were used to compare with the observed INS spectra [25,26]. Unfortunately, the quality of the calculations far outweighed that of these early INS spectra making comparison difficult.

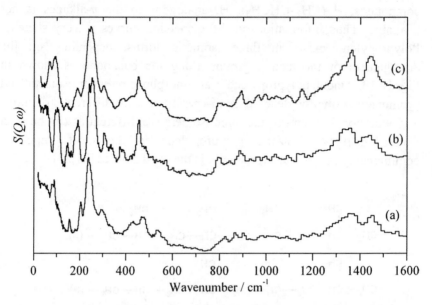

Fig. 10.14 INS spectra of the three isomeric forms of polypropylene: (a) atactic, (b) isotactic and (c) syndiotactic.

In Fig. 10.15 we show the dispersion curves [25] for isotactic polypropylene, the $S(Q,\omega)$ derived from them and the INS spectrum recorded on TOSCA. The dispersion curves were based on an erroneous assignment of 200 cm^{-1} for the methyl torsion so there is a marked discrepancy at that point. For the remainder of the spectrum, there is qualitative agreement but the detail in the INS spectrum is not reproduced. This probably stems from the neglect of the site symmetry and the intermolecular interactions. Clearly this is an area that is ripe for re-investigation with modern INS spectrometers and *ab initio* calculations.

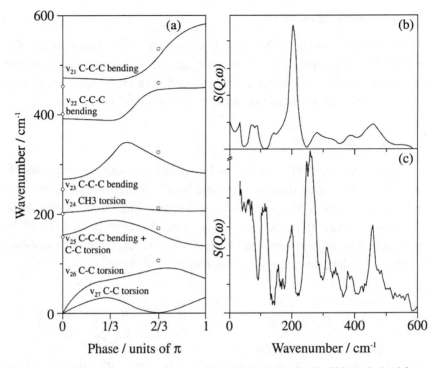

Fig. 10.15 (a) Dispersion curves for isotactic polypropylene, (b) the $S(Q,\omega)$ derived from them and (c) the INS spectrum: (a) and (b) reproduced from [25] with permission from Elsevier.

10.1.4 Nylon-6

Nylon-6, $[-(CH_2)_5-NH-C(=O)-]_n$ belongs to the important class of polyamide condensation polymers. Polyamides are characterised by the presence of secondary amides $-NHCO-$ in the backbone so hydrogen bonds are formed between neighbouring chains. These strongly influence the mechanical properties of the polymer. Nylon-6 has two crystalline forms, α and γ, which differ in the conformation of the backbone: in the α form it is planar and in the γ form it is helical. INS studies [27] of both forms of nylon-6, including oriented samples, have been made. The amide V, VI and VII modes that involve out of plane deformations of the $-NHCO-$ group were shown to depend on the crystal form. The assignments were supported by DFT calculations on model compounds.

10.1.5 Conducting polymers

A class of polymers that is the subject of active research are the conducting polymers. These are conjugated polymers that become electrically conducting when suitably doped with either electron donors such as alkali metals or electron acceptors such as iodine.

10.1.5.1 Polyacetylene

Polyacetylene, $(CH)_n$, is the archetype of conducting polymers. It consists of alternating double and single bonds. As shown in Fig. 10.16 it exists as *cis* or *trans* forms, the latter is the most important. Recent work has moved away from polyacetylene itself, because it is difficult to process and is air sensitive. However, it remains important as a test-bed for understanding the conduction mechanism in this type of polymer.

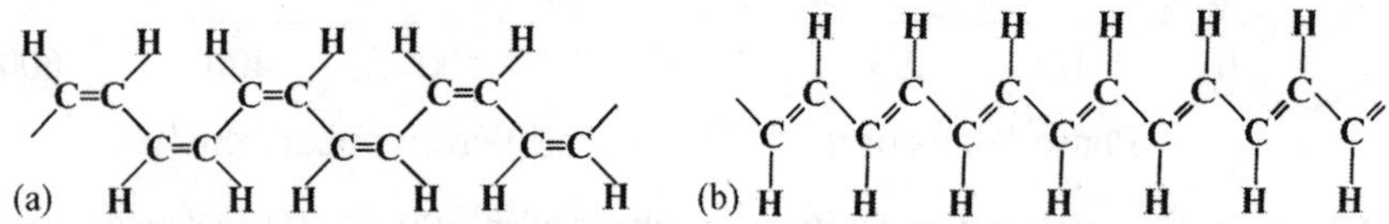

Fig. 10.16 The isomeric forms of polyacetylene: (a) *cis* and (b) *trans*.

Early work on polyacetylene was only able to cover part of the spectral range and apparently suffers from a calibration error [28]. Fig. 10.17a shows the INS spectrum of polyacetylene [29] recorded on TFXA. The spectrum in the region below 700 cm^{-1} is remarkable in that it consists of a series of terraces, each terminating in a bandhead. This is reminiscent of the v_5 mode of polyethylene and suggests that the modes are strongly dispersed. This is confirmed by the dispersion curves [30], and the resulting INS spectrum calculated from them, Fig. 10.17b and c.

The infrared and Raman spectra of polyacetylene undergo very large changes when the polymer is doped. This is because the electronic structure of the polymer has changed dramatically and infrared and Raman spectral intensities depend on the electronic properties. The metallic nature of the resulting materials also makes measuring the spectra difficult. All of these effects are irrelevant to the neutron so it might be expected that INS would be the probe of choice to study these materials. Staggeringly, this is not the case and the only work that has been done using INS is using neutron energy gain at room temperature [31,32]. As is typical of spectra recorded under these spectra difficult. conditions, the resolution is very poor and the energy transfer range is limited with little intensity above 200 cm^{-1} ($\approx k_B T$ at 300K).

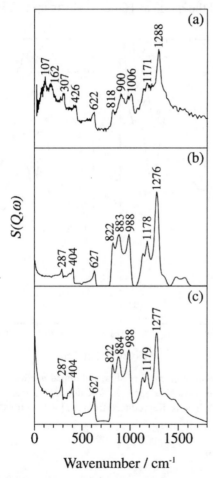

Fig. 10.17 (a) INS spectrum of *trans*-polyacetylene, (b) calculated density-of-states convoluted with a Gaussian lineshape and the instrument resolution function and (c) as (b) including the effects of the Debye-Waller factor and phonon wings. Reproduced from [29] with permission of Elsevier.

In Fig. 10.18 we compare the INS spectra of oriented films of pristine *trans*-polyacetylene and in the sodium and potassium doped states. Even with poor resolution, differences are apparent between the spectra. The most striking is the gap at 80 cm^{-1} that appears in the doped polymers. Molecular dynamics simulations show that the translational mode that peaks at 80 cm^{-1} in the pristine sample shifts down to 64 cm^{-1} in the doped samples and the mode at 160 cm^{-1} shifts up to 230 cm^{-1}.

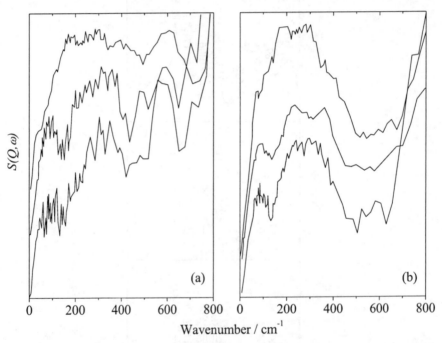

Fig. 10.18 (a) INS spectra of oriented pristine and doped *trans*-polyacetylene (IN6, ILL). Top to bottom: pristine, 14% Na doped, 14% K doped. (a) Transverse, $Q \perp$ chain axis, (b) longitudinal, $Q \parallel$ chain axis. Reproduced from [32] with permission from Elsevier.

10.1.5.2 Polyaniline

Polyaniline is the generic term which describes the materials resulting from oxidative polymerisation of aniline. These materials can be used in capacitors, displays, batteries, electromagnetic shielding and microwave absorbing systems as well as for corrosion protection.

$$2n\text{H}^+\text{X}^-\left[-\left(\text{C}_6\text{H}_4\right)-\text{NH}-\left(\text{C}_6\text{H}_4\right)-\text{NH}-\left(\text{C}_6\text{H}_4\right)-\text{N}=\left(\text{C}_6\text{H}_4\right)=\text{N}-\right]_n$$

$$\updownarrow$$

$$\left[-\left(\text{C}_6\text{H}_4\right)-\text{NH}-\left(\text{C}_6\text{H}_4\right)-\text{NH}-\left(\text{C}_6\text{H}_4\right)-\text{NH}^+=\left(\text{C}_6\text{H}_4\right)=\text{NH}^+-\right]_n 2n\text{X}^-$$

Polyaniline is a mixture of benzenoid and quinoid (*p*-phenyleneamineimine) entities. The neutral compound, emeraldine base is an equal mixture of amine and imine functionalities. Treatment with a non-oxidising acid results in protonation of the imine groups to give emeraldine salt. This causes positive charge to be transferred to the backbone and is the origin of the conductivity, which increases by 10 orders of magnitude from the base to the salt form. The polymer in the base and salt forms can be described as:

Polyaniline has been studied with INS several times [33-35]. Fig. 10.19 shows the INS spectra of the base, Fig. 10.19a, and salt forms, Fig. 10.19b, and their ring-deuterated (C_6D_4) isotopomers [33].

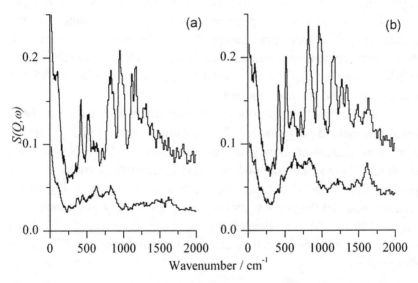

Fig. 10.19 Normalised (to sample mass) INS spectra of polyanilines [33]. Upper: all hydrogenated, lower: ring deuterated. (a) Emeraldine base, (b) emeraldine salt.

It is apparent that the spectra of the base and salt are similar, although there are differences in relative intensities, e.g. bands at 420/525 cm^{-1}, and some are split in one form and not the other e.g. 1175 cm^{-1}. What is striking from the deuterated spectra is that most of the spectral features have very little to do with the N–H functionality. The residual structure in both the ring-deuterated spectra is assigned to strongly hydrogen-bonded water trapped in the polymer.

The INS spectra in Fig 10.19 have been normalised to the sample mass in the beam; thus the intensities are directly comparable. All of the spectra sit on a continuum of intensity, attributed to the recoil of free particles with a mass of ~ 1 amu (§9.4.2). These are the hydrogens that are formally bound to the nitrogen. Unfortunately INS does not provide any information on the charge state of this species so it is not possible to distinguish H^{+}, H$^{•}$ or H^{-}.

10.1.5.3 Other conducting polymers

The commercial interest in conducting polymers means that many systems are under active study. This includes systems that are not conducting themselves but provide a solid phase medium for the conducting species. One such example is polyethyleneoxide that has had lithium trifluoromethanesulfonate dissolved in it. The charge carriers are the lithium ions but their interaction with the polymer is critical. INS can readily probe these interactions [36].

Systems that are intrinsically conducting where INS work has been carried out include polypyrrole [37] and polythiophene [38,39]. For the latter, a computational study of the variation of the INS spectrum calculated of a model compound, an isolated bithiophene molecule, as a function of the angle, ϕ, between the two rings is illustrated in Fig 10.20. It can be seen that the spectral profile changes smoothly as a function of ϕ in the 200 to 550 cm^{-1} spectral region. Thus, even using an isolated molecule approximation ϕ can be estimated to within about 10° from the INS spectrum.

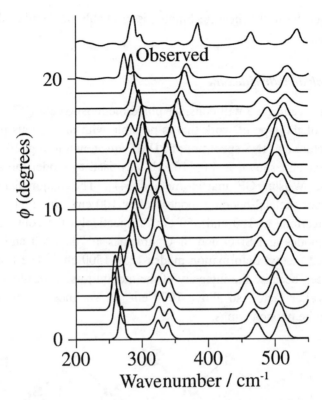

Fig. 10.20 Variation of INS spectrum of bithiophene as a function of the inter-ring angle ϕ, calculated by DFT. The measured INS spectrum is at the top. Reproduced from [39] with permission from Springer.

10.2 Amorphous polymers

While amorphous polymers constitute a significant fraction of the polymer sector, they are much less studied by vibrational spectroscopy. It is often easier to obtain the spectra than for crystalline systems but the interpretation is much more difficult. This is because the lack of local and long range order means that the group theoretical tools that are the backbone of vibrational analysis are largely inapplicable. Computational study is also difficult because the flexible nature of the polymer chains means that there are often many conformations with similar energy. In the longer term the best technique is probably molecular dynamics simulations.

10.2.1 Polydimethylsiloxane

In this section we will consider polydimethylsiloxane (PDMS) as an example of the type of work that is possible with amorphous polymers. The structure and INS spectrum of PDMS are shown in Fig. 10.21a [40]. The repeat unit shown in Fig. 10.21b was used to model the spectrum using the Wilson GF matrix method [41]. The major features are reproduced: skeletal bending modes below 100 cm^{-1}, the methyl torsion and its overtone at 180 and 360 cm^{-1} respectively, the coupled methyl rocking modes and Si–O and Si–C stretches at 700-1000 cm^{-1} and the unresolved methyl deformation modes 1250-1500 cm^{-1}. The last are not clearly seen because the intensity of the methyl torsion results in a large Debye-Waller factor, so above 1000 cm^{-1} or so, most of the intensity occurs in the phonon wings.

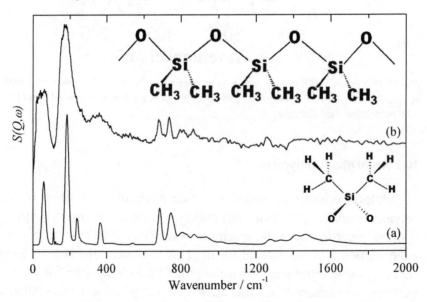

Fig. 10.21 (a) The structure and INS spectrum of polydimethylsiloxane. (b) Wilson GF fit to PDMS using just the model fragment shown.

The limitations are also clear: the integrated intensity of the methyl torsion is seriously underestimated and the relative intensities of the modes at 680 and 744 cm⁻¹ are not correct. All of these problems stem from the simplicity of the model; the methyl group sees many different environments and this results in the large width (70 cm⁻¹) of the methyl torsion. This has the potential to be used as a probe of the local environment as has been done for the ester methyl of poly(methylmethacrylate) [42].

The effect of the differing environments of the methyl group can be modelled by carrying out a DFT calculation of an oligomer of PDMS of C_1 symmetry so that the methyl groups are inequivalent. Fig. 10.22 shows a comparison of the spectrum of $H_3Si-[O-Si(CH_3)_2]_3-O-SiH_3$ with the experimental data (the silyl hydrogens are assigned zero cross section so they do not contribute to the spectrum). The agreement is better but it is only for one conformation. As stated earlier, molecular dynamics that would allow many conformations to be sampled would appear to be the way forward.

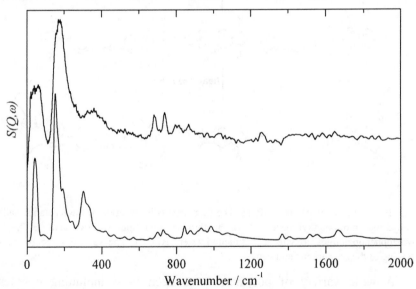

Fig. 10.22 Comparison of the INS spectrum of (a) PDMS and that of a model oligomer (b) $H_3Si-[O-Si(CH_3)_2]_3-O-SiH_3$ calculated by DFT. The silyl hydrogens are assigned zero cross section and a constant bandwidth of 7 cm⁻¹ is used.

10.2.2 Advanced composites

Advanced composites [43] are engineering materials that offer similar mechanical properties to metal alloys but are less dense. The materials consist of fibres embedded in a polymer matrix and there is a need for spectroscopic techniques that can examine the cured resins in the presence of the fibres to aid the understanding of the cure chemistry. The products are often highly cross-linked and thus insoluble, and the presence of the fibre matrix makes them difficult to study spectroscopically. INS has considerable potential in this regard since two common fibre types, glass and carbon are invisible to neutrons.

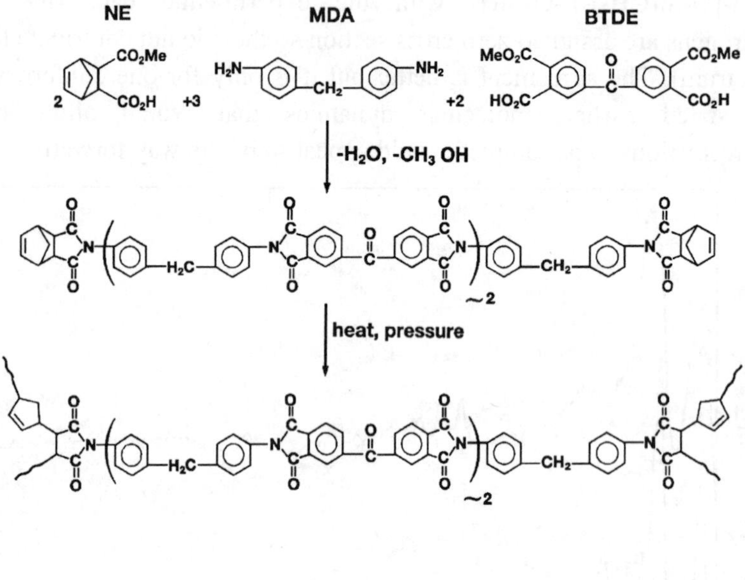

Fig. 10.23 The chemistry of PMR-15. The first stage is imidisation of the reactants, nadic anhydride monomethyl ester (NE), methylene dianiline (MDA) and 3,3',4,4'-benzophenonetetracarboxylic acid dimethyl ester (BTDE). The second stage is cross-linking of the norbornene end-caps.

A wide variety of polymers have been used including epoxies, bismaleimides and polyimides. One of the most common polyimides is PMR-15 [44]. The chemistry is complex, see Fig. 10.23, but consists essentially of two stages; imidisation to give a norbornene end-capped oligomer followed by reaction of the norbornene group to give the cross-linked polymer. The temperature at which the cross-linking reaction is carried out has a major effect on the mechanical properties of the finished product, particularly its susceptibility to microcracking. Vibrational spectroscopy has played an important role in the characterisation of polyimide/carbon fibre advanced composites, particularly the imidisation stage [45].

INS spectra of the composites cured at 270, 330 and 330°C are shown in Fig. 10.24 [46]. Differences between the three samples are apparent: bands at 1031 and 1114 cm^{-1} have diminished in intensity and there are indications of changes in the region 200—400 cm^{-1} and at 638, 720 and 1273 cm^{-1}. Comparison with the spectra of model compounds [47,48] suggests that the decrease in the 1114 cm^{-1} and the changes in the 200—400 and 600—800 cm^{-1} regions can reasonably be assigned to loss of the endcap. The 1031 cm^{-1} band does not fit this pattern and may represent a different type of cross-link.

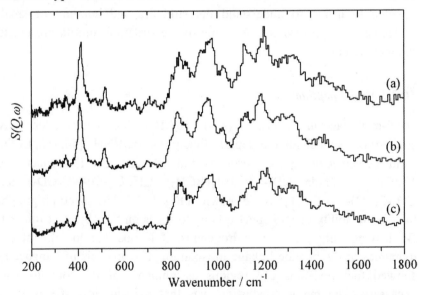

Fig. 10.24 INS spectra of PMR-15 cured at: (a) 270°C, (b) 300°C and (c) 330°C [46].

The example shown here is particularly difficult because by the cure temperatures employed here, a significant proportion of the end-groups have already reacted. In addition the composites are only ~30% by weight resin, of which only 2/7 of the molecules are the end-cap. Nonetheless, the potential of INS to study fabricated systems is real. It also highlights the possibility to study filled polymers in general [49] since most common fillers (carbon black, carbonates, silicates) are transparent to neutrons.

10.3 Biological systems

One of the most active areas in biology at present is attempting to understand how the tertiary structure (the shape) of a biomolecule arises and how this enables its function. This is important in fields such as enzyme catalysis, protein folding and DNA damage and repair. The primary structure of the biopolymer is determined by covalent bonding within, and between, the monomers but the tertiary structure is largely determined by hydrogen bonding. Thus considerable INS work has been devoted to trying to understand the structure and bonding of small molecules that possess at least some of the structural motifs present in the real systems.

10.3.1 Model peptides

N-methylacetamide, $CH_3-C(=O)-NH(CH_3)$, has been extensively studied as a model for the peptide linkage $-C(=O)-NH-$ [50,51]. In the solid the molecule forms hydrogen-bonded chains. The INS spectra of N-methylacetamide, $CD_3-C(=O)-NH(CH_3)$, $CH_3-C(=O)-NH(CD_3)$ and $CD_3-C(=O)-NH(CD_3)$ [50] are shown in Fig. 10.25. The spectra highlight the unique ability of INS spectroscopy to focus on the object of interest. By leaving only the N-H hydrogenous, it is possible to look at the dynamics of the peptide linkage in isolation. The analysis of the spectra showed that revisions of previous assignments of the torsions was necessary. The major conclusion was that existing force-fields were unable to reproduce the INS spectra. In order to model the spectra quantitatively with Wilson GF matrix method, it was necessary to decouple the proton motion from that of the rest of the molecule. This gave excellent fits to the spectra of all the isotopomers but required the startling assignment of the N-H stretch at 1600 cm^{-1} rather than the more usual ~3200 cm^{-1}. The latter was considered to be the first overtone of the stretch whose intensity in the infrared was greatly increased by anharmonicity. The 1600 cm^{-1} assignment of the N-H stretch also required that the N-H bond was considerably lengthened and that a description of -N$^{\delta-}$ ·· H$^{\delta+}$ ·· O$^{\delta-}$ rather than –N-H ·· O- was more correct.

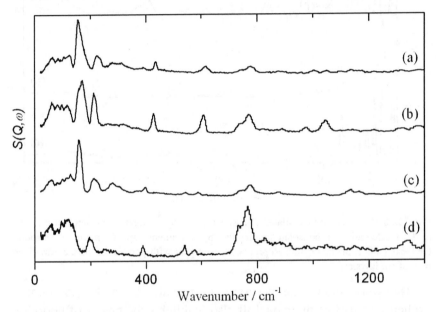

Fig. 10.25 INS spectra isotopic forms of N-methylacetamide. (a) CH_3-C(=O)-NH(CH_3), (b) CD_3-C(=O)-NH(CH_3), (c) CH_3-C(=O)-NH(CD_3) and (d) CD_3-C(=O)-NH(CD_3). Reproduced from [50] with permission from Elsevier.

The premise of the analysis in [50] was that N-methylacetamide could be treated as an isolated molecule. The crystal structure of N-methylacetamide has since been determined at 2K and the spectra revisited with the aid of periodic-DFT calculations [52], thus the hydrogen-bonding and the environment are specifically included. The structure shows a normal N-H bond length of 0.973 Å. Fig. 10.26 shows a comparison of the experimental INS spectrum, that calculated by DFT isolated molecule calculation shows poor agreement with the

experimental data. In contrast, the periodic-DFT calculation gives good
agreement and predicts a normal N-H stretch frequency of ~3200 cm⁻¹.

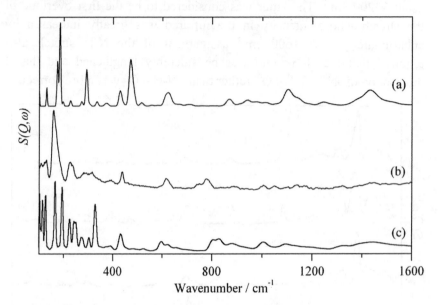

Fig. 10.26 Measured and calculated INS spectra of fully protonated N-methylacetamide.
(a) DFT calculation of the isolated molecule, (b) experimental spectrum and (c) periodic-
DFT calculation of the crystal. Reproduced from [52] with permission of the American
Institute of Physics.

The work shows the necessity to include the hydrogen-bonding
explicitly as part of the model. It also highlights the power of periodic-
DFT to analyse INS spectra. This is a technique that will become the
standard method of analysis of INS spectra
.Polyglycine, H(-CO-CH2-NH)n-H, is the simplest polypeptide. It has
two solid state structures. Form I consists of chains of polymer that are
hydrogen-bonded into two-dimensional sheets. This form (including
selectively deuterated isotopomers) has been studied by INS [53-55]. The
spectra were interpreted similarly to those of N-methylacetamide. Re-
examination of the spectra with periodic-DFT calculations is necessary.
Acetanilide, C_6H_5-NH-C(=O)-CH_3, forms hydrogen-bonded chains
similar to N-methylacetamide and as such is a potential model for phenyl
substituted peptides. However, the interest in acetanilide is that the

infrared spectrum of the amide I band at 1650 cm^{-1} has an unusual temperature dependence and overtone spectrum. This has prompted a number of suggestions including that of strong anharmonic coupling between low and high frequency modes. This gives rise to a localized vibrational excitation, a soliton, that can be both mobile and long-lived. and is potentially a means of transporting energy in biological systems.

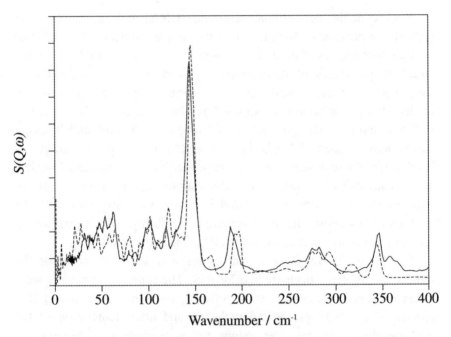

Fig.10.27 INS spectrum (solid line) and molecular dynamics simulation (dashed line) of acetanilide in the region below 400 cm^{-1}. Reproduced from [57] with permission of the American Institute of Physics.

Acetanilide, and some of its isotopomers, have been studied by INS spectroscopy [56-58]. The dispersion curves of the fully deuterated material have been measured by coherent INS [59]. A comprehensive analysis of acetanilide in the solid state was carried out with molecular dynamics simulations [57]. This includes all the lattice modes, as shown in Fig. 10.27 The simulations suggested that the barrier to the methyl torsion was enhanced when the peptide group is hydrogen-bonded and that this was a through-bond polarization effect. The methyl torsion was also markedly anharmonic (fundamental at 145 cm^{-1}, first overtone at 260 cm^{-1}). The amide I anomaly disappears on methyl deuteration so the evidence suggests that the methyl torsion is one of, if not the, mode involved in the coupling between low- and high-frequency modes.

10.3.2 Nucleic acids, nucleic acid bases, nucleotides and nucleosides

Nucleic acids, deoxyribonucleic acid (DNA) and ribonucleic acid (RNA), are polymers which take part in many biological processes such as transcription, translation, replication and catalysis. Analysis of the structural properties of these molecules helps us to understand their functions. A nucleic acid strand is formed by the repetition of nucleotides. A nucleotide is composed of a base (uracil in RNA, thymine in DNA, and cytosine, guanine and adenine in both DNA and RNA), a sugar (deoxyribose in DNA and ribose in RNA) and a phosphate group. Nucleosides (base + sugar) are also important building blocks of nucleic acids. They can adopt different conformations depending on the global structure of the nucleic acid strand in which they are involved. The analysis of ribonucleoside conformational flexibility (which depends on the force constants) is the first step in RNA structure modelling.

In Fig. 10.28 we show the structures and key references for the molecules that have been studied by INS. The initial assignments were based on a comprehensive set of infrared, resonance Raman and INS spectra of the fully protonated molecule and after deuteration of the exchangeable hydrogens. The assignments were made with the aid of *ab initio* calculations on the isolated molecules. Not unexpectedly, and as noted by the authors, the agreement was generally good for the non-hydrogen bonded parts of the molecules and poorer for those involved in the hydrogen bonding. Most of the molecules in Fig. 10.28 have been re-examined with the use of periodic-DFT calculations [66]. As shown in Fig. 10.29 for thymine, this gives much better, although not perfect, results. Fig. 10.29 also shows that for the best results, an accurate, low temperature (<100 K) structure determined by neutron diffraction (to locate the hydrogen atoms) is essential and that it is necessary to optimise both the unit cell and the atomic coordinates.

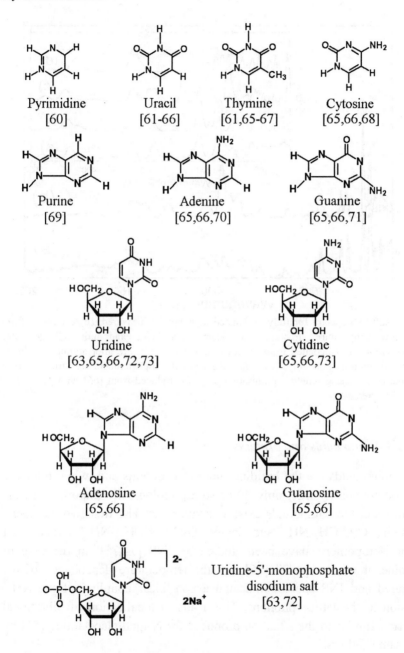

Fig. 10.28 The structures and literature references of nucleic acid bases, nucleotides and nucleosides that have been studied by INS spectroscopy.

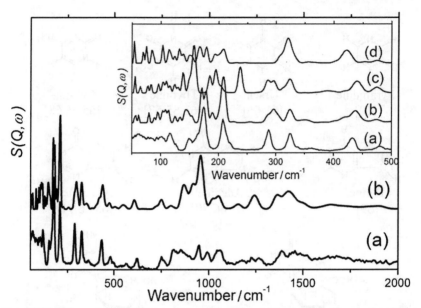

Fig. 10.29 INS spectra for fully protonated thymine: (a) measured; (b) calculated from the low temperature structure after atomic coordinate and unit cell geometry optimisation; (c) calculated from the low temperature structure after atomic coordinate geometry optimisation only; (d) calculated from the high temperature structure after atomic coordinate geometry optimisation only. Reproduced from [66] with permission from Elsevier.

10.3.3 Amino acids and proteins

Amino acids are the building blocks of proteins and, while the class of compounds is large, only 20 or so are biologically important. In the solid state, the compounds exist as zwitterions. The first two members, glycine, $-O_2C-CH_2-NH_3^+$ and alanine $-O_2C-CH(CH_3)-NH_3^+$ and some of their isotopomers, have been studied by INS [74,75]. In the case of alanine, it was proposed [75] that the temperature dependence of the infrared and INS spectra indicated a non-linear coupling of the NH_3^+ torsion to the lattice phonons. This leads to localization of vibrational energy, similar to the situation proposed for N-methylacetamide [51] by the same authors.

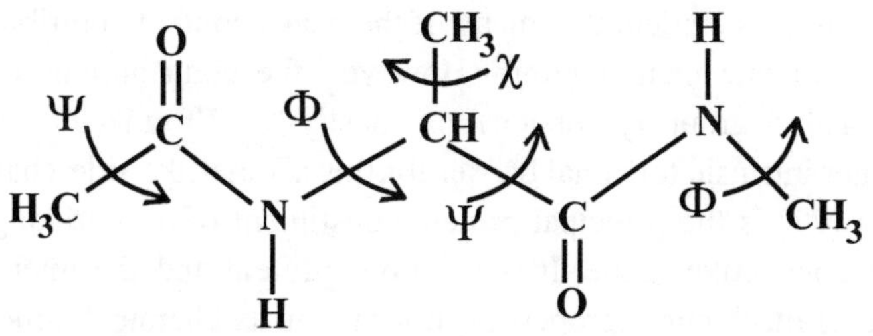

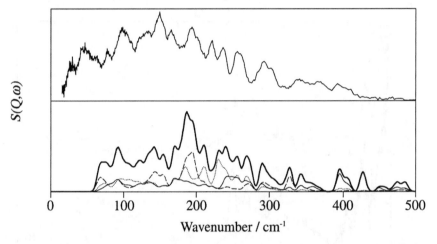

Fig. 10.30 Structure of alanine dipeptide showing the definition of the torsions. Reproduced from [76] with permission from Adenine Press.

Fig. 10.31 Observed (top) and calculated (bottom) INS spectrum of alanine dipeptide in the 0–500 cm^{-1} region. In the calculated spectrum methyl group contributions are shown: N-methyl, dotted line; terminal, dashed line; side-chain, thin solid line.. Reproduced from [76] with permission from Adenine Press.

Alanine dipeptide, see Fig. 10.30, is a useful model compound in that it contains the basic constituents of a peptide chain. Thus there are sp^3-sp^3 backbone C-C bond, backbone N-C bonds and side-chain C-C bonds. The torsions about each of these bonds are denoted as Ψ, Φ and χ respectively. The compound in the solid state has been modelled by molecular mechanics using the CHARMM force field [76]. The best results required non-zero force constants for the Ψ and Φ torsions. Fig. 10.31 compares the experimental spectrum with the calculated one. There is a uniform mismatch in frequencies of ~35 cm^{-1}, but the overall pattern is reproduced. The individual contributions of the methyl groups are also shown. It is evident that many of the peaks contain contributions from more than one methyl group. However, the contributions of the side-chain methyl group (χ torsion) are mostly > 230 cm^{-1}, consistent with the larger intrinsic torsional barrier for this alkane-like side-chain.

Collagen [77] is the principal protein constituent of a wide range of mammalian connective tissue. It is a fibrous protein and the interest is interpreting its mechanical properties in terms of its chemical structure. Properties such as elastic moduli and stress-strain curves depend on the interatomic force constants so vibrational spectroscopy is a necessary tool.

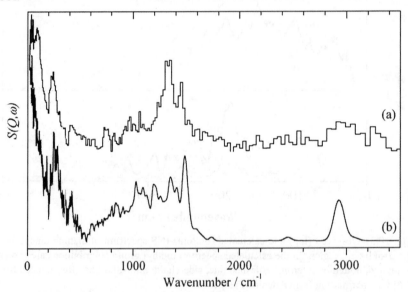

Fig. 10.32 Observed (a) and calculated (b) INS spectrum of Staphylococcal nuclease. Reproduced from [78] with permission from the American Chemical Society.

The purpose of the analysis of small molecules by molecular mechanics is to parameterise the force field so as to be able to analyse large systems. An example is the enzyme Staphylococcal nuclease. A comparison of the INS spectrum of the dried protein after exchange of labile hydrogens for deuterium, and the calculated spectrum is shown in Fig. 10.32 [78,79]. The agreement is generally good, although some details are incorrect. Examination of the displacement vectors showed that most of the modes below 1500 cm^{-1} are heavily mixed. Qualitatively, they follow the usual group frequency correlations: 350—500 cm^{-1} skeletal bending modes and torsions, 700—1000 cm^{-1} C—C stretch, CH$_2$ and CH$_3$ rock, 1000—1500 cm^{-1} CH$_2$ rock, wag, twist, scissors and CH$_3$ symmetric and asymmetric bend. The methyl torsion is calculated at 235 cm^{-1} and observed at 269 cm^{-1} showing that the parameters can be further refined. Staphylococcal nuclease has 2395 atoms, thus there are 7179 internal modes and is the largest molecule to be analysed to date.

10.3.4 Phosphate biominerals

Phosphate occurs widely in biology as 'organic' when it is bonded to a nucleic acid base or sugar and as 'inorganic' when it is present as a biomineral. As a biomineral, it can occur as both amorphous and crystalline forms.

Amorphous minerals are widely distributed in biology and occur as carbonates, phosphates and silica. Calcium and magnesium are the major cations in the phosphate deposits, but in addition, there is usually an organic component and the minerals are hydrated with up to 20% water. Such deposits are found as intracellular granules in a variety of invertebrates such as the shore crab *Carcinus maenas*. These intracellular granules, synthetic amorphous calcium phosphates and crystalline model compounds were studied [80] by infrared, Raman and INS spectroscopies to establish the protonation of the phosphates and the structure of the water in the crab granules. Hydroxyapatite, Ca$_5$(PO$_4$)$_3$OH, which has non-protonated phosphate groups and hydrogen only as part of an hydroxyl group, monetite, CaHPO$_4$, and newberyite, MgHPO$_4$.2H$_2$O, which have an acidic hydrogen and coordinated water in the latter, were used as model compounds. It was concluded that the phosphate is not protonated in the granules and that the water occurs in regions that are only loosely associated with the cations in these solids.

Crystalline hydroxyapatites are the major mineral of bones and teeth in a matrix of the protein collagen. The hydroxyls reside in channels running along the crystal *c*-axis which provide easy access to the external environment. Hydroxyapatites *in vivo* are usually described as poorly crystalline, calcium deficient and containing carbonate substitutions. The carbonate substitutions can occur at both the hydroxyl and the phosphate sites. In each case charge compensation is required, which could result in the protonation of a phosphate group to give HPO_4^{2-} or there may be a balance of substitutions with one carbonate substituting for two hydroxyl groups. Changes in the mechanical and chemical properties of bone are significant in a number of medical problems, in the ageing process and in pathological conditions caused by the incorporation of foreign ions.

Many studies have been made to establish the degree of crystallinity, crystal form and composition of the biological minerals. Fig. 10.33 shows a comparison of the infrared spectra of: highly crystalline hydroxyapatite (a), poorly crystalline hydroxyapatite (b) and ox femur after: removal of the fat (by acetone extraction), protein (by hydrazinolysis) and drying (c). In crystalline hydroxyapatite the hydroxyl group has two vibrations [81]: the O–H stretch at 3572 cm^{-1} and a twofold degenerate libration at 632 cm^{-1}. As the crystallinity decreases, (a) → (b) → (c), both modes progressively disappear. This has led to the suggestion that under physiological conditions all the hydroxyls in hydroxyapatite have been substituted, usually by carbonate [82].

Fig. 10.34 shows the INS spectrum of ox femur as the organic component is progressively removed [83]. Fig. 10.34a is very similar to that of the protein Staphylococcal nuclease, Fig 10.32, and emphasises one of the problems of working in this field: because proteins are largely made of the same monomers (amino acids), the INS spectra of very different proteins tend to look very similar. Removal of the fat results in little change in the spectrum, Fig. 10.34b. It can be seen that elimination of the protein is highly effective, Fig. 10.34c: the C–H stretching modes just below 3000 cm^{-1} and the C–H deformation modes at 1200–1500 cm^{-1} have both disappeared. There is a weak, broad peak at ~630 cm^{-1} and its overtone near 1300 cm^{-1}. For comparison, the INS spectrum of a highly crystalline reference hydroxyapatite is shown in Fig. 10.34d. The frequency match of the of the residual bone peak and that of the hydroxyapatite is exact, the width of the peak is attributed to heterogeneous broadening. The spectrum demonstrates that hydroxyl groups are still present in bone.

This is a striking example of the specificity of INS: the spectrum in Fig. 10.34c that clearly shows the presence of hydroxyls was the same sample that was used for the infrared measurements in Fig. 10.33c that showed no evidence for hydroxyls.

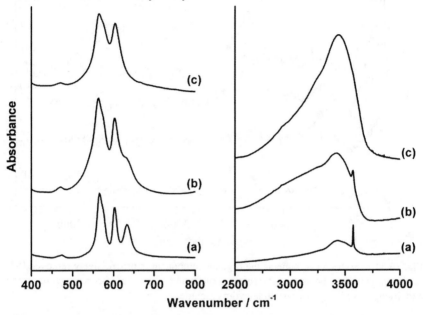

Fig. 10.33 Comparison of the infrared spectra of (a) highly crystalline hydroxyapatite (b) poorly crystalline hydroxyapatite and (c) defatted, deproteinated, dried bone showing the regions of the hydroxyl stretching vibration on the right and the hydroxyl libration on the left. Reproduced from [83] with permission from the PCCP Owner Societies.

Comparison of normalised areas of the highly crystalline hydroxyapatite and that of bone, indicated that up to ~50% of the OH groups had been substituted in bone. This conclusion was challenged [84]. It was argued that there are alternative assignments for the peak at 630 cm^{-1}, such as water or HPO_4^{2-} although comparison with model compounds does not support such assignments [80,81]. Observation of the O–H stretch mode was suggested as the definitive method. With indirect geometry spectrometers this is not possible because the resolution is degraded and, more importantly, the momentum transfer is so large, (Q ~15 Å^{-1} at 3600 cm^{-1}) that all the intensity is in the phonon wings. Direct geometry spectrometers can access transitions at large energy transfer because they have access to the low momentum transfer region.

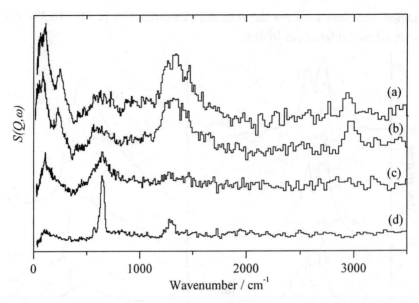

Fig. 10.34 INS spectra of bone as the organic component is progressively removed [83].
(a) Dried, powdered bone, (b) after removal of fat and drying, (c) after removal of protein
and drying and (d) highly crystalline hydroxyapatite.

In Fig. 10.35 we compare spectra obtained by different groups on the
direct geometry spectrometers LRMECS (IPNS) [84] and HET (ISIS)
[85]. The INS spectra of hydroxyapatite, Fig. 10.35a, b and e, are in good
agreement with the O–H stretch at 3585 cm^{-1}, and the combinations (O–
H stretch + libration) and (O–H stretch +2 x libration) at 4246 and 4865
cm^{-1} respectively. Fig. 10.35a and b differ in the crystallite size, the
nanometre size crystals, Fig. 10.35a, show a greater width than the
micron size crystals, Fig. 10.35b, presumably because of poorer
crystallinity. The increase in width is consistent with the greater width of
the OH libration in bone than in highly crystalline hydroxyapatite, see
Fig. 10.34c and d. The LRMECS bovine and rat cortical bone mineral
show only weak peaks in the O–H stretch region, Fig. 10.35c and d, that
do not match those expected for hydroxyapatite.In contrast, the ox femur
sample, Fig. 10.35f, shows the hydroxyapatite peaks. Comparison with a
model compound, brushite $CaHPO_4.2H_2O$, which contains both water
and HPO_4^{2-} shows no match. Thus two independent studies have arrived
at diametrically opposed conclusions: one [83,85] states that there is no

hydroxyl groups present in bone, the other [82,84] presents evidence to the contrary. The resolution of this paradox will require further INS studies: the sample preparation methods are different and LRMECS is inferior to HET both in resolution and sensitivity (proton current: IPNS ~ 15 µA, ISIS ~ 180 µA). However, recent measurements by solid-state NMR [86] of *whole* bone (organic and inorganic components) show that 21±1% of the stoichiometric quantity of OH⁻ is still present, in agreement with the ISIS results.

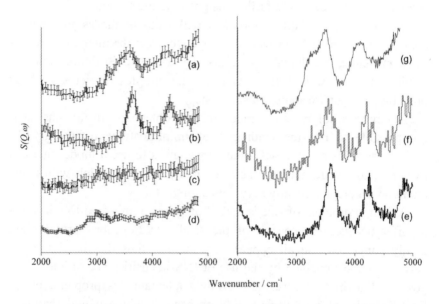

Fig. 10.35 INS spectra of bone and model compounds in the O—H stretching region, recorded by two groups of workers. Left: spectra recorded on LRMECS (IPNS) [84] (a) nanometre sized crystalline hydroxyapatite (b) micron sized crystalline hydroxyapatite, (c) bovine bone mineral, (d) bone mineral from rat. Right: spectra recorded on HET (ISIS) [85] (e) highly crystalline hydroxyapatite, (f) ox femur bone mineral, and (g) brushite $CaHPO_4.2H_2O$. The dashed vertical lines show the position of the hydroxyapatite peaks.

10.4 Conclusions

The study of soft condensed matter (polymers, biopolymers and biological materials) is likely to be the area of INS spectroscopy that sees the largest growth in the next decade. The study of model compounds is a thriving area, where the analysis follows the routes described in this book: the use of the isolated molecule approximation where intermolecular interactions are weak and *ab initio* codes for periodic systems where the interactions are significant. Re-analysis of older data is an activity that can yield further insights into the systems.

In principle, crystalline polymers will yield to the same types of analyses. In practice, significant dispersion, even in the internal modes, is often present and calculations across the complete Brillouin zone are needed. These are still rare even for the simplest systems such as polyethylene, but are an obvious next step.

Amorphous polymers are much more problematic. In this case it is not possible to use the mathematical simplifications offered by the periodic systems and the only route apparent is to use a model that can explore many conformations. This inevitably leads to models with many atoms so is computationally very demanding.

A related problem of size is that of proteins. In this area INS has less to offer because the spectra of the totally hydrogenous proteins are highly congested and all appear very similar, thus the discrimination between models enabled by the use of INS intensities is restricted. It is possible that improved resolution might alleviate this problem, but at least an order of magnitude would be needed, which would require a major innovation in instrument design. Alternatively, extensive use of novel deuteration procedures could expose the dynamics of the remaining hydrogenous components.

10.5 References

1   M. Bée (1988). *Quasielastic Neutron Scattering*, Adam Hilger, Bristol.
2   J.S. Higgins & H.C. Benoît (1994). *Polymers and Neutron Scattering*, Oxford University Press, Oxford.
3   R.J. Young & P.A. Lovell (1991). *Introduction to Polymers*, Chapman, London.

4   D.I. Bower & W.F. Maddams (1989). *The Vibrational Spectroscopy of Polymers*, Cambridge University Press, Cambridge, ch 5.

5   P.C. Painter, M.L. Coleman & J.L. Koenig (1982). *The Theory of Vibrational Spectroscopy and its Application to Polymeric Materials*, Wiley-Interscience, New York, ch 13.

6   H. Boutin & S. Yip (1968). *Molecular Spectroscopy with Neutrons*, M.I.T. Press, Cambridge (Mass), 77–84.

7   S. F. Parker (1996). J. Chem. Soc. Faraday Trans., 92, 1941–1946. Inelastic neutron scattering spectrum of polyethylene.

8   J. Barnes & B. Franconi (1978). J. Phys. Chem. Ref. Data, 7, 1309–1321. Critical review of vibrational data and force field constants for polyethylene.

9   S. Hirata & S. Iwata (1998). J. Chem. Phys., 108, 7901–7908. Density functional crystal orbital study on the normal vibrations and phonon dispersion curves of all *trans*-polyethylene.

10   T. Kitagawa & T. Miyazawa (1972). *Advances in Polymer Science,* vol. 9, Springer-Verlag, Berlin, p335. Neutron scattering and normal vibrations of polymers.

11   L. A. Feldkamp, G. Vankataraman & J.S. King (1968). *Neutron Inelastic Scattering,* Vol. 2, International Atomic Energy Authority, Vienna, p1510. Dispersion relation for skeletal vibrations in deuterated polyethylene.

12   R.G. Snyder & J.H. Schachtschneider (1965). Spectrochim. Acta, 19, 85–116 and 117–168. Vibrational analysis of the *n*-paraffins.

13   K. Okada (1965). J. Chem. Phys., 43, 2497–2510. Normal frequencies of skeletal bending vibration of planar zigzag chain with finite length.

14   D.A. Braden, S.F. Parker, J. Tomkinson & B.S. Hudson (1999). J. Chem. Phys., 111, 429–437. Inelastic neutron scattering spectra of the longitudinal acoustic modes of the normal alkanes from pentane to pentacosane.

15   H.G. Olf & B.J. Franconi (1973). J. Chem. Phys., 59, 534–544. Low frequency Raman-active lattice vibrations of *n*-paraffins.

16   S.F. Parker, J. Tomkinson, D.A. Braden & B.S. Hudson (2000). J. Chem. Soc. Chem. Commun., 165–166. Experimental test of the validity of the use of the *n*-alkanes as model compounds for polyethylene.

17   M.A. Neumann, M.R. Johnson & P.G. Radaelli (2001). Chem. Phys., 266, 53–68. The low temperature phase transition in octane and its possible generalisation to other *n*-alkanes.

18   P. Espeau, L. Robles, D. Mondieig, Y. Haget, M.A. Cuevas Diarte, H.A.J. Oonk (1996). J. de Chim. Phys. et de Physico-Chimie Bio., 93, 1217–1238. Review of the energetic and crystallographic behaviour of *n*-alkanes. 1. series from C_8H_{18} up to $C_{21}H_{44}$.

19   M. Dirand, M. Bouroukba, V. Chevallier, D. Petitjean, E. Behar, V. Ruffier-Meray (2002). J. Chem. Eng. Data, 47, 115–143. Normal alkanes, multialkane synthetic model mixtures, and real petroleum waxes: Crystallographic structures, thermodynamic properties, and crystallization.

20   J.H. Schachtschneider & R.G. Snyder (1965). Spectrochim. Acta, 21, 1527–1542. Valence force calculation of the vibrational frequencies of two forms of crystalline syndiotactic polypropylene.

21 T. Miyazawa, Y. Ideguchi & K. Fukushima (1963). J. Chem. Phys., 38, 2709–2720. Molecular vibration and structure of high polymers IV. A general method of treating degenerate normal vibrations of helical polymers and infrared-active vibrations of isotactic polypropylene.

22 R.G. Snyder & J.H. Schachtschneider (1964). Spectrochim. Acta, 20, 853–8610. Valence force calculation of the vibrational spectra of crystalline isotactic polypropylene and some deuterated polypropylenes.

23 K. Tashiro, M. Kobayashi & H. Tadokoro (1992). Polymer J., 24, 889–916. Vibrational spectra and theoretical three-dimensional elastic constants of isotactic polypropylene crystal: an important role of anharmonic vibrations.

24 G. Zerbi & L. Piseri (1968). J. Chem. Phys., 49, 3840–3844. Dispersion curves and frequency distributions in isotactic polypropylene.

25 H. Takeuchi, J.S. Higgins, A. Hill, A. Maconnachie, G. Allen & G.C. Stirling (1982). Polymer, 23, 499–504. Investigation of the methyl torsion in isotactic polypropylene—comparison between neutron inelastic scattering spectra and normal coordinate calculations.

26 G.J. Safford, H.R. Danner, H. Boutin & M. Berger (1964). J. Chem. Phys., 40, 1426–1432. Investigation of the low-frequency motions in isotactic and atactic polypropylene by neutron inelastic scattering.

27 P. Papanek, J.E Fischer & N.S. Murthy (1996). Macromol., 29, 2253–2259. Molecular vibrations in nylon 6 studied by inelastic neutron scattering. P.Papanek, J.E Fischer & N.S. Murthy (2002). Macromol., 35, 4175–4182. Low-frequency amide modes in different hydrogen-bonded forms of nylon-6 studied by inelastic neutron-scattering and density-functional calculations.

28 A. Maconnachie, A.J. Dianoux, H. Shirakawa, & M. Tasumi (1986). Synth. Met., 14, 323–327. Incoherent inelastic neutron-scattering from polyacetylenes in the 3500–400 cm^{-1} region. M. Tasumi, I. Harada, H. Takeuchi, H. Shirakawa, S. Suzuki, A. Maconnachie & A.J. Dianoux (1985). Synth. Met., 10, 293–295. Incoherent inelastic neutron-scattering from polyacetylenes in the low-frequency region.

29 S. Hirata, H. Torii, Y. Furukawa, M. Tasumi & J. Tomkinson (1996). Chem. Phys.Lett., 261, 241–245. Inelastic neutron scattering from trans-polyacetylene.

30 S. Hirata, H. Torii & M. Tasumi (1995). J. Chem. Phys., 103, 8963–8979. Vibrational analyses of *trans*-polyacetylene based on *ab initio* second-order Møller–Plesset perturbation calculations of *trans*-oligoenes.

31 A.J. Dianoux, G.R. Knellelr, J.L. Sauvajol & J.C. Smith (1994). J. Chem. Phys., 101, 634–644. Dynamics of sodium-doped polyacetylene.

32 J.L. Sauvajol, P. Papanek, J.E. Fischer, A.J. Dianoux, P.M. McNeillis, C. Mathis & B. Francois (1997). Synth. Met., 84, 941–942. Dynamics of pristine and doped conjugated polymers: A combined inelastic neutron scattering and computer simulation analysis.

33 F. Fillaux, N. Leygue, R. Baddour-Hadjean, S.F. Parker, Ph. Colomban, A. Gruger, A. Régis & L.T. Yu (1997). Chem. Phys., 216, 281–293. Inelastic neutron scattering studies of polyanilines and partially deuterated analogues.

34 D. Djurado, Y.F. Nicolau, P. Rannou, W. Luzny, E.J. Samuelsen, P. Terech, M. Bée & J.-L. Sauvajol (1999). Synth. Met., 101, 764–767. An overall view of the structure of an heterogeneous medium: The conducting polyaniline.

35 S. Folch, A. Gruger, A. Régis, R. Baddour-Hadjean & Ph. Colomban (1999). Synth. Met., 101, 795–796. Polymorphism and disorder in oligo- and polyanilines. A. El Khalki, Ph. Colomban & B. Hennion (2002). Macromol., 35, 5203–5211. Nature of protons, phase transitions, and dynamic disorder in poly- and oligoaniline bases and salts: an inelastic neutron scattering study.

36 G.J. Kearley, P. Johansson, R.G. Delaplane & J. Lindgren (2002). Solid State Ionics, 147, 237–242. Structure, vibrational-dynamics and first-principles study of diglyme as a model system for poly(ethyleneoxide).

37 F. Fillaux, S.F. Parker & L.T. Yu (2001). Solid State Ionics, 145, 451–457. Inelastic neutron scattering studies of polypyrroles and partially deuterated analogues.

38 A.D. Esposti, O. Moze, C. Taliani, J. Tomkinson, R. Zamboni & F. Zerbetto (1996). J. Chem. Phys., 104, 9704–9718. The intramolecular vibrations of prototypical polythiophenes.

39 L. van Eijck, L.D.A. Siebbeles, F.C. Grozema, I.M. de Schepper & G.J. Kearley (2002). Appl. Phys. A, 74, S496–S498. INS as a probe of inter-monomer angles in polymers.
L. van Eijck, M.R. Johnson & G.J. Kearley (2003). J. Chem. Phys., A, 107, 8980–8984. Intermolecular interactions in bithiophene as a model for polythiophene.

40 L. Jayes, A. Hard, C. Sene, S.F. Parker & U.A. Jayasooriya (2003). Anal. Chem., 75, 742–746. Vibrational spectroscopy of silicones: A Fourier transform-Raman and inelastic neutron scattering investigation.

41 G.J. Kearley (1995). Nucl. Inst. and Meth. A, 354, 53–58. A review of the analysis of molecular vibrations using INS.

42 A.J. Moreno, A. Alegrý, J. Colmenero & B. Frick (2001). Macromol., 34, 4886–4896. Methyl group dynamics in poly(methyl methacrylate): From quantum tunneling to classical hopping.

43 D. Wilson, H.D. Stenzburger & P.M. Hergenrother (Ed.) (1990) *Polyimides*. Blackie, Glasgow.

44 D.Wilson, J.K. Wells, J.N. Hay, D. Lind, G.A. Owens & F. Johnson (1987). SAMPE J., 23, 35–42. Preliminary investigations into the microcracking of PMR-15 graphite composites. 1 Effect of cure temperature.

45 S.F. Parker (1992). Vib. Spectrosc., 3, 87–104. The application of vibrational spectroscopy to the study of polyimides and their composites.

46 S.F. Parker & J.N. Hay (1995). in *Frontiers in Analytical Spectroscopy*, D.L. Andrews & A.M.C. Davies (Ed.), Royal Society of Chemistry, Cambridge, 184–188. The application of inelastic neutron scattering to advanced composites.

47 J.N Hay, J.D Boyle, S.F Parker & D. Wilson, (1989). Polymer, 30, 1032–1040. Polymerisation of N-phenylnadimide: a model for the crosslinking of PMR-15 polyimide.

48 S.F. Parker, K.P.J. Williams, D. Steele & H. Herman (2003). Phys. Chem. Chem. Phys., 5, 1508–1514. The vibrational spectra of norbornene and nadic anhydride.

49   S.F. Parker, K.P.J Williams, P. Meehan, M.A. Adams & J Tomkinson (1994). Appl. Spectrosc., 48,,669–673. The analysis of carbon black filled polymers by vibrational spectroscopy.
A.I. Nakatani, R. Ivkov, P. Papanek, H. Yang & M. Gerspacher (2000). Rubber Chem. and Technol., 73, 847–863. Inelastic neutron scattering from filled elastomers.

50   F. Fillaux, J.P. Fontaine, M.H. Baron, G.J. Kearley & J. Tomkinson (1993). Chem. Phys., 176, 249–278. Inelastic neutron scattering study of the proton dynamics in N-methylacetamide at 20 K.

51   M. Barthes, H.N. Bordallo, J. Eckert, O. Maurus, G. de Nunzio & J. Léon (1998). J. Phys. Chem. B, 6177–6183. Dynamics of crystalline N-methylacetamide: Temperature dependence on infrared and inelastic neutron scattering spectra.

52   G.J. Kearley, M.R. Johnson, M. Plazanet & E. Suard (2001). J. Chem. Phys., 115, 2614–2620. Structure and vibrational dynamics of the strongly hydrogen-bonded model peptide: N-methylacetamide.

53   F. Fillaux, J.P. Fontaine, M.H. Baron, N. Leygue, G.J. Kearley & J. Tomkinson (1994). Biophys. Chem., 53, 155–168. Inelastic neutron-scattering study of the proton transfer dynamics in polyglycine I at 20 K.

54   G.J. Kearley, F. Fillaux, M.H. Baron, S.M. Bennington & J. Tomkinson (1994). Science, 264, 1285–1289. A new look at proton transfer dynamics along the hydrogen bonds in amides and peptides.

55   F. Fillaux, M.H. Baron, N. Leygue, J. Tomkinson & G.J. Kearley (1995). Physica B, 213, 766–768. Proton transfer dynamics in polyglycine.

56   M. Barthes, R. Almairac, J.-L. Sauvajol, J. Moret, R. Currat & .J. Dianoux (1991). Phys. Rev. B, 43, 5223–5227. Incoherent neutron-scattering in acetanilide and 3 deuterated derivatives.

57   R.L. Hayward, H.D. Middendorf, U. Wanderlingh & J.C. Smith (1995). J. Chem. Phys., 102, 5525–5541. Dynamics of crystalline acetanilide: analysis using neutron scattering and computer simulation.

58   H.D. Middendorf (1997). In *Biological Macromolecular Dynamics,* (Ed.) S. Cusack, H. Büttner, M. Ferrand, P. Langan & P. Timmins, p55–67, Adenine Press, New York. Biomolecular dynamics from pulsed-source $S(Q,\omega)$ data: experiment and simulation.

59   M. Barthes, R. Almairac, J.-L. Sauvajol, R. Currat, J. Moret & J.-L. Ribet (1988). Europhys. Lett., 7, 55–60. Neutron-scattering investigation of deuterated crystalline acetanilide.

60   A. Navarro, M. Fernández-Gómez, J.J. López-González, M.P. Paz Fernández-Liencres, E. Martínez-Torres, J. Tomkinson & G.J. Kearley (1999). J. Phys. Chem. A. 103, 5833–5840. Inelastic neutron scattering spectrum and quantum mechanical calculations of the internal vibrations of pyrimidine.

61   A. Aamouche, G. Berthier, C. Coulombeau, J. P. Flament, M. Ghomi, C. Henriet, H. Jobic & P.Y. Turpin (1996). Chem. Phys., 204, 353–363. Molecular force fields of uracil and thymine, through neutron inelastic scattering experiments and scaled quantum mechanical calculations.

62 A. Aamouche, M. Ghomi, C. Coulombeau, H. Jobic, L. Grajcar, M.H. Baron, V. Baumruk, P.Y. Turpin, C. Henriet, & G. Berthier (1996). J. Phys. Chem., 100, 5224–5234. Neutron inelastic scattering, optical spectroscopies and scaled quantum mechanical force fields for analyzing the vibrational dynamics of pyrimidine nucleic acid bases. 1. Uracil.

63 M. Ghomi, A. Aamouche, H. Jobic, C. Coulombeau, & O. Bouloussa (1997). In *Biological Macromolecular Dynamics*, (Ed.) S. Cusack, H. Büttner, M. Ferrand, P. Langan & P. Timmins, p73–78. Adenine Press, New York. Neutron inelastic scattering of pyrimidine nucleic acid bases, ribonucleosides and ribonucleotides.

64 L. Grajcar, M.-H. Baron, M.-F. Lautie, S.F. Parker & F. Fillaux (1997). In *Biological Macromolecular Dynamics*, (Ed.) S. Cusack, H. Büttner, M. Ferrand, P. Langan & P. Timmins, p93–97. Adenine Press, New York. Inelastic neutron scattering study of (N)H proton dynamics in the (5, 6-D2)-uracil.

65 M.-P. Gaigeot, N. Leulliot, M. Ghomi, H. Jobic, C. Coulombeau & O. Bouloussa (2000). Chem. Phys., 261, 217–237. Analysis of the structural and vibrational properties of RNA building blocks by means of neutron inelastic scattering and density functional theory calculations.

66 M. Plazanet, N. Fukushima & M.R. Johnson (2002). Chem. Phys., 280, 53–70. Modelling molecular vibrations in extended hydrogen-bonded networks – crystalline bases of RNA and DNA and the nucleosides.

67 A. Aamouche, M. Ghomi, C. Coulombeau, ,L. Grajcar, M.H. Baron, H. Jobic & G. Berthier (1997). J. Phys. Chem., A, 101, 1808–1817. Neutron inelastic scattering, optical spectroscopies and scaled quantum mechanical force fields for analyzing the vibrational dynamics of pyrimidine nucleic acid bases. 2. Thymine.

68 A.Aamouche, M. Ghomi, L. Grajcar, M.H. Baron, F. Romain, V. Bamruk, J. Stepanek, C. Coulombeau, H. Jobic, & G. Berthier (1997). J. Phys. Chem., A, 101, 10063–10074. Neutron inelastic scattering, optical spectroscopies and scaled quantum mechanical force fields for analyzing the vibrational dynamics of pyrimidine nucleic acid bases. 3. Cytosine.

69 S.F. Parker, R. Jeans & R. Devonshire (2004). Vib. Spectrosc., 35, 173–177. Inelastic neutron scattering, Raman spectroscopy and periodic-DFT study of purine.

70 Z. Dhaouadi, M. Ghomi, J.C. Austin, R.B. Girling, R.E. Hester, P. Mojzes, L. Chinsky, P.Y. Turpin, C. Coulombeau, H. Jobic & J. Tomkinson (1993). J. Phys. Chem., 97, 1074–1084. Vibrational motions of bases of nucleic acids as revealed by neutron inelastic scattering and resonance Raman spectroscopy. 1. Adenine and its deuterated species.

71 Z. Dhaouadi, M. Ghomi, C. Coulombeau, Ce. Coulombeau, H. Jobic, P. Mojzes, L. Chinsky & P.Y. Turpin (1993). Eur. Biophys. J., 22, 225–236. The molecular force field of guanine and its deuterated species as determined from neutron inelastic scattering and resonance Raman measurements.

72 R. Bechrouri, J. Ulicny, H. Jobic, V. Baumruk, L. Bednárová, O. Bouloussa & M. Ghomi (1997). Spectroscopy of Biological Molecules: Modern Trends, 213–214. Analysis of the vibrational properties of uridine and 5'-UMP through neutron inelastic scattering, optical spectroscopies and quantum mechanical calculations.

73   N. Leulliot, M. Ghomi, H. Jobic, O. Bouloussa, V. Baumruk & C. Coulombeau
     (1999). J. Phys. Chem. B, 103, 10934–10944. Ground state properties of the nucleic
     acid constituents studied by density functional calculations. 2. Comparison between
     calculated and experimental vibrational spectra of uridine and cytidine.

74   C.L. Thaper, B.A. Dasannacharya, P.S. Goyal, R. Chakravarthy & J. Tomkinson
     (1991). Physica B, 174, 251–256. Neutron scattering from glycine and deuterated
     glycine.

75   M. Barthes, A.F. Vik, A. Spire, H.N. Bordallo & J. Eckert (2002). J. Phys. Chem. A,
     106, 5230–5241. Breathers or structural instability in solid L-alanine: A new IR and
     inelastic neutron scattering vibrational spectroscopic study.

76   J. Baudry, R.L. Hayward, H.D.M. Middendorf & J.C. Smith (1997). In *Biological
     Macromolecular Dynamics*, (Ed.) S. Cusack, H. Büttner, M. Ferrand, P. Langan &
     P. Timmins, pp. 49–54. Adenine Press, New York. Collective vibrations in
     crystalline alanine dipeptide at very low temperatures.

77   H.D. Middendorf, R.L. Hayward, S.F. Parker, J. Bradshaw & A Miller (1995).
     Biophys. J., 69, 660–673. Vibrational neutron spectroscopy of collagen and model
     polypeptides.

78   A.V. Goupil-Lamy, J.C. Smith, J. Yunoki, S.F. Parker & M. Kataoka (1997). J. Am.
     Chem. Soc., 119, 9268–9273. High-resolution vibrational inelastic neutron
     scattering: A new spectroscopic tool for globular proteins.

79   M. Kataoka, H. Kamikubo, H. Nakagawa, S.F. Parker & J.C. Smith (2003).
     Spectrosc-Int. J.,, 17, 529–535. Neutron inelastic scattering as a high-resolution
     vibrational spectroscopy: New tool for the study of protein dynamics.

80   P.C.H. Mitchell, S.F. Parker, K. Simkiss, J. Simmins & M.G. Taylor (1996). J.
     Inorg. Biochem., 62, 183–197. Hydrated sites in biogenic amorphous calcium
     phosphates: an infrared, Raman and inelastic neutron scattering study.

81   M.G. Taylor, K. Simkiss, S.F. Parker & P.C.H. Mitchell (1999). Phys. Chem. Chem.
     Phys., 1, 3141–3144. Inelastic neutron scattering studies of synthetic calcium
     phosphates.

82   C. Rey, J.L. Miquel, L. Facchini, A.P. Legrand & M.J. Glimcher (1995). Bone, 16,
     583–586. Hydroxyl groups in bone mineral.

83   M.G. Taylor, K. Simkiss, S.F. Parker & P.C.H. Mitchell (2001). Phys. Chem. Chem.
     Phys., 3, 1514–1517. Bone mineral: evidence for hydroxy groups by inelastic
     neutron scattering.

84   C.-K. Loong, C. Rey, L.T. Kuhn, C. Combes, Y. Wu, S.-H. Chen & M.J. Glimcher
     (2000). Bone, 26, 599–602. Evidence of hydroxyl-ion deficiency in bone apatites:
     An inelastic neutron-scattering study.

85   M.G. Taylor, S.F. Parker & P.C.H. Mitchell (2003). J. Mol. Struc., 651/653, 123–
     126. A study by high energy transfer inelastic neutron scattering spectroscopy of the
     mineral fraction of ox femur bone.

86   G.Y. Cho, Y.T. Wu & J.L. Ackerman (2003). Science, 300, 1123–1127. Detection
     of hydroxyl groups in bone mineral by solid-state NMR spectroscopy.

11

Non-hydrogenous Materials
and Carbon

In this chapter we consider the analysis of the spectra from non-hydrogenous materials (§11.1) with chlorine (§11.1.1) and some minerals (§11.1.2) as examples. Carbon (§11.2) in its allotropic forms of diamond (§11.2.1), graphite (§11.2.2) and the fullerenes and their derivatives (§11.2.3) has been studied extensively by INS spectroscopy.

In addition to the pure forms of carbon, there are a wide range of carbons of varying crystallinity and hydrogen content that are industrially important. These materials form a continuum from almost pure carbon to those with carbon-hydrogen ratios typical of organic compounds. The materials include amorphous hydrogenated-carbon (§11.2.4) and a range of industrial carbons (§11.2.5) such as coal, catalyst supports, catalyst coke and carbon blacks. Finally, metal carbonyl complexes (§11.2.6) are also considered.

While most INS studies exploit the large incoherent cross section of hydrogen to achieve both sensitivity and selectivity, there are a number of non-hydrogenous systems that have been successfully studied. The motivation is generally the same as for hydrogenous systems: the ease of calculation of the INS spectrum and the absence of selection rules so all the modes are observable. The theory is the same as for hydrogenous systems so the intensity is amplitude dependent. Thus modes such as out-of-plane bends and torsions give the strongest features and the lattice mode region is readily observed. These are all modes that are often difficult to observe by infrared and Raman spectroscopy.

The major difficulty with studying non-hydrogenous compounds is lack of sensitivity. This largely accounts for why there are few studies of non-hydrogenous systems. It is only in the last decade or so that neutron sources have become sufficiently intense that recording incoherent INS spectra of non-hydrogenous systems has become feasible. The only realistic way to overcome the sensitivity problem is to use large samples; sample quantities of 0.05 to 0.10 mole are needed to record a useful spectrum in 24 hours.

Deuterium is a special case. Selective deuteration is probably the most common form of sample manipulation and is often highly informative. Complete deuteration is less common, but with sufficient sample is capable of giving excellent spectra. The smaller cross section and larger mass both conspire to reduce the sensitivity but it is still larger than for virtually all other elements, so the INS spectra of fully deuterated compounds are dominated by the deuterium modes. In Fig. 11.1 we compare the spectra (normalised to one mole) of C_6H_6 and C_6D_6. The effect of the difference in cross section and amplitude of vibration is clear. (The INS spectra of partially deuterated systems are discussed along with the spectra of the parent compound).

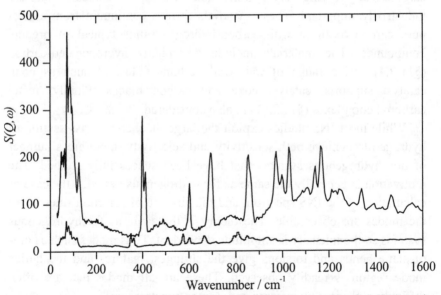

Fig. 11.1 A comparison of the INS spectra (normalised to one mole) of: C_6H_6 (upper) and C_6D_6 (lower).

It should be noted that the vast majority of non-hydrogenous samples are studied by coherent INS spectroscopy of single crystals; this is particularly the case for magnetic systems. Coherent INS is outside the scope of this book (see [1,2] for an introduction) and these systems are not considered further. There are a few coherent INS studies of non-hydrogenous and perdeuterated organic compounds; these are included in the list of compounds given in Appendix 4.

11.1 Analysis of spectra

As may be seen from Appendix 1 there are several elements with (almost) zero incoherent cross section (σ_{inc}). Carbon is a notable case in point: $\sigma_{inc} = 0.001$ barn. Thus it might be naively assumed that it would be practically impossible to obtain an INS spectrum from graphite, diamond or C_{60}. As may be seen from §11.2, this is not correct and good quality spectra are readily obtainable.

To account for this we invoke the incoherent approximation. This is discussed in more detail in §2.1.3, in essence, it treats the scattering as purely incoherent but uses the total scattering cross section rather than simply the incoherent cross section.

Hexafluorobenzene, C_6F_6 [3], provides an interesting test of the validity of the incoherent approximation. In principle, C_6F_6 should be an almost purely coherent scatterer. Fig. 11.2 shows a comparison of the experimental INS spectrum and that calculated from a DFT calculation using the incoherent approximation. It can be seen that the agreement with experiment is very good. There are discrepancies between the observed and calculated frequencies, this is because fluorine is computationally pathological.

Looking back, Fig. 11.1 provides further confirmation of the validity of the incoherent approximation. The ratio, H/D, of the normalised areas of the bands near 400 cm^{-1} is 9.8. The predicted ratio on the basis of their incoherent cross sections is 39.2 (= 80.27/2.05); in contrast, the ratio of their total scattering cross sections is 10.7 (= 82.03/7.64) in much better agreement with the experimental observation.

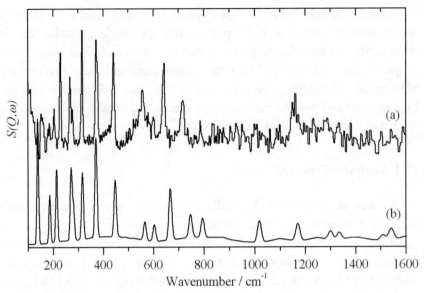

Fig. 11.2 (a) Experimental INS spectrum of C_6F_6 (0.053 mole), (b) spectrum calculated from a DFT calculation assuming the incoherent approximation.

11.1.1 Chlorine

Chlorine-35 (75.77% natural abundance) is unusual in that it has a significant incoherent cross section and a large coherent cross section. Thus natural abundance chlorine has a respectable total scattering cross section of 16.8 barn and this makes it an attractive target for incoherent INS. The relatively large atomic mass (35.5 amu) militates against this somewhat since the amplitude of motion will be small but it is still surprising that scattering from chlorine has not been more exploited.

One area that utilises the strengths of INS is that of transition metal chloro complexes. Most of the transition metals form chloro complexes and a wide variety of geometries is known. The most common are tetrahedral, square planar and octahedral. These have been extensively studied by infrared and Raman spectroscopies [4] and largely assigned. However, both O_h octahedral and D_{4h} square planar complexes have an internal mode that is infrared and Raman inactive (v_6 T_{2u} and v_5 B_{2u} respectively). There are also inactive lattice modes.

Figure 11.3 shows the INS spectra of $K_2[PtCl_6]$ [5] and $K_2[PdCl_4]$ [6]. For $K_2[PtCl_6]$, ν_6 is clearly seen at 147 cm^{-1}. A widely used empirical rule is that $\nu_6 = \nu_5/\sqrt{2}$, this incorrectly predicts $\nu_6 = 104$ cm^{-1}. The rule is based on a diagonal Wilson-type force field and is clearly an oversimplification. All of the lattice modes were also observed and assigned [5,6] for the first time.

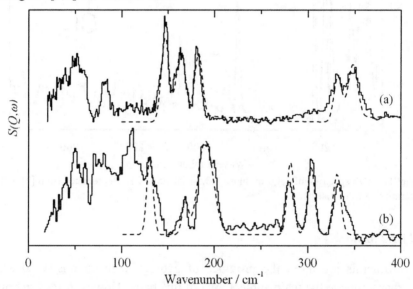

Fig. 11.3 INS spectrum of (a) $K_2[PtCl_6]$ and (b) $K_2[PdCl_4]$. The dashed lines are a fit to the internal modes using the Wilson GF matrix method.

For $K_2[PdCl_4]$ the situation was complicated by the lower symmetry, tetragonal rather than cubic. Also, all of the intensity outside the lattice mode region is quantitatively accounted for by the known modes. Careful analysis of the lattice mode region showed that the inactive ν_5 mode was at 136 cm^{-1}. This was the first time this mode had been observed in *any* homoleptic square planar complex.

Perchloro organic compounds are also amenable to study, although very few have been reported. One example is 2,4,6-trichloro-1,3,5-triazine [7], whose spectrum is shown in Fig. 11.4.

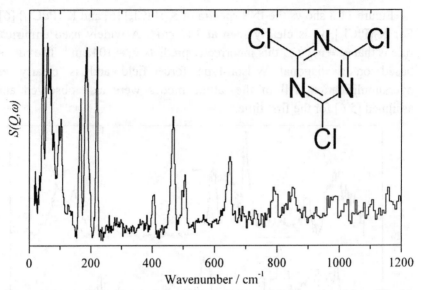

Fig. 11.4 INS spectrum of 2,4,6-trichloro-1,3,5-triazine at 5 K. Reproduced from [7] with permission from Elsevier.

11.1.2 Minerals

Minerals are more the province of geology than chemistry so are perhaps outside the main subject area of this book. However, they are an active field of study where INS has much to offer. Dispersion curves and the vibrational density of states can be calculated by lattice dynamics. These are often extrapolated to the extreme temperatures and pressures present in planetary interiors so it is essential to validate the models beforehand. As in many other areas, comparison of observed and calculated INS spectra provides a rigorous test. Two recent reviews provide a good overview of this field [8,9].

As an example, the phase diagram of the aluminosilicate Al_2SiO_5 minerals sillimanite, andalusite and kyanite is important in geothermometry and geobarometry. The transformations amongst the polymorphs involve a change in coordination of one of the aluminium atoms, which is tetrahedral in sillimanite, five-coordinated in andalusite and octahedral in kyanite. The observed (from 0.25 mol) and calculated INS spectra [10] are shown in Fig. 11.5a and 11.5b. The individual

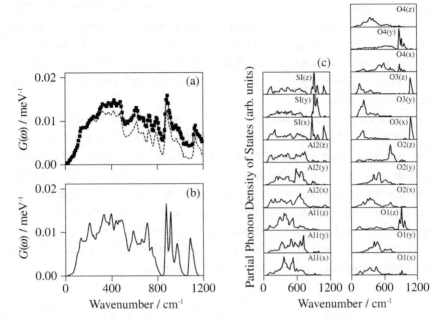

Fig. 11.5 (a) Observed INS spectrum (MARI, ISIS) of sillimanite at 15 K. The estimated one-phonon spectrum is shown by dashed lines. (b) Calculated spectrum broadened to an instrumental resolution of 16 cm^{-1}. (c) Computed partial phonon density of states of the atoms in the asymmetric unit along x, y and z in sillimanite. Reproduced from [10] with permission from the American Physical Society.

contributions of each of the seven atoms in the asymmetric unit is shown in Fig. 11.5c. Many atoms are involved in each mode and only a computational approach is capable of meaningful assignments.

The INS spectra of hydrated minerals and those containing hydroxyls, are dominated by the librational modes. The complementarity of INS to infrared and Raman spectroscopies is apparent, the optical spectra below 2000 cm^{-1} are dominated by the X–O (X = e.g. C, S, Al, Si) stretch and bend vibrations of the framework. Gypsum, $CaSO_4.2H_2O$ [11,12], provides a classic example.

11.1.2.1 Silica

Silica, SiO_2, will be considered in more detail since it is important in many different areas of technology in addition to its important role in geology and geochemistry. Silica occurs as crystalline (quartz, cristobalite) and amorphous glass forms. Besides its common use as windows and packaging, amorphous silica is used as a filler for polymers (especially in tyres) and as a catalyst support, where it can have a surface area of up to 200 m^2 g^{-1}. It is this use that is most relevant to the subject areas of this book.

The INS spectra of amorphous (vitreous) silica recorded on MARI and TOSCA at ISIS are shown in Fig. 11.6 [13]. The bands at 800–1200 cm^{-1} are Si–O stretching modes, at 300–400 cm^{-1} are the bending modes and at 100 cm^{-1} are probably optic translational modes. The detailed assignment of the spectrum has been the subject of considerable debate, but a DFT study [14] has provided some insight. Simplistically, the Si–O modes give rise to A_1 and T_2 stretching modes and the double peak in the high-frequency region results from different local modes of the tetrahedral subunits and not from LO-TO splitting.

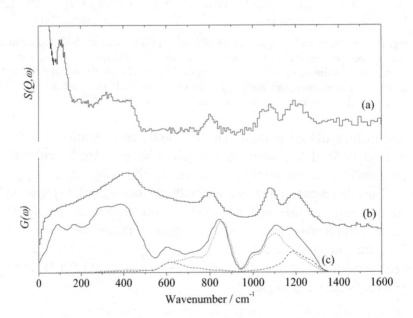

Fig. 11.6 INS spectra of amorphous silica recorded on (a) TOSCA and (b) MARI (incident energy = 1760 cm^{-1}). (c) Computed phonon density of states and the contribution of the A_1 (dashed line) and T_2 (dotted line) Si–O stretching modes. Reproduced from [14] with permission from the American Physical Society.

At very low energy, many amorphous materials show a 'Boson peak', so-called because the temperature dependence of its intensity roughly scales with the Bose-Einstein distribution. Amorphous silica is no exception and has a peak at ~40 cm^{-1}. The origin of this has been controversial but in silica it appears to be related to either transverse acoustic modes or torsions of the SiO$_4$ tetrahedra with respect to one another [15].

11.2 Carbon

Carbon has many allotropic forms: diamond, graphite, the fullerenes and their derivatives and the more readily available materials have been extensively studied by INS spectroscopy.

11.2.1 Diamond

Diamond has been studied by coherent INS [16]. Fig. 11.7 shows the dispersion curves calculated by periodic DFT; these are in excellent with the experimental data. Fig 11.8 shows a comparison of the INS spectrum derived from the dispersion curves and the TFXA experimental spectrum. Examination of the atomic displacements shows that the spectrum can be approximately described as C–C stretching modes above 1000 cm^{-1} and deformation modes below 1000 cm^{-1}. The agreement is good except for the features at 154 and 331 cm^{-1}. These are not a failure of the calculation but are a consequence of the use of graphite for the analysing crystal (§3.4.2.3) and will be discussed in §11.2.2.

11.2.2 Graphite

Graphite has been extensively studied by coherent INS [17,18], HREELS [19,20] and infrared and Raman spectroscopy [21]. Calculated dispersion curves are shown in Fig. 11.9 and the derived INS spectrum and the experimental spectrum in Fig. 11.10. As with diamond the agreement is excellent, except for the anomalously strong doublet at 112 and 128 cm^{-1} (average = 120) and the band at 324 cm^{-1}.

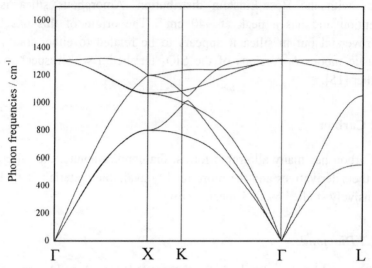

Fig. 11.7 Dispersion curves of diamond along conventional crystallographic directions as calculated by periodic DFT.

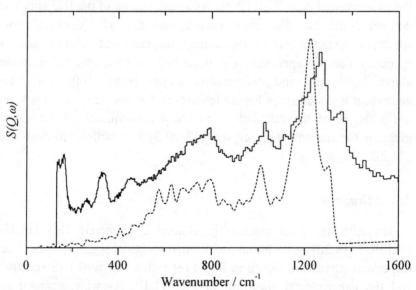

Fig. 11.8 INS spectrum of diamond derived from the dispersion curves of Fig. 11.7 (dashed line) compared with the experimental spectrum (solid line).

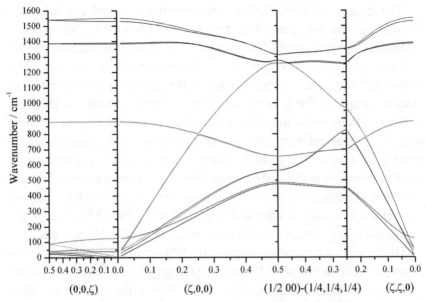

Fig. 11.9 The dispersion curves of graphite as calculated by periodic DFT.

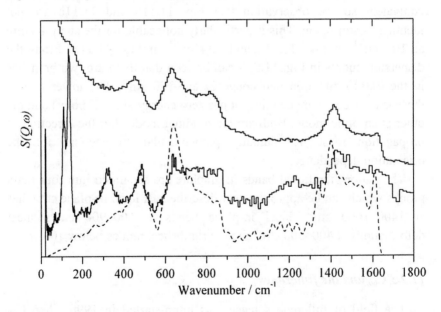

Fig. 11.10 INS spectrum of graphite derived from the dispersion curves of Fig. 11.9 (dashed line) compared with the experimental spectrum as recorded on TOSCA (lower solid line) and MARI (upper solid line).

The reasons for the differences between the observed and calculated TOSCA spectra are interesting. The 120 cm^{-1} feature has its expected strength in spectra recorded with a direct geometry instrument and the 324 cm^{-1} band does not appear, Fig. 11.10 uppermost spectrum. This shows that they are a characteristic of the instrument. Graphite is an unusual sample in that it has a large coherent cross section and gives rise to intense Bragg reflections. One explanation is that the 120 cm^{-1} features arise from coherent scattering excited in the graphite *analyser* by the (004) reflection of the graphite sample. The 324 cm^{-1} feature is even more problematic but probably arises from a hot-band transition between the states at ~100 cm^{-1}, populated at room temperature (~200 cm^{-1}), and the states that cause the peak at 476 cm^{-1} ($476 - 120 = 356 \approx 324 + E_f$). The graphite (210) reflection has the correct energy for this process.

Highly oriented pyrolytic graphite (HOPG) is a readily available form of polycrystalline graphite. It has a high degree of preferred orientation of *c*-axes of the crystallites. Fig. 11.11 shows the spectra for three different orientations. For a hexagonal crystal the *a* and *b* axes are equivalent so the observation that Fig. 11.11b and 11.11c are not identical is surprising. This is particularly noticeable for the sharp feature at 108 cm^{-1} in Fig. 11.11b and 134 cm^{-1} in Fig. 11.11c. From the dispersion curves in Fig. 11.9, it can be seen that there are two branches in the $(0,0,\xi)$ direction that converge at ~100 cm^{-1}. The lower one of these is an acoustic mode (since it has zero energy at the Γ point) and the other is an optic mode. Both are longitudinal modes but the direction of propagation must be mutually perpendicular to account for the orientation dependence.

The major vibrational bands in graphite are separated into four main groups which can be approximately described as: local in-plane stretches at 1400–1600 cm^{-1}, local in-plane bends at 600–900 cm^{-1}, sheet deformations at 400–500 cm^{-1} and the rigid sheet modes below 100 cm^{-1}.

11.2.3 C$_{60}$ and the fullerenes

The field of fullerene science was jump-started in 1990 when C$_{60}$ became available in macroscopic quantities [22]. This prompted an intense effort to understand its properties and to make new materials and compounds from C_{60}, which is still continuing. Other fullerenes, particularly C_{70}, have since become available, although in much smaller quantities. The subsequent discovery of carbon nanotubes has expanded the field even further.

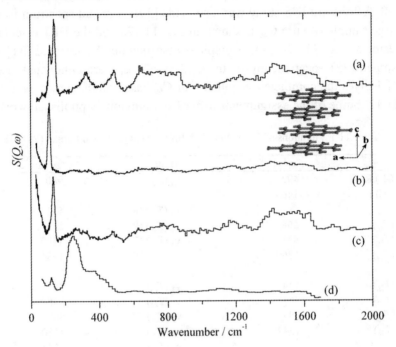

Fig. 11.11 INS spectrum of (a) polycrystalline graphite, (b) HOPG oriented with $Q \perp c$, axis, (c) HOPG rotated 90° in the a,b plane from (b) with $Q \perp c$ axis and (d) HOPG oriented with $Q \parallel c$ axis and $Q \perp a,b$ axes.

INS has played a key role in the study of C_{60} and its progeny so it will be described in some detail. A review is also available [23]. The lesser quantities of the other fullerenes has restricted INS studies to C_{70} [24,25].

11.2.3.1 C_{60}

An isolated C_{60} molecule has icosahedral, I_h, symmetry and the 174 internal modes are classified as:

$$2A_g + A_u + 3T_{1g} + 4T_{1u} + 4T_{2g} + 5T_{2u} + 6G_g + 6G_u + 8H_g + 7H_u.$$

The $2A_g$ and $8H_g$ are Raman active and the $4T_{1u}$ modes are infrared active. Thus only 14 of the 46 modes are formally observable; in contrast, *all* are INS active. This was realised early on and INS spectra from small quantities, ~1 g or less, were obtained by several groups [26-28]. A coherent INS study was also made [29]. A recent spectrum from a large sample of pure C_{60} is shown in Fig. 11.12a and the DFT calculated result in Fig. 11.12b [30]. Complementary studies by infrared [31] and Raman [32] spectroscopy of thick films of C_{60} were conducted. Table 11.1 gives the complete assignment for C_{60} based on the new data in Fig. 11.12. Some of the assignments differ from previously published work.

Table 11.1 Vibrations of C_{60} at 20K [30]. Infrared (IR) [31] and Raman (R) [32].

Mode	ω / cm^{-1}	Mode	ω / cm^{-1}
$\nu_1 (A_g)$ R	492	$\nu_1 (A_u)$	980
$\nu_2 (A_g)$ R	1466		
		$\nu_1 (T_{1u})$ IR	533
$\nu_1 (T_{1g})$	568	$\nu_2 (T_{1u})$ IR	579
$\nu_2 (T_{1g})$	833	$\nu_3 (T_{1u})$ IR	1180
$\nu_3 (T_{1g})$	1269	$\nu_4 (T_{1u})$ IR	1434
$\nu_1 (T_{2g})$	554	$\nu_1 (T_{2u})$	341
$\nu_2 (T_{2g})$	755	$\nu_2 (T_{2u})$	738
$\nu_3 (T_{2g})$	797	$\nu_3 (T_{2u})$	962
$\nu_4 (T_{2g})$	1344	$\nu_4 (T_{2u})$	1180
		$\nu_5 (T_{2u})$	1533
$\nu_1 (G_g)$	485		
$\nu_2 (G_g)$	568	$\nu_1 (G_u)$	353
$\nu_3 (G_g)$	755	$\nu_2 (G_u)$	738
$\nu_4 (G_g)$	1077	$\nu_3 (G_u)$	773
$\nu_5 (G_g)$	1315	$\nu_4 (G_u)$	962
$\nu_6 (G_g)$	1502	$\nu_5 (G_u)$	1315
		$\nu_6 (G_u)$	1434
$\nu_1 (H_g)$ R	270		
$\nu_2 (H_g)$ R	429	$\nu_1 (H_u)$	403
$\nu_3 (H_g)$ R	708	$\nu_2 (H_u)$	526
$\nu_4 (H_g)$ R	773	$\nu_3 (H_u)$	666
$\nu_5 (H_g)$ R	1101	$\nu_4 (H_u)$	755
$\nu_6 (H_g)$ R	1249	$\nu_5 (H_u)$	1217
$\nu_7 (H_g)$ R	1434	$\nu_6 (H_u)$	1344
$\nu_8 (H_g)$ R	1570	$\nu_7 (H_u)$	1570

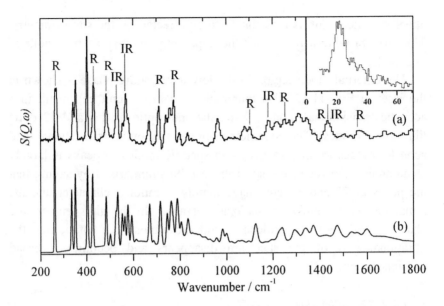

Fig. 11.12 (a) INS spectrum [30] from 0.02 mole of C_{60} at 20K and (b) calculated by DFT. The infrared (I) and Raman (R) active modes are indicated, *all* the others are forbidden. The inset shows the lattice mode region (MARI, ISIS) [28].

The modes can be roughly divided into radial ('out-of-plane' bending or torsion modes that give a buckling type of motion) below 900 cm^{-1} and tangential ('in-plane' bending or stretching modes) above 900 cm^{-1}.

The low temperature (T < 260K) crystal structure of C_{60} is cubic, space group $Pa\bar{3}$ [33], with S_6 site group symmetry. Thus the three-, four- and fivefold degenerate modes in I_h should split: $T \rightarrow A + E$, $G \rightarrow 2A + E$, $H \rightarrow A + 2E$. For most modes this is not seen, thus the intense G_u mode at 400 cm^{-1} has a spectrometer limited resolution width of 5.6 cm^{-1}, so any splitting is very small. This is the case for all the modes except the lowest internal mode at 265 cm^{-1} of H_g symmetry, which is split into two lines at 262 and 268 cm^{-1} with an intensity ratio of ~2:3. This mode is basically squashing the molecule along an axis approximately linking diagonally opposed pentagons. The weak interaction between the pentagon face and a C–C bond of a hexagon on a neighbouring molecule is responsible for the low temperature ordering of C_{60}. Only half of the pentagons are involved in these types of interaction, so a vibration that alters the interaction is most likely to be affected by the site symmetry. This may provide a reason for the apparently anomalous behaviour of this mode.

The external mode region in the low temperature phase is shown in the inset in Fig. 11.12. The four molecules in the unit cell result in 3 acoustic translational modes, 9 translational optic modes and 12 optic librational modes. The symmetry allows the optic translational and acoustic modes to mix, so assigning specific modes to peaks is largely meaningless. However, the spectrum can be summarised by saying that the peaks at 22 and 35 cm^{-1} have mainly librational character and the remainder have mainly translational character. In the high temperature phase (T > 260K), the molecules undergo almost free rotation so the librational part of the spectrum disappears and becomes a broad quasielastic line [23].

11.2.3.2 The fullerides

Reaction of C_{60} with alkali metals results in a series of ionic compounds $A_x(C_{60})$ the fullerides, that contain $(C_{60})^{x-}$ $x = 1$–6 ions [34]. Mixed metal fullerides e.g. $Na_2Rb(C_{60})_3$, are also known. Synthesis in liquid ammonia results in ammonia-containing compounds e.g. $Na_2(C_{60})(NH_3)_8$ and $Li_3(C_{60})(NH_3)_4$ [35]. Reaction with other electropositive metals also results in fulleride formation, notably $Ba_x(C_{60})$ $x = 3,4,6$. Transition metal fullerides are also known. C_{70} behaves similarly.

The interest in fullerides is that the compounds exhibit a range of phase behaviour from insulator through metallic to superconducting depending on the metal and the stoichiometry. Neutrons are an ideal probe to study such materials because the electronic nature of the phase is irrelevant to the interaction, in contrast to optical methods. Comparison of the internal mode spectrum of C_{60} with that of the alkali metal doped compounds shows that the spectrum is not drastically changed as C_{60} is reduced. The additional electrons go into the lowest unoccupied molecular orbital of T_{1u} symmetry which has mainly non-bonding character. The high energy cut-off softens slightly from 1600 cm^{-1} in C_{60} to 1534 cm^{-1} in C_{60}^{6-} and the low energy onset is largely unchanged. In the gap between the lowest internal mode and the highest external mode, weak features assigned to translational modes of the alkali metal ions are observed [36,37].

The superconductivity in the $A_3(C_{60})$ compounds is of the conventional Bardeen-Cooper-Schrieffer (BCS) type. The relatively high superconducting transition temperature (18 K for A = potassium) is ascribed to electron pair formation mediated by the high energy internal modes [38].

11.2.3.3 Dimers and polymers of C_{60}

All of the C_{60} systems considered so far have contained discrete molecules or ions. However, C_{60} undergoes a range of solid state reactions induced by the action of alkali metals, light or pressure at elevated temperature [39]. These lead to a wide variety of structural motifs including single bonded dimers (RbC_{60} when it is quenched from above 400K to 77K [40]), one- (pressure polymerized C_{60} [41-43]), two- (Na_4C_{60} [44]) and three-dimensional polymers (high temperature, high pressure polymerized C_{60} [45]). One of the motivations for the extensive work in these systems is the search for materials that are harder than diamond.

While all of these reactions lead to changes in the vibrational spectrum, the region that has been most studied by INS is the region below 300 cm^{-1}. Fig. 11.13 shows the type of spectra that are observed. In C_{60} there is a clear gap between the external and internal modes. As the structure becomes more complex the gap is progressively filled. The reason is that as polymerization occurs, translational and librational modes of the free molecules become internal modes of the polymer. Since these involve changes in covalent bonds, which are much stronger than the intermolecular forces, the new vibrations occur at higher energy than the translations and librations that they replaced.

The pattern is that increasing temperature and pressure results in progressive polymerization: C_{60} → dimers → linear polymer → 2D-polymer → 3D-polymer → 'superhard phase [46]'. The final material is interesting because hardness tests reveal that that this material is approximately two-thirds as hard as diamond but with a resistivity very similar to graphite. The INS spectrum strongly resembles that of graphite so it is proposed that the material contains graphite planes that are buckled so that they cannot slip over one another. This locks the structure and accounts for its hardness.

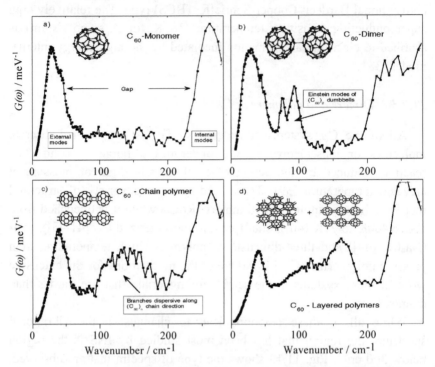

Fig. 11.13 INS spectra (IN6, ILL) obtained in neutron energy gain of the low energy region of: (a) C_{60} at 300K, (b) C_{60} dimers in the quenched phase of RbC_{60}, (c) linear chains in pressure polymerized C_{60} and (d) two-dimensional polymer sheets in pressure polymerized C_{60} (the material is a mixture of rhombohedral and tetragonal networks). Reproduced from [39] with permission from Gordon and Breach.

11.2.4 Amorphous hydrogenated carbon

Amorphous hydrogenated carbon (a-C:H) is a material that can be prepared such that it is harder, denser and more resistant to chemical attack than any other solid hydrocarbon. It is transparent across much of the infrared and has good histocompatibility. This combination of properties has lead to applications that include wear and corrosion resistant coatings on optical and electronic components such as lenses and magnetic hard disk drives. They are also used as protective coatings on plasma-facing components in experimental fusion devices in order to reduce the concentration of metal impurities in the plasma.

The material is prepared by dissociation of a gaseous hydrocarbon in saddle-field ion-beam source. Depending on the conditions, a-C:H can be a soft, polymeric material with a high hydrogen content at one extreme, through a hard or diamond-like form with intermediate hydrogen content, to a graphitic form with very low hydrogen content. The materials typically exhibit a hydrogen content between 10 and 60%. The material has been largely characterised [47–52] by a combination of neutron diffraction, which provides information on the carbon network, and INS which provides information on the hydrogen environment.

The INS spectra of a-C:H prepared under different conditions from different precursors are broadly similar. Fig. 11.14 shows [51] a typical result. The spectrum can be assigned [48] as: C–C torsions < 500 cm^{-1}, C–C stretch 875 cm^{-1}, CH_2 rock 1030 cm^{-1}, CH bend 1190 cm^{-1}, CH_2 twist 1300 cm^{-1}, CH_2 wag 1330 cm^{-1} and CH_2 scissors 1470 cm^{-1}. A more quantitative analysis of the spectra can be obtained by modelling with a

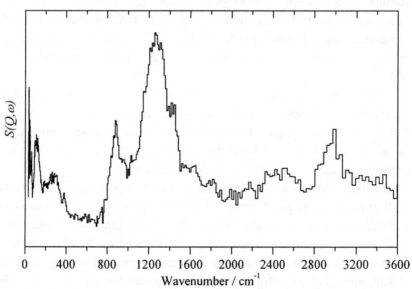

Fig. 11.14 INS spectra of amorphous hydrogenated carbon prepared from acetylene [50].

pseudo-molecule that contains all the structural elements. By varying the cross section of the different types of hydrogen so as to obtain the best fit, the relative proportions of CH, CH_2 and CH_3 groups can be determined. The analysis showed that the CH:CH_2 ratio was ~1 and that there were very few CH_3 groups.

The model that emerged [50] is of short sections (< 5 carbon atoms) of sp^3 CH_2 chains and statistically distributed CH groups in an sp^2–sp^3 carbon network with olefinic rather than aromatic (graphitic) carbons. The two environments are separated by regions of non-hydrogenated sp^2 carbons.

11.2.5 Industrial carbons

Carbon is important commercially in a wide range of forms such as coal, catalyst supports, catalyst coke, soot and carbon blacks. The last of these itself has many uses from pigments to fillers in tyre rubber. A common feature of all these materials is that they are difficult to study spectroscopically. They strongly scatter or absorb electromagnetic radiation across the visible and infrared regions making infrared and Raman spectroscopy challenging. They are insoluble and often semiconductors or contain metallic impurities making nuclear magnetic resonance studies equally challenging. All of these properties are irrelevant to INS spectroscopy and spectra are readily obtainable [53 — 66].

The hydrogen content varies widely from almost zero to values more nearly typical of organic compounds, but in general, relatively large sample sizes are needed, usually in the 10—20 g range. This has the advantage that it ensures that a macroscopic average is measured so the spectrum is representative of the bulk of the sample.

Fig. 11.15 shows a series of spectra of carbons from different sources with different hydrogen contents. Note the large sample masses used and the different ordinate scales. The spectra divide into two groups: those that resemble graphite, compare Fig. 11.10 and Fig. 11.15a, and those whose dynamics are dominated by the small amount of hydrogen present [10000 ppm H (0.01 g H/g C)corresponds to a molecular formula ~C_8H). The spectra of the hydrogenous materials bear a striking resemblance to one another in spite of their different origins. This indicates that the same basic structural motif is present in the materials. The accepted model is of a disordered graphite-like structure with hydrogen atoms decorating the edges of the planes.

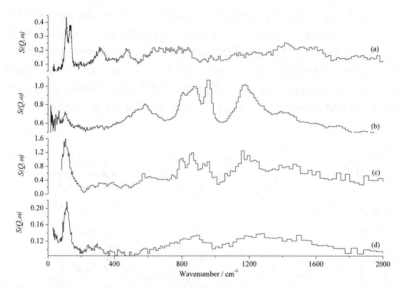

Fig. 11.15 INS spectra of carbons from different sources. (a) catalyst coke, 25.4 g, 116 ppm H [57], (b) La Mure coal, 6.6 g, 50000 ppm H (0.05 g H/g C) [53], (c) decolourising carbon, 9.7 g [59] and (d) soot from diesel engine, 4.6 g [64].

Consideration of the possible types of termination of a graphite sheet, shows that isolated hydrogens (H1), two hydrogen atoms on adjacent carbon atoms (H2) and three hydrogen atoms on adjacent carbon atoms (H3) are the most likely species and that four and five adjacent hydrogens are unlikely. Model fragments with largely H1, all H2 and largely H3 were constructed and the vibrational spectra calculated by DFT [60] as shown in Fig. 11.16.

Visualisation of the modes allow generic assignments to be made. All three spectra show a feature below 100 cm^{-1} that is assigned to torsional modes of the interior carbons. The intensity of this band comes largely from scattering by carbon. The features at ~200 and ~500 cm^{-1} are C–C torsion modes of the carbon atoms at the edges of the fragment. These result in significant proton-riding motion so are observed in the INS spectrum (§7.1). The bands at 800–1000 cm^{-1} and at 1200 cm^{-1} are the out-of-plane and the in-plane C-H bending modes and there is a distinct gap between these two types of mode, in agreement with previous work. The feature at ~1450 cm^{-1} has either been ignored or assigned to the CH$_2$ scissors mode of a small proportion of aliphatic hydrogen atoms. This new analysis shows it should be largely assigned to a coupled C–C stretch and in-plane C–H bend of the perimeter carbon atoms, accounting for its modest intensity. The C–H stretch is observed in the usual aromatic region above 3000 cm^{-1}. Thus the DFT calculations have both substantiated and extended previous assignments for these carbons.

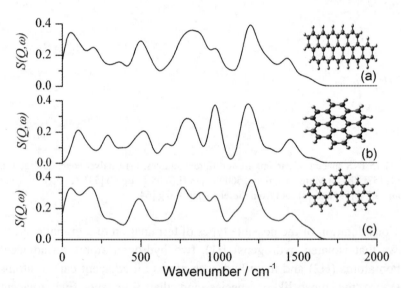

Fig. 11.16 Calculated INS spectra of model graphite fragments. (a) Mainly isolated hydrogen atoms H1, (b) coronene, two adjacent hydrogen atoms H2 and (c) mainly three adjacent hydrogen atoms. Reproduced from [60] with permission of the PCCP Owner Societies.

The model spectra show that the only region where there is any significant variation in the profile is in the C–H out-of-plane bending region around 880 cm^{-1}; exactly the region where the spectra of the carbons show variations. The INS spectrum of a single H1 or H2 or H3 unit can be examined by calculating an 'isotopic' spectrum where the mass of all the atoms, except those of interest, is set to 100 amu and their cross sections to zero. The result is that H1 contributes one band at ~770 cm^{-1}, H2 two bands at ~710 and 885 cm^{-1} and H3 three bands at ~700, 810 and 910 cm^{-1}. All the bands have approximately the same intensity since the amplitude of motion is determined by the C–H bond distance and the out-of-plane bending force constant, both of which will be very similar for similar environments as occurs here.

The calculated frequencies do not exactly match the experimental ones; the detail will depend on the particular geometry of the fragment. In addition, it is known, and confirmed by X-ray photoelectron spectroscopy and secondary ion mass spectroscopy [60], that there are a significant amount of oxygenated species present. Since these are also likely to be at the periphery of the graphite layers, they will also influence the frequencies. These are very difficult to model since the nature of these species and their relative quantities is still uncertain.

However, two conclusions can be drawn. Firstly, the variation between the carbons must reflect different distributions of H1, H2 and H3 species at the graphite edges and suggests that the distribution is characteristic of the type of carbon. This is shown by the results for a series of carbons derived from pine and shown in Fig. 11.17. The distribution of H1, H2 and H3 sites is, within errors, the same, even though the hydrogen content more than doubles. Possible explanations such as hydrogenation of oxygenated or unsaturated sites can be ruled out since the area of the aromatic C–H in-plane and out-of-plane bands increases linearly with the hydrogen content and the oxygen content is unchanged. The only consistent explanation is that the average size of the graphitic plates decreases but retains the same morphology. The large samples needed for INS are an advantage in this case because it ensures that a representative average is measured. Secondly, since the central band is usually the strongest, the major species present must be H1.

From the INS studies of coal [53–55] the dynamics are dominated by a large offset in the spectra that is proposed to be caused by free protons (§9.4.2). These are intercalated between the graphite planes and are possibly caused by sp^3 defects in the graphitic planes. The internal energy is decreased by release of free protons and delocalized electrons.

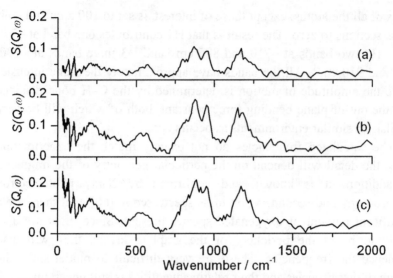

Fig. 11.17 Background subtracted INS spectra of activated carbons derived from pine. (a) 9400 ppm H, (b) 16500 ppm H and (c) 21500 ppm H. Reproduced from [60] with permission from the PCCP Owner Societies.

Other systems studied are catalyst coke [56,57] and catalyst supports [58–61]. The nature of the support has a profound effect on catalyst activity. Surprisingly, the origin of the carbon, whether peat, wood or coconut shells, is reflected at all length scales from the macroscopic to the atomic level. Carbon blacks of various types and with varying treatments have also been studied [62,64], as has diesel soot [64] and carbon anodes [66–68].

11.2.6 Transition metal carbonyls and carbonyl hydrides

Carbon monoxide is an important ligand. It coordinates to most transition metals and is implicated in many catalytic processes. Transition metal carbonyl complexes have been studied for many years. Assignments for the CO stretch region are available for many complexes but complete vibrational assignments [69] are limited to the simpler compounds such as $[M(CO)_6]$ (M = Cr, Mo, W), $[Fe(CO)_5]$ and $[Ni(CO)_4]$. Even for these, some of the fundamentals are known only from combination or overtone bands. Assignments for metal carbonyl clusters $[M_x(CO)_y]$ $(x,y > 1)$ are scarce.

INS can provide complementary information. The CO stretch region around 2000 cm^{-1} is well-suited to investigation by infrared and Raman spectroscopies. However, the M–C stretch, M–C≡O bend and C–M–C bend vibrations are much weaker and more difficult to access. Additionally, some of the modes are inactive.

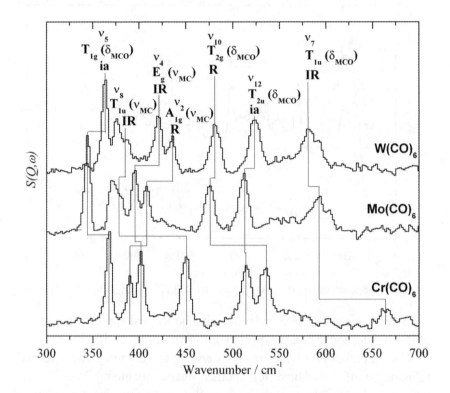

Fig. 11.18 INS spectra and assignments in O_h symmetry of: $[M(CO)_6]$ (M = Cr, Mo, W) in the M–C stretch (ν_{MC}) and M–C≡O bend (δ_{MCO}) region. The infrared (I) and Raman (R) active modes are indicated, *all* the others are forbidden (ia).

Fig. 11.18 shows the INS spectra of octahedral $[M(CO)_6]$ (M = Cr, Mo, W) [70], for the M–CO stretch and M–C–O bend region and the assignments. Half of the modes are inactive in both the infrared and Raman spectrum. Some of the *T* modes are split because the site symmetry is only C_s so the degeneracies are (formally) lifted. As is usually found, for the same type of mode, the intensities scale with the degeneracy, when Q variation is considered. For the Mo–C stretch modes at 409 (A_{1g}), 396 (E_g) and 375 cm^{-1} (T_{1u}), the relative intensities are 1 : 1.6 : 2.7 close to the 1:2:3 expected. Similarly for the Mo–C–O bending modes at 476 (T_{2g}), 512 (T_{2u}) and 596 cm^{-1} (T_{1u}), the relative intensities are: 1 : 0.98 : 0.82, close to the 1:1:1 expected.

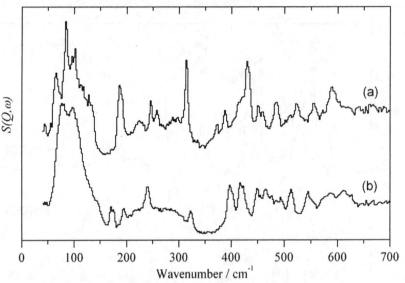

Fig. 11.19 INS spectra of (a) [Fe$_2$(CO)$_9$] (0.027 mol), (b) [Fe$_3$(CO)$_{12}$] (0.024 mol) [71].

Polynuclear metal carbonyls may also be studied and the INS spectra of [Fe$_2$(CO)$_9$] and [Fe$_3$(CO)$_{12}$] [71] are shown in Fig. 11.19.

Metal carbonyls have many derivatives obtained by (formal) substitution of a carbonyl for another ligand, including hydride. The interest in the complexes arose because they were considered as potential models for hydrogen on metal surfaces (§7.3.1). In general this is valid, in that the frequencies fall in similar ranges to those found on metal surfaces, the detailed interaction with the surface is distinctly different from that in a transition metal cluster and the analogy is imperfect.

The INS spectra of metal carbonyl hydrides [M$_y$(CO)$_z$H$_x$] ($x,y,z \geq 1$) are dominated by the hydrogen motion. However, this motion is coupled more or less strongly to the carbonyl modes and the result is a rainbow at one end of which only the hydride motion is apparent and at the other, some of the carbonyl modes are among the most intense in the spectrum.

The INS spectrum of [Co(CO)$_4$H] [72] Fig.11.20a illustrates this point. The intense band at 696 cm^{-1} is the doubly degenerate Co–H bend and the features at 330 and 430 cm^{-1} are M–C≡O bending modes that involve the axial CO. The assignments are confirmed by a DFT calculation of the isolated molecule [73], Fig. 11.20b. Fig. 11.20c is an estimate of the spectrum that would be observed on TOSCA and demonstrates the huge strides in instrument performance that have occurred in the 30 years since Fig. 11.20a was recorded on a low-bandpass spectrometer at a reactor source.

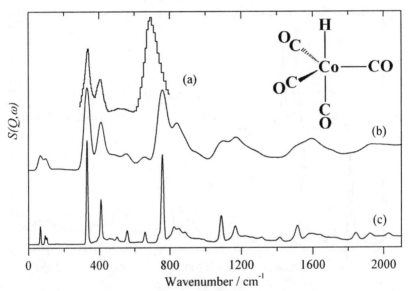

Fig. 11.20 INS spectrum of (a) [Co(CO)$_4$H] recorded on a low-bandpass spectrometer (now defunct) (b) spectrum of [Co(CO)$_4$H] calculated by DFT and displayed at the same resolution as the instrument in (a), (c) same data as (b) but displayed at TOSCA resolution. Reproduced from [72] with permission from the Royal Society of Chemistry.

The Co–H stretch occurs at 1934 cm^{-1} (2025 cm^{-1} calculated) and would not be observed. This is usual because the momentum transfer is so large that most of the intensity in this region originates from overtone and combination modes.

Polynuclear metal carbonyl hydrides have also been studied. The INS spectra of $[Os_3(CO)_{10}(\mu_2-H)_2]$ and $[Os_4(CO)_{12}H_4][74]$ are shown in Fig. 11.21. A twofold bridging hydride has three modes: symmetric and antisymmetric stretch and an out-of-plane bend. The stretches are expected to occur at about twice the frequency of the bend. Fig. 11.21a approximates to this expectation, however, there are more bands observed than expected. In particular, the observation of two bands for the out-of-plane bend at 672 and 738 cm^{-1} shows that the hydrides are coupled. This is surprising since the massive osmium atoms would mechanically decouple the hydride motions. So the coupling is likely to be electronic in nature.

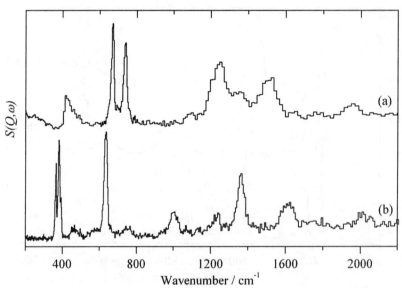

Fig. 11.21 INS spectrum of (a) $[Os_3(CO)_{10}H_2]$ and (b) $[Os_4(CO)_{12}H_4]$. Reproduced from [74] with permission from Elsevier.

The location of the hydrides in $[Os_4(CO)_{12}H_4]$ is not known with certainty, although the crystallographic data suggest that four edges of the tetrahedron formed by the osmium atoms are twofold bridged by hydrides. In this case the spectrum, Fig. 11.21b, conforms to the expectation of a simple bridging hydride, there is no evidence for coupling of the hydrides. The hydrides are coupled as shown by infrared and Raman spectroscopy [75] but the splittings are only a few wavenumbers, less than the INS resolution. The observation of the strong $Os-C\equiv O$ bending modes at 365 and 381 cm^{-1} can be accounted for by postulating a strong coupling between the hydride out-of-plane bend and some of the OsCO bending modes.

The $[Co_6(CO)_{15}H]^-$ ion is very unusual in that a neutron diffraction study of the $[\{(C_6H_5)_3P\}_2N]^+$ salt [76] showed that the cobalt atoms form a regular octahedron with a hydrogen atom located at the centre of the octahedron. An INS study of the caesium salt [77] showed one band in agreement with expectation, since the only hydrogen mode is triply degenerate motion along the Cartesian axes. Unexpectedly, a study of the potassium salt [78] showed two bands with an approximate intensity ratio of 2:1, indicating that the hydrogen atom is not located at the centre of the octahedron. A recent re-examination on TOSCA [79] of the spectra of the potassium and caesium salts is shown in Fig. 11.22. The improved resolution shows that the band at 1056 cm^{-1} in the caesium salt consists of two bands and that there are three bands in the potassium salt.

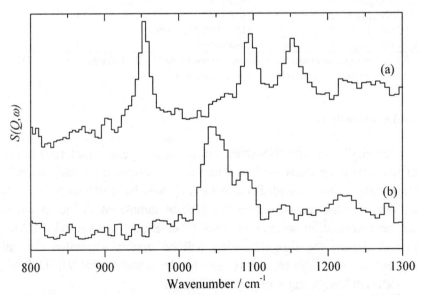

Fig. 11.22 INS spectrum of $[Co_6(CO)_{15}H]^-$ (a) potassium salt and (b) caesium salt [79].

The results suggest that the hydrogen atom is close to the centre of the octahedron in the caesium salt and occupies a different position in the potassium. The spectrum is consistent with the hydrogen having moved to one of the 'walls' and occupying a threefold bridging position. The implication is that the position of the hydrogen atom is very sensitive to the balancing cation and its local environment.

Table 11.2 lists the frequencies observed for the transition metal carbonyl hydrides that have been studied by INS.

Table 11.2 Metal–hydrogen stretch and bend frequencies of transition metal carbonyl hydrides studied by INS.

Molecule	Type (a)	ω / cm^{-1}		Reference
		Stretch	Deformation	
[Mn(CO)$_5$H]	1	1780 (b)	730	72
K[Fe(CO)$_4$H]	1		715	72
[Fe(CO)$_4$H$_2$]	1	1895 (b)	760/690	72
[Co(CO)$_4$H]	1	1934 (b)	696	72
[FeH(PF$_3$)$_4$]	1		652	72
[Mn$_3$(CO)$_{12}$H$_3$]	2		610	72
Cs[Fe$_3$(CO)$_{12}$H]	2		706	72
[Os$_3$(CO)$_{10}$H$_2$]	2	1514/1247	738/672	74
[Os$_4$(CO)$_{12}$H$_4$]	2	1611/1361	633	74
K[Co$_6$(CO)$_{15}$H]	6	1153/1096/953	(c)	79
Cs[Co$_6$(CO)$_{15}$H]	6	1090/1045	(c)	79

(a) Number of metal atoms the hydrogen is coordinated to. (b) Infrared data from [73]. (c) No M–H bending modes in this type of coordination.

11.3 Conclusions

The challenge with INS studies of non-hydrogenous materials is that of low scattering cross section, thus flux is a prerequisite and generally large samples are needed. Both of these factors have served to limit the amount of work carried out on this type of sample. With the advent of the next-generation sources at SNS (Oak Ridge, USA) and J-PARC (Tokai, Japan) the flux limitation will be greatly alleviated. In this context, the new high sensitivity, low-bandpass instrument VISION [80] at SNS will have a major impact in this area.

Direct geometry instruments also have advantages in this area. For many elements, the coherent cross section is larger than the incoherent cross section and with Q-resolved data it is possible to see additional structure [29,81] beyond that predicted by the scattering law, Eq. 2.32. These arise from coherent inelastic scattering and offer further possibilities for investigating the dynamics in such systems.

11.4 References

1   B. Dorner (1982). *Coherent Inelastic Neutron Scattering in Lattice Dynamics*, Springer Tracts In Modern Physics, vol. 93, Springer-Verlag, Berlin.

2   G. Shirane, S.M. Shapiro & J.M. Tranquada (2002). *Neutron Scattering with a Triple-Axis Spectrometer*, Cambridge University Press, Cambridge.

3   D.A. Braden & B.S. Hudson (2000). J. Phys. Chem. A, 104, 982–989. C_6F_6 and sym-$C_6F_3H_3$: *Ab initio* and DFT studies of structure, vibrations and inelastic neutron scattering spectra.

4   K. Nakamoto (1997). *Infrared and Raman Spectra of Inorganic and Coordination Compounds. Part A: Theory and Applications in Inorganic Chemistry*, 5th ed., Wiley-Interscience, New York.

5   S.F. Parker & J.B. Forsyth (1998). J. Chem. Soc. Faraday Trans., 94, 1111–1114. $K_2[MCl_6]$: (M = Pt, Ir) Location of the silent modes and forcefields.

6   S.F. Parker, H. Herman, A. Zimmerman & K.P.J. Williams (2000). Chem. Phys., 261, 261–266. The vibrational spectrum of $K_2[PdCl_4]$: first detection of the silent mode v_5.

7   G.J. Kearley, J. Tomkinson, A. Navarro, J.J. López-González, & M. Fernández-Gómez (1997). Chem. Phys., 216, 323–335. Symmetrised quantum-mechanical force-fields and INS spectra: s-triazine, trichloro-s-triazine and pyrazine.

8   M.T. Dove (2002). Eur. J. Mineral., 14, 203–224. An introduction to the use of neutron scattering methods in mineral sciences.

9   S.L. Chaplot, N. Choudhury, S. Ghose, M.N. Rao, R. Mittal & P. Goel (2002). Eur. J. Mineralogy, 14, 291–329. Inelastic neutron scattering and lattice dynamics of minerals.

10   M.N. Rao, S.L. Chaplot, N. Choudhury, K.R. Rao, R.T. Azuah, W.T. Montfrooij & S.M. Bennington (1999). Phys. Rev. B, 60, 12061–12068. Lattice dynamics and inelastic neutron scattering from sillimanite and kyanite Al_2SiO_5.

11   B. Winkler & B. Hennion (1994). Phys. Chem. Minerals, 21, 539–545. Low temperature dynamics of molecular H_2O in bassanite, gypsum and cordierite investigated by high resolution incoherent inelastic neutron scattering.

12   B. Winkler (1996). Phys. Chem. of Minerals, 23, 310–318. The dynamics of H_2O in minerals.

13   M. Arai, A.C. Hannon, A.D. Taylor, T. Otomo, A.C. Wright, R.N. Sinclair & D.L. Price (1991). Trans. Am. Cryst. Assoc., 27, 113–131. Neutron scattering law measurements for vitreous silica.

14   A. Pasquarello, J. Sarnthein & R. Car (1998). Phys. Rev. B, 57, 14133–14140. Dynamic structure factor of vitreous silica from first principles: Comparison to neutron-inelastic-scattering experiments.

15   M.J. Harris, M.T. Dove & J.M. Parker (2000). Mineral. Mag., 64, 435–440. Floppy modes and the Boson peak in crystalline and amorphous silicates: an inelastic neutron scattering study.

16   J.L. Warren, J.L. Yarnell, G. Dolling & R.A. Cowley (1967). Phys. Rev., 158, 805–808. Lattice dynamics of diamond.

17   R. Nicklow, N. Wakabayashi & H.G. Smith (1972). Phys. Rev. B, 5, 4951–4962. Lattice dynamics in pyrolytic graphite.

18   D.K. Ross (1973). J. Phys. C: Solid. State Phys., 6, 3525–3535. Inelastic neutron scattering from polycrystalline graphite up to 1920°C.

19   J.L. Wilkes, R.E. Palmer, & R.F. Willis (1987). J. Electron Spectrosc. 44, 355–360. Phonons in graphite studied by EELS.

20   C. Oshima, T. Aizawa, R. Souda, Y. Ishizawa & Y. Sumiyoshi (1988). Solid State Commun., 65, 1601–1604. Surface phonon-dispersion curves of graphite(0001) over the entire energy region.

21   G. Benedek, F. Hofmann, P. Ruggerone, G. Onida & L. Miglio (1994). Surf. Sci. Rep., 20, 3–43. Surface phonons in layered crystals - theoretical aspects.

22   W. Kratschmer, L.D. Lamb, K. Fostiropoulos & D.R. Huffmann (1990). Nature, 347, 354–358. Solid C_{60} - a new form of carbon.

23   L. Pintschovius (1996). Rep. Prog. Phys., 59, 473–510. Neutron studies of vibrations in fullerenes.

24   C. Christides, A.V. Nikolaev, T.J.S. Dennis, K. Prassides, F. Negri, G. Orlandi & F. Zerbetto (1993). J. Phys. Chem., 97, 3641–3643. Inelastic neutron scattering study of the intramolecular vibrations of the C_{70} fullerenes.

25   G. Onida, W. Andreoni, J. Kohanoff & M. Parinello (1994). Chem. Phys. Lett., 219, 1–7. Ab initio molecular dynamics of C_{70}: intramolecular vibrations and zero-point motion effects.

26   C. Coulombeau, H. Jobic, P. Bernier, C. Fabre, D. Schutz & A. Rassat (1991). Compt. Rendus Acad. Sci. Paris, 313, Serie II, 1387–1390. Spectre de diffusion inelastique des neutrons du footballene C_{60}.
     C. Coulombeau, H. Jobic, P. Bernier, C. Fabre, D. Schutz & A. Rassat (1992). J. Phys. Chem., 96, 22–24. Neutron inelastic scattering spectrum of footballene C_{60}.

27   R.L. Cappelletti, J.R.D. Copley, W.A. Kamitakahara, F. Li, J.S. Lannin, & D. Ramage (1991). Phys. Rev. Lett., 66, 3261–3264. Neutron measurements of intramolecular   vibrational   modes   in C_{60}.
     K. Prassides, T.J.S. Dennis, J.P. Hare, J. Tomkinson, H.W. Kroto, R. Taylor & D.R.M. Watkin (1991). Chem. Phys. Lett., 187, 455–458. Inelastic neutron scattering spectrum of the fullerene C_{60}.

28   C. Coulombeau, H. Jobic, C.J. Carlile, S.M. Bennington, C. Fabre & A. Rasset (1994). Fullerene Sci. and Techn., 2, 247–254. On the vibrational spectrum of C_{60} measured by neutron inelastic scattering.

29   R. Heid, L. Pinschovius & J.M. Godard (1997). Phys. Rev. B, 56, 5925–5936. Eigenvectors of internal vibrations of C_{60}: Theory and experiment.

30 S.F. Parker, J.W. Taylor, S.J. Levett, S.M. Bennington, A.J. Champion & H. Herman (2004). Phys. Rev. B, submitted for publication. Assignment of the internal modes of C_{60} by inelastic neutron scattering: one-third of present assignments are wrong.

31 K.-A. Wang, A.M. Rao, P.C. Eklund, M.S. Dresselhaus & G. Dresselhaus (1993). Phys. Rev. B, 48, 11375–11380. Observation of higher-order infrared modes in solid C_{60} films.

32 Z.-H. Dong, P. Zhou, J.M. Holden, P.C. Eklund, M.S. Dresselhaus & G. Dresselhaus (1993). Phys. Rev. B, 48, 2862–2865. Observation of higher-order Raman modes in solid C_{60} films.

33 W.I.F. David, R.I. Ibberson, J.C. Matthewman, K. Prassides, T.J.S. Dennis, J.P. Hare, H.W. Kroto, R. Taylor & D.R.M. Walton (1991). Nature, 353, 147–149. Crystal structure and bonding of ordered C_{60}.

34 S. Margadonna & K. Prassides (2002). J. Solid State Chem., 168, 639–652. Recent advances in fullerene superconductivity.

35 W.K. Fullagar (1999). Fullerene Sci. Techn., 7, 1175–1179. Molecular fullerides.

36 B. Renker, F. Gompf, H. Schober, P. Adelmann, H.J. Bornemann & R. Heid (1993). Z. Phys. B, 92, 451–455. Intermolecular vibrations in pure and doped C_{60}. An inelastic neutron scattering study.

37 B. Renker, F. Gompf, H. Schober, P. Adelmann & R. Heid (1994). J. Supercond., 7, 647–649. Intermolecular vibrations in pure and doped C_{60}: an inelastic neutron scattering study.

38 W.K. Fullagar, J.W. White & F. Trouw (1995). Physica B, 213/214, 16–21. Superconductivity, vibrational and lattice dynamics of Rb_3C_{60}.

39 H. Schober & B. Renker (1999). Neutron News, 10, 28–33. On how to do solid state chemistry with inelastic neutron scattering: the example of network formation in fullerenes.

40 H. Schober & A. Tölle (1997). Phys. Rev. B, 56, 5937–5950. Microscopic dynamics of AC_{60} compounds in the plastic, polymer, and dimer phases investigated by inelastic neutron scattering.

41 A.I. Kolesnikov, I.O. Bashkin, A.P. Moravsky, M.A. Adams, M. Prager & E.G. Ponyatovsky (1996). J. Phys.: Condens. Matter, 8, 10939–10949. Neutron scattering study of a high pressure polymeric C_{60} phase.

42 B. Renker, H. Schober, R. Heid & P. von Stein (1997). Solid State Commun., 104, 527–530. Pressure and charge induced polymerisation of C_{60}: A comparative study of lattice vibrations.

43 A. Soldatov, O. Andersson, B. Sundqvist & K. Prassides (1998). Mol. Cryst. Liq. Crys.. C, 11, 1–6. Transport and vibrational properties of pressure polymerized C_{60}.

44 H. Schober, B. Renker & R. Heid (1999). Phys. Rev. B, 60, 998–1004. Vibrational behavior of Na_4C_{60} in the monomer and two- dimensional polymer states.

45 A.V. Talyzin, L.S. Dubrovinsky, T. Le Bihan & U. Jansson (2002). J. Chem. Phys., 116, 2166–2174. *In situ* Raman study of C_{60} polymerization at high pressure high temperature conditions.

46 S.M. Bennington, N. Kitamura, M.G. Cain, M.H. Lewis & M. Arai (1999). Physica B, 263, 632–635. The structure and dynamics of hard carbon formed from C_{60} fullerene.

47 P.J.R Honeybone, R.J. Newport, W.S. Howells, J. Tomkinson S.M. Bennington & P.J Revill (1991). Chem. Phys. Lett., 180, 145–148. Inelastic neutron scattering of molecular hydrogen in amorphous hydrogenated carbon.

48 P.J.R Honeybone, R.J. Newport, J.K. Walters, W.S. Howells & J. Tomkinson (1994). Phys. Rev. B, 50, 839–845. Structural properties of amorphous hydrogenated carbon: 2. An inelastic neutron scattering study.

49 J.K. Walters, J.S. Rigden, R.J. Newport, S.F. Parker & W.S. Howells (1995). Physica Scripta, T57, 142–145. The effect of temperature on the structure of amorphous hydrogenated carbon.

50 J.K. Walters & R.J. Newport (1995). J. Phys.: Condens. Matter, 7, 1755–1769. The atomic scale structure of amorphous hydrogenated carbon.

51 J.K. Walters, R.J. Newport, S.F. Parker & W.S. Howells (1995). J. Phys.: Condens. Mat., 7, 10059–10073. A spectroscopic study of the structure of amorphous hydrogenated carbon.

52 J.K. Walters, R.J. Newport, S.F. Parker, W.S. Howells & G. Bushell-Wye (1998). J. Phys.: Condens. Mat., 10, 4161–4176. The effect of hydrogen dilution on the structure of a-C:H.

53 F. Fillaux, R. Papoular, A. Lautie & J Tomkinson (1994). Carbon, 32, 1325–1331. Inelastic neutron scattering study of the proton dynamics in carbons and coals.

54 F. Fillaux, A. Lautie, R. Papoular, S.M. Bennington & J. Tomkinson (1995). Physica B, 213, 631–633. Free proton dynamics in coal.

55 F. Fillaux, R. Papoular, A. Lautie & J. Tomkinson (1995). Fuel, 74, 865–873. Inelastic neutron scattering study of the proton dynamics in coal.

56 P. Albers, S. Bösing, G. Prescher, K. Seibold, D.K. Ross & S.F Parker (1999). Applied Catalysis A: General, 187, 233–243. Inelastic neutron scattering investigations of the products of thermally and catalytically driven catalyst coking.

57 P.W. Albers, S. Bosing, H.L. Rotgerink, D.K. Ross & S.F. Parker (2002). Carbon, 40, 1549–1558. Inelastic neutron scattering study on the influence of after-treatments on different technical cokes of varying impurity level and sp^2/sp^3 character.

58 P. Albers, R. Burmeister, K. Seibold, G. Prescher, S.F. Parker & D.K. Ross (1999). J. Catal., 181, 145–154. Investigations of palladium catalysts on different carbon supports.

59 D. Lennon, D.T. Lundie, S.D. Jackson, G.J. Kelly and S.F. Parker (2002). Langmuir, 18, 4667–4673. Characterisation of activated carbon using X-ray photoelectron spectroscopy and inelastic neutron scattering spectroscopy.

60 P. Albers, J. Pietsch, J. Krauter & S.F. Parker (2003). Phys. Chem. Chem. Phys., 5, 1941–1949. Investigations of activated carbon catalyst supports from different natural sources.

61 P.C.H. Mitchell, A.J. Ramirez-Cuesta, S.F. Parker & J. Tomkinson (2003). J. Mol. Struc., 651, 781–785. Inelastic neutron scattering in spectroscopic studies of

hydrogen on carbon-supported catalysts—experimental spectra and computed spectra of model systems.

62   P. Albers, G. Prescher, K. Seibold, D.K. Ross & F. Fillaux (1996). Carbon, 34, 903–908. Inelastic neutron-scattering study of proton dynamics in carbon-blacks.

63   P. Albers, K. Seibold, G. Prescher, B. Freund, S.F. Parker, J. Tomkinson, D.K. Ross & F. Fillaux (1999). Carbon, 37, 437–444. Neutron spectroscopic investigations on different grades of modified furnace blacks and gas blacks.

64   P.W. Albers, H. Klein, E.S. Lox, K. Seibold, G. Prescher & S.F. Parker (2000). Phys. Chem. Chem. Phys., 2, 1051–1058. INS-, SIMS- and XPS-investigation of diesel engine exhaust particles.

65   P. Albers, A. Karl, J. Mathias, D.K. Ross & S.F. Parker (2001). Carbon, 39, 1663–1676. INS-, XPS- and SIMS-investigations on the controlled postoxidation of pigment blacks -Detection of different species of strongly adsorbed water.

66   P. Zhou, P. Papanek, C. Bindra, R. Lee & J.E. Fischer (1997). J. of Power Sources, 68, 296–300. High capacity carbon anode materials: structure, hydrogen effect and stability.

67   P. Zhou, P. Papanek, R. Lee, J.E. Fischer & W.A. Kamitakahara (1997). J. Electrochem. Soc., 144, 1744–1750. Local structure and vibrational spectroscopy of disordered carbons for Li batteries: Neutron scattering studies.

68   P. Papanek, W.A. Kamitakahara, P. Zhou & J.E. Fischer (2001). J. Phys.: Condens. Mat., 13, 8287–8301. Neutron scattering studies of disordered carbon anode materials.

69   K. Nakamoto (1997). *Infrared and Raman Spectra of Inorganic and Coordination Compounds. Part B: Applications in Coordination, Organometallic and Bioinorganic Chemistry*, 5th ed., Wiley-Interscience, New York.

70   S.F. Parker & U.A. Jayasooriya, unpublished work.

71   S.F. Parker & B.S. Hudson, unpublished work.

72   J.W. White & C.J. Wright (1972). J. Chem. Soc. Faraday Trans. II, 68, 1423–1433. Neutron scattering spectra of hydridocarbonyls in the 900–200 cm^{-1} energy region.

73   V. Jonas & W. Thiel (1996). J. Chem. Phys., 105, 3636–3648. Theoretical study of the vibrational spectra of the transition-metal carbonyl hydrides [HM(CO)$_5$ (M=Mn, Re), [H$_2$M(CO)$_4$] (M=Fe, Ru, Os) and [HM(CO)$_4$] (M=Co, Rh, Ir).

74   W.C-K. Poon, P. Dallin, B.F.G. Johnson, E.A. Marseglia & J. Tomkinson (1990). Polyhedron, 9, 2759–2761. Inelastic neutron scattering study of hydride ligands in [(μ$_2$-H)$_2$Os$_3$(CO)$_{10}$]: evidence for a direct H–H interaction.

75   C.E. Anson, U.A. Jayasooriya, S.F.A. Kettle, P.L. Stanghellini & R. Rossetti (1991). Inorg. Chem., 30, 2282–2286. Vibrational spectra of the μ$_2$-hydride ligands in [H$_4$M$_4$(CO)$_{12}$] (M = Ru, Os).

76   D.W. Hart, R.G. Teller, C-Y. Wei, R. Bau, G. Longoni, S. Campanella, P. Chini & T.F. Koetzle (1981). J. Am. Chem. Soc., 103, 1458–1466. An interstitial hydrogen atom in a hexanuclear metal cluster: X-ray and neutron diffraction analysis of [(Ph$_3$P)$_2$N]$^+$[HCo$_6$(CO)$_{15}$]$^-$

77   D. Graham, J. Howard, T.C. Waddington & J. Tomkinson (1983). J. Chem. Soc. Faraday Trans. II, 79, 1713–1717. Incoherent inelastic neutron scattering studies of transition-metal hydridocarbonyls. Part 1 Cs$^+$[HCo$_6$(CO)$_{15}$]$^-$

78   J. Eckert, A. Albinati & G. Longoni (1989). Inorg. Chem., 28, 4055–4056. Inelastic
     neutron scattering study of K[HCo$_6$(CO)$_{15}$] implications for location of the hydride.
79   A. Albinati, D. Colognesi & S.F. Parker, unpublished work.
80   http://www.sns.gov/users/vol4_no2_pulse.pdf (pg 11).
81   A.C. Hannon, M. Arai & R.G. Delaplane (1995). Nucl. Instrum. and Meth., A 354,
     96 – 103. A dynamic correlation function from inelastic neutron scattering data.

12

Vibrational Spectroscopy with Neutrons— the Future

There are several themes that run through this book that provide hints to future developments. Worldwide, there are extensive programmes of new neutron source development. Taken together these provide a justification for speculating, with more than the usual degree confidence, on the future of vibrational spectroscopy with neutrons.

INS spectroscopy is currently a technique more limited by flux than resolution. It is also an expensive and scarce resource. Thus the emphasis will always be on capturing the neutrons that are available. There are two complementary approaches to this goal; capture more of the incident flux, with better guides, and capture more of the scattered flux, with larger area detectors. High index guides, $m > 2$ (§3.2.2), are becoming more commonplace and their use is almost mandatory on new instrumentation. Retrofitting existing instruments (Table 3.1) with high index guides will provide useful gains and this is an activity that will surely continue.

One of the instruments that will surely benefit from the use of such guides is the newly proposed spectrometer VISION (SNS, Oak Ridge, USA) [1]. This is a next-generation low-bandpass instrument. An innovative design coupled with the use of a guide and the higher power of a new source means that the instrument will have the equivalent of 100 times the flux of TOSCA.

Such a leap forward means that routine samples could be measured in a few minutes. It will be possible to obtain INS spectra from just a single milligram of organic sample – which will enable INS spectroscopy of biological samples to blossom. Incoherent INS spectra of non-hydrogenous samples will be routine and it will also allow INS spectra of industrially important non-hydrogenous adsorbates such as CO_x, NO_x, SO_x, to be investigated on catalysts.

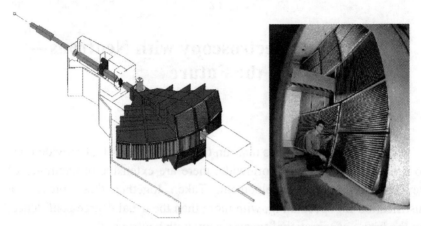

Fig. 12.1 (a) The INS direct geometry spectrometer MAPS at ISIS (Chilton, UK), (b) the position sensitive detectors on MAPS.

The MAPS spectrometer at ISIS [2], Fig. 12.1(a), is a third generation instrument that demonstrates the future direction of direct geometry instruments. The principal innovation is the use of large area, position sensitive ^{3}He detectors (§3.3.1.1). An area of 16 m^2, Fig. 12.1(b), of the sample environment tank is covered by 576 detectors that provide almost continuous coverage over a large solid angle in the forward scattering direction.

MAPS itself is optimised for the measurement of high energy magnetic excitations (> 400 cm^{-1}) in single crystals. The pixel size, when represented in terms of the Brillouin zone, is significantly smaller than the resolution volume defined by the other instrumental contributions. This type of instrument has yet to be used for incoherent scattering studies but their potential is great. While the Q-dependence is usually not useful for incoherent scattering there are cases where this information is valuable and the ability to tailor the measurement to extract it is very exciting. The disadvantage of the sophistication of this type of instrumentation is that it is much more complex to use than low-bandpass instruments. (See §3.4.4 for a discussion of direct and indirect geometry instruments.)

Beyond the instrumentation itself; better use can be made of the neutrons that are detected and more information can be extracted from the spectra. This approach is totally reliant on increasing the computing power that enables larger calculations to be undertaken. This is important in two directions: it allows the vast quantities of data from modern instruments to be treated, and the dynamics of more complex systems to be calculated. The 147456 pixels of MAPS result in 36,864 spectra that generate 0.5 Gbytes of data per measurement. While improvements and innovations in detector technology are an important part in this process, it is the continuing increase in computing power that allows such enormous data sets to be manipulated and analysed.

In the late 1990's desktop computers became powerful enough to run *ab initio* calculations of the isolated molecule in a day or two. The result was that this has become the *de facto* standard for INS analysis. It provides complete vibrational assignments and allows large molecules (~60 atoms) to be analysed. What would have been, 10 years ago, a PhD project, with no guarantee of success, can now be accomplished in a week, with certainty.

The trend is continuing: *ab initio* packages that can calculate vibrational frequencies and their atomic displacements for extended solids are now available. In addition to the analysis of new systems, this also enables older data to be mined again to provide new insights. N-methylacetamide [3,4] and thymine [5,6] are cases in point. The availability of older data is a crucial factor in this data mining and initiatives such as the INS database [7] that enable this are sure to become standard practice.

The internal modes of typical molecular crystals show little dispersion and a calculation of the frequencies at the Γ point in the Brillouin zone is usually sufficient for good agreement between observed and calculated spectra. However, there are many examples where significant dispersion is present; the alkali metal hydrides (§6.7.1), graphite (§11.2.2) and polyethylene (§10.1.1.1) being notable cases. In these instances, a calculation of the full dispersion curves are needed. In

2004, this remains state-of-the-art and is limited to systems with 10 or so atoms in the unit cell. This is a limitation that will yield to increasing computing power.

The area where *ab initio* calculations of periodic systems have yet to make their mark is that of surfaces. This has considerable potential. For an adsorbed molecule, it is conceivable to calculate a number of possible conformations and adsorption sites and compare the calculated spectra to the observed. A reason for the present paucity of examples is that vibrational spectra of adsorbed species on surfaces are relatively difficult to both calculate and measure experimentally. While single-crystal surface calculations are only an approximation to a catalyst surface they provide a useful starting point and the data from such systems can inform studies on real catalysts. The ease of extracting INS spectra from *ab initio* calculations is an area that should be of interest to the surface science community, who currently drive this area of research.

All of this discussion is focused on the crystalline state where long range order is present and can be exploited. Amorphous systems are a highly interesting and important area that is still neglected: an area that is clearly ripe for development.

In the longer term, the solution to limited flux is more, and more powerful, neutron sources. There is a considerable building programme of both reactors and spallation sources. The FRM-II reactor [8] (Munich, Germany) went critical for the first time in 2004. A replacement for the HIFAR reactor [9] (Lucas Heights, Australia) is under construction and is scheduled for operation in late 2005.

As described in Chapter 3, spallation sources have notable advantages over reactors for vibrational spectroscopy. The high flux of epithermal neutrons resulting from under-moderation of the spallation neutrons means that they are more suited to the energy range needed for vibrational spectroscopy. ISIS (Chilton, UK) will double in size by 2007 with the construction of a second target station [10]. This is optimised for neutrons below 200 cm^{-1} so will broaden the opportunities for studies of the low energy modes of much larger molecules and dihydrogen on surfaces (§6.3 and 6.4).

On a similar timescale, the SNS (Oak Ridge, USA) [11] and J-PARC

(Tokai, Japan) [12] spallation sources will be coming on line and are scheduled to reach full power around 2010. These are both 1 MW neutron sources so will deliver six times the neutron flux of ISIS.

Further into the future, J-PARC has been designed such that it is possible to upgrade it to 3 MW. There is a proposal to build a European spallation source (ESS) [13] which will have a time-averaged flux equal to that of the ILL and a peak flux 30 times that of ISIS.

Throughout the book we have stressed the two great advantages of neutrons for vibrational spectroscopy:(a) neutrons enable spectra to be obtained in circumstances where optical methods fail and (b) the ease of calculation of the spectrum from the output of modern *ab initio* calculations. There can be little doubt that this unique combination of attributes will ensure that INS spectroscopy will play an important rôle in future neutron scattering programs. The future of vibrational spectroscopy with neutrons is bright!

12.1 References

1   http://www.sns.gov/users/vol4_no2_pulse.pdf (pg 11)
2   http://www.isis.rl.ac.uk/excitations/maps/
3   F. Fillaux, J.P. Fontaine, M.H. Baron, G.J. Kearley & J. Tomkinson (1993). Chem. Phys., 176, 249–278. Inelastic neutron scattering study of the proton dynamics in N-methylacetamide and force-field calculation at 20 K.
4   G.J. Kearley, M. R. Johnson, M. Plazanet and E. Suard (2001). J. Chem. Phys., 115, 2614–2620. Structure and vibrational dynamics of the strongly hydrogen-bonded model peptide: N-methylacetamide.
5   A. Aamouche, G. Berthier, C. Coulombeau, J. P. Flament, M. Ghomi, C. Henriet, H. Jobic & P.Y. Turpin (1996). Chem. Phys., 204, 353–363. Molecular force fields of uracil and thymine, through neutron inelastic scattering experiments and scaled quantum mechanical calculations.
6   M. Plazanet, N. Fukushima & M.R. Johnson (2002). Chem. Phys., 280, 53–70. Modelling molecular vibrations in extended hydrogen-bonded networks - crystalline bases of RNA and DNA and the nucleosides.
7   http://www.isis.rl.ac.uk/insdatabase
8   http://www.frm2.tu-muenchen.de/
9   http://www.ansto.gov.au/ansto/RRR/index.html
10   http://www.isis.rl.ac.uk/TargetStation2/
11   http://www.sns.gov
12   http://jkj.tokai.jaeri.go.jp/
13   http://neutron.neutron-eu.net/n_ess

Appendix 1

Neutron Cross Sections of the Elements

Table A1.1 defines the contents of Tables A1.2 and A1.3. Table A1.2 lists the elements most important to INS. Table A1.3 is the full list; it is arranged alphabetically by chemical symbol, hence it starts with Ag for silver. Only elements up to uranium are included; if there are no data available for the element it is omitted. More complete tables which also include scattering lengths are given in [1–3].

Table A1.1 Definition of categories of Tables A1.2 and A1.3. (1 barn = 10^{-28} m^2, cross sections in parenthesis are uncertainties). Reproduced from [1] with permission from Taylor and Francis.

Column	Symbol	Unit	Quantity
1			Element name
2			Isotope
3	%		Natural abundance (a)
4	σ_{coh}	barn	Bound coherent scattering cross section
5	σ_{inc}	barn	Bound incoherent scattering cross section
6	σ_{tot}	barn	Total bound scattering cross section
7	σ_{abs}	barn	Absorption cross section for 200 cm^{-1} (2200 m s^{-1}) neutrons

(a) for radioisotopes the half-life in years is given instead

Table A1.2 Scattering and absorption cross sections of the INS important elements.

Element	Isotope	%	σ_{coh}	σ_{inc}	σ_{tot}	σ_{abs}
H			1.7568	80.26	82.02	0.3326
	1	99.985	1.7583	80.27	82.03	0.3326
	2	0.015	5.592	2.05	7.64	0.000519
C			5.551	0.001	5.551	0.0035
N			11.01	0.5	11.51	1.9
O			4.232	0.0008	4.232	0.00019

Table A1.3 Scattering and absorption cross sections of the elements.

Element	Isotope	%	σ_{coh}	σ_{inc}	σ_{tot}	σ_{abs}
Ag			4.407	0.58	4.99	63.3
	107	51.83	7.17	0.13	7.3	37.6(1.2)
	109	48.17	2.18	0.32	2.5	91.0(1.0)
Al		100	1.495	0.0082	1.503	0.231
Ar			0.458	0.225	0.683	0.675
	36	0.337	77.9	0	77.9	5.2
	38	0.063	1.5(3.1)	0	1.5(3.1)	0.8
	40	99.6	0.421	0	0.421	0.66
As		100	5.44	0.06	5.5	4.5
Au		100	7.32	0.43	7.75	98.65
B			3.54	1.7	5.24	767.(8.)
	10	20	0.144	3	3.1	3835.(9.)
	11	80	5.56	0.21	5.77	0.0055
Ba			3.23	0.15	3.38	1.1
	130	0.11	1.6	0	1.6	30.(5.)
	132	0.1	7.6	0	7.6	7
	134	2.42	4.08	0	4.08	2.0(1.6)
	135	6.59	2.74	0.5	3.2	5.8
	136	7.85	3.03	0	3.03	0.68
	137	11.23	5.86	0.5	6.4	3.6
	138	71.7	2.94	0	2.94	0.27
Be		100	7.63	0.0018	7.63	0.0076
Bi		100	9.148	0.0084	9.156	0.0338
Br			5.8	0.1	5.9	6.9
	79	50.69	5.81	0.15	5.96	11
	81	49.31	5.79	0.05	5.84	2.7
C			5.551	0.001	5.551	0.0035
	12	98.9	5.559	0	5.559	0.00353
	13	1.1	4.81	0.034	4.84	0.00137
Ca			2.78	0.05	2.83	0.43
	40	96.941	2.9	0	2.9	0.41
	42	0.647	1.42	0	1.42	0.68
	43	0.135	0.31	0.5	0.8	6.2
	44	2.086	0.25	0	0.25	0.88
	46	0.004	1.6	0	1.6	0.74
	48	0.187	0.019	0	0.019	1.09
Cd			3.04	3.46	6.5	2520.(50.)

Element	Isotope	%	σ_{coh}	σ_{inc}	σ_{tot}	σ_{abs}
	106	1.25	3.1	0	3.1(2.5)	1
	108	0.89	3.7	0	3.7	1.1
	110	12.51	4.4	0	4.4	11
	111	12.81	5.3	0.3	5.6	24
	112	24.13	5.1	0	5.1	2.2
	113	12.22	12.1	0.3	12.4	20600.(400.)
	114	28.72	7.1	0	7.1	0.34
	116	7.47	5	0	5	0.075
Ce			2.94	0.001	2.94	0.63
	136	0.19	4.23	0	4.23	7.3(1.5)
	138	0.25	5.64	0	5.64	1.1
	140	88.48	2.94	0	2.94	0.57
	142	11.08	2.84	0	2.84	0.95
Cl			11.5257	5.3	16.8	33.5
	35	75.77	17.06	4.7	21.8	44.1
	37	24.23	1.19	0.001	1.19	0.433
Co		100	0.779	4.8	5.6	37.18
Cr			1.66	1.83	3.49	3.05
	50	4.35	2.54	0	2.54	15.8
	52	83.79	3.042	0	3.042	0.76
	53	9.5	2.22	5.93	8.15	18.1(1.5)
	54	2.36	2.6	0	2.6	0.36
Cs		100	3.69	0.21	3.9	29.0(1.5)
Cu			7.485	0.55	8.03	3.78
	63	69.17	5.2	0.006	5.2	4.5
	65	30.83	14.1	0.4	14.5	2.17
Dy			35.9	54.4(1.2)	90.3	994.(13.)
	156	0.06	4.7	0	4.7	33.(3.)
	158	0.1	5.(6.)	0	5.(6.)	43.(6.)
	160	2.34	5.6	0	5.6	56.(5.)
	161	19	13.3	3.(1.)	16.(1.)	600.(25.)
	162	25.5	0.25	0	0.25	194.(10.)
	163	24.9	3.1	0.21	3.3	124.(7.)
	164	28.1	307.(3.)	0	307.(3.)	2840.(40.)
Er			7.63	1.1	8.7	159.(4.)
	162	0.14	9.7	0	9.7	19.(2.)
	164	1.56	8.4	0	8.4	13.(2.)
	166	33.4	14.1	0	14.1	19.6(1.5)
	167	22.9	1.1	0.13	1.2	659.(16.)

Element	Isotope	%	σ_{coh}	σ_{inc}	σ_{tot}	σ_{abs}
	168	27.1	6.9	0	6.9	2.74
	170	14.9	11.6	0	11.6(1.2)	5.8
Eu			6.57	2.5	9.2	4530.(40.)
	151	47.8	5.5	3.1	8.6	9100.(100.)
	153	52.2	8.5	1.3	9.8	312.(7.)
F		100	4.017	0.0008	4.018	0.0096
Fe			11.22	0.4	11.62	2.56
	54	5.8	2.2	0	2.2	2.25
	56	91.7	12.42	0	12.42	2.59
	57	2.2	0.66	0.3	1	2.48
	58	0.3	28	0	28.(26.)	1.28
Fr			0			
Ga			6.675	0.16	6.83	2.75
	69	60.1	7.8	0.091	7.89	2.18
	71	39.9	5.15	0.084	5.23	3.61
Gd			29.3	151.(2.)	180.(2.)	49700.(125.)
	152	0.2	13.(8.)	0	13.(8.)	735.(20.)
	154	2.1	13.(8.)	0	13.(8.)	85.(12.)
	155	14.8	40.8	25.(6.)	66.(6.)	61100.(400.)
	156	20.6	5	0	5	1.5(1.2)
	157	15.7	650.(4.)	394.(7.)	1044.(8.)	259000.(700.)
	158	24.8	10.(5.)	0	10.(5.)	2.2
	160	21.8	10.52	0	10.52	0.77
Ge			8.42	0.18	8.6	2.2
	70	20.5	12.6	0	12.6	3
	72	27.4	9.1	0	9.1	0.8
	73	7.8	3.17	1.5	4.7	15.1
	74	36.5	7.2	0	7.2	0.4
	76	7.8	8.(3.)	0	8.(3.)	0.16
H			1.7568	80.26	82.02	0.3326
	1	99.985	1.7583	80.27	82.03	0.3326
	2	0.015	5.592	2.05	7.64	0.000519
	3	(12.32)	2.89	0.14	3.03	0
He			1.34	0	1.34	0.00747
	3	0.00014	4.42	1.6	6	5333.(7.)
	4	99.99986	1.34	0	1.34	0
Hf			7.6	2.6	10.2	104.1
	174	0.2	15.(3.)	0	15.(3.)	561.(35.)
	176	5.2	5.5	0	5.5	23.5(3.1)

Element	Isotope	%	σ_{coh}	σ_{inc}	σ_{tot}	σ_{abs}
	177	18.6	0.1	0.1	0.2	373.(10.)
	178	27.1	4.4	0	4.4	84.(4.)
	179	13.7	7	0.14	7.1	41.(3.)
	180	35.2	21.9	0	21.9(1.0)	13.04
Hg			20.24	6.6	26.8	372.3(4.0)
	196	0.2	115.(8.)	0	115.(8.)	3080.(180.)
	198	10.1		0		2
	199	17	36.(2.)	30.(3.)	66.(2.)	2150.(48.)
	200	23.1		0		<60.
	201	13.2				7.8(2.0)
	202	29.6		0	9.828	4.89
	204	6.8		0		0.43
Ho		100	8.06	0.36	8.42	64.7(1.2)
I		100	3.5	0.31	3.81	6.15
In			2.08	0.54	2.62	193.8(1.5)
	113	4.3	3.65	0.00004	3.65	12.0(1.1)
	115	95.7	2.02	0.55	2.57	202.(2.)
Ir			14.1	0.(3.)	14.(3.)	425.(2.)
	191	37.3				954.(10.)
	193	62.7				111.(5.)
K			1.69	0.27	1.96	2.1
	39	93.258	1.76	0.25	2.01	2.1
	40	0.012	1.1	0.5	1.6	35.(8.)
	41	6.73	0.91	0.3	1.2	1.46
Kr			7.67	0.01	7.68	25.(1.)
	78	0.35		0		6.4
	80	2.25		0		11.8
	82	11.6		0		29.(20.)
	83	11.5				185.(30.)
	84	57		0	6.6	0.113
	86	17.3	8.2	0	8.2	0.003
La			8.53	1.13	9.66	8.97
	138	0.09	8.(4.)	0.5	8.5(4.0)	57.(6.)
	139	99.91	8.53	1.13	9.66	8.93
Li			0.454	0.92	1.37	70.5
	6	7.5	0.51	0.46	0.97	940.(4.)
	7	92.5	0.619	0.78	1.4	0.0454
Lu			6.53	0.7	7.2	74.(2.)
	175	97.39	6.59	0.6	7.2	21.(3.)

Element	Isotope	%	σ_{coh}	σ_{inc}	σ_{tot}	σ_{abs}
	176	2.61	4.7	1.2	5.9	2065.(35.)
Mg			3.631	0.08	3.71	0.063
	24	78.99	4.03	0	4.03	0.05
	25	10	1.65	0.28	1.93	0.19
	26	11.01	3	0	3	0.0382
Mn		100	1.75	0.4	2.15	13.3
Mo			5.67	0.04	5.71	2.48
	92	14.84	6	0	6	0.019
	94	9.25	5.81	0	5.81	0.015
	95	15.92	6	0.5	6.5	13.1
	96	16.68	4.83	0	4.83	0.5
	97	9.55	6.59	0.5	7.1	2.5
	98	24.13	5.44	0	5.44	0.127
	100	9.63	5.69	0	5.69	0.4
N			11.01	0.5	11.51	1.9
	14	99.63	11.03	0.5	11.53	1.91
	15	0.37	5.21	0.00005	5.21	0.000024
Na		100	1.66	1.62	3.28	0.53
Nb		100	6.253	0.0024	6.255	1.15
Nd			7.43	9.2	16.6	50.5(1.2)
	142	27.16	7.5	0	7.5	18.7
	143	12.18	25.(7.)	55.(7.)	80.(2.)	337.(10.)
	144	23.8	1	0	1	3.6
	145	8.29	25.(7.)	5.(5.)	30.(9.)	42.(2.)
	146	17.19	9.5	0	9.5	1.4
	148	5.75	4.1	0	4.1	2.5
	150	5.63	3.5	0	3.5	1.2
Ne			2.62	0.008	2.628	0.039
	20	90.51	2.695	0	2.695	0.036
	21	0.27	5.6	0.05	5.7	0.67
	22	9.22	1.88	0	1.88	0.046
Ni			13.3	5.2	18.5	4.49
	58	68.27	26.1	0	26.1	4.6
	60	26.1	0.99	0	0.99	2.9
	61	1.13	7.26	1.9	9.2	2.5
	62	3.59	9.5	0	9.5	14.5
	64	0.91	0.017	0	0.017	1.52
O			4.232	0.0008	4.232	0.00019
	16	99.762	4.232	0	4.232	0.0001

Element	Isotope	%	σ_{coh}	σ_{inc}	σ_{tot}	σ_{abs}
	17	0.038	4.2	0.004	4.2	0.236
	18	0.2	4.29	0	4.29	0.00016
Os			14.4	0.3	14.7	16
	184	0.02	13.(5.)	0	13.(5.)	3000.(150.)
	186	1.58	17.(5.)	0	17.(5.)	80.(13.)
	187	1.6	13.(5.)	0.3	13.(5.)	320.(10.)
	188	13.3	7.3	0	7.3	4.7
	189	16.1	14.4	0.5	14.9	25.(4.)
	190	26.4	15.2	0	15.2	13.1
	192	41	16.6	0	16.6(1.2)	2
P		100	3.307	0.005	3.312	0.172
Pa		(32800)	10.4	0.1(3.3)	10.5(3.2)	200.6(2.3)
Pb			11.115	0.003	11.118	0.171
	204	1.4	12.3	0	12.3	0.65
	206	24.1	10.68	0	10.68	0.03
	207	22.1	10.82	0.002	10.82	0.699
	208	52.4	11.34	0	11.34	0.00048
Pd			4.39	0.093	4.48	6.9
	102	1.02	7.5(1.4)	0	7.5(1.4)	3.4
	104	11.14	7.5(1.4)	0	7.5(1.4)	0.6
	105	22.33	3.8	0.8	4.6(1.1)	20.(3.)
	106	27.33	5.1	0	5.1	0.304
	108	26.46	2.1	0	2.1	8.55
	110	11.72	7.5(1.4)	0	7.5(1.4)	0.226
Pm		(2.62)	20.0(1.3)	1.3(2.0)	21.3(1.5)	168.4(3.5)
Pr		100	2.64	0.015	2.66	11.5
Pt			11.58	0.13	11.71	10.3
	190	0.01	10.(2.)	0	10.(2.)	152.(4.)
	192	0.79	12.3(1.2)	0	12.3(1.2)	10.0(2.5)
	194	32.9	14	0	14	1.44
	195	33.8	9.8	0.13	9.9	27.5(1.2)
	196	25.3	12.3	0	12.3	0.72
	198	7.2	7.6	0	7.6	3.66
Ra		(1600)	13.(3.)	0	13.(3.)	12.8(1.5)
Rb			6.32	0.5	6.8	0.38
	85	72.17	6.2	0.5	6.7	0.48
	87	27.83	6.6	0.5	7.1	0.12
Re			10.6	0.9	11.5	89.7(1.)
	185	37.4	10.2	0.5	10.7	112.(2.)

Element	Isotope	%	σ_{coh}	σ_{inc}	σ_{tot}	σ_{abs}
	187	62.6	10.9	1	11.9	76.4(1.)
Rh		100	4.34	0.3	4.6	144.8
Rn			0		12.6	
Ru			6.21	0.4	6.6	2.56
	96	5.5		0		0.28
	98	1.9		0		<8.
	99	12.7				6.9(1.0)
	100	12.6		0		4.8
	101	17				3.3
	102	31.6		0	144.8	1.17
	104	18.7		0	4.483	0.31
S			1.0186	0.007	1.026	0.53
	32	95.02	0.988	0	0.988	0.54
	33	0.75	2.8	0.3	3.1	0.54
	34	4.21	1.52	0	1.52	0.227
	36	0.02	1.1	0	1.1	0.15
Sb			3.9	0.007	3.9	4.91
	121	57.3	4.1	0.0003	4.1	5.75
	123	42.7	3.64	0.001	3.64	3.8
Sc		100	19	4.5	23.5	27.5
Se			7.98	0.32	8.3	11.7
	74	0.9	0.1	0	0.1	51.8(1.2)
	76	9	18.7	0	18.7	85.(7.)
	77	7.6	8.6	0.05	8.65	42.(4.)
	78	23.5	8.5	0	8.5	0.43
	80	49.6	7.03	0	7.03	0.61
	82	9.4	5.05	0	5.05	0.044
Si			2.163	0.004	2.167	0.171
	28	92.23	2.12	0	2.12	0.177
	29	4.67	2.78	0.001	2.78	0.101
	30	3.1	2.64	0	2.64	0.107
Sm			0.422	39.(3.)	39.(3.)	5922.(56.)
	144	3.1	1.(3.)	0	1.(3.)	0.7
	147	15.1	25.(11.)	143(19.)	39.(16.)	57.(3.)
	148	11.3	1.(3.)	0	1.(3.)	2.4
	149	13.9	63.5	137.(5.)	200.(5.)	42080.(400.)
	150	7.4	25.(11.)	0	25.(11.)	104.(4.)
	152	26.6	3.1	0	3.1	206.(6.)
	154	22.6	11.(2.)	0	11.(2.)	8.4

Element	Isotope	%	σ_{coh}	σ_{inc}	σ_{tot}	σ_{abs}
Sn			4.871	0.022	4.892	0.626
	112	1	4.5(1.5)	0	4.5(1.5)	1
	114	0.7	4.8	0	4.8	0.114
	115	0.4	4.5(1.5)	0.3	4.8(1.5)	30.(7.)
	116	14.7	4.42	0	4.42	0.14
	117	7.7	5.28	0.3	5.6	2.3
	118	24.3	4.63	0	4.63	0.22
	119	8.6	4.71	0.3	5	2.2
	120	32.4	5.29	0	5.29	0.14
	122	4.6	4.14	0	4.14	0.18
	124	5.6	4.48	0	4.48	0.133
Sr			6.19	0.06	6.25	1.28
	84	0.56	6.(2.)	0	6.(2.)	0.87
	86	9.86	4.04	0	4.04	1.04
	87	7	6.88	0.5	7.4	16.(3.)
	88	82.58	6.42	0	6.42	0.058
Ta			6	0.01	6.01	20.6
	180	0.012	6.2	0.5	7.(4.)	563.(60.)
	181	99.988	6	0.011	6.01	20.5
Tb		100	6.84	0.004	6.84	23.4
Tc		(230000)	5.8	0.5	6.3	20.(1.)
Te			4.23	0.09	4.32	4.7
	120	0.096	3.5	0	3.5	2.3
	122	2.6	1.8	0	1.8	3.4
	123	0.908	0.002	0.52	0.52	418.(30.)
	124	4.816	8	0	8	6.8(1.3)
	125	7.14	3.17	0.008	3.18	1.55
	126	18.95	3.88	0	3.88	1.04
	128	31.69	4.36	0	4.36	0.215
	130	33.8	4.55	0	4.55	0.29
Th		100	13.36	0	13.36	7.37
Ti			1.485	2.87	4.35	6.09
	46	8.2	3.05	0	3.05	0.59
	47	7.4	1.66	1.5	3.2	1.7
	48	73.8	4.65	0	4.65	7.84
	49	5.4	0.14	3.3	3.4	2.2
	50	5.2	4.8	0	4.8	0.179
Tl			9.678	0.21	9.89	3.43
	203	29.524	6.14	0.14	6.28	11.4

Element	Isotope	%	σ_{coh}	σ_{inc}	σ_{tot}	σ_{abs}
	205	70.476	11.39	0.007	11.4	0.104
Tm		100	6.28	0.1	6.38	100.(2.)
U			8.903	0.005	8.908	7.57
	233	(159000)	12.8	0.1	12.9	574.7(1.0)
	234	0.005	19.3	0	19.3	100.1(1.3)
	235	0.72	13.78	0.2	14	680.9(1.1)
	238	99.275	8.871	0	8.871	2.68
V			0.0184	5.08	5.1	5.08
	50	0.25	7.3(1.1)	0.5	7.8(1.0)	60.(40.)
	51	99.75	0.0203	5.07	5.09	4.9
W			2.97	1.63	4.6	18.3
	180	0.1	3.(4.)	0	3.(4.)	30.(20.)
	182	26.3	6.1	0	6.1	20.7
	183	14.3	5.36	0.3	5.7	10.1
	184	30.7	7.03	0	7.03	1.7
	186	28.6	0.065	0	0.065	37.9
Xe			2.96	0		23.9(1.2)
	124	0.1		0		165.(20.)
	126	0.09		0		3.5
	128	1.91		0		<8
	129	26.4				21.(5.)
	130	4.1		0		<26.
	131	21.2				85.(10.)
	132	26.9		0		0.45
	134	10.4		0		0.265
	136	8.9		0		0.26
Y		100	7.55	0.15	7.7	1.28
Yb			19.42	4	23.4	34.8
	168	0.14	2.13	0	2.13	2230.(40.)
	170	3.06	5.8	0	5.8	11.4(1.0)
	171	14.3	11.7	3.9	15.6	48.6(2.5)
	172	21.9	11.2	0	11.2	0.8
	173	16.1	11.5	3.5	15	17.1(1.3)
	174	31.8	46.8	0	46.8	69.4(5.0)
	176	12.7	9.6	0	9.6	2.85
Zn			4.054	0.077	4.131	1.11
	64	48.6	3.42	0	3.42	0.93
	66	27.9	4.48	0	4.48	0.62
	67	4.1	7.18	0.28	7.46	6.8

Element	Isotope	%	σ_{coh}	σ_{inc}	σ_{tot}	σ_{abs}
	68	18.8	4.57	0	4.57	1.1
	70	0.6	4.5	0	4.5(1.5)	0.092
Zr			6.44	0.02	6.46	0.185
	90	51.45	5.1	0	5.1	0.011
	91	11.32	9.5	0.15	9.7	1.17
	92	17.19	6.9	0	6.9	0.22
	94	17.28	8.4	0	8.4	0.0499
	96	2.76	3.8	0	3.8	0.0229

A1.1 References

1   V. F. Sears (1992). Neutron News, 3, 29–37. Neutron scattering lengths and cross sections of the elements and their isotopes.
2   A.-J.Dianoux & G. Lander , (Ed.) (2001). *Neutron Data Booklet*, Institut Laue-Langevin, Grenoble.
3   http://www.ncnr.nist.gov/resources/n-lengths/

Appendix 2

Inelastic Neutron Scattering Theory

Here we adopt a limited approach to the theory of thermal neutron scattering. The objective will be to lead the reader through the relevant mathematical steps to an understanding of the double differential scattering equation as it applies to the neutron scattering excitation of molecular vibrations in solids. Although, occasionally, some liberties will be taken with mathematical rigor, at key points our development of the equations appearing in this text will be readily identifiable with those appearing in the published literature [1, 2, 3]. To ease this recognition we have remained, generally, faithful to the use of conventional symbols.

A2.1 The neutron Schrödinger equation

This is very similar to the familiar electron Schrödinger equation, but here the particle is the neutron. It is positioned at some general coordinate, r, and has a mass, m_n. The solutions are, ψ, where:

$$\left(\frac{-\hbar^2}{2m_n}\nabla^2 + V(r)\right)\psi = E\psi \qquad \text{or} \qquad \hat{H}|\psi\rangle = E\psi \qquad (A2.1)$$

We invoke the Born approximation, which views the scattering potential, $V(r)$, as so weak a perturbation that, for incident, i, and final, f, conditions:

$$\psi = \psi_i + \psi_f \qquad \psi_i\big|_{r\to\infty} \approx \psi \qquad \text{and} \qquad E_f\big|_{r\to\infty} = E_i\big|_{r\to\infty} = E \qquad (A2.2)$$

In free space the incident, i, neutron experiences no perturbing field and at long distances from the sample:

$$\frac{-\hbar^2}{2m_n} \nabla^2 \psi_i = E \psi_i \tag{A2.3}$$

Then subtracting Eq. (A2.3) from (A2.1) and using Eq. (A2.2):

$$\frac{-\hbar^2}{2m_n} \nabla^2 \psi_f + V(r)\psi_i = E \psi_f \tag{A2.4}$$

We shall need this later to apply a well known result of quantum mechanics, involving time dependent perturbation theory, Fermi's Golden Rule [1].

Here it is convenient to recall the expression for the de Broglie wavelength, λ, of a neutron and the related neutron momentum vector, k. Readers should familiarise themselves with aspects of the manipulation of k (§2.3). The neutron velocity is v.

$$\lambda = \frac{h}{m_n v} \quad \text{or} \quad \frac{2\pi}{\lambda} = \frac{m_n v}{\hbar} \tag{A2.5}$$

$$\frac{2\pi}{\lambda} = k = |k| = \frac{m_n |v|}{\hbar} \quad \text{and} \quad E = \frac{\hbar^2 k^2}{2m_n} = \frac{m_n |v|^2}{2} \tag{A2.6}$$

Eq. (A2.4), upon substituting for E from Eq. (A2.6), rearranges to:

$$\left(\nabla^2 + k_f^2\right)\psi_f = \frac{2m_n}{\hbar^2} V(r)\psi_i \tag{A2.7}$$

A2.2 Scattering theory

When analysing the results of a neutron experiment we shall need to relate the number of neutrons seen in a detector of small area, dA, as it is positioned at different scattering angles around the sample, see Fig. 2.1. The detector subtends a small solid angle, $d\Omega$, at the sample. The number of neutrons per unit time, or flux, scattered to the detector, J_f, is directly proportional to the incident neutron flux, J_i. It is also proportional to the size of the detector but inversely proportional to the square of the distance, d_f, between the sample and the detector, also $d\Omega \equiv dA / d_f^2$. The differential cross section, $d\sigma / d\Omega$, is the constant of proportionality.

$$J_f \propto J_i \frac{dA}{d_f^2} \quad \text{thus} \quad J_f = \left(\frac{d\sigma}{d\Omega}\right) J_i \frac{dA}{d_f^2}$$

$$\left(\frac{d\sigma}{d\Omega}\right) = \frac{J_f}{J_i} \frac{d_f^2}{dA} = \frac{J_f}{J_i} \frac{1}{d\Omega}$$

(A2.8)

The flux of a beam of particles is the product of their velocity and density, $v\,\rho$, and the probability density is:

$$\rho = |\psi|^2$$

(A2.9)

The incident wave function, ψ_i, is a plane wave, conveniently written:

$$\psi_i = \exp\left(i\frac{2\pi}{\lambda_i}r\right) \qquad \rho_i = |\psi_i|^2 = 1 \quad \text{so} \quad J_i = v_i$$

(A2.10)

The scattered wave, ψ_f, is spherical since, at the energies of thermalised neutrons only (S-) spherical wave solutions need be considered. Its strength, or amplitude, is given by the parameter β.

$$\psi_f = \frac{\beta}{r}\exp\left(i\frac{2\pi}{\lambda_f}r\right)$$

(A2.11)

$$\rho_f = |\psi_f|^2 = \left(\frac{\beta}{r}\right)^2 \qquad J_f = \left(\frac{\beta}{r}\right)^2 v_f$$

In elastic scattering the incident and final velocities are equal and it is seen that the strength of the scattered wave is given by the scattering length, b, of the nucleus, where $r = d$.

$$\frac{d\sigma}{d\Omega} = \beta^2 \frac{v_f}{v_i} \frac{dA}{r^2} \frac{1}{d\Omega} = \beta^2$$

(A2.12)

$$\int_{4\pi} \frac{d\sigma}{d\Omega} = \int_{4\pi} \beta^2 = \sigma, \qquad \therefore \beta^2 = b^2 \quad \text{and} \quad \sigma = 4\pi b^2$$

Energy exchange with the sample requires the differential scattering cross section to contain energy terms, $d^2\sigma/d\Omega\,dE$. In preparation for this we define, E_0 and E_1 as the energies of the ground state and first excited state of the vibrating system of atoms.

$$\Delta E = E_1 - E_0 = \hbar\omega$$

(A2.13)

From Eq. (A2.6):

$$\frac{dE}{dk} = \left(\frac{\hbar^2}{m_n}\right)k \tag{A2.14}$$

First, however, we note that if the double differential cross section is integrated over all final energies then the simple expression for $d\sigma/d\Omega$ is recovered.

$$\int \frac{d^2\sigma}{d\Omega\,dE_f}\,dE_f = \frac{d\sigma}{d\Omega} \tag{A2.15}$$

A2.2.1 The transition rate—Fermi's Golden Rule

The transition rate between the initial and the final state of a system depends both on the strength of the coupling between those states and on the number of ways the transition can happen, the density of the final states. A transition occurs more often as the coupling, or scattering strength, becomes stronger. The Golden Rule rate, $W_{i\to f}$, (first used by Fermi) at which neutrons scatter from a system and change its energy state, the number of transitions per second, is given by:

$$W_{i\to f} = \left(\frac{2\pi}{\hbar}\right)|\langle final\ \ states|V(r)|initial\ \ states\rangle|^2\,\rho_f \tag{A2.16}$$

where ρ_f is the density of final states. The transition probability is proportional to the square of the integral of the interaction over all of the space appropriate to the problem. The *matrix element*, Eq. (A2.17), for the interaction is the coupling term and its bra-ket notation can be written explicitly in integral form.

$$\langle final\ \ states|V(r)|initial\ \ states\rangle =$$
$$\iint \psi_{n,f}^* \psi_{N,f}^*\ \ V(r)\ \ \psi_{n,i}\psi_{N,i}\,dv_N\,dv_n \tag{A2.17}$$

Here, the sample, of N atoms, is represented by the wavefunction, ψ_N, and the N volume elements dv_N contain the individual atoms, whilst the single volume dv_n contains the neutron. The integral is taken over all the $N+1$ volume elements.

We can now proceed to determine the specific scattering cross section for, say a $(1\leftarrow0)$ transition, i→f. Substituting into Eq. (A2.8):

$$\left(\frac{d\sigma}{d\Omega}\right)_{i\rightarrow f} = \frac{J_f}{J_i\,d\Omega} = \frac{W_{i\rightarrow f}}{J_i\,d\Omega} = \frac{2\pi}{\hbar} \frac{\left|\left\langle \psi^*_{n,f}\psi^*_{N,f} \left|V(r)\right| \psi_{n,f}\psi_{N,f} \right\rangle\right|^2}{J_i} \frac{\rho_f}{d\Omega} \quad (A2.18)$$

There only remains the straightforward task of finding appropriate expressions for the incident flux and density of final states for substitution into the cross section equation. Unfortunately, although deriving these expressions is straightforward it is also tedious and long-winded. Both expressions must be derived in terms of a normalisation constant, this is usually achieved through a standard mathematical device, Box-normalisation [1]. A normalised volume is defined and this permits the number of objects within the volume to be enumerated. By this process densities can be readily determined. The procedure adds little to our understanding of the scattering process and the normalisation factors cancel in the final equation, therefore this is omitted, for details see [1]. The normalised incident flux then reduces to the velocity of the incident neutrons, see Eq. (A2.10).

$$J_i = \left(\frac{\hbar}{m_n}\right)k_i \quad (A2.19)$$

The normalised density of final momentum states becomes:

$$\rho_f = \left(\frac{k_f^2}{(2\pi)^3}\right)\frac{dk_f}{dE_f}d\Omega = \left(\frac{k_f^2}{(2\pi)^3}\right)\frac{m_n}{\hbar^2}\frac{1}{k_f}d\Omega \quad (A2.20)$$

And the normalised neutron wavefunction is:

$$\psi_n = \exp(i\,\boldsymbol{k}.\boldsymbol{r}) \quad (A2.21)$$

Substituting Eqs. (A2.19) and (A2.20) into Eq. (A2.18) provides:

$$\left(\frac{d\sigma}{d\Omega}\right)_{i\rightarrow f} = \frac{2\pi}{\hbar} \frac{\left|\left\langle \psi^*_{n,f}\psi^*_{N,f} \left|V(r)\right| \psi_{n,f}\psi_{N,f} \right\rangle\right|^2}{J_i} \frac{\left(\frac{k_f}{(2\pi)^3}\right)\frac{m_n}{\hbar^2}d\Omega}{d\Omega} \quad (A2.22a)$$

$$\left(\frac{d\sigma}{d\Omega}\right)_{i\to f} = \frac{2\pi}{\hbar} \frac{\left|\left\langle \psi_{n,f}^* \psi_{N,f}^* \left|V(r)\right| \psi_{n,f} \psi_{N,f} \right\rangle\right|^2}{J_i} \left(\frac{k_f}{(2\pi)^3}\right)\frac{m_n}{\hbar^2} \qquad (A2.22)$$

$$= \frac{2\pi}{\hbar}\frac{m_n}{\hbar}\frac{m_n}{\hbar^2}\frac{1}{(2\pi)^3}\frac{k_f}{k_i}\left|\left\langle \psi_{n,f}^* \psi_{N,f}^* \left|V(r)\right| \psi_{n,f} \psi_{N,f} \right\rangle\right|^2$$

This cross section is for neutrons that are scattered into a solid angle of $d\Omega$ that lies in the direction given by the final wavevector, k_f. In the experiment the incident neutron energy is fixed and the initial and final states of the sample are also fixed, even if not known. The total energy of the system is conserved and that which is lost by the neutron is gained by the sample, so:

$$E_f - E_i + \hbar\omega = 0 \qquad (A2.23)$$

Since the energy is conserved, the final energy of the neutrons is also fixed. All of the scattered neutrons appear with precisely the same energy, which is conveniently represented by a delta-function in energy. The delta-function has several useful mathematical properties, including:

$$\int \delta(x)dx = \begin{cases} 1 & x=0 \\ 0 & x\neq 0 \end{cases} \qquad (A2.24)$$

Substituting Eq. (A2.23) into (A2.24) gives the energy delta function. Multiplying the differential cross section of Eq. (A2.22) by the energy delta function provides a function equivalent to the double differential cross section. As can be seen from their integrals

$$\int \left(\frac{d\sigma}{d\Omega}\right)_{i\to f} \delta(E_f - E_i + \hbar\omega)\,dE_f = \left(\frac{d\sigma}{d\Omega}\right) = \int\left(\frac{d^2\sigma}{d\Omega\,dE_f}\right)dE_f \qquad (A2.25)$$

Eq. (A2.22) becomes, after the multiplication and with rearrangement:

$$\frac{d^2\sigma}{dE_f\,d\Omega}\bigg|_{i\to f} = \frac{k_f}{k_i}\left(\frac{m_n}{2\pi\hbar^2}\right)^2\left|\left\langle \psi_{n,f}^* \psi_{N,f}^* \left|V(r)\right| \psi_{n,f} \psi_{N,f} \right\rangle\right|^2 \delta(E_f - E_i + \hbar\omega)$$

$$(A2.26)$$

There are *four terms* in Eq. (A2.26), *first* is the ratio of the incident and final neutron momenta. The *second* term groups the fundamental constants and the *final* term ensures that the difference between the incident and final neutron energies equals the difference between quantised energy states of the system (or zero for elastic scattering). The *third* term describes how the initial states are related to the final states through the scattering potential, $V(r)$.

A2.2.2 The form of the scattering potential

The mathematical form of the potential used in the derivation of the double differential cross section is the Fermi pseudo-potential, $V_F(r_n)$. The total potential is represented as the sum of the non penetrable parts of space occupied by the atomic nuclei, labelled l, at positions r_l.

$$V_F(r_n) = \frac{2\pi\hbar^2}{m_n} \sum_l b_l \, \delta(r_n - r_l) \qquad (A2.27)$$

Only when the neutron is at an atomic position, $r_n - r_l = 0$, does it encounter the scattering potential of strength b_l. In the Born approximation only this form for $V_F(r_n)$ can successfully reproduce S-wave scattering from bound nuclei.

We shall now evaluate the matrix-element Eq. (A2.17) for the Fermi pseudo-potential representing a simple monatomic Bravais lattice. Here, the distance of the neutron's position, r_n, from the lth atom's position, r_l, is $x_l = r_n - r_l$,

$$\left\langle \psi_{n,f}^* \psi_{N,f}^* \left| V_F(x_l) \right| \psi_{n,f} \psi_{N,f} \right\rangle$$

Substituting the neutron wavefunction, from Eq. (A2.10)

$$= \sum_l \int \psi_{N,f}^* \exp[-i k_f.(x_l + r_l)] V_F(x_l) \psi_{N,i} \exp[i k_i.(x_l + r_l)] dr_l \, dx_l$$

$$(A2.28)$$

Also, $Q = k_i - k_f$ and we recall that when the neutron encounters the atom at $r_n = r_l$, $\delta(x_l) = 1$. Then, using [1]:

$$\int V_F(x_l)\exp[i\,Q.x_l]dx_l = V_F(Q) = \frac{2\pi\hbar^2}{m_n}b_l \tag{A2.29}$$

the matrix element Eq. (A2.28) becomes:

$$= \sum_l \frac{2\pi\hbar^2}{m_n} b_l \langle \psi_{N,f}^* \left| \exp[i\,Q.r_l] \right| \psi_{N,i} \rangle$$

Substituting this into Eq. (A2.26), gives:

$$\frac{d^2\sigma}{dE_f\,d\Omega}\bigg|_{i\to f} = \frac{k_f}{k_i}\left(\frac{m_n}{2\pi\hbar^2}\right)^2$$
$$\left| \sum_l \frac{2\pi\hbar^2}{m_n} b_l \langle \psi_{N,f}^* \left| \exp[i\,Q.r_l] \right| \psi_{N,i} \rangle \right|^2 \delta(E_f - E_i + \hbar\omega) \tag{A2.30}$$

or

$$\frac{d^2\sigma}{dE_f\,d\Omega}\bigg|_{i\to f} = \frac{k_f}{k_i}\left| \sum_l b_l \langle \psi_{N,f}^* \left| \exp[i\,Q.r_l] \right| \psi_{N,i} \rangle \right|^2 \delta(E_f - E_i + \hbar\omega) \tag{A2.31}$$

which can be compared to Eq. (2.40) of [1]. The summation runs over all l, the index that numbers each atom, of which there are N. Expanding the quadratic there are thus N^2 terms, labelled $l'\,l$, each of which is itself a quadratic contribution to the overall value of the matrix element, they all have the general form:

$$b_{l'}^* \langle \psi_{N,f} \left| \exp[i\,Q.r_{l'}] \right| \psi_{N,i} \rangle^* b_l \langle \psi_{N,f} \left| \exp[i\,Q.r_l] \right| \psi_{N,i} \rangle \tag{A2.32}$$

However, since:

$$\langle f|A|i \rangle^* = \langle i|A^*|f \rangle \tag{A2.33}$$

and scattering lengths are real, then each contribution resolves to

$$= b_{l'}\,b_l \langle \psi_{N,f} \left| \exp[-i\,Q.r_{l'}] \right| \psi_{N,i} \rangle \langle \psi_{N,f} \left| \exp[i\,Q.r_l] \right| \psi_{N,i} \rangle \tag{A2.34}$$

The integral expression for the energy delta function is (see Appendix A of [1]).

$$\delta\left(E_{\mathrm{i}}-E_{\mathrm{f}}+\hbar\omega\right)=\frac{1}{2\pi\hbar}\int_{-\infty}^{\infty}\exp\left(\frac{\mathrm{i}\left(E_{\mathrm{f}}-E_{\mathrm{i}}\right)t}{\hbar}\right)\exp\left(-\mathrm{i}\,\omega t\right)\mathrm{d}t \qquad (A2.35)$$

On substituting Eqs. (A2.34) and (A2.35) into Eq. (A2.31), we obtain:

$$\left.\frac{\mathrm{d}^2\sigma}{\mathrm{d}E_{\mathrm{f}}\,\mathrm{d}\Omega}\right|_{\mathrm{i}\to\mathrm{f}}=\frac{k_{\mathrm{f}}}{k_{\mathrm{i}}}\sum_{l,l'}b_{l'}\,b_l\left\langle\psi_{N,\mathrm{i}}\left|\exp\left[-\mathrm{i}\,\boldsymbol{Q}.\boldsymbol{r}_{l'}\right]\right|\psi_{N,\mathrm{f}}\right\rangle\left\langle\psi_{N,\mathrm{f}}\left|\exp\left[\mathrm{i}\,\boldsymbol{Q}.\boldsymbol{r}_l\right]\right|\psi_{N,\mathrm{i}}\right\rangle$$

$$\times\frac{1}{2\pi\hbar}\int_{-\infty}^{\infty}\exp\left(\frac{\mathrm{i}\left(E_{\mathrm{f}}-E_{\mathrm{i}}\right)t}{\hbar}\right)\exp\left(-\mathrm{i}\,\omega t\right)\mathrm{d}t \qquad (A2.36)$$

Also, [1]

$$\exp\left(\frac{-\mathrm{i}\hat{H}t}{\hbar}\right)|\psi\rangle=\exp\left(\frac{-\mathrm{i}Et}{\hbar}\right)|\psi\rangle \qquad (A2.37)$$

and so:

$$\left.\frac{\mathrm{d}^2\sigma}{\mathrm{d}E_{\mathrm{f}}\,\mathrm{d}\Omega}\right|_{\mathrm{i}\to\mathrm{f}}=\frac{k_{\mathrm{f}}}{k_{\mathrm{i}}}\frac{1}{2\pi\hbar}\sum_{l,l'}b_{l'}\,b_l\left\langle\psi_{N,\mathrm{i}}\left|\exp\left[-\mathrm{i}\,\boldsymbol{Q}.\boldsymbol{r}_{l'}\right]\right|\psi_{N,\mathrm{f}}\right\rangle$$

$$\left\langle\psi_{N,\mathrm{f}}\left|\exp\left(\frac{\mathrm{i}\hat{H}t}{\hbar}\right)\exp\left[\mathrm{i}\,\boldsymbol{Q}.\boldsymbol{r}_l\right]\exp\left(\frac{-\mathrm{i}\hat{H}t}{\hbar}\right)\right|\psi_{N,\mathrm{i}}\right\rangle\exp\left(-\mathrm{i}\,\omega t\right)\mathrm{d}t \qquad (A2.38)$$

Using the closure relationship for any operators A and B allows us to account for the contributions from all final states of the system (that conserve energy and momentum).

$$\sum_{\mathrm{f}}\left\langle\psi_{\mathrm{i}}|A|\psi_{\mathrm{f}}\right\rangle\left\langle\psi_{\mathrm{f}}|B|\psi_{\mathrm{i}}\right\rangle=\left\langle\psi_{\mathrm{i}}|AB|\psi_{\mathrm{i}}\right\rangle \qquad (A2.39)$$

The double differential cross section becomes

$$\left.\frac{\mathrm{d}^2\sigma}{\mathrm{d}E_{\mathrm{f}}\,\mathrm{d}\Omega}\right|_{\mathrm{i}\to\mathrm{all\,f}}=\frac{k_{\mathrm{f}}}{k_{\mathrm{i}}}\frac{1}{2\pi\hbar}\sum_{l,l'}b_{l'}\,b_l\int_{-\infty}^{\infty}\exp\left[-\mathrm{i}\,\omega t\right]\mathrm{d}t\times \qquad (A2.40)$$

$$\left\langle\psi_{N,\mathrm{i}}\left|\exp\left[-\mathrm{i}\,\boldsymbol{Q}.\boldsymbol{r}_{l'}\right]\underbrace{\exp\left(\frac{\mathrm{i}\hat{H}t}{\hbar}\right)\exp\left[\mathrm{i}\,\boldsymbol{Q}.\boldsymbol{r}_l\right]\exp\left(\frac{-\mathrm{i}\hat{H}t}{\hbar}\right)}_{\text{time dependent Heisenberg operator}}\right|\psi_{N,\mathrm{i}}\right\rangle$$

time dependent Heisenberg operator

Conveniently we can write the Heisenberg operator in a more compact form

$$\exp\left(\frac{i\hat{H}t}{\hbar}\right)\exp[i\boldsymbol{Q}.\boldsymbol{r}_l]\exp\left(\frac{-i\hat{H}t}{\hbar}\right) \equiv \exp[i\boldsymbol{Q}.\boldsymbol{r}_l(t)] \tag{A2.41}$$

and this allows us to identify the time independent term in the same form but evaluated at time zero, $t = 0$.

$$\exp[-i\boldsymbol{Q}.\boldsymbol{r}_{l'}] \equiv \exp[-i\boldsymbol{Q}.\boldsymbol{r}_{l'}(0)] \tag{A2.42}$$

(The two operators, $r_l(0)$ and $r_l(t)$, do not commute and their order in the integral must be preserved.)

The double differential of Eq. (A2.40) is still specific with respect to the initial states. All the populated initial states will contribute to the experimental observable in proportion to the probability of their being occupied.

$$\frac{d^2\sigma}{dE_f\,d\Omega} = \sum_i P_i \left(\frac{d^2\sigma}{dE_f\,d\Omega}\right)_{i\rightarrow \text{all}\,f} \tag{A2.43}$$

The factor P_i represents the probability of finding a particular initial state of the system occupied. If the sample is at a given temperature, T, this is the well known Boltzmann distribution.

$$P_i = \frac{\exp\left(\frac{-E_i}{k_B T}\right)}{\sum_i \exp\left(\frac{-E_i}{k_B T}\right)} \tag{A2.44}$$

Using Eq. (A2.43) and substituting Eqs. (A2.41) and (A2.42) into Eq. (A2.40) gives

$$\frac{d^2\sigma}{dE_f\,d\Omega} = \frac{k_f}{k_i}\frac{1}{2\pi\hbar}\sum_{l,l'} b_{l'}\,b_l \int_{-\infty}^{\infty}\exp[-i\omega t]dt\sum_i P_i \tag{A2.45}$$

$$\langle \psi_{N,i}|\exp[-i\mathbf{Q}.\mathbf{r}_{l'}(0)]\exp[i\mathbf{Q}.\mathbf{r}_l(t)]|\psi_{N,i}\rangle$$

The scattering experiments considered in this work are exclusively incoherent in nature (§2.1) and neutrons are scattered by individual atoms. The dynamics that can be probed by this technique refer to the motions of those atoms taken individually, no information about their motions relative to other atoms is available from the results. The incoherent cross section is therefore appropriate, Eq. (2.3).

$$\sigma_{\text{inc}} = 4\pi\left(\langle b^2 \rangle - \langle b \rangle^2\right) \tag{A2.46}$$

Thus the time dependent Heisenberg operator required above is that which represents the relative motion of a single target atom as a function of time. The neutron is probing how the present position of the scattering atom, at time $t > 0$, correlates with its previous position measured at an arbitrary earlier time, whose value is chosen as $t = 0$. Molecular vibrations involve repeated cyclic trajectories of the atoms and the correlation of the position of a given atom at different times is straightforward. (In the technique of coherent inelastic neutron scattering the correlated motion of pairs of atoms is observed.)

The thermal average of an operator A at a temperature T is written compactly as

$$\sum_i P_i \langle \psi_i | A | \psi_i \rangle = \langle A \rangle_T \tag{A2.47}$$

Substituting Eqs. (A2.46) and (A2.47) into Eq. (2.45) gives, see also Eq. (2.28)

$$\frac{d^2\sigma}{dE_f \, d\Omega} = \frac{\sigma_{\text{inc}}}{4\pi} \frac{k_f}{k_i} \frac{1}{2\pi\hbar} \tag{A2.48}$$

$$\sum_{l=1}^{N} \int \langle \exp[-i\boldsymbol{Q}.\boldsymbol{r}_l(0)] \exp[i\boldsymbol{Q}.\boldsymbol{r}_l(t)] \rangle_T \exp[-i\omega t] \, dt$$

The introduction of the incoherent cross section reduces the sum to run over only the single index, l, (compare Eq. (2.69) of [1]).

A2.2.3 The scattering law

It is convenient to rewrite this equation in terms of a van Hove response function, S, which emphasises the structure and dynamics of the sample, also called the 'incoherent Scattering Law', which has units of per energy. Since all of the work reported here involves only incoherent scattering we suppress specific indications of its incoherent nature.

$$S(\boldsymbol{Q},\omega) = \frac{4\pi}{\sigma} \frac{k_i}{k_f} \left(\frac{d^2\sigma}{dE_f \, d\Omega} \right) \qquad (A2.49)$$

The Scattering Law is the Fourier transform of the intermediate scattering function, $\Im(\boldsymbol{Q},t)$, which is often used as the starting point to interpret a broad range of experimental results.

$$S(\boldsymbol{Q},\omega) = \frac{1}{2\pi\hbar} \int \Im(\boldsymbol{Q},t) \exp(-i\omega t) dt \qquad (A2.50)$$

A2.3 Scattering from vibrating molecules

We shall express the Scattering Law in terms of the internal and external dynamics of the molecules in crystals that form the backbone of the work discussed in this book. If the molecules consist of N_{atom} atoms each and altogether there are N_{mol} molecules in a crystal, then the total number of atoms in the system is $N = N_{atom} \times N_{mol}$. We will consider the dynamics of the molecule and the crystal separately.

If $r(t)$ is the time dependent atomic position vector, taken with respect to the origin of the crystallographic cell, it can be expressed as the sum of two terms, the molecular centre of mass, this is a time dependent vector, $u(t)_{ext}$, that allows a description of the vibrations of the crystal, the phonons. The second term is the position vector of the atom, $u(t)_{int}$, given by a Cartesian coordinate system with its origin at the molecular centre of mass.

$$r(t) = u(t)_{ext} + u(t)_{int} \qquad (A2.51)$$

So, the Heisenberg operator in Eq. (A2.41) becomes, see also Eq. (2.29):

$$\exp(i\,\boldsymbol{Q}.r(t)) = \exp(i\,\boldsymbol{Q}.u(t)_{ext})\exp(i\,\boldsymbol{Q}.u(t)_{int}) \qquad (A2.52)$$

and since the Fourier transform of a product is a convolution, $\otimes$, we may separate the Scattering Law, Eq. (A2.49), as:

$$S(\boldsymbol{Q},\omega)_{total} = S(\boldsymbol{Q},\omega)_{int} \otimes S(\boldsymbol{Q},\omega)_{ext} \qquad (A2.53)$$

This convolution will have significant consequences for the predicted spectral shapes but, for the present, it will be ignored as we develop the response predicted for the internal molecular vibrations alone. (This allows us to assume that the molecule is alone in a gas-like phase but is not free to recoil.) From the definition of the scattering function given above the internal dynamics are represented by:

$$S(\mathbf{Q}, \omega)_{int} = \frac{1}{2\pi\hbar} \sum_{l=1}^{N} \int \langle \exp[-i\mathbf{Q}.\mathbf{r}_l(0)_{int}] \exp[i\mathbf{Q}.\mathbf{r}_l(t)_{int}] \rangle_T \exp[-i\omega t] \, dt$$

(A2.54)

It is convenient develop the total internal atomic displacements into their individual vibrations in Cartesian components. The time-displacement of a particular atom l is then the sum of the displacements of the atom in each of the internal modes, labelled $v = 1, 2, \ldots 3N_{atom} - 6$.

$$\mathbf{u}_l(t)_{int} = \sum_{v} {}^{v}\mathbf{u}_l(t)$$

(A2.55)

$$= {}^{1}\mathbf{u}_l(t)_x + {}^{1}\mathbf{u}_l(t)_y + {}^{1}\mathbf{u}_l(t)_z + {}^{2}\mathbf{u}_l(t)_x + {}^{2}\mathbf{u}_l(t)_y + \cdots$$

We shall now focus on the dynamics of a single atom and momentarily suppress the subscript l. The dynamics of the molecule will eventually be the sum over the dynamics of all of its atoms taken individually. This approach is chosen because, in general, molecules are of low symmetry and contain many different types of atom. In condensed matter physics texts the dynamics is developed in terms of a simple monatomic lattice and the subscript l is removed by summation, which leads to an extra factor N outside the integral [1].

Then in one dimension, say along the x direction, the scattering law for the chosen atom is:

$$S(\mathbf{Q}, \omega) = \frac{1}{2\pi\hbar} \int \left\langle \exp\left[-i\mathbf{Q}.\sum_{v} {}^{v}u(0)_x\right] \exp\left[i\mathbf{Q}.\sum_{v} {}^{v}u(t)_x\right] \right\rangle_T$$

(A2.56)

$$\exp(-i\omega t) \, dt$$

We expand the summations over the internal modes and recognise that in the harmonic approximation the modes are dynamically decoupled and no cross terms can result. The sum of the arguments in the exponential becomes a product of exponentials and Eq. (A2.56) becomes, after suppressing the Cartesian indicator, x.

$$S(\boldsymbol{Q},\omega) = \frac{1}{2\pi\hbar} \prod_{\nu=1}^{\nu=3N_{atom}-6} \int \left\langle \exp\left[-i\boldsymbol{Q}.^{\nu}\boldsymbol{u}(0)\right]\exp\left[i\boldsymbol{Q}.^{\nu}\boldsymbol{u}(t)\right]\right\rangle_{T} \qquad (A2.57)$$

$$\exp(-i\omega t)dt$$

In proceeding further we shall treat each mode individually and, in the manner of Eq. (A2.50) rewrite each term of Eq. (A2.57) as:

$$S(\boldsymbol{Q},\omega_\nu) = \frac{1}{2\pi\hbar} \int \mathfrak{I}_\nu(\boldsymbol{Q},t)\exp(-i\omega t)dt \qquad (A2.58)$$

This defines $\mathfrak{I}_\nu(\boldsymbol{Q}, t)$ as:

$$\mathfrak{I}_\nu(\boldsymbol{Q},t) = \left\langle \exp\left[-i\boldsymbol{Q}.^{\nu}\boldsymbol{u}(0)\right]\exp\left[i\boldsymbol{Q}.^{\nu}\boldsymbol{u}(t)\right]\right\rangle_{T} \qquad (A2.59)$$

The position of the nucleus, as a function of time, is given in terms of the creation, $\hat{a}^{+}$, and annihilation, $\hat{a}^{-}$, operators by [2], see also Appendix E of [1].

$$^{\nu}\boldsymbol{u}(t) = ^{\nu}\boldsymbol{u}\left\{\hat{a}^{+}\exp(i\omega t) + \hat{a}^{-}\exp(-i\omega t)\right\} \qquad (A2.60)$$

This is equivalent to, but more convenient than, the alternative representation.

$$^{\nu}\boldsymbol{u}(t) = ^{\nu}\boldsymbol{u}\cos\omega t + \frac{p_{atom}}{m\omega}\sin\omega t \qquad (A2.61)$$

Where $^{\nu}\boldsymbol{u}$ is the time independent maximum amplitude of the vibration, p_{atom} is the atom's momentum and m is its mass. We also use the following identity relating thermal averages.

$$\left\langle \exp A \exp B \right\rangle_{T} = \exp\left\langle A^{2}\right\rangle_{T} \exp\left\langle A B \right\rangle_{T} \qquad (A2.62)$$

Eq. (A2.59) becomes:

$$\mathfrak{I}_\nu = \exp\left\langle -\left[\boldsymbol{Q}.^{\nu}\boldsymbol{u}(0)\right]^{2}\right\rangle_{T} \exp\left\langle \left[\boldsymbol{Q}.^{\nu}\boldsymbol{u}(0)\right]\left[\boldsymbol{Q}.^{\nu}\boldsymbol{u}(t)\right]\right\rangle_{T} \qquad (A2.63)$$

We now proceed to take the terms of Eq. (A2.63) individually. We recognise that, for a given mode ν, only the energy levels 0, 1, 2, 3… are available and given by the quantum number, n. The thermal average of the argument of first term is, see Eq.(A2.47) and [2].

$$\left\langle -\left[\boldsymbol{Q}.^{\nu}\,\boldsymbol{u}(0)\right]^2\right\rangle_T = \sum_n P_n\langle n|-\left[\boldsymbol{Q}.^{\nu}\,\boldsymbol{u}(0)\right]^2|n\rangle = -\left[\boldsymbol{Q}.^{\nu}\,\boldsymbol{u}\right]^2\left(2\langle n\rangle_T + 1\right) \quad \text{(A2.64)}$$

The expectation value of the quantum number, n, is:

$$\langle n\rangle_T = \sum_n P_n\langle \psi_{N,\text{i}}|n|\psi_{N,\text{f}}\rangle = \sum_n n P_n\langle \psi_{N,\text{i}}|\psi_{N,\text{f}}\rangle = \sum_n n P_n \quad \text{(A2.65)}$$

We can rewrite Eq. (A2.44) more compactly if we define 2ß as:

$$P_n = \frac{\exp\left(-\dfrac{n\hbar\omega_\nu}{k_{\mathrm{B}}T}\right)}{\sum_n \exp\left(-\dfrac{n\hbar\omega_\nu}{k_{\mathrm{B}}T}\right)} = \frac{\exp(-n2\text{ß})}{\sum_n \exp(-n2\text{ß})} \quad \text{(A2.66)}$$

We also have:

$$\sum_n \exp(-n2\text{ß}) = \frac{1}{1-\exp(-2\text{ß})}$$
$$\sum_n n\exp(-n2\text{ß}) = \frac{\exp(-2\text{ß})}{\left(1-\exp(-2\text{ß})\right)^2} \quad \text{(A2.67)}$$

Which gives, upon substitution into Eq. (A2.65):

$$\langle n\rangle_T = \sum_n n\,\frac{\exp(-n2\text{ß})}{\sum_n \exp(-n2\text{ß})} = \frac{\exp(-2\text{ß})}{1-\exp(-2\text{ß})} = \frac{1}{\exp(2\text{ß})-1} \quad \text{(A2.68)}$$

We substitute this result into Eq. (A2.64), using the hyperbolic cotangent, and obtain.

$$\left\langle -\left[\mathbf{Q}.^{\nu}\,\mathbf{u}(0)\right]^2\right\rangle_T = -\left[\mathbf{Q}.^{\nu}\,\mathbf{u}\right]^2\left(\frac{1+\exp(2\text{ß})}{\exp(2\text{ß})-1}\right)\left(\frac{1}{2}\right)$$

$$= -\left[\mathbf{Q}.^{\nu}\,\mathbf{u}\right]^2 \coth(\text{ß})$$

$$= -\left[\mathbf{Q}.^{\nu}\,\mathbf{u}\right]^2 \coth\left(\frac{\hbar\omega_\nu}{2k_{\mathrm{B}}T}\right) \quad \text{(A2.69)}$$

We complete our treatment of the first term by recalling its form in Eq. (A2.63)

$$\exp\left\langle -\left[\underline{\varrho}\cdot^{\nu} u(0)\right]^{2}\right\rangle_{T} = \exp\left\{ -\left[\underline{\varrho}\cdot^{\nu} u\right]^{2} \coth(\beta)\right\} \tag{A2.70}$$

The second term in Eq. (A2.63), is

$$\exp\left\langle \left[\underline{\varrho}\cdot^{\nu} u(0)\right]\left[\underline{\varrho}\cdot^{\nu} u(t)\right]\right\rangle_{T}$$
$$= \exp\left\{ \left[\underline{\varrho}\cdot^{\nu} u\right]^{2} \left[(\langle n\rangle_{T} +1)\exp(-i\omega t)+\langle n\rangle_{T} \exp(i\omega t)\right]\right\} \tag{A2.71}$$

Again substituting from Eq. (A2.68) and rearranging Eq.(A2.71):

$$= \exp\left[\left(\left[\underline{\varrho}\cdot^{\nu} u\right]^{2} \frac{\sqrt{\exp(2\beta)}}{\exp(2\beta)-1}\right)\left\{\frac{\sqrt{\exp(2\beta)}}{\exp(i\omega t)} + \frac{\exp(i\omega t)}{\sqrt{\exp(2\beta)}}\right\}\right] \tag{A2.72}$$

We recognise the form of Eq. (A2.72) in its relationship to that of the modified Bessel function of the first kind, I.

$$\sum_{n=-\infty}^{\infty} y^{n} I_{n}\{x\} = \exp\left[\left(\frac{x}{2}\right)\left\{y+\frac{1}{y}\right\}\right] \tag{A2.73}$$

Where x and y are:

$$x = 2\left[\underline{\varrho}\cdot^{\nu} u\right]^{2} \frac{\sqrt{\exp(2\beta)}}{\exp(2\beta)-1} \qquad y = \frac{\exp(i\omega t)}{\sqrt{\exp(2\beta)}} \tag{A2.74}$$

Substituting Eq. (A2.73) into Eq. (A2.72), and using the hyperbolic sine function allows us to regroup terms from Eqs. (A2.65), (A2.71), (A2.72), (A2.73).

$$\exp\left\langle \left[\underline{\varrho}\cdot^{\nu} u(0)\right]\left[\underline{\varrho}\cdot^{\nu} u(t)\right]\right\rangle_{T} = \sum_{n=-\infty}^{+\infty} \exp[n\beta]\exp(in\omega t)I_{n}\left\{\frac{\left[\underline{\varrho}\cdot^{\nu} u\right]^{2}}{\sinh(\beta)}\right\} \tag{A2.75}$$

During an experiment the sample is cold, *ca* 20K (= 14 cm^{-1}). The lowest internal vibrations are typically about 300 cm^{-1} and the hyperbolic sine function will, except at the very lowest energies, have an argument greater than ten. The argument of the Bessel function is, therefore, less than 10^{-3} and it can be safely represented by the first term of its power series expansion. Where, for an arbitrary argument x:

$$I_n\{x\} \approx \frac{1}{|n|!}\left(\frac{x}{2}\right)^{|n|} \tag{A2.76}$$

From the definition of the hyperbolic sine function Eq. (A2.75) becomes:

$$= \sum_n \exp[n\beta]\, \frac{\exp(in\omega t)}{|n|!}\left\{\frac{2\left[\mathbf{Q}.^v\mathbf{u}\right]^2}{(\exp(\beta)-\exp(-\beta))}\frac{1}{2}\right\}^{|n|} \tag{A2.77}$$

We shall require the intensities of the orders taken individually, thus we suppress the summation and recognise that the result is specific to a given n, also for cold samples $\exp(-\beta) \approx 0$ and simplifying Eq. (A2.77):

$$\exp\left\langle\left[\mathbf{Q}.^v\mathbf{u}(0)\right]\left[\mathbf{Q}.^v\mathbf{u}(t)\right]\right\rangle_T^n = \exp(in\omega t)\frac{\exp[n\beta]}{\exp(|n|\beta)}\frac{\left\{\left[\mathbf{Q}.^v\mathbf{u}\right]^2\right\}^{|n|}}{|n|!} \tag{A2.78}$$

The order, n, labels the final state of the sample and $n = 1$, represents the (0–1) transition; $n = -1$, is (1–0); and so on. The $n = 0$ solution refers to (0–0), elastic scattering. Substituting for neutron energy loss processes eliminates the temperature factors and allows us to write, $|n|!$ and $|n|$ simply as $n!$ and n. This completes the treatment of the second term of Eq. (A2.63).

We generalise the result over three dimensions (x, y, z), substitute the results of Eqs. (A2.70) and (A2.78) into Eq. (A2.57), through Eqs. (A2.63), (A2.59) and (A2.58). The individual Fourier components of the product in Eq. (A2.57) each contribute a delta function to the spectrum, one at each normal mode frequency and its overtones, or orders.

The time independent product of exponentials in Eq. (A2.70) becomes the exponential of a sum over all modes. Thus Eq. (A2.57) becomes for the individual v^{th} normal mode, excited to its nth harmonic, in its low temperature approximation,

$$S(\mathbf{Q},\omega_v)^n = \frac{\left[\mathbf{Q}.^v\mathbf{u}\right]^{2n}}{n!}\exp\left(-\left(\mathbf{Q}.\sum_v {}^v\mathbf{u}\right)^2\right)\delta(E_i - E_f + n\hbar\omega_v) \tag{A2.79}$$

We now reintroduce the atomic index, l, and include the atomic cross section, σ. Then, for the scattering intensity contribution to a given spectral band, that arising from the lth atom is,

$$S^{\bullet}(Q,\omega_v)_l^n = \frac{\sigma_l}{4\pi} \frac{\left[Q.^v u_l\right]^{2n}}{n!} \exp\left(-\left(Q.\sum_v {}^v u_l\right)^2\right)$$

(A2.80)

The delta function can be omitted since the expression is only evaluated at specific transitions, where its value is unity. This is the basis of Eq. (2.32), in Chapter 2, where the significance of $S^{\bullet}$ is explained. The spectrum consists of transitions, appearing as a sequence of sharp bands at neutron energy loss values corresponding to the energies of all the internal excitations of the molecule.

As described in Chapter 4, for the case of simple molecules the displacement vectors can be calculated by hand but it is laborious and the output of the normalised displacement vectors, ${}^v L$, from commercial programs is usually to be preferred. These are related to the displacement vectors through:

$$\begin{aligned}
{}^v u_l^2 &= {}^v L \left(\frac{\hbar}{2 m_l \omega_v}\right) \coth\left(\frac{\hbar \omega_v}{2 k_B T}\right) \\
&= \frac{{}^v L}{m_l} \left(\frac{h}{2(2\pi)^2 c \tilde{v}_v}\right) \coth\left(\frac{h c \tilde{v}_v}{2 k_B T}\right)
\end{aligned}$$

(A2.81)

(Where the relationship of our conventional symbol, ω, is given in respect of the common spectroscopic symbol $\tilde{v}$.) Substituting for the fundamental constants, see below for the tensor notation B.

$$ {}^v B_l = {}^v u_l^2 = \frac{16.9}{{}^v \mu_l \omega_v} \coth\left(\frac{1.47 \omega_v}{T}\right) $$

(A2.82)

With the transition at ω, in cm^{-1}, the sample temperature T, in K, and the reduced mass μ, in atomic mass units, then the atomic mean square displacement is in $Å^2$.

A2.4 Debye-Waller factor

This is the name given to the exponential factor in the Scattering Law and it is often found written, $\exp(-2W)$. It should be noted that the argument, $-2W$, is specifically not the Debye-Waller factor despite the occasional solecism found in the literature. It is often found written:

$$\exp(-2W) = \exp\left(-\left(\boldsymbol{Q}.\boldsymbol{U}_l\right)^2\right)$$

$$\boldsymbol{U}_l = \sum_v {}^v\boldsymbol{u}_l$$

(A2.83)

The Debye-Waller factor always decreases the observed intensity in neutron spectroscopy and significant experimental efforts are made to reduce its impact. To follow this more closely we retrace our steps and, from Eq. (A2.70), write the full expression for $2W$, instead of the low temperature approximation used in Eq.(A2.80).

$$2W = \left(\boldsymbol{Q}.\sum_v \left\{ {}^v\boldsymbol{u}_l \coth\left(\frac{\hbar\omega_v}{2k_B T}\right)\right\}\right)^2$$

(A2.84)

As can be appreciated, the lower the temperature the closer the root mean square displacement, ${}^v\boldsymbol{u}_l$, approaches its minimum value, the zero-point motion. A reduction in the neutron momentum transfer will also increase the Debye-Waller factor but this parameter is not always open to variation (§3.4.4).

A2.5 Powder averaging

The average value of the Scattering Law taken with respect to all available directions in space is required to compare its calculated value to the experimental results obtained form powders.

There have been several approaches including, analytical, numerical and semi-analytical. These have all been discussed in the literature. Each approach has its own advantages and disadvantages. The analytical approach is very accurate but lacks generality, being applicable to only spherical local geometry, O_h, or cylindrical local geometry, $C_{\infty v}$, or $D_{\infty h}$ [4]. The numerical approach is quite general but, to be accurate, involves many calculations, a slow process [5]. We shall focus on a semi-analytical approach that is general, fast to compute and sufficiently accurate [6].

We start by developing the quadratic terms in Eq. (A2.80), for a given atom. The vector components along the Cartesian coordinate scheme used to generate the root mean square displacement vectors are developed (a superscripted T denotes matrix transpose).

$$^{v}B = {}^{v}u^{T} \; {}^{v}u = {}^{v}\left(u_x, \; u_y, \; u_z\right)^{T} \; {}^{v}\left(u_x, \; u_y, \; u_z\right)$$

$$= \begin{pmatrix} \begin{pmatrix} u_x \\ u_y \\ u_z \end{pmatrix} {}^{v}\begin{pmatrix} u_x & u_y & u_z \end{pmatrix} \end{pmatrix} = {}^{v}\begin{pmatrix} u_x u_x & u_x u_y & u_x u_z \\ u_y u_x & u_y u_y & u_y u_z \\ u_z u_x & u_z u_y & u_z u_z \end{pmatrix} \qquad (A2.85)$$

The mode specific vibrational root mean squared displacement vector is thus replaced by the mean square displacement tensor of that atom in that specific mode, it has units of $Å^2$. The tensor is represented in Eq. (A2.85) by a matrix, in more compact notation it is written, for a specific vibrational mode and atom $^{v}B_l$. The colon-operator (:) or tensor contraction operation is found by taking the trace (Tr) of the product of the two tensors, see Eq. (2.50). Further we write the total mean square displacement tensor of an atom as the sum over all the individual vibrational contributions.

$$A_l = \sum_v {}^{v}B_l \qquad (A2.86)$$

The powder averaged version of Eq. (A2.80) for an atom in a given fundamental mode, $n = 1$, is the average taken over all directions of Q.

$$S(Q, \omega)_l = \frac{1}{4\pi} \int QQ^{T} {}^{:v}B_l \exp(-QQ^{T} : A_l) \, dQ \qquad (A2.87)$$

This is obtained by expanding the exponential as a power series [6].

$$\exp(-QQ : A) = 1 - (QQ : A) + \frac{1}{2!}(QQ : A)(QQ : A) -$$

$$=1-\boldsymbol{QQ}:\boldsymbol{A}+\frac{1}{2!}\boldsymbol{QQQQ}\;\vdots\;\boldsymbol{AA}-\frac{1}{3!}\boldsymbol{QQQQQQ}:\boldsymbol{AAA}+ \qquad \text{(A2.88)}$$

If e_Q is the unit vector in the direction of $\boldsymbol{Q}$ then, for example, the second (indices i and j) and fourth rank (indices i, j, k and l) tensors are [6]

$$\frac{1}{4\pi}\int\boldsymbol{QQ}\,\mathrm{d}Q=\frac{Q^2}{4\pi}\int e_Q e_Q \mathrm{d}e_Q=\frac{Q^2}{3}\delta_{ij} \qquad \text{(A2.89)}$$

$$\frac{1}{4\pi}\int\boldsymbol{QQQQ}\,\mathrm{d}Q=\frac{Q^4}{4\pi}\int e_Q e_Q e_Q e_Q\,\mathrm{d}e_Q=\frac{Q^4}{15}(\delta_{ij}\delta_{kl}+\delta_{ik}\delta_{jl}+\delta_{il}\delta_{jk}) \qquad \text{(A2.90)}$$

Similar expressions exist for the tensors of higher rank and can be used to rewrite Eq. (A2.87) term by term

$$S(Q,\omega)\big|_{1\text{st term}}=\frac{1}{4\pi}\int\boldsymbol{QQ}^{\mathrm{T}}:{}^{v}\boldsymbol{B}\,\mathrm{d}Q=\left(\frac{Q^2}{3}\delta_{ij}:{}^{v}\boldsymbol{B}\right)$$
$$=\frac{Q^2}{3}{}^{\mathrm{Tr}}\!\left({}^{v}\boldsymbol{B}\right) \qquad \text{(A2.91)}$$

$$S(Q,\omega)\big|_{2\text{nd term}}=\frac{1}{2!}\frac{1}{4\pi}\int\boldsymbol{QQ}^{\mathrm{T}}:{}^{v}\boldsymbol{B}\,\boldsymbol{QQ}^{\mathrm{T}}:\boldsymbol{A}\,\mathrm{d}Q=\boldsymbol{QQQQ}\;\vdots\;{}^{v}\boldsymbol{B}\boldsymbol{A} \qquad \text{(A2.92)}$$

$$=\left(\frac{Q^4}{15}(\delta_{ij}\delta_{kl}+\delta_{ik}\delta_{jl}+\delta_{il}\delta_{jk}):{}^{v}\boldsymbol{B}\boldsymbol{A}\right)=\frac{Q^4}{15}\left\{{}^{\mathrm{Tr}}\boldsymbol{A}^{\mathrm{Tr}}\!\left({}^{v}\boldsymbol{B}\right)+{}^{v}\boldsymbol{B}:\boldsymbol{A}\right\}$$

An exponential power series is recuperated by dividing through by the first term, Eq. (A2.91). Where the total mean square displacement tensor A is a spheroid that deviates little from the isotropic, i.e. one that is almost a sphere, then terms linear in A will be adequate [6]. This is the 'almost isotropic' approximation and Eq. (A2.87) becomes Eq. (2.41) for $n=1$.

$$S(Q,\omega)=\frac{Q^2}{3}{}^{\mathrm{Tr}}\!\left({}^{v}\boldsymbol{B}\right)\left[1-\frac{Q^2}{2!\,5}\left\{{}^{\mathrm{Tr}}\boldsymbol{A}+2\left(\frac{{}^{v}\boldsymbol{B}:\boldsymbol{A}}{{}^{\mathrm{Tr}}\left({}^{v}\boldsymbol{B}\right)}\right)\right\}+\frac{Q^4}{3!\,35}\left\{\;\right\}^2-....\right]$$

$$\text{(A2.93a)}$$

$$=\frac{Q^2}{3}{}^{\text{Tr}}\left({}^{\nu}B\right)\exp(-Q^2\alpha^{\nu}) \qquad (A2.93b)$$

$$\alpha^{\nu}=\frac{1}{5}\left\{{}^{\text{Tr}}A+2\left(\frac{{}^{\nu}B:A}{{}^{\text{Tr}}\left({}^{\nu}B\right)}\right)\right\}$$

The argument of the Debye-Waller factor is mode specific since the contraction of A with the B of a specific mode generates a particular response depending on the orientation of the atomic displacement in that mode. The case of two quantum transitions, $n = 2$, covers $(2\leftarrow0)$ and $(1\leftarrow0)(1'\leftarrow0)$, then [4]:

$$S(Q,\omega)^{\nu}\big|(0-2)$$

$$=\frac{Q^4}{30}\left\{{}^{\text{Tr}}\left({}^{\nu}B\right)^{\text{Tr}}\left({}^{\nu}B\right)+2\ {}^{\nu}B{:}^{\nu}B\right\}\exp\left(-Q^2\beta^{\nu}\right) \qquad (A2.94)$$

$$S(Q,\omega)^{\nu}\bigg|\begin{matrix}(0-1')(0-1)\\(0-1)(0-1')\end{matrix} \qquad (A2.95)$$

$$=\frac{Q^4}{15}\left\{{}^{\text{Tr}}\left({}^{\nu}B\right)^{\text{Tr}}\left({}^{\nu'}B\right)+\left({}^{\nu}B{:}^{\nu'}B\right)+\left({}^{\nu'}B{:}^{\nu}B\right)\right\}\exp\left(-Q^2\ \beta^{\nu}\right)$$

where

$$\beta^{\nu}=\frac{{}^{\text{Tr}}(A)}{3} \qquad (A2.96)$$

Here we have made the isotropic approximation for the Debye-Waller factor, although a full mode specific expression for β has been derived it is complex and adds little to the accuracy of the final result [6].

A2.6 References

1 G.L Squires (1996). *Introduction to the Theory of Thermal Neutron Scattering*, Dover Publications Inc., New York.

2 D.D. Fitts (1999). *Principles of Quantum Mechanics as Applied to Chemistry and Chemical Physics*, Cambridge University Press, Cambridge.

3 S.W. Lovesey (1986). *Condensed Matter Physics: Dynamic Correlations*, The Benjamin/Cummins Publishing Company Inc., Menlo Park.

4   J. Tomkinson, M. Warner & A.D. Taylor (1984). Mol. Phys., 51, 381-385. Powder averages for neutron spectroscopy of anisotropic molecular oscillators.

5   B.S. Hudson, A. Warshel & R.G. Gordon (1974). J. Chem. Phys., 61, 2929-2939. Molecular inelastic neutron scattering: computational methods using consistent force fields.

6   T.C. Waddington, J. Howard, K.P. Brierley & J. Tomkinson (1982). Chem. Phys., 64, 193-201. Inelastic neutron scattering spectra of the alkali metal (Na,K) bifluorides: the harmonic overtone of v_3.

Appendix 3

The Resolution Function of Crystal Analyser Spectrometers

In this Appendix we derive the analytical expression for the energy resolution of a low-bandpass spectrometer like TOSCA (§3.1) (also known as crystal analyser spectrometers) and describe two key features of the design (§3.2), time focussing (§3.2.1) and the Marx principle (§3.2.2) that improve the resolution at high and low energy transfer respectively.

A3.1 The resolution function

The resolution function for crystal analyser instruments is complex but can be summarised as [1,2]:

$$\frac{\Delta E}{E_t} = \frac{1}{E_t}\sqrt{\left\{\left(\Delta t_t\right)^2 + \left(\Delta d_i\right)^2 + \left(\Delta E_f\right)^2 + \left(\Delta d_f\right)^2\right\}} \tag{A3.1}$$

where $\Delta E/E_t$ is the relative energy resolution at energy transfer E_t, Δt_t is the total timing uncertainty (largely determined by the pulse width from the moderator), Δd_i is the total uncertainty due to the initial flight path, ΔE_f and Δd_f are the total uncertainties due to the final energy and the final flight path respectively. These both depend on the thickness of the sample (d_s), graphite analyser (d_g) and detector (d_d). The form of Eq. (A3.1), where the errors are added in quadrature, assumes that each of the resolution components has a Gaussian distribution and it is the variance (the square of the standard deviation, Γ^2) of the quantity that is needed. For components that have a Gaussian distribution, the derived quantity is the full width at half maximum (FWHM).

In this case Γ = (FWHM)/2.35. (For components that are better approximated with a boxcar distribution, e.g. detector and sample width, Γ = (FWHM)/($2\sqrt{3}$).)

Eq. (A3.1) is a summary of the full expression:

$$\frac{\Delta E}{E_{tr}} = \frac{1}{E_{tr}} \left\{ \left(\left[\frac{\partial E_{tr}}{\partial t_t} \right] \Delta t_t \right)^2 + \left(\left[\frac{\partial E_{ta}}{\partial d_i} \right] \Delta d_i \right)^2 + \left(\left[\frac{\partial E_{tr}}{\partial E_f} \right] \Delta E_f \right)^2 + \left(\left[\frac{\partial E_{tr}}{\partial d_f} \right] \Delta d_f \right)^2 \right\}^{1/2} \quad \text{(A3.2)}$$

The quadratic terms will be dealt with in the order they appear in Eq. (A3.2).

A3.1.1 The time dependent term

The derivatives are calculated from the basic equation for time-of-flight INS spectroscopy, where m_n is the neutron mass:

$$t_t = t_i + t_f = \frac{d_i}{v_i} + \frac{d_f}{v_f}$$

$$= \left(d_i \Big/ \sqrt{\frac{2E_i}{m_n}} \right) + \left(d_f \Big/ \sqrt{\frac{2E_f}{m_n}} \right) \quad \text{(A3.3)}$$

$$= \left(\sqrt{\frac{m_n d_i^2}{2(E_{tr} + E_f)}} \right) + \left(\sqrt{\frac{m_n d_f^2}{2E_f}} \right)$$

Solving for the energy transfer gives:

$$E_{tr} = \frac{L_i^2}{\left[\frac{2t_t^2}{m_n} + \frac{d_f^2}{E_f^2} - 2t_t \left(\frac{2d_f^2}{m_n E_f} \right)^{1/2} \right]} - E_f = E_i - E_f \quad \text{(A3.4)}$$

Differentiating Eq. (A3.4), with respect to time:

$$\frac{\partial E_{tr}}{\partial t_t} = \frac{d_i^2}{\left[\frac{2t_t^2}{m_n} + \frac{d_f^2}{E_f^2} - 2t_t \left(\frac{2d_f^2}{m_n E_f} \right)^{1/2} \right]^2} \left[2t_t \left(\frac{2}{m_n} \right) - 2 \left(\frac{2d_f^2}{m_n E_f} \right)^{1/2} \right] \quad \text{(A3.5)}$$

substituting for E_i from Eq. (A3.4)

$$= \frac{E_i^2}{d_i^2}\left[2t_t\left(\frac{2}{m_n}\right) - 2\left(\frac{2d_f^2}{m_n E_f}\right)^{1/2}\right] \tag{A3.6}$$

and substituting for t_t from Eq. (A3.3)

$$= \frac{E_i^2}{d_i^2}\left[2\left(\frac{2}{m_n}\right)\left\{\left(\frac{m_n d_i^2}{2E_i}\right)^{1/2} + \left(\frac{m_n d_f^2}{2E_f}\right)^{1/2}\right\} - 2\left(\frac{2E_f}{m_n d_f^2}\right)^{1/2}\right] \tag{A3.7}$$

$$= 2\left(\frac{2}{m_n}\right)^{1/2}\frac{E_i^2}{d_f^2}\left[\left(\frac{d_i^2}{E_i}\right)^{1/2} + \left(\frac{d_f^2}{E_f}\right)^{1/2} - \left(\frac{d_f^2}{E_f}\right)^{1/2}\right] \tag{A3.8}$$

$$= 2\left(\frac{2}{m_n}\right)^{1/2}\frac{E_i^{3/2}}{d_i} \tag{A3.9}$$

Also

$$\Delta t_t = \left\{(\Delta t_m)^2 + (\Delta t_{ch})^2\right\}^{1/2}$$

$$= \left\{\left(A\left[\frac{\hbar^2}{2m_n E_i}\right]^{1/2}\right)^2 + (\Delta t_{ch})^2\right\}^{1/2} \tag{A3.10}$$

Where Δt_m is the time width of the moderator pulse in microseconds. Its energy dependence is specific to a given neutron source and is described empirically, A is the moderator constant, (the units of $\hbar^2$ in Eq. (A3.10) are quite exceptional, see Table A3.1). The time channel uncertainty, Δt_{ch}, in microseconds, is usually negligible in comparison to the moderator pulse width.

A3.1.2 The incident flight path dependent term

Differentiating the incident flight path dependent term in Eq. (A3.2), with respect to d_i:

$$\frac{\partial E_{tr}}{\partial d_i} = \frac{2d_i}{\left[\dfrac{2t_t^2}{m_n} + \dfrac{d_f^2}{E_f^2} - 2t_t\left(\dfrac{2d_f^2}{m_n E_f}\right)^{1/2}\right]}$$

$$= \frac{2E_i}{d_i}$$

(A3.11)

The total uncertainty in the incident flightpath, Δd_i, includes contributions from $\Delta d_i'$ the real length uncertainty and sample thickness, Δd_s, both are very small:

$$\Delta d_i = \left\{ \left(\Delta d_i'\right)^2 + \left(\Delta d_s\right)^2 \right\}^{1/2}$$

(A3.12)

This term is usually ignored.

A3.1.3 The final energy dependent term

Differentiating the final energy dependent term in Eq. (A3.2), with respect to E_f:

$$\frac{\partial E_{tr}}{\partial E_f} = -1 + \frac{d_i^2}{\left[\dfrac{2t_t^2}{m_n} + \dfrac{d_f^2}{E_f} - 2t_{tot}\left(\dfrac{2d_f^2}{m_n E_f}\right)^{1/2}\right]^2}$$

(A3.13)

$$\times (-1)\left[\frac{-d_f^2}{E_f^2} - 2t_t\left(\frac{2d_f^2}{m_n}\right)^{1/2}\left(\frac{-1}{2}\right)\left(\frac{1}{E_f}\right)^{3/2}\right]$$

$$= -1 + \frac{E_i^2}{d_i^2}\left[\frac{d_f^2}{E_f^2} - t_t\left(\frac{2d_f^2}{m_n}\right)^{1/2}\left(\frac{1}{E_f}\right)^{3/2}\right]$$

(A3.14)

$$= -1 + \frac{E_i^2}{d_i^2}\left[\frac{d_f^2}{E_f^2} - \left(\frac{2d_f^2}{m_n}\right)^{1/2}\left(\frac{1}{E_f}\right)^{3/2}\left\{\left(\frac{m_n d_i^2}{2E_i}\right)^{1/2} + \left(\frac{m_n d_f^2}{2E_f}\right)^{1/2}\right\}\right]$$

(A3.15)

$$= -1 + \frac{E_i^2}{d_i^2} \left[\frac{d_f^2}{E_f^2} - \left(\frac{2d_f^2}{m_n} \right)^{1/2} \left(\frac{1}{E_f} \right)^{3/2} \left(\frac{m_n d_i^2}{2E_i} \right)^{1/2} - \frac{d_f^2}{E_f^2} \right] \tag{A3.16}$$

$$= -\left\{ 1 + \left(\frac{d_f}{d_i} \right) \left(\frac{E_i}{E_f} \right)^{3/2} \right\} \tag{A3.17}$$

Also the uncertainty in the final energy is

$$\Delta E_f = \left\{ \left(2E_f \frac{d_t}{d_A} \right)^2 + \left(2E_f \, \Delta\theta_B \cot\theta_B \right)^2 \right\}^{1/2} \tag{A3.18}$$

where d_A is the vertical distance from the sample to the graphite analyser planes, see Figure A3.1. Also d_B is the vertical distance from the graphite analyser planes to the detector.

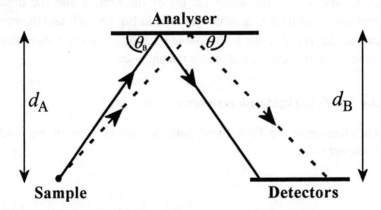

Fig. A3.1 Schematic of the time focusing effect on a crystal analyser spectrometer.

For simplicity, we assume that $d_A = d_B$ hence:

$$d_A = \frac{d_f}{2} \sin\theta_B \tag{A3.19}$$

Also

$$d_t = \left\{ (d_s)^2 + (2d_A)^2 + (d_d)^2 \right\}^{1/2} \tag{A3.20}$$

θ_B is the Bragg angle at the analyser, see Fig. A3.1. $\Delta\theta_B$ is given by [3]:

$$\Delta\theta_B = \frac{\left\{\alpha_2^2\alpha_3^2 + \alpha_2^2\eta_g^2 + \alpha_3^2\eta_g^2\right\}^{1/2}}{\left\{\alpha_2^2 + \alpha_3^2 + \left(2\eta_g\right)^2\right\}^{1/2}} \tag{A3.21}$$

Where α_2 is the divergence of the beam scattered from the sample, α_3 is the divergence of the beam diffracted by the analyser, as seen at the detector and η_g is the mosaic spread of the graphite analyser.

$$\alpha_2 = \frac{w_s}{2}\frac{\left[1+\left(\cot\theta_B\right)^2\right]}{d_g} \tag{A3.22}$$

and

$$\alpha_3 = \frac{w_d}{2}\frac{\left[1+\left(\cot\theta_B\right)^2\right]}{d_g} \tag{A3.23}$$

Here, w_s and w_d are the width (in m) of the sample and the detector respectively. The thickness of the graphite in Eq. (A3.20) and the mosaic spread in Eq. (A3.21) are both doubled since the neutron traverses the graphite twice as it enters and leaves the analyser.

A3.1.4 The final flight path dependent term

Differentiating the final flight path dependent term in Eq. (A3.2), with respect to d_f:

$$\frac{\partial E_{tr}}{\partial d_f} = \frac{d_i^2}{\left[\frac{2t_t^2}{m_n}+\frac{d_f^2}{E_f}-2t_t\left(\frac{2d_f^2}{m_n E_f}\right)^{1/2}\right]^2}\left[2t_t\left(\frac{2}{m_n E_f}\right)^{1/2}+\frac{2d_f}{E_f^2}\right] \tag{A3.24}$$

$$= \frac{-E_i^2}{d_i^2}\left[\frac{2d_f}{E_f^2}-2t_t\left(\frac{2}{m_n E_f}\right)^{1/2}\right] \tag{A3.25}$$

$$= \frac{-E_i^2}{d_i^2} \left[\frac{2d_f}{E_f^2} - 2\left(\frac{2}{m_n E_f}\right)^{1/2} \left\{ \left(\frac{m_n d_i^2}{2E_i}\right)^{1/2} + \left(\frac{m_n d_f^2}{2E_f}\right)^{1/2} \right\} \right] \quad \text{(A3.26)}$$

$$= \frac{-E_i^2}{d_i^2} \left[\frac{2d_f}{E_f^2} - 2\left(\frac{2}{m_n E_f}\right)^{1/2} \left(\frac{m_n d_i^2}{2E_i}\right)^{1/2} - \frac{2d_f}{E_f^2} \right] \quad \text{(A3.27)}$$

$$= \frac{2}{d_f} \left(\frac{E_i^3}{E_f}\right)^{1/2} \quad \text{(A3.28)}$$

Also

$$\Delta d_f = \frac{2d_t}{\sin \theta_B} \quad \text{(A3.29)}$$

Table A3.1 Parameters for TFXA, TOSCA-1 and TOSCA at ISIS.

Parameter (units)		TFXA	TOSCA1	TOSCA	
				Back	Forward
Primary flightpath (m)	d_i	12.13	12.264	17.00	17.00
Average secondary flightpath (m)	d_f	0.671	0.7456	0.6279	
Average final energy (meV)	E_f	3.909	3.51	3.32	3.35
Average Bragg angle for graphite (°)	θ_B	43.00	46.03	47.73	47.45
ISIS water moderator constant (μs)	A	44	44	44	44
Time channel uncertainty (μs)	Δt_{ch}	2	2	2	2
Sample thickness (m)	d_s	0.002	0.002	0.002	0.002
Graphite thickness (m)	d_g	0.002	0.002	0.002	0.002
Detector thickness (m)	d_d	0.006	0.0025	0.0025	0.0025
Uncertainty in primary flightpath (m)	Δd_i	0.0021	0.0021	0.0021	0.0021
Sample width (m)	w_s	0.020	0.020	0.040	0.040
Detector width (m)	w_d	0.012	0.012	0.12	0.12
Mosaic of the graphite analyser (°)	η_g	2.5	2.5	2.5	2.5
Constant of Eq. (A3.10) (meV Å)	$\hbar^2$		4.18019		

A3.1.5 The resolution function

Collecting the individual terms, Eqs. (A3.9), (A3.10), (A3.11), (A3.12), (A3.17), (A3.18), (A3.20), (A3.28), (A3.29) and substituting in Eq. (A3.2) gives:

$$\frac{\Delta E}{E_{tr}} = \frac{1}{E_{tr}} \left\{ \left(\left[\frac{\partial E_{tr}}{\partial t_t} \right] \Delta t_t \right)^2 + \left(\left[\frac{\partial E_{tr}}{\partial d_i} \right] \Delta d_i \right)^2 + \left(\left[\frac{\partial E_{tr}}{\partial E_f} \right] \Delta E_f \right)^2 + \left(\left[\frac{\partial E_{tr}}{\partial d_f} \right] \Delta d_f \right)^2 \right\}^{\frac{1}{2}} \qquad (A3.2)$$

$$= \frac{1}{E_{tr}} \left\{ \begin{array}{l} \left(\left[\left(\frac{2}{m_n} \right)^{\frac{1}{2}} \frac{E_i^{\frac{3}{2}}}{d_i} \right] \left[A^2 \left(\frac{\hbar^2}{2m_n E_i} \right) + (\Delta t_{ch})^2 \right] \right)^2 \\[2mm] + \left(\left[\frac{2E_i}{d_i} \right] \left[(\Delta d_i)^2 + (\Delta d_s)^2 \right]^{\frac{1}{2}} \right)^2 \\[2mm] + \left[\left(1 + \left(\frac{d_f}{d_i} \right) \left(\frac{E_i}{E_f} \right)^{\frac{3}{2}} \right) \left(2E_f \frac{\left((d_s)^2 + (2d_g)^2 + (d_d)^2 \right)^{\frac{1}{2}}}{d_A} \right)^2 + (2E_f \Delta\theta_B \cot\theta_B) \right]^{\frac{1}{2}} \right)^2 \\[2mm] + \left(\frac{2}{d_f} \left[\frac{E_i^3}{E_f} \right]^{\frac{1}{2}} \frac{2d_t}{\sin\theta_B} \right)^2 \end{array} \right\}^{\frac{1}{2}}$$

$$(A3.30)$$

There are several factors worth noting:

The most significant terms are the first, third and fourth. The incident flightpath, d_i, appears in the denominator of two of these, thus increasing d_i gives an immediate improvement in resolution as found for TOSCA over TFXA. However, unless other action is also taken, the incident flux will fall off rapidly from solid angle considerations.

It is possible to obtain a limited improvement in resolution by reducing the thickness of the sample, graphite and analyser as seen for TOSCA1 relative to TOSCA. If the these are reduced too much then the other terms in the resolution function will dominate and no further improvement will be seen. The price of reducing these terms is that the detected flux decreases, so compensating action is also required here.

Increasing the Bragg angle, θ_B, also improves the resolution. The cost is that the volume available for the analysers and detectors becomes severely limited.

Since the incident energy E_i always appears in the numerator of the terms it follows that the resolution will degrade to larger energy transfer.

Eq. (A3.30) gives a reasonable description of the resolution function and qualitatively reproduces the behaviour well. In detail, it is found to underestimate the resolution observed experimentally. The problem stems from Eq. (A3.21) which predicts a divergence in the analyser that is larger than observed. The reason for this is that all the variables are assumed to be independent and this is only an approximation. Recent work on the development of a TOSCA-like spectrometer on the SNS source VISION (Oak Ridge, USA) has lead to a significant improvement in the mathematical representation of the resolution elements of these types of spectrometer. At the time of going to press this work was unpublished [4].

A3.2 Design elements

All crystal analyser spectrometers of the TOSCA type have two design features that improve their resolution. These are the use of time focussing and the Marx principle (also known as energy focussing). Incorporation of these features improves the resolution at high and low energy transfer respectively.

A3.2.1 Time focusing

Time focussing was first introduced in the LAM-D spectrometer at KENS [1] and has been adopted in all subsequent crystal analyser instruments. The key feature is that the planes of the sample, analyser and detector are parallel as shown in Fig. A3.1. For simplicity, a point sample is assumed. From Fig A3.1 for the ray shown by the solid line it can be seen that the length of the final flight path from sample to detector, d_f, is given by:

$$d_f = \frac{d_A}{\sin \theta_B} + \frac{d_B}{\sin \theta_B} = \frac{(d_A + d_B)}{\sin \theta_B} \tag{A3.31}$$

Where d_A and d_B are the vertical distances from the sample to analyser and analyser to detector respectively. In the present case we consider the more general condition where $d_A \neq d_B$. The wavelength of any scattered neutron, λ_f, is given by:

$$\lambda_f = \frac{h}{m_n v_f} = \left(\frac{h}{m_n}\right)\left(\frac{t_f}{d_f}\right) \tag{A3.32}$$

Where t_f is the flight time along d_f. Substituting (A3.31) and (A3.32) into Bragg's law: here the planes of the graphite analyser are separated by d_{PG}

$$\lambda_f = 2d_{PG} \sin \theta_B \tag{A3.33}$$

$$\left(\frac{h}{m_n}\right)\left(\frac{t_f}{d_f}\right) = 2d_{PG} \frac{(d_A + d_B)}{d_f}$$

$$\therefore t_f = 2d_{PG}(d_A + d_B)\left(\frac{m_n}{h}\right) = \text{constant}$$

Thus the travel time around the final flightpath is independent of the neutron's energy. Thus the time uncertainty in the final flightpath is the same for all neutrons irrespective of their initial energy. At large energy transfer, t_f is a significant contribution to the total flight time, so by making the uncertainty in t_f constant the relative contribution to the total uncertainty is diminished and the resolution is improved.

Qualitatively Eq. (A3.33) can be understood by considering the two scattered neutrons shown by the solid and dashed lines in Fig. A3.1. These have scattering angles from the analyser of θ_A and θ, with $\theta_A > \theta$. The total flightpath for the neutron scattering at θ_A is clearly less than that scattering at θ. From Bragg's law the wavelength of the neutron scattering at θ_A must be longer than that scattering at θ, hence its energy, and thus its velocity, must be less than that scattering at θ. So the greater velocity of the neutron scattering at θ compensates for the greater pathlength and the travel time is the same for both neutrons.

A3.2.2 The Marx principle

The Marx principle applies when irrespective of their divergence; the design of the spectrometer requires all neutrons of a given energy to reach a particular focal point. This is illustrated in Fig. A3.2. For a point sample, reflection in the analyser produces a series of focal points for a range of energies and these fall on a plane parallel to the analyser. For a given energy, the distance from the analyser to the focal plane is the same as the distance from the sample to the analyser. As shown in Fig. A3.2, divergent rays from the sample of the same energy (wavelength) find crystallites in the analyser of the correct orientation to reflect on to the detector. The mosaicity of the analyser is being used as a means of focussing the scattered neutrons. This emphasises the need for significant mosaicity in the analyser, this is easy to achieve with pyrolytic graphite and is one of the reasons for its use.

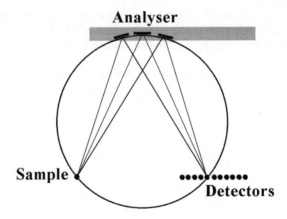

Fig. A3.2 Illustration of the Marx principle. Scattered neutrons of the same energy are reflected by differently oriented crystallites in the analyser and focused on to the same detector. The focal circle for the fifth detector is shown.

The Marx principle improves the resolution at low energy transfer. The reason for this is that all scattered neutrons of the *same* energy have the same t_f thus reducing the uncertainty in the final energy, which is a large contribution at low energy transfer.

Note that the Marx principle is more restrictive than time focussing; in the latter all that is necessary is that the sample, analyser and detector planes are parallel, in the former the planes must not only be parallel but the sample and detector planes must also be co-planar. Time and energy focussing are so successful that all current and planned crystal analyser spectrometers employ it. The only exception is the forward scattering bank on TOSCA where limitations on space meant that the detectors were placed slightly downstream from the sample plane. In practice, the displacement is small enough that the resolution is essentially the same in both the forward and backscattering detector banks.

A3.3 References

1   S. Ikeda & N. Watanabe (1984). Nucl. Instrum. and Meth., 221, 571-576. High resolution tof crystal analyzer spectrometer for large energy transfer incoherent neutron scattering.
2   J. Penfold & J. Tomkinson (1986). Rutherford Appleton Laboratory Report RAL-86-019. The ISIS time focused crystal spectrometer, TFXA.
3   C.G. Windsor (1981). *Pulsed Neutron Scattering*, Taylor and Francis, London, p336.
4   J.Z. Larese, University of Tennessee, private communication.

Appendix 4

Systems Studied by INS

The purpose of this Appendix is to provide a starting point for searching the literature. The spectra selected are generally those that show well-resolved vibrational features and are largely from the mid-1980s onwards. Some older work of particular significance is also included. We recognise that the references will become out-dated, however, they can be used to search backwards, since they include references to previous work (if any), and forwards, since they can be used in a citation search. For these reasons, for any given system we have not attempted to provide comprehensive references; instead, we have selected what appears to us to be the most complete analysis of the data. Other useful sources are the websites for the instruments, since many of these include a list of publications. The INS database at www.isis.rl.ac.uk/insdatabase also includes references to where the spectra were published.

The publications are listed in two different ways; Table A4.1 is alphabetical by common or systematic name, thus acetone not propanone, propylene not propene. The formula is given in the usual style, thus $Fe(CO)_5$ not FeC_5O_5. Catalyst studies are listed in two ways: under the name of the adsorbate, with the entry starting with &. Thus the next entry after thiophene is & MoS_2, showing that thiophene on MoS_2 has been studied. Catalyst studies are also listed under the name of the catalyst; so there is an entry for molybdenum disulfide with & H_2 and & thiophene. The two types of entry allow the reader to find all the molecules that have been studied on a particular catalyst or all the catalysts a particular molecule has been studied on. Hydrogen is sub-divided into systems where H_2 was the species studied e.g. H_2 on Ru/C or whether it was H atoms (H_{ads}), thus entries for Pt/C and RuS_2 occur. Minerals appear under their common name and their chemical name, thus newberyite, $MgHPO_4.3H_2O$, is also listed as magnesium hydrogenphosphate trihydrate.

We show in the column headed H/D where isotopomers were studied (Y). The spectrometer on which a spectrum was recorded is also reported; most of these are described in Table 3.2. If the system has been studied by coherent INS, the references are also included and TAS (triple axis spectroscopy) appears in the instrument column. Much of the early work was carried out on low-bandpass spectrometers that are now defunct, these were generally of the beryllium filter type and are indicated as 'LBS'.

Hydrogen-in-metal systems (§6.6) constitute a large fraction of the materials that have been studied by INS. The spectra of many of these are very similar; consisting of broad lines (because of dispersion) often recorded at low resolution. The field has been extensively documented [1–5] so only the hydrides of relevance to catalysis or those with unusual spectra are included. Note that no hydrogen-in-platinum system is known.

Table A4.2 includes the same information as Table A4.1 but it is listed by formula, with the elements listed alphabetically. Thus Zeises's salt $K[Pt(C_2H_4)Cl_3]$ appears as $C_2Cl_3H_4KPt$ and HCl as ClH. Water of hydration is included so gypsum, $CaSO_4.2H_2O$, is listed as CaH_4O_6S. Catalysts are listed by the metal, so hydrogen on Pt/C is found under Pt and CoAlPO under Co. Catalysts with more than one metal, are listed alphabetically by the metal, so PtRu/C is found as PtRu. Adsorbed dihydrogen, H_2, and its complexes are listed under H_2. Polymers are listed as (repeat unit)$_n$ so polyethylene is $(C_2H_4)_n$ and polyacetylene is $(C_2H_2)_n$, the n is ignored and they are listed by the repeat unit. Condensation polymers *e.g.* Nylon 6,6, $-(NH(CH_2)_6NHCO(CH_2)_6CO)_n-$ are treated similarly and occur as: $(C_{14}H_{26}N_2O_2)_n$. Polymers are best located from Table A4.1.

Table A4.1 Alphabetical list of systems studied by high resolution vibrational INS spectroscopy.

Name	Formula	H/D	Instrument	Refs.
Acenapthene	$C_{12}H_{10}$		TFXA	7
Acetanilide	C_8H_9NO	Y	TFXA	8,9
Acetonitrile	CH_3CN		TFXA	10,11
& Ni			TFXA, IN1BeF	11,286
Acetonitriledichloro(ethene)-platinum(II)	$[Pt(CH_3CN)(C_2H_4) - Cl_2]$	Y	LBS	298
Acetylacetone	$CH_3COCH_2COCH_3$	Y	TOSCA	12
Acetylenedecacarbonyltriosmium	$[Os_3(\mu_2\text{-}CO)\text{-}(CO)_9(C_2H_2)]$		TFXA	201
Acetylenehexacarbonyldicobalt	$[Co_2(CO)_6(C_2H_2)]$		IN1BeF	13
Adenine	$C_5H_5N_5$		TXFA	6,14,15
Adenosine	$C_{10}H_{15}N_5O_4$		TFXA	6,15
l-Alanine	$H_2NCH_2CH_2CO_2H$		FDS	16
Alanine dipeptide	$CH_3CONHCH(CH_3)\text{-}CONHCH_3$		TFXA	17
Alumina	Al_2O_3		TAS	259
Aluminium trihydride	AlH_3		TFXA	19
2-Amino-3-methyl-1-butanol	$(CH_3)_2CHCH(NH_2)C H_2OH$		IN1BeF	20
2-Aminopyrazine-3-carboxylic acid	$C_5H_5N_3O_2$		NERA-PR	21
Amminedichloro(ethene)platinum(II)	$[Pt(C_2H_4)Cl_2(NH_3)]$	Y	LBS	298
Ammoniated titanium disulfide	$(NH_3)_xTiS_2$		FDS	22,23
Ammonium benzoate	$NH_4C_6H_5CO_2$	Y	TFXA	24
Ammonium bromide	NH_4Br		TFXA	26,27,28
Ammonium chloride	NH_4Cl		TFXA	26,27
Ammonium hydrogensulphate	NH_4HSO_4		TFXA	29
Ammonium iodide	NH_4I		TFXA	26
Ammonium nitrate	NH_4NO_3		TFXA	26
Ammonium tantalum tungstate	NH_4TaWO_6		TFXA	306
Ammonium tetrafluoroaluminate(III)	NH_4AlF_4		TFXA	30
Ammonium trifluorocadmate(II)	$NH_4[CdF_3]$		TFXA	25
Ammonium trifluoromanganate(II)	$NH_4[MnF_3]$		TFXA	25
Ammonium trifluorozincate(II)	$NH_4[ZnF_3]$		TFXA	25
Aniline	$C_6H_5NH_2$		TFXA	31
Anilinedichloro(ethene)platinum(II)	$[Pt(C_2H_4)(C_6H_5N_2)\text{-}Cl_2]$	Y	LBS	298
Anilinium bromide	$(C_6H_5NH_3)Br$	Y	LBS	293
Anilinium chloride	$(C_6H_5NH_3)Cl$	Y	LBS	293

Name	Formula	H/D	Instrument	Refs.
Anilinium iodide	$(C_6H_5NH_3)I$	Y	LBS	293
Anthracene	$C_{14}H_{10}$		TFXA	7,32
Anthrone	$C_{14}H_{10}O$		TFXA	33
Apophyllite	$Ca_4H_{17}O_{29}Si_8K$		TFXA	34
Augelite	$Al_2(OH)_3(PO_4)$		TOSCA	35
Barium imide	$BaNH$		TFXA	36
Barium chlorate monohydrate	$Ba(ClO_3)_2.H_2O$		TFXA	304
Barium tetrahydridopalladate(0)	$Ba_2[PdH_4]$		TFXA	37
Benzene	C_6H_6	Y	TFXA,TAS IN1BeF,	38,39, 271
& Ni			IN1BeF	272,273
& Pt			IN1BeF	274
& Zeolite Y			IN1BeF	38,271
Benzenetricarbonylchromium	$[Cr(CO)_3C_6H_6]$		IN1BeF, IN4	40
Benzoic acid	C_6H_5COOH	Y	TFXA	41
1,4-Benzoquinone	$C_6H_4O_2$		TFXA, NERA-PR	42, 43
Benzylic amide [2] catenane	$C_{64}H_{56}N_8O_8$		TFXA	45
Bianthrone	$C_{28}H_{16}O_2$		TFXA	33
Bikitaite	$Li_2Al_2Si_4O_{14}H_4$		TFXA	34
Bis(allyl)dichlorodipalladium	$[\{Pd(\eta^3-C_3H_5)\}_2(\mu-Cl)_2]$		TFXA	18
Bis(dcype)trihydridodiplatinum tetraphenylborate (dcype = 1,2-bis(dicyclohexylphosphanyl)ethane	$[Pt_2(dcype)_2(H)_3][BPh_4]$		FDS	215
Bis(dicarbonylcyclopentadienyliron)	$[\{Fe(CO)_2(C_5H_5)\}_2]$		TFXA	54
1,8-Bis(dimethylamino)naphthalene	$C_{10}H_6\{N(CH_3)_2\}_2$		NERA-PR	44
Bis(4-methylpyridine)cobalt(II) dichloride	$[Co(C_6H_7N)_2]Cl_2$		TFXA	56
Bis(4-methylpyridine)manganese(II) dichloride	$[Mn(C_6H_7N)_2]Cl_2$		TFXA	56
Bis(4-methylpyridine)zinc dichloride	$[Zn(C_6H_7N)]Cl_2$		TFXA	56
Borane-ammonia complex	$H_3B:NH_3$		TOSCA	317
Borneole	$C_{10}H_{18}O$	Y	NERA-PR	52
Brushite	$CaHPO_6H_4$		TFXA	46
Buckminsterfullerene	C_{60}		MARI, TOSCA	238
Buckminsterfullerene polymer	$(C_{60})_n$		MARI, IN6 TOSCA	240-243
1-Butanol	C_4H_9OH		TOSCA	81
2-Butyne	$CH_3-C{\equiv}C-CH_3$		TOSCA	47

Name	Formula	H/D	Instrument	Refs.
C_{70} fullerene	C_{70}		TFXA	239
Caesium dihydrogenarsenate	$Cs(H_2AsO_4)$		TFXA	258
Caesium dihydrogenphosphate	$Cs(H_2PO_4)$		TFXA	305
Caesium dodecaborane	$Cs_2(B_{12}H_{12})$		TOSCA	48
Caesium fulleride	CsC_{60}		IN6	249,250
Caesium hydride	CsH		TOSCA	303
Caesium hydrogencarbonate	$Cs(HCO_3)$		TFXA	49
Caesium hydrogensulphate	$Cs(HSO_4)$		TFXA	256
Caesium methanesulfonate	$Cs(CH_3SO_3)$		LBS	292
Caesium pentadecacarbonylhydrido-hexacobaltate(1−)	$Cs[HCo_6(CO)_{15}]$		IN1BeF	252
Caesium undecacarbonylhydridoferrate(1−)	$Cs[HFe_3(CO)_{11}]$		LBS	254
Calcium hydride	CaH_2		IN1BeF	221
Calcium hydrogenphosphate	$CaHPO_4$		TFXA	50
Calcium hydrogenphosphate dihydrate	CaH_5PO_6		TFXA	46
Calcium hydroxide	$Ca(OH)_2$		TFXA, NERA-PR	51
Camphor	$C_{10}H_{16}O$	Y	NERA-PR	52
Chloroform	$CHCl_3$		FDS	53
3-Chlorostyrene	$Cl-C_4H_7-CH=CH_2$		TOSCA	327
Chromous acid (chromium(III) oxide hydroxide)	$CrOOH$		FDS	55
Coal	C_x		TFXA, MARI	7,229,230
Collagen 25% hydration			TFXA	59
Copper(II) formate	$Cu(HCO_2)_2$		TFXA	60
Copper(II) formate tetrahydrate	$Cu(HCO_2)_2.4H_2O$		TFXA TOSCA	60,61
Coronene	$C_{24}H_{12}$		TOSCA	62
Cubane	C_8H_8		FANS, TAS	63,64
Cyclohexanedione	$C_6H_8O_2$		TOSCA, FANS	315
Cyclohexanedione benzene complex	$(C_6H_8O_2)_6C_6H_6$	Y	TOSCA, FANS	315
Cyclohexene	C_6H_{10}		IN1BeF	275
& Ni			IN1BeF	275
Cytidine	$C_9H_{11}N_3O_5$		TFXA	6,15,65
Cytosine	$C_4H_5N_3O$		TFXA	6,15,66
Decacarbonyldihydridotriosmium	$[Os_3(CO)_{10}(\mu_2-H)_2]$		TFXA	72
Decane	$C_{10}H_{22}$		TFXA	67
1,8-Diaminonaphthalene	$C_{10}H_6(NH_2)_2$		NERA-PR	68

Name	Formula	H/D	Instrument	Refs.
Diamond	C		TFXA,TAS	89,223
1,3-Diaminopropane	$C_3H_{10}N_2$	Y	TFXA	70
Dicarbonyl(cyclopentadienyl)-thiopheneiron tetrafluoroborate	$[Fe(CO)_2(C_4H_4S)-(C_5H_5)][BF_4]$		TFXA	54
Dichloro(dimethylformamide)ethene-platinum(II)	$[Pt(C_2H_4)Cl_2-HCON(CH_3)_2]$	Y	LBS	298
Dichloro(dimethylsulfoxide)ethene-platinum(II)	$[Pt(CH_3)_2SO(C_2H_4)-Cl_2]$	Y	LBS	298
Diglyme	$C_6H_{14}O_3$		TFXA	71
2,5-Dihydroxy-1,4-benzoquinone	$(HO)_2C_6H_2(=O)_2$		NERA-PR	73
2,3-Dimethylnorbornane	C_9H_{16}		IN1BeF	134
Disodium 12-tungstohydrogenphosphate hydrate	$Na_2PW_{12}O_{53}H_{26}$		TFXA	74
Dithiophene	$C_8H_6S_2$		TFXA	193,324
Docosane	$C_{22}H_{46}$		TFXA	67
Dodecacarbonyltetrahydrido-tetraosmium	$[Os_4(CO)_{12}H_4]$		TFXA	72
Dodecacarbonyltrihydrido-trimanganese	$[Mn_3(CO)_{12}H_3]$		LBS	254
Dodecahedrane	$C_{20}H_{20}$		TOSCA	76
Dodecane	$C_{12}H_{26}$		TFXA	67
Durene	$C_{10}H_{14}$	Y	TFXA	77
Eicosane	$C_{20}H_{42}$		TFXA	67
Emodin	$C_{15}H_{10}O_5$		TFXA	312
Ethene	C_2H_4		TFXA	79,80
Ethanol	C_2H_5OH		TFXA	82
Ethylene oxide	C_2H_4O		IN1BeF	326
Ethylidyne tricobalt nonacarbonyl (Nonacarbonyl(ethylidyne)tricobalt)	$[Co_3(CCH_3)(CO)_9]$	Y	TFXA	83
Ferrocene	$[Fe(C_5H_5)_2]$		TFXA	84
3-Fluorostyrene	$3-FC_8H_7$		TOSCA	85
4-Fluorostyrene	$4-FC_8H_7$		TOSCA	86
Formamide	$HCONH_2$		CHEX	87
Formic acid	HCO_2H	Y	TFXA	60,212
Glycine	$H_2NCH_2CO_2H$	Y	TFXA	88
Graphite	C		TFXA, TAS	89,224, 225
Guanidinium dithionate	$[C(NH_2)_3]_2^+ [SO_3SO_3]^{2-}$		FANS	90
Guanidinium methanesulfonate	$[C(NH_2)_3]^+[CH_3SO_3]^-$		FANS	90
Guanidinium triflate	$[C(NH_2)_3]^+[CF_3SO_3]^-$		FANS	90
Guanine	$C_5H_5N_5O$		TFXA	6,15,91
Guanosine	$C_{10}H_{12}N_5O_5$		TFXA	6,15

Name	Formula	H/D	Instrument	Refs.
Gypsum	$CaSO_4.2H_2O$		TFXA	92,93
Heneicosane	$C_{21}H_{44}$		TFXA	67
Heptadecane	$C_{17}H_{36}$		TFXA	67
Heptane	C_7H_{16}		TFXA	67
Hexachloroethane	C_2Cl_6		IN4	297
Hexadecane	$C_{16}H_{34}$	Y	TFXA	67,69
Hexafluorobenzene	C_6F_6		TOSCA	94
Hexakis-μ-acetatotri(aqua)-μ₃-oxo-trichromium(III) chloride hydrate	$[Cr_3(H_2O)_3O-(O_2CCH_3)_6]Cl.H_2O$		TFXA	307,308
Hexakis-μ-acetatotri(aqua)-μ₃-oxo-triiron(III,III,II) dihydrate	$[Fe_3(H_2O)_3O-(O_2CCH_3)_6].2H_2O$		TFXA	307
Hexakis-μ-acetatotri(aqua) -μ₃-oxo-triiron(III) chloride hydrate	$[Fe_3(H_2O)_3O-(O_2CCH_3)_6]Cl.H_2O$		TFXA	307
Hexamethylenetetramine	$C_6H_{12}N_4$		IN1BeF	95
Hexane	C_6H_{14}		TFXA	67
Hexathiophene	$C_{24}H_{14}S_6$		TFXA	193
Hydridotetra(trifluorophosphine)cobalt	$[CoH(PF_3)_4]$		LBS	254
Hydrogarnet	$Sr_3Al_2(O_4H_4)_3$		HRMECS	96
Hydrogen	H_2	Y	TOSCA	97,98
& CoAlPO			TOSCA	99
& MoS₂			TFXA	107,108
& MoS₂/Al₂O₃			TFXA	107
& nanotubes			TOSCA	100
& Ru/C			TOSCA	101
& WS₂			TFXA	108
& $[Fe(H)_2(\eta_2\text{-}H_2)(PEtPh_2)_3]$			FDS	328
& $[Fe(H)(\eta_2\text{-}H_2)\text{-}(PPh_2CH_2CH_2PPh_2)_2][BF_4]$			FDS	329
& $[RhH_2(H_2)Tp^{3,5\text{-Me}}]$			FDS	214
& $[W(CO)_3(\eta^2\text{-}H_2)(PCy_3)_2]$			FDS	213
& $[W(CO)_3(\eta^2\text{-}H_2)(Pi\text{-}Pr_3)_2]$			FDS	330
& Zeolite A			BT4,IN4	265-267
& Zeolite X			IN4	267
Hydrogen adsorbed	H_{ads}			
& Au/TiO₂			FDS	334
& Ni			IN1BeF	262,263
& Pd			IN1BeF	264
& Pt			IN1BeF BT4,MARI TOSCA	278-282
& Pt/C			TOSCA	102,103
& PtRu/C			TOSCA	103,282

Name	Formula	H/D	Instrument	Refs.
& Pt/SiO$_2$			IN1BeF	278
& Ru/C			TOSCA	103
& RuS$_2$			TFXA	104
Hydrogen chloride	HCl		TOSCA	342
Hydrogen sulfide	H$_2$S		TOSCA	105,342
Hydroxyapatite	Ca$_5$(PO$_4$)$_3$OH		TFXA	46
9-Hydroxyphenalenone	C$_{13}$H$_8$O$_2$	Y	TOSCA	322
8-Hydroxyquinoline N-oxide	C$_9$H$_7$NO	Y	NERA-PR	336
Ice Ic	H$_2$O		TFXA	110
Ice Ih	H$_2$O	Y	TOSCA	110
Ice II	H$_2$O		TFXA	110
Ice V	H$_2$O		TFXA	110
Ice VII	H$_2$O		TFXA	110
Ice VIII	H$_2$O		TFXA	110
Ice IX	H$_2$O		TFXA	110
Ice XI	H$_2$O	Y	TFXA	110
Ice low-density amorphous	H$_2$O		TFXA	110
Imidazole	C$_3$H$_4$N$_2$		TFXA,TAS	111,226
Indium hydrogensulfate	In(HSO$_4$)(SO$_4$).H$_2$O		TFXA	309
Iron(III) hydrogensulfate	Fe(HSO$_4$)(SO$_4$).H$_2$O		TFXA	309
Isoleucine	C$_6$H$_{13}$NO$_2$	Y	NERA-PR	340
Kaolinite	Si$_2$Al$_2$O$_5$(OH)$_4$		FDS	112
Lanthanum carbide hydride	La$_2$C$_3$H$_{1.5}$		IN1BeF	113
Lead hydrogenphosphate	PbHPO$_4$		TFXA	302
Leucine	C$_6$H$_{13}$NO$_2$	Y	NERA-PR	339
Lithium acetate dideuterate	Li(C$_2$H$_3$O$_2$).2D$_2$O	Y	TFXA	114
Lithium dihydridopalladate(0)	Li$_2$[PdH$_2$]		TOSCA	37
Lithium hydride	LiH		TOSCA, TAS	313,314
Lithium methanesulfonate	Li[CH$_3$SO$_3$]		LBS	292
Magnesium hydride	MgH$_2$		TOSCA	222
Magnesium hexahydridoferrate(II)	Mg$_2$[FeH$_6$]	Y	TFXA	115
Magnesium hydrogenphosphate trihydrate	MgHPO$_4$.3H$_2$O		TFXA	50
Magnesium pentahydridocobaltate(I)	Mg$_2$[CoH$_5$]	Y	TFXA	116
Magnesium tetrahydridonickelate(0)	Mg$_2$[NiH$_4$]	Y	TFXA	117
Maleimide	C$_4$H$_3$NO$_2$		TOSCA	321
Maleic anhydride	C$_4$H$_2$O$_3$		TFXA	118
Manganese hydride	MnH$_{0.07}$	Y	TFXA, MARI, IN6	4,119
Methane	CH$_4$			

Name	Formula	H/D	Instrument	Refs.
& Ru/Al$_2$O$_3$			FDS	332,333
Methanol	CH$_3$OH		TOSCA	82
Methoxy-benzylidene-butyl-aniline	CH$_3$(CH$_2$)$_3$C$_6$H$_4$N=C HC$_6$H$_4$OCH$_3$		IN4	120
N-Methylacetamide	C$_3$H$_7$NO		TFXA, FDS	121-123
N-Methylformamide	HCONHCH$_3$		CHEX	124
N-Methylmaleimide	C$_5$H$_5$NO$_2$		TFXA	125
2-Methylnorbornane	C$_8$H$_{14}$		IN1BeF	133
Methyltrioxorhenium(VII)	[Re(CH$_3$)O$_3$]		TOSCA	126
Molybdenum disulfide	MoS$_2$			
& Thiophene	C$_4$H$_4$S		TFXA	54
& Hydrogen	H$_2$		TFXA	107,108
Monetite	CaHPO$_4$		TFXA	50
Napthalene	C$_{10}$H$_8$		TAS	109,228
Natrolite	Na$_2$Al$_2$Si$_3$O$_{12}$H$_4$		TFXA	34
Neopentane	C$_5$H$_{12}$		IN4, IN1BeF	128
Newberyite	MgHPO$_4$.3H$_2$O		TFXA	50
Nickel				
& Acetonitrile			TFXA, IN1BeF	11,286
& Benzene			IN1BeF	272,273
& Cyclohexene			IN1BeF	275
Nickel hydride	NiH		IN1BeF	4,211
Niobium hydride	NbH		FDS,HET	129,217
Nitroxide radical 2,2,6,6-tetramethyl -4-oxopiperidine	C$_9$H$_{14}$NO$_2$		TFXA	137
Nonacarbonyl(ethylidyne)tricobalt	[Co$_3$(CCH$_3$)(CO)$_9$]	Y	TFXA	83
Nonadecane	C$_{19}$H$_{40}$		TFXA	67
Nonane	C$_9$H$_{20}$		TFXA	67
Norbornane	C$_7$H$_{12}$		IN1BeF, NERA-PR	130,131
Norbornene	C$_7$H$_{10}$		TOSCA	132
Nylon-6	(HN-(CH$_2$)$_6$-CO)$_n$		FANS	135,136
Octadecane	C$_{18}$H$_{38}$		TFXA	67,138
Octahydrosilasesquioxane	H$_8$Si$_8$O$_{12}$		TFXA	139
Octane	C$_8$H$_{18}$		TFXA	67,325
Oxamide	C$_2$H$_4$N$_2$O$_2$		TOSCA	140
Pagodane	C$_{20}$H$_{20}$		TOSCA	318
Palladium copper hydride	PdCuH		TFXA	141
Palladium hydride	PdH	Y	TFXA,	142-144

Name	Formula	H/D	Instrument	Refs.
			TAS	
Pentacarbonyhydridomanganese	[[Mn(CO)$_5$H]		LBS	254
Pentacarbonylmethylmanganese	[MnCH$_3$(CO)$_5$]		FDS	145
Pentachlorophenol	C$_6$Cl$_5$OH		NERA-PR	146
Pentacosane	C$_{25}$H$_{52}$		TFXA	67
Pentadecane	C$_{15}$H$_{32}$		TFXA	67
Pentane	C$_5$H$_{12}$		TFXA	67
Perylene	C$_{20}$H$_{12}$		TFXA,TAS	7,227
Phenanthrene	C$_{12}$H$_{10}$		TFXA	7
Platinum				
& Benzene			IN1BeF	274
Polyacetylene	(CH)$_n$		TFXA	147
Polyaniline	(C$_6$H$_4$NH)$_n$		TFXA, TAS	148-150
Polydimethylsiloxane	(SiO$_2$(CH$_3$)$_2$)$_n$		TOSCA	151
Polyethylene	(C$_2$H$_4$)$_n$	Y	TFXA,TAS IN1BeF	152-155
Polyglycine	(COCH$_2$NH)$_n$		TFXA	156
Polymethylmethacrylate	(C$_5$H$_8$O$_2$)$_n$		TOSCA	157
Polypropylene	(CH$_3$CHCH$_2$)$_n$		LBS	158
Polypyrrole	(C$_4$H$_2$N)$_n$		TFXA	159
Potassium aquapentachloroferrate(III)	K$_2$[FeCl$_5$H$_2$O]		TFXA	160
Potassium borohydride	K[BH$_4$]		TOSCA	316
Potassium dichlorohydrido-disulphitoplatinate(II)	K$_3$[PtCl$_2$H(SO$_3$)$_2$]		TFXA	162
Potassium dihydrogenphosphate	KH$_2$PO$_4$		TFXA	163
Potassium formate	HCO$_2$K		TFXA	60
Potassium fulleride	KC$_{60}$		IN6	249
Potassium fulleride	K$_3$C$_{60}$		TFXA	244,245
Potassium hexachloroiridate(IV)	K$_2$[IrCl$_6$]		TFXA	164
Potassium hexachloroplatinate(IV)	K$_2$[PtCl$_6$]		TFXA	164
Potassium dihydrogenarsenate	KH$_2$AsO$_4$		TFXA	258
Potassium hydrogendichloromaleate	K Cl$_2$C$_4$HO$_4$		IN1BeF	295
Potassium hydrogenmaleate	K C$_4$H$_3$O$_4$	Y	TFXA	165,294
Potassium hydrogenmalonate	K C$_4$H$_5$O$_4$	Y	IN1BeF, CAS	296
Potassium hydrogen monochloromaleate	K ClC$_4$H$_2$O$_4$		IN1BeF	294
Potassium hydrogenphthalate	K C$_8$H$_5$O$_4$		TOSCA	166
Potassium methanesulfonate	K CH$_3$SO$_3$		LBS	292
Potassium oxalate monohydrate	K C$_2$O$_4$H.H$_2$O		TFXA	255
Potassium pentadecacarbonyl-	K[Co$_6$(CO)$_{15}$H]		FDS	253

Name	Formula	H/D	Instrument	Refs.
hydridohexacobaltate(1−)				
Potassium tetrachloropalladate(II)	$K_2[PdCl_4]$		TFXA	167
Potassium tetrahydridopalladate(II)	$K_2[PdH_4]$		TOSCA	37
Progesterone	$C_{21}H_{30}O_2$		NERA-PR	335
l-Proline	$C_5H_6NO_2$		NERA-PR	168
1-Propanol	C_3H_7OH		TOSCA	81
Propene	C_3H_6		TFXA	80
p-tert-butylcalix[8]arene(1:1)C_{60} complex			TFXA	169
Purine	$C_5H_4N_4$		TFXA	206
Pyrazine	$C_4H_4N_2$		TFXA	170
Pyridazine	$C_4H_4N_2$		TFXA	171
Pyridine	C_5H_5N		TFXA	78,341
Pyridine N-oxide semiperchlorate	$C_5H_5NO.0.5HClO_4$	Y	TFXA	172
Pyrimidine	$C_4H_4N_2$		TFXA	106
Pyrrole	C_4H_5N		TFXA	173
& Faujasite			IN1BeF	331
Pyrope	$[Mg_{0.92}Fe_{0.05}Ca_{0.03}]_3$ $Al_2Si_3O_{12}$		TFXA	174
Rubidium aquapentachloroferrate(III)	$Rb_2[FeCl_5H_2O]$		TFXA	160
Rubidium dihydrogenarsenate	RbH_2AsO_4		TFXA	258
Rubidium fulleride	RbC_{60}		IN6	249
Rubidium fulleride	Rb_3C_{60}		IN6,TFXA	245,246
Rubidium fulleride	Rb_6C_{60}		TFXA	247
Rubidium hexahydridoplatinate(IV)	Rb_2PtH_6	Y	TOSCA, MARI	161
Rubidium hydride	RbH		TOSCA	303
Rubidium methanesulfonate	$Rb(CH_3SO_3)$		LBS	292
Rubidium pentahydridozincate(II)	$Rb_3[ZnH_5]$	Y	TFXA	117
Ruthenium sulfide				
& H_2			TFXA	104,277
& NH_3			TFXA	276
Scolecite	$CaAl_2Si_3O_{10}.3H_2O$		TFXA	34
1-Serine	$C_3H_5O_3(NH_2)$	Y	TOSCA	175
Silica	SiO_2		MARI,TAS	260,261
Sodium barium trihydridopalladate(0)	$NaBa[PdH_3]$		TFXA	37
Sodium bifluoride	$NaHF_2$		TFXA	176
Sodium borohydride	$Na[BH_4]$		TOSCA	316
Sodium dihydridopalladate(0)	$Na_2[PdH_2]$		TFXA	37
Sodium formate	$NaHCO_2$		TFXA	60
Sodium fulleride	Na_4C_{60}		IN6	248

Name	Formula	H/D	Instrument	Refs.
Sodium hydride	NaH		TOSCA	303
Sodium hydrogencarbonate	$NaHCO_3$		TFXA	49
Sodium methanesulfonate	$Na(CH_3SO_3)$		LBS	292
Sodium rubidium fulleride	Na_2RbC_{60}		TFXA,IN6	251
Sodium zeolite X	NaX		IN4, IN1BeF	178,179
Sodium zeolite Y	NaY		IN4, IN1BeF	178,179
Sodium zeolite ZSM5	NaZSM5		IN4, IN1BeF	178,179
Staphyloccal nuclease			TFXA	180
Strontium diazenide	SrN		TFXA, MARI	181
Strontium diazenide nitride	SrN_2		TFXA, MARI	181
Sulfur	S		TAS	182
Tantalum hydride	TaH		FDS	129
Terbium nickel aluminate	$TbNiAlH_{1.4}$		FDS	203
Testosterone	$C_{19}H_{28}O_2$		NERA-PR	335
Tetrabromoethene	C_2Br_4		TFXA	184
Tetracarbonyldihydridoiron	$[Fe(CO)_4H_2]$		LBS	254
Tetracarbonylhydridocobalt	$[Co(CO)_4H]$		LBS	254
Tetrachloropyrimidine	$C_4Cl_4N_2$		TFXA	185
Tetracosane	$C_{24}H_{50}$		TFXA	67
Tetracyanoethene	C_6N_4		TAS	186
Tetracyanoquinodimethane	$C_{12}H_4N_4$		NERA-PR	337
Tetradecane	$C_{14}H_{30}$		TFXA	67
Tetrahydrofuran	C_4H_8O	Y	TFXA	187,188
Tetrakisimidazolezinc(II) [11]B-tetrafluoroborate	$C_{12}H_{16}N_8B_2F_8Zn$		TFXA	208
Tetramethylammonium trichlorogermanate(II)	$N(CH_3)_4[GeCl_3]$		TFXA	189
Tetramethyltin	$[Sn(CH_3)_4]$		TFXA	190
Tetrathiafulvalene	$C_6H_4S_4$		TOSCA	323
Tetrathiophene	$C_{16}H_{10}S_4$		TFXA	193
Thallium dihydrogenarsenate	TlH_2AsO_4		TFXA	257
Thallium dihydrogenphosphate	TlH_2PO_4	Y	TFXA	192
Thiophene	C_4H_4S		TFXA	127
& MoS_2			TFXA	54
Tricarbonylthiophenechromium	$[Cr(CO)_3C_4H_4S]$		TFXA	54
Tricarbonylthiophenemanganese(I) trifluoromethylsulphonate	$[Mn(CO)_3(C_4H_4S)]$ (CF_3SO_3)		TFXA	54
l-Threonine	$C_4H_9NO_3$	Y	TFXA	194

Name	Formula	H/D	Instrument	Refs.
Thymine	$C_5H_6N_2O_2$		TFXA	6,216
Tin hydrogenphosphate	$Sn(HPO_4)_2.H_2O$		TFXA	309
Titanium hydride	$TiH_{0.74}$		TFXA	220
Triaminotrinitrobenzene	$C_6H_6N_6O_6$		FDS	195
Triazine	$C_3H_3N_3$		TFXA	170
Tribromomesitylene	$C_9H_9Br_3$		TFXA	196
Tricaesium hydrogenselenate	$Cs_3H(SO_4)_2$		TFXA	197
Tricalcium silicate	$3CaO.SiO_2$		FANS	301
2,4,6-Trichloro-1,3,5-triazine	$C_3Cl_3N_3$		TFXA	170
Trichloroethene	$ClCH=CCl_2$		FDS	53
Trichloromesitylene	$C_6Cl_3(CH_3)_3$		TFXA	198
Tricosane	$C_{23}H_{48}$		TFXA	67
Tridecane	$C_{13}H_{28}$		TFXA	67
1,3,5-Trifluorobenzene	$C_6H_3F_3$		TOSCA	94
1,3,5-Trichloro-2,4,6-trifluorobenzene	$C_3Cl_3F_3$		TAS	199
Trimethylaluminium	$[Al(CH_3)_3]$	Y	TOSCA	183
Trimethylgallium	$[Ga(CH_3)_3]$	Y	TOSCA	191
Trimethyloxosulfonium iodide	$(CH_3)_3SOI$		IN4	200
Trirubidium hydrogenselenate	$Rb_3H(SO_4)_2$		TFXA	197
Trisethylenediaminecobalt(III) chloride (en = $H_2NCH_2CH_2NH_2$)	$[Co(en)_3]Cl_3$		TFXA	57
Trisethylenediaminecobalt(III) chloride dihydrate	$[Co(en)_3]Cl_3.2H_2O$		TFXA	57
Trisodium 12-tungstophosphate hydrate	$Na_3PW_{12}O_{40}.12H_2O$		TFXA	74
Trispropylenediaminecobalt(III) chloride (pn = $H_2NCH_2CH_2CH_2NH_2$)	$[Co(pn)_3]Cl_3$		TFXA	58
Trispropylenediaminecobalt(III) chloride dihydrate	$[Co(pn)_3]Cl_3.2H_2O$		TFXA	58
12-Tungstohydrogenphosphate hydrate disodium salt	$Na_2PW_{12}O_{40}.13H_2O$		TFXA	74
12-Tungstophosphoric acid	$H_3PW_{12}O_{40}$		IN1BeF	75
12- Tungstophosphoric acid hemihydrate	$H_3PW_{12}O_{40}.0.5H_2O$		TFXA	74
12-Tungstophosphoric acid hexahydrate	$H_3PW_{12}O_{40}6H_2O$		TFXA	74
12-Tungstophosphoric acid monohydrate	$H_3PW_{12}O_{40}.H_2O$		TFXA	74
12-Tungstophosphate hydrate trisodium salt	$Na_3PW_{12}O_{40}.12H_2O$		TFXA	74
Uracil	$C_4H_4N_2O_2$	Y	TFXA	6,202
Uranium nickel aluminium hydride	$UNiAlH_2$		FDS	203
Urea	H_2NCONH_2		TOSCA	204

Name	Formula	H/D	Instrument	Refs.
Urea pimelonitrile co-crystal	$C_8H_{14}N_4O$		TOSCA	48
Uridine	$C_9H_{12}N_2O_6$		TFXA	6,65
Uridine-5'-monophosphate disodium salt	$C_9H_{11}N_2Na_2O_9P$		TFXA	205
Valine	$C_5H_{11}NO_2$	Y	NERA-PR	338
Wairakite	$CaAl_2Si_4O_{14}H_4$		TFXA	34
Wardite	$NaAl_3(OH)_4(PO_4)_2.$ $2H_2O$		TOSCA	320
Zeise's salt	$K[Pt(C_2H_4)Cl_3]$		IN1BeF	207
Zeolite rho	Rho zeolite		BT4	299,300
Zeolite X			IN4, IN1BeF	178,179
& 1-Butanol			TOSCA	81
& Ethanol			TOSCA	81
& Methanol			TOSCA	81,82
& Propan-1-ol			TOSCA	81
& Water			TOSCA	319
Zeolite Y			IN4, TFXA	178,179
			IN1BeF	268-270
& Benzene			IN1BeF	271
& Water			TOSCA	319
Zeolite ZSM5			IN4, IN1BeF	178,179
Zirconium hydride	ZrH_x	Y	HET,TFXA KDSOG-M	217-219, 310,311
Zirconium titanium hydride	$ZrTi_2H_{3.6}$		TFXA	209

Table A4.2 List of systems studied by high resolution vibrational INS spectroscopy, ordered alphabetically by formula.

Formula	Name	H/D	Instrument	Refs.
AlC_3H_9	Trimethylaluminium	Y	TFXA	183
AlH_2NiU	Uranium nickel aluminate		FDS	203
AlH_3	Aluminium trihydride		TFXA	19
AlH_4F_4N	Ammonium tetrafluoroaluminate(III)		TFXA	30
$Al_2Ca_{0.09}Fe_{0.15}Mg_{2.76}O_{12}Si_3$	Pyrope		TFXA	174
$Al_2CaH_4O_{14}Si_4$	Wairakite		TFXA	34
$Al_2CaH_6O_{13}Si_3$	Scolecite		TFXA	34
$Al_2H_3O_7P$	Augelite		TFXA	35
$Al_2H_4Li_2O_{14}Si_4$	Bikitaite		TFXA	34
$Al_2H_4Na_2O_{12}Si_3$	Natrolite		TFXA	34
$Al_2H_4O_9Si_2$	Kaolinite		FDS	112
$Al_2H_{12}O_{12}Sr_3$	Hydrogarnet		HRMECS	96
Al_2O_3	Alumina		TAS	259
$Al_3H_8NaO_{14}P_2$	Wardite		TOSCA	320
$AsCsH_2O_4$	Caesium dihydrogenarsenate		TFXA	258
AsH_2KO_4	Potassium dihydrogenarsenate		TFXA	258
AsH_2KO_4Rb	Rubidium dihydrogenarsenate		TFXA	258
AsH_2O_4Tl	Thallium dihydrogenarsenate		TFXA	257
$BC_{11}F_4FeH_9O_2S$	Dicarbonyl(cyclopentadienyl)-thiopheneiron tetrafluoroborate		TFXA	54
$B_2C_{12}F_8H_{16}N_8Zn$	Tetrakisimidazolezinc(II) tetrafluoroborate		TFXA	208
$B_{12}Cs_2H_{12}$	Caesium dodecaborane		TOSCA	48
BH_4Na	Sodium borohydride		TOSCA	316
BH_4K	Potassium borohydride		TOSCA	316
BH_6N	Borane-ammonia complex		TOSCA	317
$Ba_2Cl_2H_2O_8$	Barium chlorate monohydrate		TFXA	304
$BaHN$	Barium imide		TFXA	36
BaH_3NaPd	Sodium barium trihydridopalladate(0)		TFXA	37
Ba_2H_4Pd	Barium tetrahydridopalladate(0)		TFXA	37
BrC_6H_8N	Anilinium bromide	Y	LBS	293
C	Graphite		TFXA,TAS	89,224, 225
C	Diamond		TAS	89,223
C_{60}	Buckminsterfullerene		MARI, TOSCA	238

Formula	Name	H/D	Instrument	Refs.
$(C_{60})_n$	Buckminsterfullerene polymer		MARI, IN6	240-
			TOSCA	243
C_{70}	C_{70} fullerene		TOSCA	239
C_x	Coal		TFXA,	7,229,
			MARI	230
C_x	Carbon—industrial forms		TOSCA,	232-
			FANS	237
a-C:H	Amorphous carbon hydrogen		TFXA	231
$CCsHO_3$	Caesium hydrogencarbonate		TFXA	49
$CCsH_3O_3S$	Caesium methanesulfonate		LBS	292
$C_{60}Cs$	Caesium fulleride		IN6	249,
				250
$CHCl_3$	Chloroform		FDS	53
$CHNaO_3$	Sodium hydrogencarbonate		TFXA	49
CHO_2K	Potassium formate		TFXA	60
CHO_2Na	Sodium formate		TFXA	60
CH_2O_2	Formic acid	Y		60,212
CH_3KO_3S	Potassium methanesulfonate		LBS	292
CH_3LiO_3S	Lithium methanesulfonate		LBS	292
CH_3NO	Formamide		CHEX	87
CH_3NaO_3S	Sodium methanesulfonate		LBS	292
CH_3O_3RbS	Rubidium methanesulfonate		LBS	292
CH_3O_3Re	Methyltrioxorhenium(VII)		TOSCA	126
CH_4	Methane			
	& Ru/Al$_2$O$_3$		FDS	332,
				333
CH_4N_2O	Urea		TOSCA	204
CH_4O	Methanol		TOSCA	81,82
C_2Br_4	Tetrabromoethene		TFXA	184
$C_2Cl_2H_7NPt$	Amminedichloro(ethene)platinum(II)	Y	LBS	298
$C_2Cl_3H_4KPt$	Zeise's salt		IN1BeF	207
C_2Cl_6	Hexachloroethane		IN4	297
$C_2CuH_2O_4$	Copper formate		TFXA	60
$C_2CuH_{10}O_8$	Copper formate tetrahydrate		TFXA	60,61
			TOSCA	
C_2HCl_3	Trichloroethene		FDS	53
$(C_2H_2)_n$	Polyacetylene		TFXA	147
$C_2H_3KO_5$	Potassium oxalate monohydrate		TFXA	255
C_2H_3N	Acetonitrile		TFXA	10,11
	& Ni		TFXA,	11,286
			IN1BeF	

Formula	Name	H/D	Instrument	Refs.
$(C_2H_3NO)_n$	Polyglycine		TFXA	156
C_2H_4	Ethene (ethylene)		TFXA	79,80
$(C_2H_4)_n$	Polyethylene	Y	TFXA,TAS IN1BeF	152-155
$C_2H_4N_2O_2$	Oxamide		TOSCA	140
C_2H_4O	Ethylene oxide		IN1BeF	326
C_2H_5NO	N-methylformamide		CHEX	124
$C_2H_5NO_2$	Glycine	Y	TFXA	88
C_2H_6O	Ethanol		TOSCA	81
$(C_2H_6SiO_2)_n$	Polydimethylsiloxane		TOSCA	151
$C_2H_7LiO_4$	Lithium acetate dideuterate	Y	TFXA	114
$C_3Cl_5N_3$	2,4,6-Trichloro-1,3,5-triazine		TFXA	170
$C_3F_3H_{12}N_6O_3S$	Guanidinium triflate		FANS	90
C_3GaH_9	Trimethylgallium	Y	TOSCA	191
$C_3H_{1.5}La_2$	Lanthanium carbide hydride		IN1BeF	113
$C_3H_4N_2$	Imidazole		TFXA	111, 226
$C_3H_5N_3$	Triazine		TFXA	170
C_3H_6	Propene		TFXA	79
$(C_3H_6)_n$	Polypropylene		LBS	158
C_3H_7NO	N-Methylacetamide		TFXA, FDS	121-123
$C_3H_7NO_2$	*l*-Alanine		FDS	16
$C_3H_7NO_3$	*l*-Serine	Y	TOSCA	175
C_3H_8O	1-Propanol		TOSCA	81
C_3H_9IOS	Trimethyloxosulfonium iodide		IN4	200
$C_3H_{10}N_2$	1,3-Diaminopropane		TFXA	70
$C_3H_{12}N_6O_6S_2$	Guanidinium dithionate		FANS	90
$C_3H_{15}N_6O_3S$	Guanidinium methanesulfonate		FANS	90
$C_4ClH_2KO_4$	Potassium hydrogenmonochloromaleate		IN1BeF	294
$C_4Cl_2HKO_4$	Potassium hydrogendichloromaleate		IN1BeF	295
$C_4Cl_2H_7NPt$	Acetonitriledichloro(ethene)platinum(II)	Y	LBS	298
$C_4Cl_2H_{10}OPtS$	Dichloro(dimethylsulfoxide)ethene-platinum(II)	Y	LBS	298
$C_4Cl_3GeH_{12}N$	Tetramethylammonium trichlorogermanate		TFXA	189
$C_4Cl_4N_2$	Tetrachloropyrimidine		TFXA	185
C_4CoHO_4	Tetracarbonylhydridocobalt		LBS	254
$C_4FeH_2O_4$	Tetracarbonyldihydridoiron		LBS	254
$(C_4H_2N)_n$	Polypyrroles		TFXA	159

Formula	Name	H/D	Instrument	Refs.
$C_4H_2O_3$	Maleic anhydride		TFXA	118
$C_4H_3NO_2$	Maleimide		TOSCA	321
$C_4H_3O_4K$	Potassium hydrogenmaleate		TFXA	165, 294
$C_4H_5O_4K$	Potassium hydrogenmalonate	Y	IN1BeF, CAS	296
$C_4H_4N_2$	Pyrazine		TFXA	170
$C_4H_4N_2$	Pyrimidine		TFXA	106
$C_4H_4N_2$	Pyridazine		TFXA	171
$C_4H_4N_2O_2$	Uracil	Y	TFXA	6,15,202
C_4H_4S	Thiophene		TFXA	127
& MoS_2			TFXA	54
C_4H_5N	Pyrrole		TFXA	173
& Faujasite			IN1BeF	331
$C_4H_5N_3O$	Cytosine		TFXA	6,15,66
C_4H_6	2-Butyne		TOSCA	47
C_4H_8O	Tetrahydrofuran	Y	TFXA	187, 188
$C_4H_9NO_3$	*l*-Threonine		TFXA	194
$C_4H_{10}O$	1-Butanol		TOSCA	81
$C_4H_{12}Sn$	Tetramethyltin		TFXA	190
$C_5Cl_2H_{11}NOPt$	Dichloro(dimethylformamide)ethene-platinum(II)	Y	LBS	298
C_5HMnO_5	Pentacarbonylhydridomanganese		LBS	254
C_5H_5N	Pyridine		TFXA	78,341
$C_5H_4N_4$	Purine		TFXA	206
$C_5H_5NO_2$	N-Methylmaleimide		TFXA	125
$C_5H_5N_3O_2$	2-Aminopyrazine-3-carboxylic acid		NERA-PR	21
$C_5H_5N_5$	Adenine		TFXA	6,14
$C_5H_5N_5O$	Guanine	Y	TFXA	6,15,91
$C_5H_6NO_2$	*l*-Proline		NERA-PR	168
$C_5H_6N_2O_2$	Thymine		TFXA	6,216
$(C_5H_8O_2)_n$	Polymethylmethacrylate		TOSCA	157
$C_5H_{11}NO_2$	Valine	Y	NERA-PR	338
C_5H_{12}	Pentane		TFXA	67
C_5H_{12}	Neopentane		IN4, IN1BeF	128
$C_5H_{13}NO$	2-Amino-3-methyl-1-butanol		IN1BeF	20
C_6ClH_8N	Anilinium chloride	Y	LBS	293

Formula	Name	H/D	Instrument	Refs.
$C_6Cl_6H_{10}Pd$	Bis(allyl)dichlorodipalladium		TFXA	18
C_6Cl_5HO	Pentachlorophenol		NERA-PR	146
$C_6F_3H_3$	1,3,5-Trifluorobenzene		TFXA	94
C_6F_6	Hexafluorobenzene		TOSCA	94
$C_6H_3MnO_5$	Pentacarbonylmethylmanganese		FDS	145
$C_6H_4O_2$	1,4-Benzoquinone		NERA-PR	42,43
$C_6H_4O_4$	2,5-Dihydroxy-1,4-benzoquinone		NERA-PR	73
$C_6H_4S_4$	Tetrathiafulvalene		TOSCA	323
$(C_6H_5N)_n$	Polyaniline		TFXA, TAS	148-150
C_6H_6	Benzene	Y	TFXA, IN1BeF, TAS	38,39
& Ni			IN1BeF	272
& Pt			IN1BeF	274
& Zeolite Y			IN1BeF	271
$C_6H_6N_6O_6$	Triaminotrinitrobenzene		FDS	195
C_6H_7N	Aniline		TFXA	31
C_6H_8IN	Anilinium iodide	Y	LBS	293
$C_6H_8O_2$	1,3-Cyclohexanedione		TOSCA, FANS	315
C_6H_{10}	Cyclohexene		IN1BeF	275
& Ni			IN1BeF	275
$C_6H_{12}N_2O_2$	Alanine dipeptide		TFXA	17
$C_6H_{12}N_4$	Hexamethylenetetramine		IN1BeF	95
$C_6H_{13}NO_2$	Leucine	Y	NERA-PR	339
$C_6H_{13}NO_2$	Isoleucine	Y	NERA-PR	340
C_6H_{14}	Hexane		TFXA	67
$C_6H_{14}O_3$	Diglyme		TFXA	71
$C_6Cl_3CoH_{24}N_6$	Tris(ethylenediamine)cobalt(III) trichloride		TFXA	57
$C_6Cl_3CoH_{28}N_6O_2$	Tris(ethylenediamine)cobalt(III) trichloride dihydrate		TFXA	57
C_6N_4	Tetracyanoethene		TAS	186
$C_7H_4CrO_3S$	Tricarbonylthiophenechromium		TFXA	54
$C_7H_6O_2$	Benzoic acid		TFXA	41
$C_7H_9NO_2$	Ammonium benzoate		TFXA	24
C_7H_{10}	Norbornene		TFXA	132
C_7H_{12}	Norbornane		IN1BeF, NERA-PR	130, 131

Formula	Name	H/D	Instrument	Refs.
$(C_7H_{13}NO)_n$	Nylon-6		FANS	135, 136
C_7H_{16}	Heptane		TFXA	67
C_8ClH_7	3-Chlorostyrene		TOSCA	327
$C_8Cl_2H_{11}NPt$	Ethylenedichlorodimethylaniline-platinum(II)	Y	LBS	298
$C_8Co_2H_2O_6$	Acetylenehexacarbonyldicobalt		IN1BeF	13
C_8FH_7	3-Fluorostyrene		TOSCA	85
C_8FH_7	4-Fluorostyrene		TOSCA	86
$C_8F_3H_4MnO_6S_2$	Tricarbonylthiophenemanganese(0) trifluoromethylsulphonate		TFXA	54
$C_8H_5KO_4$	Potassium hydrogenphthalate	Y	TOSCA	166
$C_8H_6S_2$	Dithiophene		TFXA	193, 324
C_8H_8	Cubane		FANS, TAS	63,64
C_8H_9NO	Acetanilide	Y	TFXA	8,9
$C_8H_{14}N_4O$	Urea pimelonitrile co-crystal		TOSCA	48
C_8H_{14}	2-Methylnorbornane		IN1BeF	133
C_8H_{18}	Octane		TFXA	67,325
$C_9Br_3H_9$	Tribromomesitylene		TFXA	196
$C_9Cl_3H_9$	Trichloromesitylene		TFXA	198
$C_9CrH_6O_3$	Benzenetricarbonylchromium		IN1BeF, IN4	40
C_9H_7NO	8-Hydroxyquinoline N-oxide	Y	NERA-PR	336
$C_9H_{11}N_2Na_2O_9P$	Uridine-5'-monophosphate disodium salt		TFXA	205
$C_9H_{11}N_3O_5$	Cytidine		TFXA	6,15,65
$C_9H_{12}N_2O_6$	Uridine		TFXA	6,15,65
$C_9H_{14}NO_2$	Nitroxide radical 2,2,6,6-tetramethyl-4-oxopiperidine		TFXA	137
C_9H_{16}	2,3-Dimethylnorbornane		IN1BeF	134
C_9H_{20}	Nonane		TFXA	67
$C_9H_{30}Cl_3CoN_6$	Trispropylenediaminecobalt(III) chloride		TFXA	578
$C_9H_{34}Cl_3CoN_6O_2$	Trispropylenediaminecobalt(III) chloride dihydrate		TFXA	58
$C_{10}FeH_{10}$	Ferrocene		TFXA	84
$C_{10}H_2O_{10}Os_3$	Decacarbonyldihydridotriosmium		TFXA	72
$C_{10}H_{10}N_2$	1,8-Diaminonaphthalene		NERA-PR	68
$C_{10}H_8$	Napthalene		TAS	109, 228
$C_{10}H_{12}N_5O_5$	Guanosine		TFXA	6,15

Formula	Name	H/D	Instrument	Refs.
$C_{10}H_{14}$	Durene	Y	TFXA	77
$C_{10}H_{15}N_5O_4$	Adenosine		TFXA	6,15
$C_{10}H_{16}O$	Camphor	Y	NERA-PR	52
$C_{10}H_{18}O$	Borneole	Y	NERA-PR	52
$C_{10}H_{22}$	Decane		TFXA	67
$C_{11}Co_3H_3O_9$	Nonacarbonyl(ethylidyne)tricobalt	Y	TFXA	83
$C_{11}CsFe_3HO_{12}$	Caesium undecacarbonyl-hydridoferrate(1−)		LBS	254
$C_{12}ClCr_3H_{26}O_{15}$	Hexakis-μ-acetatotri(aqua)-μ₃-oxo-trichromium(III) chloride hydrate		TFXA	307, 308
$C_{12}ClFe_3H_{26}O_{15}$	Hexakis-μ-acetatotri(aqua)-μ₃-oxo-triiron(III) chloride hydrate		TFXA	307
$C_{12}FeH_{10}O_4$	Bis(dicarbonylcyclopentadienyliron)		TFXA	54
$C_{12}Fe_3H_{28}O_{16}$	Hexakis-μ-acetatotri(aqua)-μ₃-oxo-triiron(III,III,II) dihydrate		TFXA	307
$C_{12}H_2O_{10}Os_3$	Acetylenedodecacarbonyltriosmium		TFXA	201
$C_{12}H_3Mn_3O_{12}$	Dodecacarbonyltrihydridotrimanganese		LBS	254
$C_{12}H_4N_4$	Tetracyanoquinodimethane		NERA-PR	337
$C_{12}H_4O_{12}Os_4$	Tetrahydridotetraosmium dodecacarbonyl		TFXA	72
$C_{12}H_{10}$	Acenapthene		TFXA	7
$C_{12}H_{10}$	Phenanthrene			7
$C_{12}Cl_2CoH_{14}N_2$	Bis(4-methylpyridine)cobalt(II) chloride		TFXA	56
$C_{12}H_{14}Cl_2N_2Mn$	Bis(4-methylpyridine)manganese(II) chloride		TFXA	56
$C_{12}H_{14}Cl_2N_2Zn$	Bis(4-methylpyridine)zinc(II) chloride		TFXA	56
$C_{12}H_{26}$	Dodecane		TFXA	67
$C_{13}H_8O_2$	9-Hydroxyphenalenone	Y	TOSCA	322
$C_{13}H_{28}$	Tridecane		TFXA	67
$C_{14}H_{10}$	Anthracene		TFXA	7,32
$C_{14}H_{10}O$	Anthrone		TFXA	32
$C_{14}H_{18}N_2$	1,8-Bis(dimethylamino)napthalene		NERA-PR	44
$C_{14}H_{30}$	Tetradecane		TFXA	67
$C_{15}H_{10}O_4$	Emodin		TFXA	312
$C_{15}Co_6CsHO_{15}$	Caesium pentadecacarbonyl-hydridohexacobaltate(1−)		IN1BeF	252
$C_{15}Co_6HKO_{15}$	Potassium pentadecacarbonyl-hydridohexacobaltate(1−)		FDS	253
$C_{15}H_{32}$	Pentadecane		TFXA	67
$C_{16}H_{10}S_4$	Tetrathiophene		TFXA	193
$C_{16}H_{34}$	Hexadecane	Y	TFXA	67,69

Formula	Name	H/D	Instrument	Refs.
$C_{17}H_{36}$	Heptadecane		TFXA	67
$C_{18}H_{21}NO$	Methoxy-benzylidene-butyl-aniline		IN4	120
$C_{18}H_{38}$	Octadecane		TFXA	138
$C_{19}H_{28}O_2$	Testosterone		NERA-PR	335
$C_{19}H_{40}$	Nonadecane		TFXA	67
$C_{20}H_{12}$	Perylene	Y	TFXA,TAS	7,227
$C_{20}H_{20}$	Dodecahedrane		TOSCA	76
$C_{20}H_{20}$	Pagodane		TOSCA	318
$C_{20}H_{42}$	Eicosane		TFXA	67
$C_{21}H_{30}O_2$	Progesterone		NERA-PR	335
$C_{21}H_{44}$	Heneicosane		TFXA	67
$C_{22}H_{46}$	Docosane		TFXA	67
$C_{23}H_{48}$	Tricosane		TFXA	67
$C_{24}H_{12}$	Coronene		TOSCA	62
$C_{24}H_{14}S_6$	Hexathiophene	Y	TOSCA	193
$C_{24}H_{50}$	Tetracosane		TFXA	67
$C_{25}H_{52}$	Pentacosane		TFXA	67
$C_{28}H_{16}O_2$	Bianthrone		TFXA	33
$C_{60}Cs$	Caesium fulleride		IN6	249, 250
$C_{60}K$	Potassium fulleride		IN6	249
$C_{60}K_3$	Potassium fulleride		TFXA, IN6	244, 245
$C_{60}Na_4$	Sodium fulleride		IN6	248
$C_{60}Na_2Rb$	Sodium rubidium fulleride		TFXA,IN6	251
$C_{60}Rb_3$	Rubidium fulleride		IN6, TFXA	245, 246
$C_{60}Rb$	Rubidium fulleride		IN6	249
$C_{60}Rb_6$	Rubidium fulleride		TFXA	247
$C_{64}H_{56}N_8O_8$	Benzylic amide [2] catenane		TFXA	45
$CaHPO_4$	Calcium hydrogenphosphate		TFXA	50
$CaHPO_4$	Monetite		TFXA	50
CaH_2	Calcium hydride		IN1BeF	221
CaH_2O_2	Calcium hydroxide		TFXA, NERA-PR	51
CaH_4O_6S	Gypsum		TFXA	92,93
CaH_5PO_6	Calcium hydrogenphosphate dihydrate		TFXA	46
CaH_5PO_6	Brushite		TFXA	46
Ca_3O_5Si	Tricalcium silicate		FANS	301

Formula	Name	H/D	Instrument	Refs.
$Ca_4H_{17}O_{29}Si_8K$	Apophyllite		TFXA	34
$Ca_5HO_{13}P_3$	Hydroxyapatite		TFXA	46
CdF_3H_4N	Ammonium cadmium trifluoride		TFXA	25
ClH	Hydrogen chloride		TOSCA	342
$Cl_2HK_3O_6PtS_2$	Potassium dichlorohydrido-disulphitoplatinate(II)		TFXA	162
Cl_4K_2Pd	Potassium tetrachloropalladate(II)		TFXA	167
$Cl_5FeH_2K_2O$	Potassium aquapentachloroferrate(II)		TFXA	160
$Cl_5FeH_2ORb_2$	Rubidium aquopentachloroferrate(II)		TFXA	160
Cl_6IrK_2	Potassium hexachloroiridate(IV)		TFXA	164
Cl_6K_2Pt	Potassium hexachloroplatinate(IV)		TFXA	164
$CoHF_{12}P_4$	Tetrakis(trifluorophosphine)-hydridocobalt		LBS	254
CoH_5Mg_2	Magnesium pentahydridocobaltate(I)	Y	TFXA	116
$CrHO_2$	Chromous acid (chromium(III) oxide hydroxide)		FDS	55
CsH	Caesium hydride		TOSCA	303
$CsHO_4S$	Caesium hydrogensulphate		TFXA	256
CsH_2O_4P	Caesium dihydrogenphosphate		TFXA	305
$Cs_3HO_8S_2$	Tricaesium hydrogenselenate		TFXA	197
CuHPd	Palladium copper hydride		TFXA	141
F_2HNa	Sodium bifluoride		TFXA	176
F_3H_4MnN	Ammonium trifluoromanganate(II)		TFXA	25
F_3H_4NZn	Ammonium trifluorozincate(II)		TFXA	25
FeH_6Mg_2	Magnesium hexahydridoferrate(II)	Y	TFXA	115
H_{ads}	Hydrogen adsorbed			
	& Au/TiO_2		FDS	334
	& Pt		IN1BeF, BT4, TOSCA	278-281
	& Pt/C		TOSCA, MARI	102, 103, 281, 282
	& PtRu/C		TOSCA, MARI	103, 282
	& Pt/SiO_2		IN1BeF	278
	& Ru/C		TOSCA	103
	& RuS_2		TOSCA	104, 277
	& TiO_2		FDS	334

Formula	Name	H/D	Instrument	Refs.
H_2	Hydrogen	Y	TOSCA	97,98
	& CoAlPO		TOSCA	99
	& NaA		IN4,BT4	265-267
	& nanotubes		TOSCA	100
	& Ru/C		TOSCA	101
	& $[Fe(H)_2(\eta_2\text{-}H_2)(PEtPh_2)_3]$		FDS	328
	& $[Fe(H)(\eta_2\text{-}H_2)\text{-}$ $(PPh_2CH_2CH_2PPh_2)_2][BF_4]$		FDS	329
	& $[RhH_2(H_2)Tp^{3,5\text{-}Me}]$		FDS	214
	& $[W(CO)_3(\eta^2\text{-}H_2)(PCy_3)_2]$		FDS	213
	& $[W(CO)_3(\eta^2\text{-}H_2)(Pi\text{-}Pr_3)_2]$		FDS	330
$H_{0.74}Ti$	Titanium hydride		TFXA	220
HFe	Iron hydride		IN1BeF	4
HLi	Lithium hydride		TOSCA, TAS	313, 314
$HMn_{0.07}$	Manganese hydride		TFXA, IN6 MARI,	119
HNa	Sodium hydride		TOSCA	303
HNi	Nickel hydride		IN1BeF	211
$HO_8Rb_3Se_2$	Trirubidium hydrogenselenate		TFXA	197
$HPPbO_4$	Lead hydrogenphosphate		TFXA	302
HPd	Palladium hydride	Y	TFXA, TAS	142-144
H_xNb	Niobium hydride		FDS,HET	129, 217
HRb	Rubidium hydride		TOSCA	303
H_xTa	Tantalum hydride		FDS	129
H_xZr	Zirconium hydride	Y	HET,TFXA KDSOG-M	217-219, 310, 311
HF_2K	Potassium bifluoride		TFXA	176
HF_2Na	Sodium bifluoride		TFXA	176
H_2KO_4P	Potassium dihydrogenphosphate		TFXA	163
H_2Li_2Pd	Lithium dihydridopalladate(0)		TOSCA	37
H_2Mg	Magnesium hydride		TOSCA	222
H_2Na_2Pd	Sodium dihydridopalladate(0)		TFXA	37
H_2O	Ice	Y	TFXA	110
H_2O_4PTl	Thallium dihydrogenphosphate	Y	TFXA	192
H_2S	Hydrogen sulfide		TOSCA	105

Formula	Name	H/D	Instrument	Refs.
H_3FeO_9S	Iron hydrogensulphate		TFXA	309
H_3InO_9S	Indium hydrogensulphate		TFXA	309
$(H_3N)_xS_2Ti$	Ammoniated titanium disulfide		FDS	22,23
H_3N	Ammonia			
& RuS_2			TFXA	276
$H_3O_{40}PW_{12}$	12-Tungstophosphoric acid		IN1BeF	75
$H_{3\cdot6}Ti_2Zr$	Zirconium titanium hydride		TFXA	209
H_4K_2Pd	Potassium tetrahydridopalladate(II)		TOSCA	37
H_4Mg_2Ni	Magnesium tetrahydridonickelate(0)	Y	TFXA	117
H_4NO_6TaW	Ammonium tantalum tungstate		TFXA	306
$H_4O_9P_2Sn$	Tin hydrogen sulphate		TFXA	309
$H_4O_{40.5}PW_{12}$	12-Tungstophosphoric acid hemihydrate		TFXA	74
$H_5O_{41}PW_{12}$	12-Tungstophosphoric acid monohydrate		TFXA	74
H_5Rb_3Zn	Rubidium pentahydridozincate(II)	Y	TFXA	117
H_6PtRb_2	Rubidium hexahydridoplatinate(IV)	Y	TOSCA, MARI	161
H_7MgO_7P	Magnesium hydrogenphosphate trihydrate		TFXA	50
H_7MgO_7P	Newberyite		TFXA	50
$H_8O_{12}Si_8$	Octahydrosilasesquioxane	Y	TFXA	139
$H_{15}O_{46}PW_{12}$	12-Tungstophosphoric acid hexahydrate		TFXA	74
$H_{24}Na_3O_{52}PW_{12}$	Trisodium 12-tungstophosphate hydrate		TFXA	74
$H_{26}Na_2O_{53}PW_{12}$	12-tungstohydrogenphosphatehydrate disodium salt		TFXA	74
MoS_2	Molybdenum disulfide			
& C_4H_4S	& Thiophene		TFXA	54
& H_2	& Hydrogen		TFXA	107, 108
NSr	Strontium diazenide nitride		TFXA	181
N_2Sr	Strontium diazenide		TFXA	181
O_2Si	Silica		MARI, TAS	260, 261
Ni				
& CH_3CN			TFXA	11,286
& C_6H_6			IN1BeF, BT4,	274, 283
& C_6H_{10}			IN1BeF	275
& H_{ads}		Y	IN1BeF, BT4	262, 263, 284, 285

Formula	Name	H/D Instrument	Refs.
Pd			
	& H_{ads}	IN1BeF, BT4	287-291
Pt			
	& C_2H_2	BT4	283
	& C_2H_4	BT4	283
	& C_6H_6	IN1BeF, BT4,	274, 283
	& H_{ads}	IN1BeF, BT4, TOSCA	278-281
Pt/C			
	& H_{ads}	TOSCA, MARI	281, 282
PtRu/C			
	& H_{ads}	TOSCA, MARI	103, 281, 282
Pt/SiO$_2$			
	& H_{ads}	IN1BeF	278
Ru/C			
	& H_{ads}	TOSCA	103
	& H_2	TFXA	101
RuS$_2$	Ruthenium sulfide		
	& H_2	TFXA	104,277
	& NH_3	TFXA	276
S	Sulfur	TAS	182
S$_2$W	Tungsten disulfide		
	& H_2	TFXA	108
Zeolite rho	Zeolite rho	BT4	299, 300
Zeolite X	Zeolite X	IN4, IN1BeF	178, 179
	& 1-Butanol	TOSCA	81
	& Ethanol	TOSCA	81
	& Methanol	TOSCA	81,82
	& 1-Propanol	TOSCA	81
	& Water	TOSCA	319
Zeolite Y	Zeolite Y	IN4, TFXA IN1BeF	178, 179, 268-270

Formula	Name	H/D Instrument	Refs.
	& Water	TOSCA	319
Zeolite ZSM5	Zeolite ZSM5	IN4,	178,
		IN1BeF	179

A4.1 References

1   T. Springer (1978). In *Topics in Applied Physics 28. Hydrogen in Metals I. Basic Properties*, (Ed.) G. Alefield & J. Völkl, pp.75–100, Springer-Verlag, Berlin. Investigation of vibrations in metal hydrides by neutron spectroscopy.

2   D.K. Ross (1997). In *Topics in Applied Physics 73. Hydrogen in Metals III. Properties and Applications*, (Ed.) H. Wipf, pp.153–214, Springer-Verlag, Berlin. Neutron scattering studies of metal hydrogen systems.

3   Y. Fukai (1993). *The Metal-Hydrogen System*, Springer-Verlag, Berlin.

4   A.I. Kolesnikov, V.E. Antonov, V.K. Fedotov, G. Grosse, A.S. Ivanov & F.E. Wagner (2002). Physica B, 316, 158–161. Lattice dynamics of high-pressure hydrides of the group VI–VIII transition metals.

5   I. Anderson (1994). In *Neutron Scattering from Hydrogen in Materials*, (Ed.) A. Furrer, pp. 142–167, World Scientific, Singapore. The dynamics of hydrogen in metals studied by inelastic neutron scattering.

6   M. Plazanet, N. Fukushima & M.R. Johnson (2002). Chem. Phys., 280, 53–70. Modelling molecular vibrations in extended hydrogen-bonded networks – crystalline bases of RNA and DNA and the nucleosides.

7   F. Fillaux, R. Papoular, A. Lautie & J. Tomkinson (1995). Fuel, 74, 865–873. Inelastic neutron scattering study of the proton dynamics in coal.

8   R.L. Hayward, H.D. Middendorf, U. Wanderlingh & J.C. Smith (1995). J. Chem. Phys., 102, 5525–5541. Dynamics of crystalline acetanilide: analysis using neutron scattering and computer simulation.

9   M. Barthes, R. Almairac, J.–L. Sauvajol, R. Currat & J.–L. Ribet (1988). Europhys. Lett., 7, 55–60. Neutron-scattering investigation of deuterated crystalline acetanilide.

10 P.H. Gamlen, W.J. Stead, J. Tomkinson & J.W. White (1991). J. Chem. Soc. Faraday Trans., 87, 539–545. Dynamics of acetonitrile crystals and clusters.

11 F. Hochard, H. Jobic, G. Clugnet, A.J. Renouprez & J. Tomkinson (1993). Catalysis Letters, 21, 381–389. Inelastic neutron scattering study of acetonitrile adsorbed on Raney nickel.

12 M.R. Johnson, N.H. Jones, A. Geis, A.J. Horsewill, & H.P. Trommsdorff (2002). J. Chem. Phys., 116, 5694–5700. Structure and dynamics of the keto and enol forms of acetylacetone in the solid state.

13 H. Jobic, C.C. Santini & C. Coulombeau (1991). Inorg. Chem., 30, 3088–3090. Vibrational modes of acetylenedicobalt hexacarbonyl studied by neutron spectroscopy.

14 Z. Dhaouadi, M. Ghomi, J.C. Austin, R.B. Girling, R.E. Hester, P. Mojzes, L. Chinsky, R.Y. Turpin, C. Coulombeau, H. Jobic & J. Tomkinson (1993). J. Phys.

Chem., 97, 1074–1084. Vibrational motions of bases of nucleic acids as revealed by neutron inelastic scattering and resonance Raman spectroscopy. 1. Adenine and its deuterated species.

15   M.-P. Gaigeot, N. Leulliot, M. Ghomi, H. Jobic, C. Coulombeau & O. Bouloussa (2000). Chem. Phys., 261, 217–237. Analysis of the structural and vibrational properties of RNA building blocks by means of neutron inelastic scattering and density functional theory calculations.

16   M. Barthes, A.F. Vik, A. Spire, H.N. Bordallo & J. Eckert (2002). J. Phys. Chem. A, 106, 5230–5241. Breathers or structural instability in solid L-alanine: a new IR and inelastic neutron scattering vibrational spectroscopic study.

17   J. Baudry, R.L. Hayward, H.D. Middendorf & J.C. Smith (1997). In *Biological Macromolecular Dynamics*, (Ed.) S. Cusack, H. Büttner, M. Ferrand P. Langan & P. Timmins, pp. 49–54. Adenine Press, New York. Collective vibrations in crystalline alanine dipeptide at very low temperatures.

18   P.C.H. Mitchell, M. Bowker, N. Price, S. Poulston, D. James & S.F. Parker (2000). Topics in Catalysis, 11/12, 223–227. Iron antimony oxide selective oxidation catalysts—an inelastic neutron scattering study.

19   A.I. Kolesnikov, M.A. Adams, V.E. Antonov, N.A. Chirin, E.A. Goremychkin, G.G. Inikova, Yu.E. Markushkin, M. Prager & I.L. Sashin (1996). J. Phys.: Condens. Matter, 8, 2529–2538. Neutron spectroscopy of aluminium trihydride.

20   K. Molvinger, J. Court & H. Jobic (2001). J. Mol. Catal. A-Chem., 174, 245–248. Evidence for the anchoring of 2-amino-3-methyl-1-butanol at the surface of NiB_2 agglomerate by inelastic neutron spectroscopy.

21   A. Pawlukojć, I. Natkaniec, Z. Malarski & J. Leciejewicz (2000). J. Mol. Struct., 516, 7–14. The dynamical pattern of the 2-aminopyrazine-3-carboxylic acid molecule by inelastic and incoherent neutron scattering, Raman spectroscopy and ab initio calculations.

22   W.S. Glaunsinger, M.J. McKelvy, E.M. Larson, R.B. Vondreele, J. Eckert & N.L. Ross (1989). Solid State Ionics, 34, 281–286. Incoherent inelastic neutron-scattering investigation of ammoniated titanium disulfide.

23   E.W. Ong, J. Eckert, L.A. Dotson & W.S. Glaunsinger (1994). Chemistry of Materials, 6, 1946–1954. Nature of guest species within alkaline-earth ammonia intercalates of titanium disulfide.

24   J. Tomkinson & G.J. Kearley (1992). Physica B, 180–181, 665–667. The effect of recoil of chemically distinct molecular ions on inelastic neutron scattering spectra of molecular vibrations.

25   J. Rubín, J. Bartolomé & J. Tomkinson (1995). J. Phys.: Condens. Matter, 7, 8723–8740. The dynamics of NH_4^+ in the NH_4MF_3 perovskites: II inelastic neutron scattering study.

26   J. Tomkinson & G.J. Kearley (1989). J. Chem. Phys., 91, 5164–5169. Phonon wings in inelastic neutron scattering spectroscopy: the harmonic approximation.

27   P.S. Goyal, B.C. Boland, J. Penfold, A.D. Taylor & J. Tomkinson (1987). In *Dynamics of Molecular Crystals*, (Ed.) J. Lascombe, Elsevier, Amsterdam, 429–434. Higher librational modes of the NH_4^+ ion in ammonium halides: high resolution incoherent inelastic scattering spectra.

28 M.A. Adams & J. Tomkinson (1992). Physica B, 180–181, 694–696. Incoherent inelastic neutron scattering spectrum of ammonium bromide at a pressure of 15kbar.

29 A.V. Belushkin, T. Mhiri & S.F. Parker (1997). Physica B, 234–236, 92–94. Inelastic neutron-scattering study of $Cs_{1-x}(NH_4)_xHSO_4$ mixed crystals.

30 J. Rubín, E. Palacios, J. Bartolomé, J. Tomkinson & J.L. Fourquet (1992). Physica B, 180–181, 723–725. NH_4^+ Translational and librational spectrum in NH_4AlF_4.

31 M.H. Herzog-Cance, D.J. Jones, R.El. Mejjad, J. Roziere & J. Tomkinson (1992). J. Chem. Soc. Faraday Trans., 88, 2275–2281. Study of ion exchange and intercalation of organic bases in layered substrates by vibrational spectroscopy. Inelastic neutron scattering, infrared and Raman spectroscopies of aniline inserted alpha and gamma zirconium hydrogen phosphates.

32 B. Dorner, E.L. Bokhenkov, S.L. Chaplot, J. Kalus, I. Natkaniec, G.S. Pawley, U. Schmelzer & E.F. Sheka (1982). J. Phys. C: Solid State Phys., 15, 2353–2365. The 12 external and the 4 lowest internal phonon dispersion branches in d_{10}-anthracene at 12K.

33 J. Ulicny, M. Ghomi, H. Jobic, P. Miskovsky & A. Aamouche (1997). J. Mol. Struct., 410, 497–501. Neutron inelastic scattering and density functional calculations for analysing the force fields of hypericine model compounds: anthrone and bianthrone.

34 C.M.B. Line & G.J. Kearley (2000). J. Chem. Phys., 112, 9058–9067. An inelastic incoherent neutron scattering study of water in small-pored zeolites and other water-bearing minerals.

35 D.K. Breitinger, J. Mohr, D. Colognesi, S.F. Parker, H. Schukow & R.G. Schwab (2001). J. Mol. Struct., 563, 377–382. Vibrational spectra of augelites $Al_2(OH)_3(XO_4)$ (X = P, As, V).

36 R. Essmann, H. Jacobs & J. Tomkinson (1993). J. Alloys Cmpds., 191, 131–134. Neutron vibrational spectroscopy of imide ions (NH^{2-}) in bariumimide (BaNH).

37 M. Olofsson-Mårtensson, U. Häussermann, J. Tomkinson, & D. Noréus (2000). J. Am. Chem. Soc., 122, 6960–6970. Stabilization of electron-dense palladium-hydrido complexes in solid-state hydrides.

38 H. Jobic & A.N. Fitch (1996). Studies in Surface Science and Catalysis, 105, 559–566. Vibrational study of benzene adsorbed in NaY zeolite by neutron spectroscopy.

39 B.M. Powell, G. Dolling & H. Bonadeo (1978). J. Chem. Phys., 69, 2428–2433. Intermolecular dynamics of deuterated benzene.

40 H. Jobic, J. Tomkinson & A.J. Renouprez (1980). Mol. Phys., 39, 4, 989–999. Neutron inelastic scattering spectrum and valence force field for benzenetricarbonylchromium.

41 M. Plazanet, N. Fukushima, M.R. Johnson, A.J. Horsewill & H.P. Trommsdorff (2001). J. Chem. Phys., 115, 3241–3248. The vibrational spectrum of crystalline benzoic acid: Inelastic neutron scattering and density functional theory calculations.

42 J. Penfold & J. Tomkinson (1986). J. Chem. Phys., 85, 6246–6247. First observation of the optically inactive modes in solid 1,4-benzoquinone.

43 A. Pawlukojć, I. Natkaniec, I. Majerz, L. Sobczyk & E. Grech (2001). Chem. Phys. Lett., 346, 112–116. Inelastic neutron scattering (INS) studies on low frequency vibrations of 1,4-benzoquinone.

44   A. Pawlukojć, I. Natkaniec, E. Grech, J. Baran, Z. Malarski & L. Sobczyk (1998). Spectrochim. Acta, 54A, 439–448. Incoherent inelastic neutron scattering, Raman and IR absorption studies on 1,8-bis(dimethylamino)naphthalene and its protonated forms.

45   R. Caciuffo, A.D. Esposti, M.S. Deleuze, D.A. Leigh, A. Murphy, B. Paci, S.F. Parker & F. Zerbetto (1998). J. Chem. Phys., 109, 11094–11100. Inelastic neutron scattering of large molecular systems: The case of the original benzylic amide [2]catenane.

46   M.G. Taylor, K. Simkiss, S.F. Parker & P.C.H. Mitchell (1999). Phys. Chem. Chem. Phys., 1, 3141–3144. An inelastic neutron scattering study of synthetic calcium phosphates.

47   O. Kirstein, M. Prager, M.R. Johnson & S.F. Parker (2002). J. Chem. Phys., 117, 1313–1319. Lattice dynamics and methyl rotational excitations of 2-butyne.

48   B.S. Hudson (2001). J. Phys. Chem. A, 105, 3949–3960. Inelastic neutron scattering: a tool in molecular vibrational spectroscopy and a test of ab initio methods.

49   F. Fillaux & J. Tomkinson (1992). J. Mol Struct., 270, 339–349. Proton transfer dynamics in the hydrogen bond. Inelastic neutron scattering spectra of Na, Rb and Cs hydrogen carbonates at low temperature.

50   P.C.H. Mitchell, S.F. Parker, K. Simkiss, J. Simmins, & M.G. Taylor (1996). J. Inorg. Biochem., 62, 183–197. Hydrated sites in biogenic amorphous calcium phosphates: an infrared, Raman and inelastic neutron scattering study.

51   R. Baddour-Hadjean, F. Fillaux, N. Floquet, A.V. Belushkin, I. Natkaniec, L. Desgranges & D. Grebille (1995). Chem. Phys., 197, 81–90. Inelastic neutron-scattering study of proton dynamics in $Ca(OH)_2$ at 20 K.

52   K. Holderna-Natkaniec, I. Natkaniec, S. Habrylo & J. Mayer (1994). Physica B, 194, 369–370. Comparative neutron-scattering study of molecular ordering in d-camphor and dl-borneole.

53   A.M. Davidson, C.F. Mellot, J. Eckert & A.K. Cheetham (2000). J. Phys. Chem. B, 104, 432–438. An inelastic neutron scattering and NIR–FT Raman spectroscopy study of chloroform and trichloroethylene in faujasites.

54   P.C.H. Mitchell, D.A. Green, E. Payen, J. Tomkinson & S.F. Parker (1999). Phys. Chem. Chem. Phys., 1, 3357–3363. Interaction of thiophene with a molybdenum disulfide catalyst: an inelastic neutron scattering study.

55   J. Tomkinson, A.D. Taylor, J. Howard, J. Eckert & J.A. Goldstone (1985). J. Chem. Phys., 82, 1112–1114. The inelastic neutron scattering spectrum of chromous acid at high energy transfers.

56   F. Fillaux, C.J. Carlile & G.J. Kearley (1993). Mol. Phys., 80, 671–683. Rotational tunnelling and the rotational potential of CH_3-groups in the zinc, manganese and cobalt chloride salts of 4-methylpyridine.

57   A.J. Ramirez-Cuesta, P.C.H. Mitchell, A.P. Wilkinson, S.F. Parker & P.M. Rodger (1998). J. Chem. Soc. Chem. Comm., 2653–2654. Dynamics of water in a templated aluminophosphate: molecular dynamics simulation of inelastic neutron scattering spectra.

58 A.J. Ramirez-Cuesta, P.C.H. Mitchell, S.F. Parker & P.M. Rodger (1999). Phys. Chem. Chem. Phys., 1, 5711–5715. Dynamics of water and template molecules in the interlayer space of a layered aluminophosphate. Experimental inelastic neutron scattering spectra and molecular dynamics simulated spectra.

59 H.D. Middendorf, R.L. Hayward, S.F. Parker, J. Bradshaw & A. Miller (1995). Biophysical Journal, 69, 660–673. Vibrational neutron spectroscopy of collagen and model polypeptides.

60 P.C.H. Mitchell, R.P. Holroyd, S. Poulston, M. Bowker & S.F. Parker (1997). J. Chem. Soc. Faraday Trans., 93, 2569–2577. Inelastic neutron scattering of model compounds for surface formates. Potassium formate, copper formate and formic acid.

61 T. Omura, K. Itoh & S.F. Parker (2004). Incoherent inelastic neutron study of crystalline antiferroelectric $Cu(HCOO)_2.4H_2O$. J. Neutron. Res., accepted for publication.

62 P.C.H. Mitchell, A.J. Ramirez-Cuesta, S.F. Parker & J. Tomkinson (2003). J. Mol. Struct., 651–653, 781–785. Inelastic neutron scattering in spectroscopic studies of hydrogen on carbon-supported catalytsts-experimental spectra and computed spectra of model systems.

63 T. Yildirim, C. Kilic, S. Ciraci, P.M. Gehring, D.A. Neumann, P.E. Eaton & T. Emrick (1999). Chem. Phys. Lett., 309, 234–240. Vibrations of the cubane molecule: inelastic neutron scattering study and theory.

64 P.M. Gehring, D.A. Neumann, W.A. Kamitakahara, J.J. Rush, P.E. Eaton & D.P. Vanmeurs (1995). J. Phys. Chem., 99, 4429–4434. Neutron-scattering study of the lattice modes of solid cubane.

65 N. Leulliot, M. Ghomi, H. Jobic, O. Bouloussa, V. Baumruk & C. Coulombeau (1999). J. Phys. Chem. B, 103, 10934–10944. Ground state properties of the nucleic acid constituents studied by density functional calculations. 2. Comparison between calculated and experimental vibrational spectra of uridine and cytidine.

66 A. Aamouche, M. Ghomi, L. Grajcar, M.H. Baron, F. Romain, V. Baumruk, J. Stepanek, C. Coulombeau, H. Jobic & G. Berthier (1997). J. Phys. Chem. A, 101, 10063–10074. Neutron inelastic scattering, optical spectroscopies and scaled quantum mechanical force fields for analyzing the vibrational dynamics of pyrimidine nucleic acid bases. 3. Cytosine.

67 D.A. Braden, S.F. Parker, J. Tomkinson & B.S. Hudson (1999). J. Chem. Phys., 111, 429–437. Inelastic neutron scattering spectra of the longitudinal acoustic modes of the normal alkanes from pentane to pentacosane.

68 I. Majerz, A. Pawlukojć, L. Sobczyk, E. Grech & J. Nowicka-Scheibe (2002). Polish J. Chem., 76 409–417. Dimerization of 1,8-diaminonaphthalene. DFT theoretical, infra-red, Raman and inelastic neutron scattering studies.

69 S.F. Parker, J. Tomkinson, D.A. Braden & B.S. Hudson (2000). J. Chem. Soc. Chem. Comm., 165–166. Experimental test of the validity of the use of the n-alkanes as model compounds for polyethylene.

70 M.P.M. Marques, L.A.E.B. de Carvalho & J. Tomkinson (2002). J. Phys. Chem. A, 106, 2473–2482. Study of biogenic alpha and omega-polyamines by combined

inelastic neutron scattering and Raman spectroscopies and by ab initio molecular orbital calculations.

71   G.J. Kearley, P. Johansson, R.G. Delaplane & J. Lindgren (2002). Solid State Ionics, 147, 237–242. Structure, vibrational-dynamics and first-principles study of diglyme as a model system for poly(ethyleneoxide).

72   W.C.-K. Poon, P. Dallin, B.F.G. Johnson, E.A. Marseglia & J. Tomkinson (1990). Polyhedron, 9, 2759–2761. Inelastic neutron scattering study of hydride ligands in $[(\mu_2\text{-}H)_2Os_3(CO)_{10}]$: evidence for a direct H–H interaction.

73   A. Pawlukojć, I. Natkaniec, J. Nowicka-Scheibe, E. Grech & L. Sobczyk (2003). Spectrochim. Acta, 59A, 537–542. Inelastic neutron scattering (INS) studies on 2,5-dihydroxy-1,4-benzoquinone (DHBQ).

74   U.B. Mioc, Ph. Colomban, M. Davidovic & J. Tomkinson (1994). J. Mol. Struct., 326, 99–107. Inelastic neutron scattering study of protonic species during the thermal dehydration of 12-tungtophosphoric hexahydrate.

75   N. Essayem, Y.Y. Tong, H. Jobic & J.C. Vedrine (2000). Appl. Cat. A–General, 194, 109–122. Characterization of protonic sites in $H_3PW_{12}O_{40}$ and $Cs_{1.9}H_{1.1}PW_{12}O_{40}$: a solid-state H-1, H-2, P-31 MAS-NMR and inelastic neutron scattering study on samples prepared under standard reaction conditions.

76   B.S. Hudson, D.A. Braden, S.F. Parker & H. Prinzbach (2000). Angew. Chem. Int. Ed. Engl., 39, 514–516. The vibrational inelastic neutron scattering spectrum of dodecahedrane: experiment and DFT simulation.

77   M. Plazanet, M.R. Johnson, J.D. Gale, T. Yildirim, G.J. Kearley, M.T. Fernández-Diaz, D. Sánchez-Portal, E. Artacho, J.M. Soler, P. Ordejón, A. Garcia & H.P. Trommsdorff (2000). Chem. Phys., 261, 189–203. The structure and dynamics of crystalline durene by neutron scattering and numerical modelling using density functional methods.

78   A. Navarro, M. Fernández-Gómez, J.J. López-González, F. Partal, J. Tomkinson & G.J. Kearley (1999). In *Neutrons and Numerical Methods — N₂M*, (Ed.) M.R. Johnson, G.J. Kearley & H. G. Büttner, AIP Conf. Proc. Vol. 479, pp. 172–178. Density functional theory and *ab initio* methods applied to the analysis of inelastic neutron scattering spectra.

79   H. Jobic (1984). Chem. Phys. Lett., 106, 321–324. Neutron inelastic-scattering from ethylene—an unusual spectrum.

80   D. Lennon, J. McNamara, J.R. Phillips, R.M. Ibberson & S.F. Parker (2000). Phys. Chem. Chem. Phys., 2, 4447–4451. An inelastic neutron scattering spectroscopic investigation of the adsorption of ethene and propene on carbon.

81   R. Schenkel, A. Jentys, S.F. Parker & J.A. Lercher (2004). J. Phys. Chem. B, 108, 15013–15026. INS and IR and NMR spectroscopic study of C_1-C_4 alcohols adsorbed on alkali metal-exchanged zeolite X.

82   R. Schenkel, A. Jentys, S.F. Parker & J.A. Lercher (2004). J. Phys. Chem. B, 108, 7902–7910. Investigation of the adsorption of methanol on alkali metal cation exchanged zeolite X by inelastic neutron scattering.

83   S.F. Parker, N.A. Marsh, L.M. Camus, M.K. Whittlesey, U.A. Jayasooriya & G.J. Kearley (2002). J. Phys. Chem. A, 106, 5797–5802. Ethylidyne tricobalt nonacarbonyl: infrared, FT-Raman and inelastic neutron scattering spectra.

84   E. Kemner, I.M. de Schepper, G.J. Kearley & U.A. Jayasooriya (2000). J. Chem. Phys., 112, 10926–10929. The vibrational spectrum of solid ferrocene by inelastic neutron scattering.

85   J.M. Granadino-Roldán, M. Fernández-Gomez, A. Navarro, L.M. Camus & U.A. Jayasooriya (2002). Phys. Chem. Chem. Phys., 4, 4890–4901. Refined, scaled and canonical force fields for the cis- and trans-3-fluorostyrene conformers. An interplay between theoretical calculations, IR/Raman and INS data.

86   J.M. Granadino-Roldán , M. Fernández-Gómez , A. Navarro & U.A. Jayasooriya (2003). Phys. Chem. Chem. Phys., 5, 1760–1768. The molecular force field of 4-fluorostyrene: an insight into its vibrational analysis using inelastic neutron scattering, optical spectroscopies (IR/Raman) and theoretical calculations.

87   C.N. Tam, P. Bour, J. Eckert & F.R. Trouw (1997). J. Phys. Chem. A, 101, 5877–5884. Inelastic neutron scattering study of hydrogen-bonded solid formamide.

88   C.L. Thaper, B.A. Dasannacharya, P.S. Goyal, R. Chakravarthy & J. Tomkinson (1991). Physica B, 174, 251–256. Neutron scattering from glycine and deuterated glycine.

89   J.K. Walters, R.J. Newport, S.F. Parker & W.S. Howells (1995). J. Phys.: Condens. Matter, 7, 10059–10073. A spectroscopic study of the structure of amorphous hydrogenated carbon.

90   A.M. Pivovar, M.D. Ward, T. Yildirim & D.A. Neumann (2001). J. Chem. Phys., 115, 1909–1915. Vibrational mode analysis of isomorphous hydrogen-bonded guanidinium sulfonates with inelastic neutron scattering and density-functional theory.

91   Z. Dhaouadi, M. Ghomi, Ce. Coulombeau, C. Coulombeau, H. Jobic, P. Mojzes, L. Chinsky and P.Y. Turpin (1993). Eur. Biophys. J., 22, 225–236. The molecular force field of guanine and its deuterated species as determined from neutron inelastic scattering and resonance Raman measurements.

92   B. Winkler & B. Hennion (1994). Phys. Chem. Minerals, 21, 539–545. Low tempearature dynamics of molecular H_2O in Bassanite, Gypsum and Cordierite investigated by high resolution incoherent inelastic neutron scattering.

93   B. Winkler (1996). Phys. Chem. Minerals, 23, 310–318. The dynamics of H_2O in minerals.

94   D.A. Braden & B.S. Hudson (2000). J. Phys. Chem. A, 104, 982–989. C_6F_6 and sym-$C_6F_3H_3$: Ab initio and DFT studies of structure, vibrations and inelastic neutron scattering spectra.

95   H. Jobic & H.J. Lauter (1988). J. Chem. Phys. 88, 5450–5456. Calculation of the effect of the Debye-Waller factor on the intensities of molecular modes measured by neutron inelastic scattering. Application to hexamethylenetetramine.

96   G.A. Lager, J.C. Nipko & C.K. Loong (1997). Physica B, 241–243, 406–408. Inelastic neutron scattering study of the (O_4H_4) substitution in garnet.

97   M. Celli, D. Colognesi & M. Zoppi (2002). J. Low Temp. Phys., 126, 585–590. The microscopic dynamics of liquid and solid parahydrogen.

98   M. Celli, D. Colognesi & M. Zoppi (2002). Phys. Rev. E, 66, art. no. 021202. Direct experimental access to microscopic dynamics in liquid hydrogen.

99   A.J. Ramirez-Cuesta, P.C.H. Mitchell & S.F. Parker (2001). J. Mol. Cat. A: Chemical, 167, 217–224. An inelastic neutron scattering study of the interaction of dihydrogen with the cobalt site of a cobalt aluminophosphate catalyst. Two-dimensional quantum rotation of adsorbed dihydrogen.

100 P.A. Georgiev, D.K. Ross, A. DeMonte, U. Montaretto-Marullo, R.A.H. Edwards, A.J. Ramirez-Cuesta and D. Colognesi (2004). J. Phys.: Condens. Matter, 16, L73–L78. Hydrogen site occupancies in single-walled carbon nanotubes studied by inelastic neutron scattering.

101 P.C.H. Mitchell, S.F. Parker, J. Tomkinson & D. Thompsett (1998). J. Chem. Soc. Faraday Trans., 94, 1489–1493. Adsorbed states of dihydrogen on a carbon supported ruthenium catalyst: an inelastic neutron scattering study.

102 P. Albers, E. Auer, K. Ruth & S.F. Parker (2000). J. Catal., 196, 174–179. Inelastic neutron scattering investigation of the nature of surface sites occupied by hydrogen on highly dispersed platinum on commercial carbon black supports.

103 P.C.H. Mitchell, A.J. Ramirez-Cuesta, S.F. Parker, J. Tomkinson & D. Thompsett. (2003). J. Phys. Chem. B, 107, 6838–6845. Hydrogen spillover on carbon-supported metal catalysts studied by inelastic neutron scattering. Surface vibrational states and hydrogen riding modes.

104 H. Jobic, G. Clugnet, M. Lacroix, S.B. Yuan, C. Mirodatos & M. Breysse (1993). J. Am. Chem. Soc., 115, 3654–3657. Identification of new hydrogen species absorbed on ruthenium sulfide by neutron spectroscopy.

105 C. Andreani, E. Degiorgi, R. Senesi, F. Cilloco, D. Colognesi, J. Mayers, M. Nardone & E. Pace (2001). J. Chem. Phys., 114, 387–398. Single particle dynamics in fluid and solid hydrogen sulphide: An inelastic neutron scattering study.

106 A. Navarro, M. Fernández-Gómez, J.J. López-González, M. Paz Fernández-Liencres, E. Martínez-Torres, J. Tomkinson & G.J. Kearley (1999). J. Phys. Chem. A, 103, 5833–5840. Inelastic neutron scattering spectrum and quantum mechanical calculations of the internal vibrations of pyrimidine.

107 P. Sundberg, R.B. Moyes & J. Tomkinson (1991). Bulletin Des Societes Chimiques Belges, 100, 967–976. Inelastic neutron-scattering spectroscopy of hydrogen adsorbed on powdered- MoS_2, MoS_2-alumina and nickel-promoted MoS_2.

108 P.N. Jones, E. Knozinger, W. Langel, R.B. Moyes & J. Tomkinson (1988). Surf. Sci., 207, 159–176. Adsorption of molecular hydrogen at high pressures and temperatures on MoS_2 and WS_2 observed by inelastic neutron scattering.

109 I. Natkaniec, E.L. Bokhenkov, B. Dorner, J. Kalus, G.A. Mackenzie, G.S. Pawley, U. Schmelzer & E.F. Sheka (1980). J. Phys. C: Solid State Phys., 13, 4265–4283. Phonon dispersion in d_8-naphthalene crystal at 6K.

110 J.-C. Li (1996). J. Chem. Phys., 105, 6733–6755. Inelastic neutron scattering studies of hydrogen bonding in ices.

111 P.W. Loeffen, R.F. Pettifer, F. Fillaux & G.J. Kearley (1995). J. Chem. Phys., 103, 8444–8455. Vibrational force field of solid imidazole from inelastic neutron scattering.

112 C.T. Johnston, D.L. Bish, J. Eckert & L.A. Brown (2000). J. Phys. Chem. B, 104, 8080–8088. Infrared and inelastic neutron scattering study of the 1.03-and 0.95-nm kaolinite-hydrazine intercalation complexes.

113 G. Auffermann, A. Simon, Th. Gulden, G.J. Kearley & A. Ivanov (2001). Z. Anorg. Allg. Chem., 627, 307–311. Location and vibrations of hydrogen in $La_2C_3H_{1.5}$.

114 B. Nicolai, G.J. Kearley, M.R. Johnson, F. Fillaux & E. Suard (1998). J. Chem. Phys., 109, 9062–9074. Crystal structure and low-temperature methyl-group dynamics of cobalt and nickel acetates.

115 S.F. Parker, K.P.J. Williams, M. Bortz & K. Yvon (1997). Inorg. Chem., 36, 5218–5221. Inelastic neutron scattering, infrared and Raman spectroscopic studies of Mg_2FeH_6 and Mg_2FeD_6.

116 S.F. Parker, J.C. Sprunt, U.A. Jayasooriya, M. Bortz & K. Yvon (1998). J. Chem. Soc. Faraday Trans., 94, 2595–2599. Inelastic neutron scattering, infrared and Raman spectroscopic studies of Mg_2CoH_5 and Mg_2CoD_5.

117 S.F. Parker, K.P.J. Williams, T. Smith, M. Bortz, B. Bertheville & K. Yvon (2002). Phys. Chem. Chem. Phys., 4, 1732–1737. Vibrational spectroscopy of tetrahedral ternary metal hydrides: Mg_2NiH_4, Rb_3ZnH_5 and their deuterides.

118 S.F. Parker, C.C. Wilson, J. Tomkinson, D.A. Keen, K. Shankland, A.J. Ramirez-Cuesta, P.C.H. Mitchell, A.J. Florence & N. Shankland (2001). J. Phys. Chem. A, 105, 3064–3070. Structure and dynamics of maleic anhydride.

119 A.I. Kolesnikov, V.E. Antonov, S.M. Bennington, B. Dorner, V.K. Fedotov, G. Grosse, J.-C. Li, S.F. Parker & F.E. Wagner (1999). Physica B, 263–264, 421–423. The vibrational spectrum and giant tunnelling effect of hydrogen dissolved in α-Mn.

120 P. Derollez, M. Bée & H. Jobic (1992). Spectrochim. Acta 48A, (5) 743–748. Inelastic neutron scattering in the glassy and metastable phases of methoxy-benzylidene-butyl-aniline.

121 F. Fillaux, J.P. Fontaine, M.H. Baron, G.J. Kearley & J. Tomkinson (1993). Chem. Phys., 176, 249–278. Inelastic neutron scattering study of the proton dynamics in N-methylacetamide and force-field calculation.

122 M. Barthes, H.N. Bordallo, J. Eckert, O. Maurus, G. de Nunzio & J. Léon (1998). J. Phys. Chem. B, 6177–6183. Dynamics of crystalline N-methylacetamide: temperature dependence on infrared and inelastic neutron scattering spectra.

123 G.J. Kearley, M.R. Johnson, M. Plazanet & E. Suard (2001). J. Chem. Phys., 115, 2614–2620. Structure and vibrational dynamics of the strongly hydrogen-bonded model peptide: N-methylacetamide.

124 P. Bour, C.N. Tam, J. Sopková & F.R. Trouw (1998). J. Chem. Phys., 108, 351–358. Measurement and ab initio modeling of the inelastic neutron scattering of solid N-methylformamide.

125 S.F. Parker (1995). Spectrochim. Acta, 51A, 2067–2072. Vibrational spectroscopy of N-methylmaleimide.

126 S.F. Parker & H. Herman (2000). Spectrochim. Acta, 56A, 1123–1129. The vibrational spectra of methyltrioxorhenium(VII).

127 G.D. Atter, D.M. Chapman, R.E. Hester, D.A. Green, P.C.H. Mitchell & J. Tomkinson (1997). J. Chem. Soc. Faraday Trans., 93, 2977–2980. Refined ab initio inelastic neutron scattering spectrum of thiophene.

128 H. Jobic, S. Sportouch & A. Renouprez (1983). J. Mol. Spec., 99, 47–55. Neutron inelastic-scattering spectrum and valence force-field for neopentane.

129 J. Eckert, J.A. Goldstone, D. Tonks & D. Richter (1983). Phys. Rev. B, 27, 1980–1990. Inelastic neutron-scattering studies of vibrational excitations of hydrogen in Nb and Ta.

130 Y. Brunel, Ce. Coulombeau, C. Coulombeau, M. Moutin & H. Jobic (1983). J. Am. Chem. Soc., 105, 6411–6416. Optical and neutron inelastic-scattering study of norbornane—a new assignment of vibrational frequencies.

131 K. Holderna-Natkaniec, I. Natkaniec & V.D. Khravryutchenko (2003). Phase Transitions, 76, 275–279. Neutron spectroscopy of norbornane.

132 S.F. Parker, K.P.J. Williams, D. Steele & H. Herman (2003). Phys. Chem. Chem. Phys., 5, 1508–1514. The vibrational spectra of norbornene and nadic anhydride.

133 Y. Brunel, Ce. Coulombeau, C. Coulombeau & H. Jobic (1985). J. Phys. Chem., 89, 937–943. Optical and neutron inelastic-scattering study of 2-methylnorbornanes.

134 Y. Brunel, Ce. Coulombeau, C. Coulombeau & H. Jobic (1986). J. Phys. Chem., 90, 2008–2015. Optical and neutron inelastic-scattering study of 2,3-dimethylnorbornanes.

135 P. Papanek, J.E. Fischer & N.S. Murthy (1996). Macromol., 29, 2253–2259. Molecular vibrations in nylon 6 studied by inelastic neutron scattering.

136 P. Papanek, J.E. Fischer & N.S. Murthy (2002). Macromol., 35, 4175–4182. Low-frequency amide modes in different hydrogen-bonded forms of nylon-6 studied by inelastic neutron-scattering and density-functional calculations.

137 M. Bée, J. Combet, F. Guillaume, N.D. Morelon, M. Ferrand, D. Djurado & A.J. Dianoux (1996). Physica B, 226, 15–27. Neutron-scattering studies of linear-chains in an organic inclusion compound.

138 S.F. Parker, D.A. Braden, J. Tomkinson & B.S. Hudson (1998). J. Phys. Chem. B, 102, 5955–5956. Full longitudinal acoustic mode (LAM) spectrum of an n-alkane: comparison of observed and computed incoherent inelastic neutron scattering spectrum of n-octadecane.

139 C. Marcolli, P. Lainé, R. Bühler, G. Calzaferri & J. Tomkinson (1997). J. Phys. Chem. B, 101, 1171–1179. Vibrations of $H_8Si_8O_{12}$, $D_8Si_8O_{12}$ and $H_{10}Si_{10}O_{15}$ as determined by INS, IR and Raman experiments.

140 B.S. Hudson, J. Tse, M.Z. Zgierski, S.F. Parker, D.A. Braden & C. Middleton (2000). Chem. Phys., 261, 249–260. The inelastic incoherent neutron spectrum of crystalline oxamide: experiment and simulation of a solid.

141 A.I. Kolesnikov, V.E. Antonov, A.M. Balgurov, S.M. Bennington & M. Prager (1994). J. Phys.: Condens. Matter, 6, 9001–9008. Neutron scattering studies of the structure and dynamics of the PdCuH ordered phase produced under a high hydrogen pressure.

142 D.K. Ross, V.E. Antonov, E.L. Bokhenkov, A.I. Kolesnikov, E.G. Ponyatovsky & J. Tomkinson (1998). Phys. Rev. B, 58, 2591–2595. Strong anisotropy in the inelastic neutron scattering from PdH at high energy transfer.

143 P. Albers, M. Poniatowski, S.F. Parker & D.K. Ross (2000). J. Phys.: Condens. Matter, 12, 4451–4463. Inelastic neutron scattering on different grades of palladium of varying pretreatment.

144 J.M. Rowe, J.J. Rush, H.G. Smith, M. Mostelle & H.E. Flotow (1974). Phys. Rev. Lett., 33, 1297–1300. Lattice dynamics of a single crystal of $PdD_{0.63}$.

145 M.A. Andrews, J. Eckert, J.A. Goldstone, L. Passell & B. Swanson (1983). J. Am. Chem. Soc., 105, 2262–2269. Incoherent inelastic neutron-scattering, infrared, Raman, and X-Ray diffraction studies of pentacarbonylmethylmanganese, $Mn(CO)_5CH_3$.

146 A. Pawlukojć, I. Natkaniec, I. Majerz & L. Sobczyk (2001). Spectrochim. Acta, 57A, 2775–2779. Inelastic neutron scattering studies on low frequency vibrations of pentachlorophenol.

147 S. Hirata, H. Torii, Y. Furukawa, M. Tasumi & J. Tomkinson (1996). Chem. Phys. Lett., 261, 241–245. Inelastic neutron scattering from *trans*-polyacetylene.

148 F. Fillaux, N. Leygue, R. Baddour-Hadjean, S.F. Parker, Ph. Colomban, A. Gruger, A. Régis & L.T. Yu (1997). Chem. Phys., 216, 281–293. Inelastic neutron scattering studies of polyanilines and partially deuterated analogues.

149 S. Folch, A. Gruger, A. Régis, R. Baddour-Hadjean & Ph. Colomban (1999). Synth. Met., 101, 795–796. Polymorphism and disorder in oligo- and polyanilines.

150 A. El. Khalki, Ph. Colomban & B. Hennion (2002). Macromol., 35, 5203–5211. Nature of protons, phase transitions, and dynamic disorder in poly- and oligoaniline bases and salts: an inelastic neutron scattering study.

151 L. Jayes, A.P. Hard, C. Sene, S.F. Parker & U.A. Jayasooriya (2003). Anal. Chem., 75, 742–746. Vibrational spectroscopy of silicones: an FT-Raman and inelastic neutron scattering investigation.

152 J.F. Twisleton, J.W. White & P.A. Reynolds (1982). Polymer, 23, 578–588. Dynamical studies of fully oriented deuteropolyethylene by inelastic neutron scattering.

153 S.F. Parker (1996). J. Chem. Soc. Faraday Trans., 92, 1941–1946. Inelastic neutron scattering spectra of polyethylene.

154 H. Jobic (1982). J. Chem. Phys., 76, 2693–2696. Neutron inelastic scattering from oriented and polycrystalline polyethylene: observation and polarization properties of the optical phonons.

155 S.F. Parker (1997). Macromolecular Symposia, 119, 227–234. Inelastic neutron scattering of polyethylene and n-alkanes.

156 F. Fillaux, J.P. Fontaine, M.H. Baron, N. Leygue, G.J. Kearley & J. Tomkinson (1994). Biophysical Chemistry, 53, 155–168. Inelastic neutron-scattering study of the proton transfer dynamics in polyglycine I at 20K.

157 J. Moreno, A. Alegria, J. Colmenero & B. Frick (2001). Macromol., 34, 4886–4896. Methyl group dynamics in poly(methyl methacrylate): from quantum tunneling to classical hopping.

158 H. Takeuchi, J.S. Higgins, A. Hill, A. Machonnachie, G. Allen & G.C. Stirling (1982). Polymer, 23, 499–504. Investigation of the methyl torsion in isotactic polypropylene – comparison between neutron inelastic scattering spectra and normal coordinate calculations.

159 F. Fillaux, S.F. Parker & L.T. Yu (2001). Solid State Ionics, 145, 451–457. Inelastic neutron scattering studies of polypyrroles and partially deuterated analogues.

160 S.F. Parker, K. Shankland, J.C. Sprunt & U.A. Jayasooriya (1997). Spectrochim. Acta, 53A, 2333–2339. The nine modes of complexed water.

161 S.F. Parker, S.M. Bennington, A.J. Ramirez-Cuesta, G. Auffermann, W. Bronger, H. Herman, K.P.J. Williams & T. Smith (2003). J. Am. Chem. Soc., 125, 11656–11661. Inelastic neutron scattering, Raman spectroscopy and periodic-DFT studies of Rb_2PtH_6 and Rb_2PtD_6.

162 D.K. Breitinger, G. Bauer, M Raidel, R. Breiter & J. Tomkinson (1992). J. Mol. Struct., 267, 55–60. Vibrations of hydrogen bonds in hydrogendisulfito and hydrogensulfito complexes.

163 A.V. Belushkin & M.A. Adams (1997). Physica B, 234–236, 37–39. Lattice dynamics of KH_2PO_4 at high pressure.

164 S.F. Parker & J.B. Forsyth (1998). J. Chem. Soc. Faraday Trans., 94, 1111–1114. K_2MCl_6: (M = Pt, Ir) Location of the silent modes and forcefields.

165 F. Fillaux, N. Leygue, J. Tomkinson, A. Cousson & W. Paulus (1999). Chem. Phys., 244, 387–403. Structure and dynamics of the symmetric hydrogen bond in potassium hydrogen maleate: a neutron scattering study.

166 D. Colognesi, M. Celli, F. Cilloco, R.J. Newport, S.F. Parker, V. Rossi-Albertini, F. Sacchetti, J. Tomkinson & M. Zoppi (2002). Appl. Phys. A, 74 [Suppl.], S64–S66. TOSCA neutron spectrometer; the final configuration.

167 S.F. Parker, H. Herman, A. Zimmerman & K.P.J. Williams (2000). Chem. Phys., 261, 261–266. The vibrational spectrum of K_2PdCl_4: first detection of the silent mode v_5.

168 A. Pawlukojć, J. Leciejewicz, I. Natkaniec & J. Nowicka-Scheibe (2003). Polish J. Chem., 77, 75–85. Neutron spectroscopy, IR, Raman and ab initio study of L-proline.

169 D. Paci, G. Amoretti, G. Arduini, G. Ruani, S. Shinkai, T. Suzuki, F. Ugozzoli & R. Caciuffo (1997). Phys. Rev. B, 55, 5566–5569. Vibrational spectrum of C_{60} in the p-tert-butylcalix[8]arene(1:1)C_{60} complex.

170 G.J. Kearley, J. Tomkinson, A. Navarro, J.J. López-González, & M. Fernández-Gómez (1997). Chem. Phys., 216, 323–335. Symmetrised quantum-mechanical force-fields and INS spectra: s-triazine, trichloro-s-triazine and pyrazine.

171 A. Navarro, J. Vaquez, M. Montejo, J.J.L. Gonzalez & G.J. Kearley (2002). Chem. Phys. Lett., 361, 483–491. A reinvestigation of the v_7 and v_{10} modes of pyridazine on the basis of the inelastic neutron scattering spectrum analysis.

172 J. Wasicki, M. Jaskólski, Z. Pajak, M. Szafran, Z. Dega-Szafran, M.A. Adams & S.F. Parker (1999). J. Mol. Struct., 476, 81–95. Crystal structure and molecular motion in pyridine N-oxide semiperchlorate.

173 E. Geidel, H. Jobic & S.F. Parker (1999). *Proceedings of the 12th International Zeolite Conference,* (Ed). M.M.J. Treacy, B.K. Marcus, M.E. Bisher & J.B. Higgins), Materials Research Society, Pittsburgh, pp.2609–2614. Vibrational spectroscopic investigations of pyrrole adsorption in faujasites: studies by infrared, Raman and neutron spectroscopy.

174 A. Pavese, G. Artioli & O. Moze (1998). European J. Mineralogy, 10, 59–69. Inelastic neutron scattering from pyrope powder: experimental data and theoretical calculations.

175 A. Pawlukojc´, J. Leciejewicz, J. Tomkinson & S.F. Parker (2002). Spectrochim. Acta, 58A, 2897–2904. Neutron spectroscopy study of hydrogen bonds dynamics in *l*-serine.

176 G.J. Kearley, J. Tomkinson & J. Penfold (1987). Zeitschrift Fur Physik B, 69, 63–67. New constraints for normal-mode analysis of inelastic neutron-scattering spectra: application to the HF_2^- ion.

177 J. Tomkinson & G.J. Kearley (1989). J. Chem. Phys., 91, 5164–5169. Phonon wings in inelastic neutron scattering spectroscopy: the harmonic approximation.

178 H. Jobic (2000). Physica B, 276, 222–225. Inelastic scattering of organic molecules in zeolites.

179 H. Jobic, K.S. Smirnov & D. Bougeard (2001). Chem. Phys. Lett., 344, 147–153. Inelastic neutron scattering spectra of zeolite frameworks – experiment and modeling.

180 A.V. Goupil-Lamy, J.C. Smith, J. Yunoki, F. Tokunaga, S.F. Parker & M. Kataoka (1997). J. Am. Chem. Soc., 119, 9268–9273. High-resolution vibrational inelastic neutron scattering: A new spectroscopic tool for globular proteins.

181 G. Auffermann, Y. Prots, R. Kniep, S.F. Parker & S.M. Bennington (2002). Chem. Phys. Chem., 3, 815–817. Inelastic neutron scattering spectroscopy of diazenides: detection of the N=N stretch.

182 R.P. Rinaldi & G.S. Pawley (1975). J. Phys. C: Solid State Phys., 8, 599–616. Investigation of the intermolecular modes in orthorhombic sulphur.

183 M. Prager, H. Grimm, S.F. Parker, R. Lechner, A. Desmedt, S. McGrady & E. Koglin (2002). J. Phys.: Condens. Matter, 14, 1833–1845. Methyl group rotation in trimethylaluminium.

184 R. Mukhopadhyay & S.L. Chaplot (2002). Chem. Phys. Lett., 358, 219–223. Phonon density of states in tetrabromoethylene: lattice dynamic and inelastic neutron scattering study.

185 A. Navarro, M. Fernández-Gómez, M.P. Fernández-Liencres, C. A. Morrison, D.W.H. Rankin & H.E. Robertson (1999). Phys. Chem. Chem. Phys., 1, 3453–3460. Tetrachloropyrimidine: molecular structure by electron diffraction, vibrational analysis by infrared, Raman and inelastic neutron scattering spectroscopies and quantum mechanical calculations.

186 S.L. Chaplot, A. Mierzejewski, G.S. Pawley, J. Lefebvre & T. Luty (1983). J. Phys. C: Solid State Phys., 16, 625–644. Phonon dispersion of the external and low-frequency internal vibrations in monoclinic tetracyanoethylene at 5K.

187 B. Cadioli, E. Gallinella, C. Coulombeau, H. Jobic & G. Berthier (1993). J. Phys. Chem., 97, 7844–7856. Geometric structure and vibrational spectrum of tetrahydrofuran.

188 C. Coulombeau & H. Jobic (1995). J. Mol. Struct. (Theochem), 330, 127–130. Contribution to the vibrational normal modes analysis of d_8-THF by neutron inelastic scattering.

189 B. Winkler, I. Kaiser, M. Chall, G. Coddens, B. Hennion & R. Kahn (1997). Physica B, 234–236, 70–71. Dynamics of $N(CH_3)_4GeCl_3$.

190 G.J. Kearley & J. Tomkinson (1992). Physica B, 180–181, 700–702. The inelastic neutron spectrum of $Sn(CH_3)_4$.

191 M. Prager, J. Combet, S.F. Parker, A. Desmedt & R.E. Lechner (2002). J. Phys.: Condens. Matter, 14, 10145–10157. The methyl rotational potentials of $Ga(CH_3)_3$ derived by neutron spectroscopy.

192 B. Pasquier, N. LeCalvé, S. Alhomsiteiar & F. Fillaux (1993). Chem. Phys., 171, 203–220. Vibrational study of various crystalline phases of thallium dihydrogen phosphate TlH_2PO_4 and its deuterated analog TlD_2PO_4.

193 A.D. Esposti, O. Moze, C. Taliani, J. Tomkinson, R. Zamboni & F. Zerbetto (1996). J. Chem. Phys., 104, 9704–9718. The intramolecular vibrations of prototypical polythiophenes.

194 A. Pawlukojc´, J. Leciejewicz, J. Tomkinson & S.F. Parker (2001). Spectrochim. Acta, 57A, 2513–2523. Neutron scattering, infra red, Raman spectroscopy and ab initio study of L-threonine.

195 S.K. Satija, B. Swanson, J. Eckert & J.A. Goldstone (1991). J. Phys. Chem., 95, 10103–10109. High pressure Raman scattering and inelastic neutron scattering studies of triaminotrinitrobenzene.

196 F. Boudjada, J. Meinnel, A. Cousson, W. Paulus, M. Mani & M. Sanquer (1999). In *Neutrons and Numerical Methods — N₂M*, (Ed.) M.R. Johnson, G.J. Kearley & H. G. Büttner, AIP Conf. Proc. Vol. 479, pp. 217–222. Tribromomesitylene structure at 14 K methyl conformation and tunnelling.

197 A.V. Belushkin, J. Tomkinson & L.A. Shuvalov (1993). J. Physique II (France), 3, 217–225. Inelastic neutron scattering study of proton dynamics in $Cs_3H(SeO_4)_2$ and $Rb_3H(SeO_4)_2$.

198 J. Meinnel, W. Häusler, B. Mani, M. Tazi, M. Nusimovici, M. Sanquer, B. Wyncke, A. Heidemann, C.J. Carlile, J. Tomkinson & B. Hennion (1992). Physica B, 180–181, 711–713. Methyl tunnelling in trihalogeno-trimethyl-benzenes.

199 M.T. Dove, B.M. Powell, G.S. Pawley, S.L. Chaplot & A. Mierzejewski (1989). J. Chem. Phys., 90, 1918–1923. Inelastic neutron scattering determination of phonon dispersion curves in the molecular crystal sym-$C_6F_3Cl_3$.

200 C. Sourisseau, M. Bée, A. Dworkin & H. Jobic (1985). J. Raman Spec., 16, 44–56. Infrared, Raman and inelastic neutron-scattering study of phase-transitions in trimethyloxosulfonium iodide, $(CH_3)_3SOI$.

201 S.F. Parker, P.H. Dallin, B.T. Keiller, C.E. Anson & U.A. Jayasooriya (1999). Phys. Chem. Chem. Phys., 1, 2589–2592. An inelastic neutron scattering study and re-assignment of the vibrational spectrum of $[Os_3(CO)_9(\mu_2\text{-}CO)(\mu_3\text{-}\eta^2\text{-}C_2H_2)]$, a model compound for chemisorbed ethyne.

202 A. Aamouche, M. Ghomi, C. Coulombeau, H. Jobic, L. Grajcar, M.H. Baron, V. Baumruk, P.Y. Turpin, C. Henriet, & G. Berthier (1996). J. Phys. Chem., 100, 5224–5234. Neutron inelastic scattering, optical spectroscopies and scaled quantum mechanical force fields for analyzing the vibrational dynamics of pyrimidine nucleic acid bases. 1. Uracil.

203 H.N. Bordallo, A.I. Kolesnikov, A.V. Kolomiets, W. Kalceff, H. Nakotte & J. Eckert (2003). J. Phys.: Condens. Matter, 15, 2551–2559. Inelastic neutron scattering studies of $TbNiAlH_{1.4}$ and $UNiAlH_2$ hydrides.

204 M.R. Johnson, K. Parlinski, I. Natkaniec & B.S. Hudson (2003). Chem. Phys., 291, 53–60. Ab initio calculations and INS measurements of phonons and molecular vibrations in a model peptide compound—urea.

205 R. Bechrouri, J. Ulicny, H. Jobic, V. Baumruk, L. Bednárová, O. Bouloussa & M. Ghomi (1997). Spectroscopy of Biological Molecules: Modern Trends, 288, 213–214. Analysis of the vibrational properties of uridine and 5'-UMP through neutron inelastic scattering, optical spectroscopies and quantum mechanical calculations.

206 S.F. Parker, R. Jeans & R. Devonshire (2004). Vib. Spec., 35, 173–177. Inelastic neutron scattering, Raman and periodic-DFT studies of purine.

207 H. Jobic (1985). J. Mol. Struct., 131, 167–175. A new inelastic neutron scattering study of Zeise's salt, $K[Pt(C_2H_4)Cl_3]$, and a more confident assignment of the vibrational frequencies.

208 P.W. Loeffen, R.F. Pettifer, & J. Tomkinson (1996). Chem. Phys., 208, 403–420. Vibrational force field of zinc tetraimidazole from inelastic neutron scattering.

209 J.F. Fernandez, M. Kemali & D.K. Ross (1997). J. Alloys Cmpds., 253, 248–251. Incoherent inelastic neutron scattering from the C-15 Laves phase $ZrTi_2H_{3.6}$.

210 V.E. Antonov, M. Baier, B. Dorner, V.K. Fedotov, G. Grosse, A.I. Kolesnikov, E.G. Ponyatovsky, G. Schneider & F.E. Wagner (2002). J. Phys. Condens.: Matter, 14, 6427–6445. High-pressure hydrides of iron and its alloys.

211 V.E. Antonov, V.K. Fedotov, B.A. Gnesin, G. Grosse, A.S. Ivanov, A.I. Kolesnikov & F.E. Wagner (2000). Europhys. Lett., 51, 140–146. Anisotropy in the inelastic neutron scattering from fcc NiH.

212 C.V. Berney & J.W. White (1977). J. Am. Chem. Soc., 99, 6878–6880. Selective deuteration in neutron-scattering spectroscopy: formic acid and deuterated derivatives.

213 B.R. Bender, G.J. Kubas, L.H. Jones, B.I. Swanson, J. Eckert, K.B. Capps & C.D. Hoff (1997). J. Am. Chem. Soc., 119, 9179–9190. Why does D_2 bind better than H_2? A theoretical and experimental study of the equilibrium isotope effect on H_2 binding in a $M(\eta^2\text{-}H_2)$ complex. Normal coordinate analysis of $W(CO)_3(PCy_3)_2(\eta^2\text{-}H_2)$.

214 J. Eckert, C.E. Webster, M.B. Hall, A. Albinati & L.M. Venanzi (2002). Inorganica Chimica Acta, 330, 240–249. The vibrational spectrum of $Tp^{3,5\text{-Me}}RhH_2(H_2)$: a computational and inelastic neutron scattering study.

215 A.L. Bandini, G. Banditelli, M. Manassero, A. Albinati, D. Colognesi & J. Eckert (2003). Eur. J. Inorg. Chem., 21, 3958–3967. Binuclear hydridoplatinum(II): One-pot synthesis, INS spectra and X-ray crystal structure of $[Pt_2(dcype)_2(H)_3][BPh_4]$ {dcype=1,2-bis(dicyclohexylphosphanyl)ethane}.

216 A. Aamouche, M. Ghomi, C. Coulombeau, L. Grajcar, M.H. Baron, H. Jobic, & G. Berthier (1997). J. Phys. Chem. A, 101, 1808–1817. Neutron inelastic scattering, optical spectroscopies and scaled quantum mechanical force fields for analyzing the vibrational dynamics of pyrimidine nucleic acid bases. 2. Thymine.

217 S. Ikeda, M. Furusaka, T. Fukunaga & A.D. Taylor (1990). J. Phys.: Condens. Matter, 2, 4675–4684. Hydrogen wave-functions in the metal-hydrides ZrH_2 and $NbH_{0.3}$.

218 A.I. Kolesnikov, I.O. Bashkin, A.V. Belushkin, E.G. Ponyatovsky & M. Prager (1994). J. Phys.: Condens Matter, 6, 8989–9000. Inelastic neutron scattering study of ordered γ-ZrH.

219 A.I. Kolesnikov, A.M. Balgurov, I.O. Bashkin, A.V. Belushkin, E.G. Ponyatovsky & M. Prager (1994). J. Phys.: Condens. Matter, 6, 8977–8988. Neutron scattering studies of ordered γ-ZrD.

220 I.O. Bashkin, A.I. Kolesnikov, V.Yu Malyshev, E.G. Ponyatovsky, M. Prager & J. Tomkinson (1993). Zeitschrift für Physikalische Chemie, 179, 335–342. Hydrogen interaction and bound multiphonon states in vibrational spectra of titanium hydrides.

221 P. Morris, D.K. Ross, S. Ivanov, D.R Weaver & O. Serot (2004). J. Alloys and Compds., 363, 88–92. Inelastic neutron scattering study of the vibration frequencies of hydrogen in calcium dihydride.

222 H.G. Schimmel, M.R. Johnson, G.J. Kearley, A.J. Ramirez-Cuesta, J. Huot & F.M. Mulder (2004). Materials Science and Engineering B–Solid State Materials for Advanced Technology, 108, 38–41. The vibrational spectrum of magnesium hydride from inelastic neutron scattering and density functional theory.

223 J.L. Warren, J.L. Yarnell, G. Dolling & R.A. Cowley (1967). Phys. Rev., 158, 805–808. Lattice dynamics of diamond.

224 R. Nicklow, N. Wakabayashi & H.G. Smith (1972). Phys. Rev. B, 5, 4951–4962. Lattice dynamics of pyrolytic graphite.

225 D.K. Ross (1973). J. Phys. C: Solid State Phys., 6, 3525–3535. Inelastic neutron scattering from polycrystalline graphite at temperatures up to 1920°C.

226 K.H. Link, H. Grimm, B. Dorner, H. Zimmermann, H. Stiller & P. Bleckmann (1985). J. Phys. Chem. Solids, 46, 135–142. Determination of the lattice vibrations of imidazole by neutron scattering.

227 J. Schleifer, J. Kalus, U. Schmelzer & G. Eckold (1989). Phys. Stat. Sol. B, 154, 153–166. Phonon dispersion in an α-perylene D_{12}-crystal at 10K.

228 E.F. Sheka, E.L. Bokhenkov, B. Dorner, J. Kalus, G.A. Mackenzie, I. Natkaniec, G.S. Pawley & U. Schmelzer (1984). J. Phys. C-Solid State Phys., 17, 5893–5914. Anharmonicity of phonons in crystalline naphthalene.

229 F. Fillaux, R. Papoular, A. Lautie & J. Tomkinson (1994). Carbon, 32, 1325–1331. Inelastic neutron scattering study of the proton dynamics in carbons and coals.

230 F. Fillaux, A. Lautie, R. Papoular & S.M. Bennington (1995). Physica B, 213, 631–633. Free proton dynamics in coal.

231 J.K. Walters, R.J. Newport, S.F. Parker & W.S. Howells (1995). J. Phys.: Condens. Matter, 7, 10059–10073. A spectroscopic study of the structure of amorphous hydrogenated carbon.

232 P. Albers, G. Prescher, K. Seibold, D.K. Ross & F. Fillaux (1996). Carbon, 34, 903–908. Inelastic neutron scattering study of proton dynamics in carbon blacks.

233 P. Albers, S. Bösing, G. Prescher, K. Seibold, D.K. Ross & S.F Parker (1999). Applied Catalysis A: General, 187, 233–243. Inelastic neutron scattering investigations of the products of thermally and catalytically driven catalyst coking.

234 D. Lennon, D.T. Lundie, S.D. Jackson, G.J. Kelly and S.F. Parker (2002). Langmuir, 18, 4667–4673. Characterisation of activated carbon using X-ray photoelectron spectroscopy and inelastic neutron scattering spectroscopy.

235 P.W. Albers, J. Pietsch, J. Krauter & S.F. Parker (2003). Phys. Chem. Chem. Phys., 5, 1941–1949. Investigations of activated carbon catalyst supports from different natural sources.

236 P. Albers, K. Seibold, G. Prescher, B. Freund, S.F. Parker, J. Tomkinson, D.K. Ross & F. Fillaux (1999). Carbon, 37, 437–444. Neutron spectroscopic investigations on different grades of modified furnace blacks.

237 P. Papanek, W.A. Kamitakahara, P. Zhou & J.E. Fischer (2001). J. Phys.: Condens. Matter, 13, 8287–8301. Neutron scattering studies of disordered carbon anode materials.

238 S.F. Parker, A.J. Champion, S.M. Bennington, S.J. Levett, J.W. Taylor and H. Herman (2004). Phys. Rev. B, submitted for publication. Assignment of the internal modes of C_{60} by inelastic neutron scattering spectroscopy – one-third of accepted assignments are wrong!

239 C. Christides, A.V. Nikolaev, T.J.S. Dennis, K. Prassides, F. Negri, G. Orlandi & F. Zerbetto (1993). J. Phys. Chem., 97, 3641–3643. Inelastic neutron scattering study of the intramolecular vibrations of C_{70}.

240 A.I. Kolesnikov, I.O. Bashkin, A.P. Moravsky, M.A. Adams, M. Prager & E.G. Ponyatovsky (1996). J. Phys.: Condens. Matter, 8, 10939–10949. Neutron scattering study of a high pressure polymeric C_{60} phase.

241 B. Renker, H. Schober, R. Heid & P. von Stein (1997). Solid State Comm., 104, 527–530. Pressure and charge induced polymerisation of C_{60}: A comparative study of lattice vibrations.

242 A. Soldatov, O. Andersson, B. Sundqvist & K. Prassides (1998). Molecular Crystals and Liquid Crystals Science and Technology C Molecular Materials, 11, 1–6. Transport and vibrational properties of pressure polymerized C_{60}.

243 S.M. Bennington, N. Kitamura, M.G. Cain, M.H. Lewis & M. Arai (1999). Physica B, 263, 632–635. The structure and dynamics of hard carbon formed from C_{60} fullerene.

244 K. Prassides, J. Tomkinson, C. Christides, M.J. Rosseinsky, D.W. Murphy & R.C. Haddon (1991). Nature, 354, 462–463. Vibrational spectroscopy of superconducting fulleride K_3C_{60} by inelastic neutron scattering.

245 B. Renker, F. Gompf, H. Schober, P. Adelmann, H.J. Bornemann & R. Heid (1993). Zeitschrift für Physik B–Condensed Matter, 92, 451–455. Intermolecular vibrations in pure and doped C_{60} – an inelastic neutron-scattering study.

246 W.K. Fullagar, J.W. White & F.R. Trouw (1995). Physica B, 213–214, 16–21. Superconductivity, vibrational and lattice dynamics of Rb_3C_{60}.

247 K. Prassides, C. Christides, M.J. Rosseinsky, J. Tomkinson, D.W. Murphy & R.C. Haddon (1992). Europhys. Lett., 19, 629–635. Neutron spectroscopy and electron-phonon coupling in alkali-metal doped fullerides.

248 H. Schober, B. Renker & R. Heid (1999). Phys. Rev. B, 60, 998–1004. Vibrational behavior of Na_4C_{60} in the monomer and two- dimensional polymer states.

249 H. Schober, A. Tölle, B. Renker, R. Heid & F. Gompf (1997). Phys. Rev. B, 56, 5937–5950. Microscopic dynamics of AC_{60} compounds in the plastic, polymer, and dimer phases investigated by inelastic neutron scattering.

250 L. Cristofolini, C.M. Brown, A.J. Dianoux, M. Kosaka, K. Prassides, K. Tanigaki & K. Vavekis (1996). J. Chem. Soc. Chem. Comm., 2465–2466. Interfullerene vibrations in the polymeric fulleride CsC_{60}.

251 K. Prassides, C.M. Brown, S. Margadonna, K. Kordatos, K. Tanigaki, E. Suard, A.J. Dianoux & K.D. Knudsen (2000). J. Mat. Chem., 10, 1443–1449. Powder diffraction and inelastic neutron scattering studies of the Na_2RbC_{60} fulleride.

252 D. Graham, J. Howard, T.C. Waddington & J. Tomkinson (1983). J. Chem. Soc. Faraday Trans. II, 79, 1713–1717. Incoherent inelastic neutron scattering studies of transition-metal hydridocarbonyls. Part 1 $Cs^+[HCo_6(CO)_{15}]^-$.

253 J. Eckert, A. Albinati & G. Longoni (1989). Inorg. Chem., 28, 4055–4056. Inelastic neutron scattering study of $K[HCo_6(CO)_{15}]$ implications for location of the hydride.

254 J.W. White & C.J. Wright (1972). J. Chem. Soc. Faraday Trans. II, 68, 1423–1433. Neutron scattering spectra of hydridocarbonyls in the 900–200 cm^{-1} energy region.

255 R. G. Delaplane, H. Kuppers, H. Noreland & S.F. Parker (2002). Appl. Phys. A [Suppl], 74, S1366—S1367. Dynamics of the water molecule in potassium oxalate monohydrate as studied by single crystal inelastic neutron scattering.

256 A.V. Belushkin, M.A. Adams, A.I. Kolesnikov & L.A. Shuvalov (1994). J. Phys.: Condens. Matter, 6, 5823–5832. Lattice dynamics and effects of anharmonicity in different phases of cesium hydrogen sulphate.

257 Z. Ouafik, N. Le Calvé & B. Pasquier (1995). Chem. Phys., 194, 145–158. Vibrational study of various crystalline phases of thallium dihydrogen arsenate TlH_2AsO_4. Comparison with thallium dihydrogen phosphate TlH_2PO_4.

258 N. Le Calvé, B. Pasquier & Z. Ouafik (1997). Chem. Phys., 222, 299–313. Vibrational study by inelastic neutron scattering, infrared adsorption and Raman scattering of potassium, rubidium and cesium dihydrogenarsenate crystals, comparison with thallium dihydrogenarsenate.

259 H. Schober, D. Strauch & B. Dorner (1993). Zeitschrift fur Physik B–Condensed Matter, 92, 273–283. Lattice-dynamics of sapphire (Al_2O_3).

260 M. Arai, A.C. Hannon, A.D. Taylor, T. Otomo, A.C. Wright, R.N. Sinclair & D.L. Price (1991). Trans. Am. Cryst. Assoc., 27, 113–131. Neutron scattering law measurements for vitreous silica.

261 D. Strauch & B. Dorner (1993). J. Phys.: Condens. Matter, 5, 6149–6154. Lattice-dynamics of alpha-quartz: 1. experiment.

262 H. Jobic & A.J. Renouprez (1984). J. Chem. Soc. Faraday Trans. I, 80, 1991–1997. Inelastic neutron scattering spectroscopy of hydrogen adsorbed on Raney nickel.

263 H. Jobic, G. Clugnet & A.J. Renouprez (1987). J. Electron Spectroscopy and Related Phenomena, 45, 281–290. Neutron inelastic spectroscopy of hydrogen adsorbed at different pressures on a Raney nickel catalyst.

264 H. Jobic, J.P. Candy, V. Perrichon & A. Renouprez (1985). J. Chem. Soc. Faraday Trans. I, 81, 1955–1961. Neutron-scattering and volumetric study of hydrogen adsorbed and absorbed on Raney palladium.

265 J.M. Nicol, J. Eckert & J. Howard (1988). J. Phys. Chem., 92, 7117–7121. Dynamics of molecular hydrogen adsorbed in CoNa-A zeolite.

266 J. Eckert, J.M. Nicol, J. Howard & F.R. Trouw (1996). J. Phys. Chem., 100, 10646–10651. Adsorption of hydrogen in Ca-exchanged Na-A zeolites probed by inelastic neutron scattering spectroscopy.

267 J. Eckert, F. Trouw, A.L.R. Bug & R. Lobo (1999). In *Proceedings of the 12th International Zeolite Conference*, (Ed.) M.M.J. Treacy, B.K. Marcus, M.E. Bisher and J.B. Higgins, pp.119–125, Materials Research Society, Pittsburgh. Adsorption sites in zeolites A and X probed by competitive adsoorption of H_2 with N_2 or O_2: implications for N_2/O_2 separation.

268 H. Jobic (1991). J. Catal., 131, 289–293. Observation of the fundamental bending vibrations of hydroxyl groups in HNaY zeolite by neutron inelastic scattering.

269 W.P.J.H. Jacobs, H. Jobic, J.H.M.C. van Wolput & R.A. van Santen (1992). Zeolites, 12, 315–319. Fourier-transform infrared and inelastic neutron scattering study of HY zeolites.

270 W.P.J.H. Jacobs, R.A. van Santen & H. Jobic (1994). J. Chem. Soc. Faraday Trans., 90, 1191–1196. Inelastic neutron scattering study of NH_4Y zeolites.

271 H. Jobic, A. Renouprez, A.N. Fitch & H.J. Lauter (1987). J. Chem. Soc. Faraday Trans. I, 83, 3199–3205. Neutron spectroscopic study of polycrystalline benzene and of benzene adsorbed in Na-Y Zeolite.

272 H. Jobic, J. Tomkinson, J.P. Candy, P. Fouilloux & A.J. Renouprez (1980). Surf. Sci., 95, 496–510. The structure of benzene chemisorbed on Raney nickel: a neutron inelastic spectroscopy determination.

273 A.J. Renouprez, G. Clugnet & H. Jobic (1982). J. Catal., 74, 296–306. The interaction between benzene and nickel—a neutron inelastic spectroscopy study.

274 H. Jobic & A.J. Renouprez (1981). Surf. Sci., 111, 53–62. Neutron inelastic spectroscopy of benzene chemisorbed on Raney platinum.

275 J.P. Candy, H. Jobic & A.J. Renouprez (1983). J. Phys. Chem., 87, 1227–1230. Chemisorption of cyclohexene on nickel—a volumetric and neutron inelastic spectroscopy study.

276 H. Jobic, M. Lacroix, T. Decamp & M. Breysse (1995). J. Catal., 157, 414–422. Characterization of ammonia adsorption on ruthenium sulfide: identification of amino species by inelastic neutron-scattering.

277 C. Dumonteil, M. Lacroix, C. Geantet, H. Jobic & M. Breysse (1999). J. Catal., 187, 464–473. Hydrogen activation and reactivity of ruthenium sulfide catalysts: influence of the dispersion.

278 A.J. Renouprez & H. Jobic (1988). J. Catal., 113, 509–516. Neutron-scattering study of hydrogen adsorption on platinum catalysts.

279 J.J. Rush, R.R. Cavanagh, R.D. Kelley & J.M. Rowe (1985). J. Chem. Phys., 83, 5339–5341. Interaction of vibrating H-atoms on the surface of platinum particles by isotope-dilution neutron spectroscopy.

280 T.J. Udovic, R.R. Cavanagh & J.J. Rush (1988). J. Am. Chem. Soc., 110, 5590–5591. Neutron spectroscopic evidence for adsorbed hydroxyl species on platinum black.

281 P.W. Albers, M. Lopez, G. Sextl, G. Jeske & S.F. Parker (2004). J. Catal., 223, 44–53. Inelastic neutron scattering investigation on the site occupation of atomic hydrogen on platinum particles of different size.

282 S.F. Parker, J.W. Taylor, P. Albers, M. Lopez, G. Sextl, D. Lennon, A.R. McInroy & I.W. Sutherland (2004). Vib. Spec., 35, 179–182. Inelastic neutron scattering studies of hydrogen on fuel cell catalysts.

283 R.R. Cavanagh, J.J. Rush, R.D. Kelley & T.J. Udovic (1984). J. Chem. Phys., 80, 3478–3484. Adsorption and decomposition of hydrocarbons on platinum black—vibrational-modes from NIS.

284 R.R. Cavanagh, R.D. Kelley & J.J. Rush (1982). J. Chem. Phys., 77, 1540–1547. Neutron spectroscopy of hydrogen and deuterium on Raney nickel.

285 R.D. Kelley, R.R. Cavanagh & J.J. Rush (1983). J. Catal., 83, 464–468. Co-adsorption and reaction of H_2 and CO on Raney nickel: neutron vibrational spectroscopy.

286 F. Hochard, H. Jobic, J. Massardier & A.J. Renouprez (1995). J. Mol. Cat. A–Chem., 95, 165–172. Gas-phase hydrogenation of acetonitrile on Raney-nickel catalysts—reactive hydrogen.

287 J.M. Nicol, J.J. Rush & R.D. Kelley (1987). Phys. Rev. B, 36, 9315–9317. Neutron spectroscopic evidence for subsurface hydrogen in palladium.

288 J.M. Nicol, J.J. Rush & R.D. Kelley (1988). Surf. Sci., 197, 67–80. Inelastic neutron-scattering studies of the interaction of hydrogen with palladium black.

289 J.M. Nicol, T.J. Udovic, J.J. Rush & R.D. Kelley (1988). Langmuir, 4, 294–297. Isotope-dilution neutron spectroscopy—a vibrational probe of hydrogen-deuterium adsorbate interactions on palladium black.

290 H. Jobic, J.P. Candy, V. Perrichon & A.J. Renouprez (1985). J. Chem. Soc. Faraday Trans. I, 81, 1955–1961. Neutron-scattering and volumetric study of hydrogen adsorbed and absorbed on Raney palladium.

291 H. Jobic & A. Renouprez (1987). J. Less-Common Met., 129, 311–316. Formation of hydrides in small particles of palladium supported in Y-zeolite.

292 C.I. Ratcliffe, T.C. Waddington & J. Howard (1982). J. Chem. Soc. Faraday Trans. II, 78, 1881–1893. Inelastic neutron-scattering studies and barriers to methyl rotation in alkali-metal methanesulfonate salts.

293 K.P. Brierley, J. Howard, K. Robson, T.C. Waddington & C.I. Ratcliffe (1982). J. Chem. Soc. Faraday Trans. II, 78, 1101–1119. Inelastic neutron-scattering studies of the torsional and librational modes of the anilinium halides, $C_6H_5NH_3X$.

294 J. Tomkinson, I.J. Braid, J. Howard & T.C. Waddington (1982). Chem. Phys., 64, 151–157. Hydrogen-bonding in potassium hydrogen maleate and some simple derivatives studied by inelastic neutron-scattering spectroscopy.

295 J. Howard, T.C. Waddington & J. Tomkinson (1982). Chem. Phys. Lett., 85, 428–429. An inelastic neutron-scattering study of hydrogen-bonding in potassium hydrogen dichloromaleate.

296 T. Brun, J. Howard & J. Tomkinson (1986). Spectrochim. Acta, 42A, 1209–1216. A single-crystal and polycrystalline study of potassium hydrogen malonate using inelastic neutron-scattering.

297 J. Howard & T.C. Waddington (1981). Chem. Phys. Lett., 83, 211–214. Incoherent inelastic neutron-scattering studies of the torsional modes of hexachloroethane.

298 J. Howard, K. Robson & T.C. Waddington (1982). Spectrochim. Acta, 38A, 903–911. Inelastic neutron-scattering studies of ethylene-containing square-planar platinum complexes.

299 M.J. Wax, R.R. Cavanagh, J.J. Rush, G.D. Stucky, L. Abrams & D.R. Corbin (1986). J. Phys. Chem., 90, 532–534. A neutron-scattering study of zeolite-rho.

300 T.J. Udovic, R.R. Cavanagh, J.J. Rush, M.J. Wax, G.D. Stucky, G.A. Jones & D.R. Corbin (1987). J. Phys. Chem., 91, 5968–5973. Neutron-scattering study of NH_4^+ dynamics during the deammoniation of NH_4-rho zeolite.

301 J.J. Thomas, J.J. Chen, H.M. Jennings & D.A. Neumann (2003). Chem. Mater., 15, 3813–3817. Ca-OH Bonding in the C-S-H gel phase of tricalcium silicate and white Portland cement pastes measured by inelastic neutron scattering.

302 B. Pasquier, F. Fillaux & J. Tomkinson (1995). Physica B, 213–214, 658–660. Proton dynamics in lead hydrogen phosphate crystal.

303 G. Auffermann, G.D. Barrera, D. Colognesi, G. Corradi, A.J. Ramirez-Cuesta & M. Zoppi (2004). J. Phys.: Condens. Matter, 16, 5731–5743. Hydrogen dynamics in heavy alkali metal hydrides through inelastic neutron scattering.

304 B.A. Dasannacharya, P.S. Goyal, C.L. Thaper, J. Penfold & J. Tomkinson (1989). Physica B, 156–157, 135–136. Dynamics of H_2O in $Ba(ClO_3)_2.H_2O$.

305 F. Fillaux, B. Marchon, A. Novak & J. Tomkinson (1989). Chem. Phys., 130, 257–270. Proton dynamics in the hydrogen bond, inelastic neutron scattering by single crystals of CsH_2PO_4 at 20 K.

306 A.V. Powell & P.G. Dickens (1990). Appl. Phys. A, 51, 226–230. An incoherent inelastic neutron scattering study of NH_4TaWO_6.

307 R.P. White, N.C. Sa-ard, S.K. Bollen, R.D. Cannon, U.A. Jayasooriya, S.T. Robertson, U. Steigenberger & J. Tomkinson (1990). Spectrochim. Acta, 46A, 903–910. Vibrational spectroscopy of oxo-centred trinuclear metal complexes: an inelastic neutron scattering study.

308 S.K. Bollen, N.C. Sa-ard, E. Karu, R.P. White, R.D. Cannon, U.A. Jayasooriya, C.J. Carlile, S.T. Robertson, U. Steigenberger & J. Tomkinson (1990). Spectrochim. Acta, 46A, 911–915. Methyl proton tunnelling in the $[Cr_3O(OOCCH_3)_6(H_2O)_3]^+$ ion: a neutron scattering study.

309 D.J. Jones, J. Roziere, J. Tomkinson & J. Penfold (1989). J. Mol. Struct., 197, 113–121. Incoherent inelastic neutron scattering studies of proton conducting materials: $Sn(HPO_4)_2.H_2O$ and $HM(SO_4)_2.H_2O$, M=Fe,In.

310 A. Radelescu, I. Padureanu, S.N. Rapeanu & A. Beldiman (2000). Phys. Stat. Sol. B, 217, 777–784. On the high-frequency limit of lattice vibrations in zirconium hydrides and deuterides.

311 A. Radelescu, I. Padureanu, S.N. Rapeanu, Zh.A. Kozlov & V.A. Semenov (2000). Phys. Stat. Sol. B, 222, 445–450. On the dynamics and site occupation of D atoms in δ and ε deuterides of Zr.

312 J. Ulicny, N. Leulliot, L. Grajcar, M.H. Baron, H. Jobic & M. Ghomi (1999). In *Neutrons and Numerical Methods — N_2M*, (Ed.) M.R. Johnson, G.J. Kearley & H. G. Büttner, AIP Conf. Proc. Vol. 479, pp. 183–186. NIS, IR and Raman spectra with quantum mechanical calculations for analysing the force field of hypericin model compounds.

313 D. Colognesi, A.J. Ramirez-Cuesta, M. Zoppi, R. Senesi & T. Abdul-Redah (2004). Physica B, 350, E983–E986. Extraction of the density of phonon states in LiH and NaH.

314 J.L. Varble, J.L. Warren and J.L. Yarnell (1968). Phys. Rev., 168, 980–989. Lattice dynamics of lithium hydride.

315 B.S. Hudson, D.A. Braden, D.G. Allis, T. Jenkins, S. Baronov, C. Middleton, R. Withnall & C.M. Brown (2004). J. Phys. Chem. A, 108, 7356–7363. The crystalline enol of 1,3-cyclohexanedione and its complex with benzene: vibrational spectra, simulation of structure and dynamics and evidence for cooperative hydrogen bonding.

316 D.G. Allis & B.S. Hudson (2004). Chem. Phys. Lett., 385, 166–172. Inelastic neutron scattering spectra of $NaBH_4$ and KBH_4 reproduction of anion mode shifts via periodic DFT.

317 D.G. Allis, M.E. Kosmowski & B.S. Hudson (2004). J. Am. Chem. Soc., 126, 7756–7757. The inelastic neutron scattering spectrum of $H_3B:NH_3$ and the reproduction of its solid-state features by periodic DFT.

318 D.G. Allis, H. Prinzbach & B.S. Hudson (2004). Chem. Phys. Lett., 386, 356–363. Inelastic neutron scattering spectra of pagodane: experiment and DFT calculations.

319 I.A. Beta, H. Böhlig & B. Hunger (2004). Phys. Chem. Chem. Phys., 6, 1975–1981. Structure of adsorption complexes of water in zeolites of different types studied by infrared spectroscopy and inelastic neutron scattering.

320 D.K. Breitinger, H.-H. Belz, L. Hajba, V. Komlósi, J. Mink, G. Brehm, D. Colognesi, S.F. Parker & R.G. Schwab (2004). J. Mol. Struct., 706, 95–99. Combined vibrational spectra of natural wardite

321 C. Corsaro & S.F. Parker (2004). Physica B, 350, e591–e593. Vibrational spectroscopy of maleimide.

322 R. Weihrich, M.H. Limage, S.F. Parker & F. Fillaux (2004). J. Mol. Struct., 700, 147–149. Proton tunnelling in the intramolecular hydrogen bond of 9-hydroxyphenalenone.

323 G.R. Burns, F. Favier, D.J. Jones, J. Rozière & G.J. Kearley (2003). J. Chem. Phys., 119, 4929–4933. Potential model for tetrathiafulvalene based on inelastic neutron scattering and Raman spectra.

324 L. van Eijck, M.R. Johnson & G.J. Kearley (2003). J. Phys. Chem. A, 107, 8980–8984. Intermolecular interactions in bithiophene as a model for polythiophene.

325 M.A. Neumann, M.R. Johnson & P.G. Radaelli (2001). Chem. Phys., 266, 53–68. The low temperature phase transition in octane and its possible generalisation to other n-alkanes.

326 C. Coulombeau & H. Jobic (1988).). J. Mol. Struct., 176, 213–222. Neutron inelastic spectroscopy of ethylene-oxide and partial reassignment of the vibrational frequencies.

327 J.M. Granadino-Roldán, M. Fernández-Gomez, A. Navarro, T.P Ruiz & U.A. Jayasooriya (2004). Phys. Chem. Chem. Phys., 6, 1133–1143. An approach to the structure and vibrational analysis of cis- and trans-3-chlorostyrene through IR/Raman and INS spectroscopies and theoretical ab initio/DFT calculations.

328 L.S. Vandersluys, J. Eckert, O. Eisenstein, J.H. Hall, J.C. Huffman, S.A. Jackson, T.F. Koetzle, G.J. Kubas, P.J. Vergamini & K.G. Caulton (1990). J. Am. Chem. Soc., 112, 4831–4841. An attractive 'cis-effect' of hydride on neighbor ligands—experimental and theoretical-studies on the structure and intramolecular rearrangements of $Fe(H)_2(\eta_2-H_2)(PEtPh_2)_3$.

329 J. Eckert, H. Blank, M.T. Bautista & R.H. Morris (1990). Inorg. Chem., 29, 747–750. Dynamics of molecular hydrogen in the complex trans-$[Fe(\eta_2-H_2)(H)(PPh_2CH_2CH_2PPh_2)_2]$ BF_4 in the solid-state as revealed by neutron scattering experiments.

330 J. Eckert, G.J. Kubas, J.H. Hall, P.J. Hay & C.M. Boyle (1990). J. Am. Chem. Soc., 112, 2324–2332. Molecular hydrogen complexes. 6. The barrier to rotation of η^2-H_2 in $M(CO)_3(PR_3)_2(\eta^2-H_2)$ (M = W, Mo; R = Cy, i-Pr). Why does D_2 bind better than H_2? Inelastic neutron scattering theoretical and molecular mechanics studies.

331 H. Förster, H. Fuess, E. Geidel, B. Hunger, H. Jobic, C. Kirschhock, O. Klepel & K. Krause (1999). Phys. Chem. Chem. Phys., 1, 593–603. Adsorption of pyrrole derivatives in alkali metal cation-exchanged faujasites: comparative studies by surface vibrational techniques, X-ray diffraction and temperature-programmed desorption augmented with theoretical studies Part I. Pyrrole as probe molecule.

332 S. Chinta, T.V. Choudhary, L.L. Daemen, J. Eckert & D.W. Goodman (2002). Angew. Chem. Int. Ed. Engl., 41, 144–146. Characterization of C_2 (C_xH_y) intermediates from adsorption and decomposition of methane on supported metal catalysts by in situ INS vibrational spectroscopy.

333 T.V. Choudhary & D.W. Goodman (2002). Topics in Catalysis, 20, 35–42. Methane activation on ruthenium: the nature of the surface intermediates.

334 S. Chinta, T.V. Choudhary, L.L. Daemen, J. Eckert & D.W. Goodman (2002). J. Am. Chem. Soc., 126, 38–39. The nature of the surface species formed on Au/TiO_2 during the reaction of H_2 and O_2: and inelastic neutron scattering study.

335 K. Holderna-Natkaniec, A. Szyczewski, I. Natkaniec, V.D. Khavryutchenko & A. Pawlukojc (2002). Appl. Phys. A [Suppl.], 74, S1274–S1276. Progesterone and testosterone studies by neutron-scattering methods and quantum chemistry calculations.

336 T. Dziembowska, M. Szafran, E. Jagodzinska, I. Natkaniec, A. Pawlukojc, J.S. Kwiatkowski & J. Baran (2003). Spectrochim. Acta, 59A, 2175–2189. DFT studies of the structure and vibrational spectra of 8- hydroxyquinoline N-oxide.

337 A. Pawlukojc, I. Natkaniec, G. Bator, L. Sobczyk & E. Grech (2003). Chem. Phys. Lett., 378, 665–672. Inelastic neutron scattering (INS) spectrum of tetracyanoquinodimethane (TCNQ).

338 A. Pawlukojc, L. Bobrowicz, I. Natkaniec & J. Leciejewicz (1995). Spectrochim. Acta, 51A, 303–308. The IINS spectroscopy of amino-acids—l-valine and dl-valine.

339 A. Pawlukojc, J. Leciejewicz & I. Natkaniec (1996). Spectrochim. Acta, 52A, 29–32. The INS spectroscopy of amino acids: L-leucine.

340 A. Pawlukojc, K. Bajdor, J.C. Dobrowolski, J. Leciejewicz & I. Natkaniec (1997). Spectrochim. Acta, 53A, 927–931. The IINS spectroscopy of amino acids: L-isoleucine.

341 F. Partal, M. Fernández-Gómez, J.J. López-González, A. Navarro & G.J. Kearley (2000). Chem. Phys., 261, 239–247. Vibrational analysis of the inelastic neutron scattering spectrum of pyridine.

342 D. Colognesi, C. Andreani & E. Degiorgi (2003). J. Neutron Res., 11,123–143. Phonon density of states from a crystal analyzer inverse-geometry spectrometer: a study on ordered solid hydrogen sulfide and hydrogen chloride.

Index

For a complete listing of compounds and systems studied by INS see Appendix 4.

Substances are indexed under the IUPAC systematic stoichiometric name (i.e. with structural descriptors omitted) or the name generally used in the neutron scattering literature (even if obsolete or wrong) and cross referenced as needed.

The hydrogen molecule is generally indexed under dihydrogen; the hydrogen atom, free or combined, under hydrogen.

Coordination compounds and organometallics are indexed both on the central metal and the ligand of principal interest.

RETURN TO: PHYSICS LIBRARY

351 LeConte Hall   510-642-3122

LOAN PERIOD 1 1-MONTH	2	3
4	5	6

ALL BOOKS MAY BE RECALLED AFTER 7 DAYS.
Renewable by telephone.

DUE AS STAMPED BELOW.

This book will be held in PHYSICS LIBRARY until **JAN 2 4 2006**		
FEB 2 3 2006		
5/08/2007		
U.C. BERKELEY SENT ON ILL		
JAN 0 9 2007		
3 MONTHS LOAN		
JUN 0 1 2010		

FORM NO. DD 22
500   4-03

UNIVERSITY OF CALIFORNIA, BERKELEY
Berkeley, California 94720–6000